Summary of Formulas

Effective Interest Rate per Payment Period

- Discrete compounding $i = [(1 + r/(CK)]^C - 1$
- Continuous compounding $i = e^{r/K} - 1$

where i = effective interest rate per payment period
  r = nominal interest rate or APR
  C = number of interest periods per payment period
  K = number of payment periods per year
  r/K = nominal interest rate per payment period

Market Interest Rate

$$i = i' + \bar{f} + i'\bar{f}$$

where i = market interest rate
  i' = inflation-free interest rate
  $\bar{f}$ = general inflation rate

Present Value of Perpetuities

$$P = \frac{A}{i}$$

Capital Recovery with Return

$$CR(i) = (I - S)(A/P, i, N) + iS$$

Book Value

$$BV_n = I - \sum_{j=1}^{n} D_j$$

Straight-Line Depreciation

$$D_n = \frac{(I - S)}{N}$$

Declining Balance Depreciation

$$D_n = \alpha I (1 - \alpha)^{n-1}$$

where α = declining balance rate, and $0 < \alpha \le \dfrac{2}{N}$

Cost of Equity

$$i_e = r_f + \beta[r_M - r_f]$$

where i_e = cost of equity
  r_f = risk-free interest rate

Modified ACRS Factors				
	Recovery Period (Years)			
Year	**3**	**5**	**7**	**10**
1	33.33	20.00	14.29	10.00
2	44.45	32.00	24.49	18.00
3	14.81	19.20	17.49	14.44
4	7.41	11.52	12.49	11.52
5		11.52	8.93	9.22
6		5.76	8.92	7.37
7			8.93	6.55
8			4.46	6.55
9				6.55
10				6.56
11				3.28

  β = market related risk index
  r_M = market rate of return

Cost of Debt

$$i_d = \left(\frac{c_s}{c_d}\right)k_s(1 - t_m) + \left(\frac{c_b}{c_d}\right)k_b(1 - t_m)$$

where i_d = cost of debt
  c_s = the amount of term loan
  c_b = the amount of bond financing
  c_d = total debt = $c_s + c_b$
  k_s = the before-tax interest rate on the term loan
  k_b = the before-tax interest rate on the bond
  t_m = the firm's marginal tax rate

Weighted—Average Cost of Capital

$$k = \frac{i_d c_d}{V} + \frac{i_e c_e}{V}$$

where k = cost of capital
  c_e = total equity capital
  $V = c_d + c_e$

Fundamentals of Engineering Economics

Fundamentals
of Engineering
Economics

Chan S. Park

Department of Industrial and Systems Engineering
Auburn University

PEARSON
Prentice
Hall

UPPER SADDLE RIVER, NJ 07458

Library of Congress Cataloging-in-Publication Data on File

Vice President and Editorial Director, ECS: *Marcia J. Horton*
Acquisitions Editor: *Dorothy Marrero*
Vice President and Director of Production and Manufacturing, ESM: *David W. Riccardi*
Executive Managing Editor: *Vince O'Brien*
Managing Editor: *David A. George*
Production Editor: *Scott Disanno*
Director of Creative Services: *Paul Belfanti*
Creative Director: *Carole Anson*
Art Director and Cover Manager: *Maureen Eide*
Managing Editor, AV Management and Production: *Patricia Burns*
Art Editor: *Xiaohong Zhu*
Manufacturing Manager: *Trudy Pisciotti*
Manufacturing Buyer: *Lisa McDowell*
Marketing Manager: *Holly Stark*

© 2004 by Pearson Education, Inc.
Upper Saddle River, New Jersey 07458

Excel is a registered trademark of the Microsoft Coporation, One Microsoft Way, Redmond WA 98052-6399.

10 9 8 7 6 5 4 3 2

ISBN 0-13-030791-2

Pearson Education Ltd., *London*
Pearson Education Australia Pty. Ltd. *Sydney*
Pearson Education Singapore, Pte. Ltd.
Pearson Education North Asia Ltd., *Hong Kong*
Pearson Education Canada, Inc., *Toronto*
Pearson Educatión de Mexico, S.A. de C.V.
Pearson Education—Japan, *Tokyo*
Pearson Education Malaysia, Pte. Ltd.
Pearson Education, Inc., *Upper Saddle River, New Jersey*

To my wife: Inkyung (Kim)

Table of Contents

Evaluating Business and Engineering Assets

5 Present-Worth Analysis

•IV

Special Topics in Engineering Economics 393

Preface

Why Fundamentals of Engineering Economics?

Engineering economics is one of the most practical subject matters in the engineering curriculum, but it is always challenging and an ever-changing discipline. *Contemporary Engineering Economics* (CEE) was first published in 1993, and since then we have tried to reflect changes in the business world in each new edition, along with the latest innovations in education and publishing. These changes have resulted in a better, more complete textbook, but one that is much longer than it was originally intended. This may present a problem: today, covering the textbook in a single term is increasingly difficult. Therefore, we decided to create *Fundamentals of Engineering Economics* (FEE) for those who like *Fundamentals* but think a smaller, more concise textbook would better serve their needs.

Goals of the Text

This text aims not only to provide sound and comprehensive coverage of the concepts of engineering economics, but also to address the practical concerns of engineering economics. More specifically, this text has the following goals:

1. To build a thorough understanding of the theoretical and conceptual basis upon which the practice of financial project analysis is built.
2. To satisfy the very practical needs of the engineer toward making informed financial decisions when acting as a team member or project manager for an engineering project.
3. To incorporate all critical decision-making tools—including the most contemporary, computer-oriented ones that engineers bring to the task of making informed financial decisions.
4. To appeal to the full range of engineering disciplines for which this course is often required: industrial, civil, mechanical, electrical, computer, aerospace, chemical, and manufacturing engineering, as well as engineering technology.

Intended Market and Use

This text is intended for use in the introductory engineering economics course. Unlike the larger textbook (CEE), it is possible to cover FEE in a single term, and perhaps even to supplement it with a few outside readings or cases. Although the chapters in FEE are arranged logically, they are written in a flexible, modular format, allowing instructors to cover the material in a different sequence.

Steps Taken to Streamline the Textbook

We decided to streamline the textbook by retaining the depth and level of rigor in CEE, while eliminating some less critical topics in each chapter. This resulted in reducing the total number of chapters by four chapters in two steps. Such core topics as the time value of money, measures of investment worth, development of project cash flows, and the relationship between risk and return are still discussed in great detail.

- First, we eliminated the three chapters on cost accounting, principles of investing, and capital budgeting. We address these issues in other parts of the textbook, but in less depth than was contained in the deleted chapters.

- Second, we consolidated the two chapters on depreciation and income taxes into one chapter, thus eliminating one more chapter. This consolidation produced some unexpected benefits—students understand depreciation and income taxes in the context of project cash flow analysis, rather than a separate accounting chapter.

- Third, moving the inflation material from late in the textbook to the end of the equivalence chapters enables students to understand better the nature of inflation in the context of time value of money.

- Fourth, the project cash flow analysis chapter (Chapter 9) is significantly streamlined—it begins with the definitions and classifications of various cost elements that will be a part of a project cash flow statement. Then, it presents the income tax rate to use in developing a project cash flow statement. It also presents the appropriate interest rate to use in after-tax economic analysis. Finally, it illustrates how to develop a project cash flow statement considering (1) operating activities, (2) investing activities, and (3) financing activities.

- Fifth, the handling project uncertainty chapter (Chapter 10) has been consolidated by introducing the risk-adjusted discount rate approach and investment strategies under uncertainty, but eliminating the decision-tree analysis.

- Finally, the chapter on understanding financial statements has been moved to the end of the book as a capstone chapter, illustrating that a corporation does not make a large-scale investment decision on an engineering project based on just profitability alone. It considers both the financial impact on the bottom-line of business as well as the market value of the corporation.

FEE is significantly different from CEE, but most of the chapters will be familiar to users of CEE. Although we pruned some material and clarified, updated, and otherwise improved all of the chapters, FEE should still be regarded as an alternative version of CEE.

Features of the Book

Although FEE is a streamlined version of CEE, we did retain all of the pedagogical elements and supporting materials that helped make CEE so successful. Some of the features are:

- Each chapter opens with a real economic decision describing how an individual decision maker or actual corporation has wrestled with the issues discussed in the chapter. These opening cases heighten students' interest by pointing out the real-world relevance and applicability of what might otherwise seem to be dry technical material.
- There are a large number of end-of-chapter problems and exam-type questions varying in level of difficulty; these problems thoroughly cover the book's various topics.
- Most chapters contain a section titled "Short Case Studies with Excel," enabling students to use Excel to answer a set of questions. These problems reinforce the concepts covered in the chapter and provide students with an opportunity to become more proficient with the use of an electronic spreadsheet.

Taking Advantage of the Internet

The integration of computer use is another important feature of *Fundamentals of Engineering Economics*. Students have greater access to and familiarity with the various spreadsheet tools, and instructors have greater inclination either to treat these topics explicitly in the course or to encourage students to experiment independently. A remaining concern is that the use of computers will undermine true understanding of course concepts. This text does not promote the trivial or mindless use of computers as a replacement for genuine understanding of and skill in applying traditional solution methods. Rather, it focuses on the computer's productivity-enhancing benefits for complex project cash flow development and analysis. Specifically, *Fundamentals of Engineering Economics* includes:

- A robust introduction to computer automation in the form of the Cash Flow Analyzer problem, which can be accessed from the book's website (http:// www.prenhall.com/park).
- An introduction to spreadsheets using Microsoft Excel examples. For spreadsheet coverage, the emphasis is on demonstrating complex concepts that can be resolved much more efficiently on a computer than by traditional long-hand solutions.

Book Website

The companion website (http://www.prenhall.com/park) has been created and maintained by the author. This text takes advantage of the Internet as a tool that has become an increasingly important resource medium used to access a variety of information on the Web. This website contains a variety of resources for both instructors and students, including sample test questions, supplemental problems, and various on-line financial calculators. As you type the address and click the open button, you will see the *Fundamentals of Engineering Economics* Home Page (Figure P.1). As you will note from the figure, several menus are available. Each menu item is explained as follows:

- **Study Guides**. Click this menu to find out what resource materials are available on the website. This site includes (1) sample text questions, (2) solutions to chapter problems, (3) interest tables, and (4) computer notes with Excel files of selected example problems in the text.

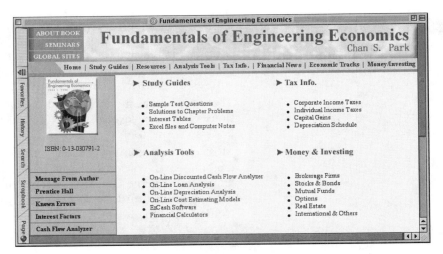

FIGURE P.1

- **Analysis Tools**. This section includes (1) a Cash Flow Analyzer and (2) a collection of various financial calculators available on the Internet. The Cash Flow Analyzer is an integrated computer software package written in Java. The software includes the most frequently used methods of economic analysis. It is menu driven for convenience and flexibility, and it provides (1) a flexible and easy-to-use cash flow editor for data input and modifications, and (2) an extensive array of computational modules and user selected graphic outputs.

- **Instructor Resources**. This section includes information useful to instructors who teach an engineering economic course with the *Fundamentals of Engineering Economics* text is found in this location. Typically, course outlines based on a quarterly as well as a semester system are provided as an aid to instructors who adopt this text for the first time. A collection of well-designed engineering economic case studies is listed at this location. Initially, only a few case problems will be provided, but as new ones are developed or found, the case library will be expanded in volume as well as in variety. You will also find lecture notes developed by the author.

- **Tax Information**. This section will serve as a clearinghouse in terms of disseminating ever-changing tax information, including personal as well as corporate income taxes. Links are provided to various tax sites on the Web, so you will find the most up-to-date information on depreciation schedules as well as capital gains taxes.

- **Financial News**. This section provides access to various financial news outlets on the Web. The site divides news outlets into on-line news, and daily, weekly, and monthly publications.

- **Economic Tracks**. This section includes cost and price information as well as the most recent interest rate trends. In particular, the consumer price indices, productivity figures, and employment cost indices are some of the representative economic data provided.

- **Money and Investing**. This section provides a gateway to a variety of information useful to conducting engineering economic analysis. For example, a direct link is provided to the most up-to-date stock prices, options, and mutual funds performances.

What Supplements are Available to Students and Instructors?

FEE includes several ancillary materials designed to enhance the student's learning experience, while making it easier for the instructor to prepare for and conduct classes. The ancillaries are described below.

For Students

- *Excel for Engineering Economics* (supplement), containing information on how to use Excel for engineering economic studies and various Excel applications.
- *Study Guides for Fundamentals of Engineering Economics* (supplement), which contains more than 200 completely worked out solutions and guides on how to take the FE exam on engineering economics and sample test questions.

For Instructors

- A comprehensive *Instructor's Manual* that includes answers to end-of-chapter problems and Excel solutions to all complex problems and short case studies.
- PowerPoint slides for lecture notes, the entire contents of the Instructor's Manual in Word format, test questions, and Excel spreadsheet files are available in a passcode protected area in the Instructor Resources section of the Book Website.

ACKNOWLEDGMENTS

This book reflects the efforts of a great many individuals over a number of years. In particular, I would like to recognize the following individuals, whose reviews and comments have contributed to this edition. Once again, I would like to thank each of them:

Richard V. Petitt, United States Military Academy; James R. Smith, Tennessee Technological University; Bruce Hartsough, University of California at Davis; Iris V. Rivero, Texas Tech University; Donald R. Smith, Texas A&M University; Bruce McCann, University of Texas at Austin; Dolores Gooding, University of South Florida; and Stan Uryasev, University of Florida.

Personally, I wish to thank the following individuals for their additional input to the new edition: Michael D. Park, McKinsey & Company, who read the entire manuscript and offered numerous and critical comments to improve the content of the book; Luke Miller, Yeji Jung and Edward Park, who helped me in preparing the Instructor's Manual; Junmo Yang, who helped me in developing the book website; Dorothy Marrero, my editor at Prentice Hall, who assumed responsibility for the overall project; and Scott Disanno, the production editor, who oversaw the entire book production.

CHAN S. PARK
Auburn, Alabama

Understanding Money and Its Management

.1

Bose Packs Concert Acoustics into Home-Speaker Systems[1]

Dr. Amar G. Bose, an MIT professor and chairman of speaker manufacturer Bose Corporation, defied the conventional wisdom of consumer electronics. Dr. Bose grew up poor in Philadelphia, where his father emigrated from India and worked as an importer until he lost the business during World War II. While his mother worked as a teacher, Dr. Bose set up a radio-repair business at the age of 14 in the basement; this business soon became the family's main support. He entered MIT and never left, earning a doctoral degree in 1956. As a reward for finishing his research, he decided to buy himself a stereo system. Although he had done his homework on the hi-fi's engineering specifications, he was profoundly disappointed with his purchase. Mulling over why something that looked good on paper sounded bad in the open air, Dr. Bose concluded that the answer involved directionality. In a concert hall, sound waves radiate outward from the instruments and bounce back at the audience from the walls. However, home stereo speakers aimed sound only forward. Therefore, Dr. Bose began tinkering to develop a home speaker that could reproduce the concert experience.

In 1964, he formed Bose Corporation,[2] and four years later he introduced his first successful speaker, the 901. Based on the principle of reflected sound, the speaker bounces sounds off walls and ceilings in order to surround the listener. In 1968, Dr. Bose pioneered the use of "reflected sound" in an effort to bring concert-hall quality to home-speaker systems. A decade later, he convinced General Motors Corporation to let his company design a high-end speaker system for the Cadillac Seville, helping to push car stereos beyond the mediocre. Recently, he introduced a compact radio system that can produce rich bass sound. In the process, Bose has become the world's number-one speaker maker, with annual sales of more than $700 million, and one of the few U.S. firms that beats the Japanese in consumer electronics. Dr. Amar G. Bose was inducted into the Radio Hall of Fame in 2000. In 2002, his success vaulted Dr. Bose into *Forbes* magazine's list of the 400 wealthiest Americans (he was 288th), with a net worth estimated at $800 million.[3]

[1]William M. Bulkeley, "Bose Packs Concert Acoustics into Home-Speaker Systems," *The Wall Street Journal*, December 31, 1996.
[2]Courtesy of Bose Corporation—History of Company on its website (http://www.bose.com).
[3]"400 Richest Americans," Forbes.com (http://www.Forbes.com), February 12, 2003.

Engineering Economic Decisions

The story of how Dr. Bose got motivated to invent a directional home speaker and eventually transformed his invention into a multimillion-dollar business is not an uncommon one in today's market. Companies such as Dell, Microsoft, and Yahoo all produce computer-related products and have market values of several billion dollars. These companies were all started by highly motivated young college students just like Dr. Bose. Another thing that is common to all these successful businesses is that they have capable and imaginative engineers who constantly generate good ideas for capital investment, execute them well, and obtain good results. You might wonder about what kind of role these engineers play in making such business decisions. In other words, what specific tasks are assigned to these engineers, and what tools and techniques are available to them for making such capital-investment decisions? In this book, we will consider many investment situations, personal as well as business. The focus, however, will be on evaluating engineering projects on the basis of economic desirability and in light of the investment situations that face a typical firm.

1-1   The Rational Decision-Making Process

We, as individuals or businesspersons, constantly make decisions in our daily lives. We make most of them automatically, without consciously recognizing that we are actually following some sort of a logical decision flowchart. Rational decision making can be a complex process that contains a number of essential elements. Instead of presenting some rigid rational decision making processes, we will provide examples of how two engineering students approached their financial as well as engineering design problems. By reviewing these examples, we will be able to identify some essential elements common to any rational decision-making process. The first example illustrates how a student named Monica narrowed down her choice between two competing alternatives when buying an automobile. The second example illustrates how a typical class-project idea evolves and how a student named Sonya approached the design problem by following a logical method of analysis.

1.1.1   How Do We Make Typical Personal Decisions?

For Monica Russell, a senior at the University of Washington, the future holds a new car. Her 1993 Honda Civic has clocked almost 110,000 miles, and she wants to replace it soon. But how to do it—buy or lease? In either case, "car payments would be difficult," said the engineering major, who works as a part-time cashier at a local supermarket. "I have never leased before, but I am leaning toward it this time to save on the down payment. I also don't want to worry about major repairs," she said. For Monica, leasing would provide the warranty protection she wants, along with a new car every three years. On the other hand, she would be limited to driving only a specified number of miles, usually 12,000 per year, after which she would have to pay 20 cents or more per mile. Monica is well aware that choosing the right vehicle is an important decision, and so is choosing the best possible financing. Yet, at this point, Monica is unsure of the implications of buying versus leasing.

Establishing the Goal or Objective   Monica decided to survey the local papers and the Internet for the latest lease programs, including factory-subsidized "sweetheart" deals and special incentive packages. Of the cars that were within her budget, the 2003 Saturn ION.3 and the 2003 Honda Civic DX coupe appeared to be equally attractive in terms of style, price, and options. Monica finally decided to visit the dealers' lots to see how both models looked and to take them for a test drive. Both cars gave her very satisfactory driving experiences. Monica thought that it would be important to examine carefully many technical as well as safety features of the automobiles. After her examination, it seemed that both models were virtually identical in terms of reliability, safety features, and quality.

Evaluation of Feasible Alternatives   Monica figured that her 1993 Honda could be traded in at around $2,000. This amount would be just enough to make any down payment required for leasing the new automobile. Through her research, Monica also learned that there are two types of leases: open end and closed end. The most popular by far was closed end, because open-end leases expose the consumer to possible higher payments at the end of the lease if the car depreciates faster than expected. If Monica were to take a closed-end lease, she could just return the vehicle at the end of the lease and "walk away" to lease or buy another vehicle. However, she would have to pay for extra mileage or excess wear or damage. She thought that since she would not be a "pedal-to-the-metal driver," lease-end charges would not be a problem for her.

To get the best financial deal, Monica obtained some financial facts from both dealers on their best offers. With each offer, she added all the costs of leasing, from the down payment to the disposition fee due at the end of the lease. This sum would determine the total cost of leasing that vehicle, not counting routine items such as oil changes and other maintenance. See Table 1.1 for a comparison of the costs of both offers. It appeared that, with the Saturn ION.3, Monica could save about $622 in total lease payments [(47 months × $29 monthly lease payment savings) −$741 total due at signing (including the first month's lease payment savings)], over the Honda Civic, plus $250 on the disposition fee (which the Saturn did not have), for a total savings of $872.[4] However, if she were to drive any additional miles over the limit, her savings would be reduced by five cents (the difference between the two cars' mileage surcharges) for each additional mile. Monica would need to drive

TABLE 1.1 Financial Data for Auto Leasing: Saturn versus Honda

Auto Leasing	Saturn	Honda	Difference Saturn − Honda
1. Manufacturer's suggested retail price (MSRP)	$15,573	$15,810	−$273
2. Lease length	48 months	48 months	
3. Allowed mileage	48,000 miles	48,000 miles	
4. Monthly lease payment	$219	$248	−$29
5. Mileage surcharge over 36,000 miles	$0.20 per mile	$0.15 per mile	+$0.05 per mile
6. Disposition fee at lease end	$0	$250	$250
7. Total due at signing:			
• First month's lease payment	$219	$248	
• Down payment	$1,100	$800	
• Administrative fee	$495	$0	
• Refundable security deposit	$200	$225	
Total	$2,014	$1,273	+$741

• Models compared: The 2003 Saturn ION.3 with automatic transmission and A/C and the 2003 Honda Civic DX coupe with automatic transmission and A/C.
• Disposition fee: This is a paperwork charge for getting the vehicle ready for resale after the lease end.

[4]If Monica considered the time value of money in her comparison, the amount of actual savings would be less than $872, which we will demonstrate in Chapter 2.

about 17,440 extra miles over the limit in order to lose all the savings. Because she could not anticipate her driving needs after graduation, her conclusion was to lease the Honda Civic DX. Certainly, any monetary savings would be important, but she preferred having some flexibility in her future driving needs.

Knowing Other Opportunities   If Monica had been interested in buying the car, it would have been even more challenging to determine precisely whether she would be better off buying than leasing. To make a comparison of leasing versus buying, Monica could have considered what she likely would pay for the same vehicle under both scenarios. If she would own the car for as long as she would lease it, she could sell the car and use the proceeds to pay off any outstanding loan. If finances were her only consideration, her choice would depend on the specifics of the deal. But beyond finances, she would need to consider the positives and negatives of her personal preferences. By leasing, she would never experience the "joy" of the last payment—but she would have a new car every three years.

Review of Monica's Decision-Making Process   Now we may revisit the decision-making process in a more structured way. The analysis can be thought of as including the six steps as summarized in Figure 1.1.

These six steps are known as the "rational decision-making process." Certainly, we do not always follow these six steps in every decision problem. Some decision problems may not require much of our time and effort. Quite often, we even make our decisions solely on emotional reasons. However, for any complex economic decision problem, a structured decision framework such as that outlined here proves to be worthwhile.

1.1.2   How Do We Approach an Engineering Design Problem?

The idea of design and development is what most distinguishes engineering from science, the latter being concerned principally with understanding the world as it is. Decisions made during the engineering design phase of a product's development determine the majority of the costs of manufacturing that product. As design and manufacturing processes become more complex, the engineer increasingly will be called upon to make decisions that involve money. In this section, we provide an

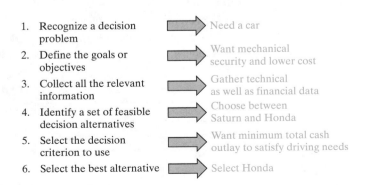

1. Recognize a decision problem   → Need a car
2. Define the goals or objectives   → Want mechanical security and lower cost
3. Collect all the relevant information → Gather technical as well as financial data
4. Identify a set of feasible decision alternatives → Choose between Saturn and Honda
5. Select the decision criterion to use → Want minimum total cash outlay to satisfy driving needs
6. Select the best alternative   → Select Honda

FIGURE 1.1 Logical steps to follow in a car-leasing decision

example of how engineers get from "thought" to "thing." The story we relate of how an electrical engineering student approached her design problem and exercised her judgment has much to teach us about some of the fundamental characteristics of the human endeavor known as engineering decision making.[5]

Getting an Idea: Necessity is the Mother of Invention

Most consumers abhor lukewarm beverages, especially during the hot days of summer. Throughout history, necessity has been the mother of invention. So, several years ago, Sonya Talton, an electrical engineering student at Johns Hopkins University, had a revolutionary idea—a self-chilling soda can!

Picture this: It's one of those sweltering, hazy August afternoons. Your friends have finally gotten their acts together for a picnic at the lake. Together, you pull out the items you brought with you: blankets, a radio, sunscreen, sandwiches, chips, and soda. You wipe the sweat from your neck, reach for a soda, and realize that it's about the same temperature as the 90°F afternoon. Great start! Everyone's just dying to make another trip back to the store for ice. Why can't someone come up with a soda container that can chill itself, anyway?

Setting Design Goals and Objectives

Sonya decided to take on the topic of a soda container that can chill itself as a term project in her engineering graphics and design course. The professor stressed innovative thinking and urged students to consider practical, but novel, concepts. The first thing Sonya needed to do was to establish some goals for the project:

- Get the soda as cold as possible in the shortest possible time.
- Keep the container design simple.
- Keep the size and weight of the newly designed container similar to that of the traditional soda can. (This factor would allow beverage companies to use existing vending machines and storage equipment.)
- Keep the production costs low.
- Make the product environmentally safe.

Evaluating Design Alternatives

With these goals in mind, Sonya had to think of a practical, yet innovative, way of chilling the can. Ice was the obvious choice—practical, but not innovative. Sonya had a great idea: what about a chemical ice pack? Sonya asked herself what would go inside such an ice pack. The answer she came up with was ammonium nitrate (NH_4NO_3) and a water pouch. When pressure is applied to the chemical ice pack, the water pouch breaks and mixes with the NH_4NO_3, creating an endothermic reaction (the absorption of heat). The NH_4NO_3 draws the heat out of the soda, causing it to chill. (See Figure 1.2.) How much water should go in the water pouch? The first amount Sonya tried was 135 mL. After several trials involving different amounts of water, Sonya found that she could chill the soda can from 80°F to 48°F in a three-minute period. The required amount of water was about 115 mL.

---

[5]Background materials from 1991 Annual Report, GWC Whiting School of Engineering, Johns Hopkins University (with permission).

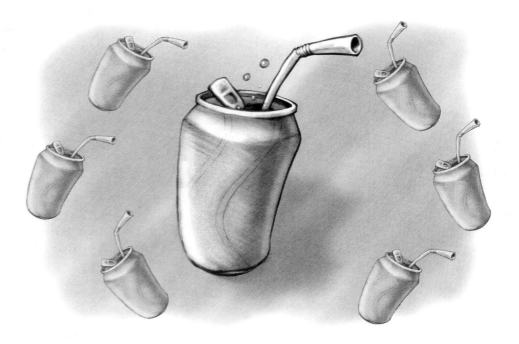

FIGURE 1.2 Conceptual design for self-chilling soda can

At this point, she needed to determine how cold a refrigerated soda gets, as a basis for comparison. She put a can in the fridge for two days and found that it chilled to 41° F. Sonya's idea was definitely feasible. But was it economically marketable?

Gauging Product Cost and Price   In Sonya's engineering graphics and design course, the topic of how economic feasibility plays a major role in the engineering design process was discussed. The professor emphasized the importance of marketing surveys and benefit–cost analyses as ways to gauge a product's potential. To determine the marketability of her self-chilling soda can, Sonya surveyed approximately 80 people. She asked them only two questions: their age and how much would they be willing to pay for a self-chilling can of soda. The under-21 group was willing to pay the most, 84 cents on average. The 40-plus bunch wanted to pay only 68 cents on average. Overall, the surveyed group would be willing to shell out 75 cents for a self-chilling soda can. (This poll was hardly a scientific market survey, but it did give Sonya a feel for what would be a reasonable price for her product.)

The next hurdle was to determine the existing production cost of one traditional can of soda. Also, how much more would it cost to produce the self-chiller? Would it be profitable? She went to the library, and there she found the bulk cost of the chemicals and materials she would need. Then she calculated how much she would require for one unit of soda. She couldn't believe it! It costs only 12 cents to manufacture one can of soda, including transportation. Her can of soda would cost 2 or 3 cents more. That wasn't bad, considering that the average consumer was willing to pay up to 25 cents more for the self-chilling can than for the traditional one.

Considering Green Engineering The only two constraints left to consider were possible chemical contamination of the soda and recyclability. Theoretically, it should be possible to build a machine that would drain the solution from the can and recrystallize it. The ammonium nitrate could then be reused in future soda cans; in addition, the plastic outer can could be recycled. Chemical contamination of the soda, the only remaining restriction, was a big concern. Unfortunately, there was absolutely no way to ensure that the chemical and the soda would never come in contact with one another inside the cans. To ease consumer fears, Sonya decided that a color or odor indicator could be added to alert the consumer to contamination if it occurred.

What is Sonya's conclusion? The self-chilling beverage container (can) would be an incredible technological advancement. The product would be convenient for the beach, picnics, sporting events, and barbecues. Its design would incorporate consumer convenience while addressing environmental concerns. It would be innovative, yet inexpensive, and it would have an economic as well as social impact on society.

1.1.3 What Makes Economic Decisions Differ from Other Design Decisions?

Economic decisions differ in a fundamental way from the types of decisions typically encountered in engineering design. In a design situation, the engineer uses known physical properties, the principles of chemistry and physics, engineering design correlations, and engineering judgment to arrive at a workable and optimal design. If the judgment is sound, the calculations are done correctly, and we ignore technological advances, the design is time invariant. In other words, if the engineering design to meet a particular need is done today, next year, or in five years time, the final design will not change significantly.

In considering economic decisions, the measurement of investment attractiveness, which is the subject of this book, is relatively straightforward. However, information required in such evaluations always involves predicting or forecasting product sales, product selling price, and various costs over some future time frame— 5 years, 10 years, 25 years, etc.

All such forecasts have two things in common. First, they are never completely accurate when compared with the actual values realized at future times. Second, a prediction or forecast made today is likely to be different than one made at some point in the future. It is this ever-changing view of the future that can make it necessary to revisit and even change previous economic decisions. Thus, unlike engineering design outcomes, the conclusions reached through economic evaluation are not necessarily time invariant. Economic decisions have to be based on the best information available at the time of the decision and a thorough understanding of the uncertainties in the forecasted data.

1-2 The Engineer's Role in Business

What role do engineers play within a firm? What specific tasks are assigned to the engineering staff, and what tools and techniques are available to it to improve a firm's profits? Engineers are called upon to participate in a variety of decision-making processes, ranging from manufacturing and marketing to financing decisions. We will restrict our focus, however, to various economic decisions related to engineering projects. We refer to these decisions as **engineering economic decisions**.

1.2.1 Making Capital-Expenditure Decisions

In manufacturing, engineering is involved in every detail of producing goods, from conceptual design to shipping. In fact, engineering decisions account for the majority (some say 85%) of product costs. Engineers must consider the effective use of capital assets such as buildings and machinery. One of the engineer's primary tasks is to plan for the acquisition of equipment (**capital expenditure**) that will enable the firm to design and produce products economically. (See Figure 1.3.)

With the purchase of any fixed asset, equipment for example, we need to estimate the profits (more precisely, the cash flows) that the asset will generate during its service period. In other words, we have to make capital-expenditure decisions based on predictions about the future. Suppose, for example, that you are considering the purchase of a deburring machine to meet the anticipated demand for hubs and sleeves used in the production of gear couplings. You expect the machine to last 10 years. This purchase decision thus involves an implicit 10-year sales forecast for the gear couplings, which means that a long waiting period will be required before you will know whether the purchase was justified.

An inaccurate estimate of asset needs can have serious consequences. If you invest too much in assets, you incur unnecessarily heavy expenses. Spending too little on fixed assets is also harmful, for then your firm's equipment may be too obsolete to produce products competitively, and without an adequate capacity, you may lose a portion of your market share to rival firms. Regaining lost customers involves heavy marketing expenses and may even require price reductions or product improvements, both of which are costly.

Engineering Economic Decisions

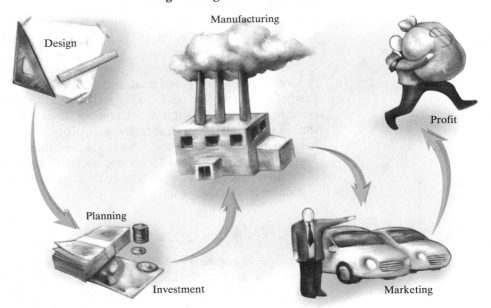

FIGURE 1.3 One of the primary functions of engineers: making capital-budgeting decisions

1.2.2 Large-Scale Engineering Economic Decisions

The economic decisions that engineers make in business differ very little from those made by Sonya, except for the scale of the concern. Let us consider a real-world engineering decision problem of a much larger scale. Public concern about poor air quality is increasing, particularly regarding pollution caused by gasoline-powered automobiles. With requirements looming in a number of jurisdictions for automakers to produce electric vehicles, General Motors Corporation has decided to build an advanced electric car to be known as GEN II–EV1,[6] or just EV1 for short. The biggest question remaining about the feasibility of the vehicle concerns its battery.[7] With its current experimental battery design, EV1's monthly operating cost would be roughly twice that of a conventional automobile.[8] The primary advantage of the design, however, is that EV1 does not emit any pollutants, a feature that could be very appealing at a time when government air-quality standards are becoming more stringent and consumer interest in the environment is ever growing.

Engineers at General Motors have stated that the total annual demand for EV1 would need to be 100,000 cars in order to justify production. Although the management of General Motors has already decided to build the battery-powered electric car, the engineers involved in making the engineering economic decision are still debating about whether the demand for such a car would be sufficient to justify its production.

Obviously, this level of engineering economic decision is more complex and more significant to the company than a decision about when to purchase a new lathe. Projects of this nature involve large sums of money over long periods of time, and it is difficult to predict market demand accurately. (See Figure 1.4.) An erroneous forecast of product demand can have serious consequences: with any overexpansion, unnecessary expenses will have to be paid for unused raw materials and finished products. In the case of EV1, if an improved battery design that lowers the car's monthly operating cost never materializes, demand may remain insufficient to justify the project.

1.2.3 Impact of Engineering Projects on Financial Statements

Engineers must also understand the business environment in which a company's major business decisions are made. It is important for an engineering project to generate profits, but it also must strengthen the firm's overall financial position. How do we measure General Motors's success in the EV1 project? Will enough

[6]Official GM's website for EV1: http://www.gmev.com.

[7]The EV1 with a high-capacity lead-acid pack has an estimated "real-world" driving range of 55 to 95 miles, depending on terrain, driving habits, and temperature. The range with the nickel-metal hydride (NiMH) battery pack is even greater. Again, depending on terrain, driving habits, temperature, and humidity, estimated real-world driving range will vary from 75 to 130 miles. Certainly, the EV1 isn't practical for long trips, simply because it is not designed for that purpose. However, battery technology is currently being developed that might make those trips possible in the near future.

[8]The manufacturer suggested retail price (MSRP) for the EV1 ranges from $33,995 to $43,995, depending on the model year and the battery pack. The monthly lease payment ranges from $350 to $575. The start-up costs of the EV1 could be very expensive to some, as home electric wiring must be 220 V compatible. It would cost about $1,000 for the home charging unit and its installation. However, the car is a pollution-free, low maintenance vehicle that only costs about 2 cents per mile to operate.

GM's Electric Car Project

- Requires a large sum of investment

- Takes a long time to see the financial outcomes

- Difficult to predict the revenue and cost streams

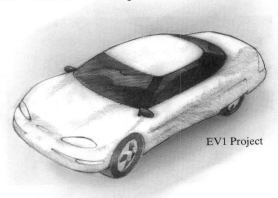

EV1 Project

FIGURE 1.4 A large-scale engineering project: GM's EV1 project

EV1 models be sold, for example, to generate sufficient profits? While the EV1 project will provide comfortable, reliable pollution-free driving for its customers, the bottom line is its financial performance over the long run.

Regardless of a business's form, each company has to produce basic financial statements at the end of each operating cycle (typically, a year). These financial statements provide the basis for future investment analysis. In practice, we seldom make investment decisions based solely on an estimate of a project's profitability, because we must also consider the project's overall impact on the financial strength and position of the company. For example, some companies with low cash flow may be unable to bear the risk of a large project like EV1, even if it is profitable. (See Figure 1.5.)

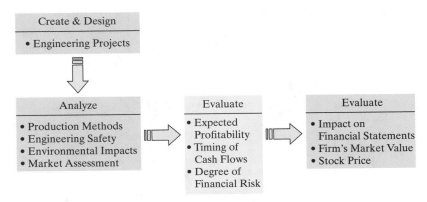

FIGURE 1.5 How a successful engineering project affects a firm's market value

Suppose that you are the president of the GM Corporation. Further suppose that you hold some shares in the company, which makes you one of the company's many owners. What objectives would you set for the company? One of your objectives should be to increase the company's value to its owners (including yourself) as much as possible. While all firms are in business in hopes of making a profit, what determines the market value of a company are not profits per se, but rather cash flows. It is, after all, available cash that determines the future investments and growth of the firm. The market price of your company's stock to some extent represents the value of your company. Many factors affect your company's market value: present and expected future earnings, the timing and duration of these earnings, and the risks associated with the earnings. Certainly, any successful investment decision will increase a company's market value. Stock price can be a good indicator of your company's financial health and may also reflect the market's attitude about how well your company is managed for the benefit of its owners.

If investors like the new electric car, the result will be an increased demand for the company's stock. This increased demand, in turn, will cause stock prices, and hence shareholder wealth, to increase. Any successful investment decision on EV1's scale will tend to increase a firm's stock prices in the marketplace and promote long-term success. Thus, in making a large-scale engineering project decision, we must consider its possible effect on the firm's market value.

1-3 Types of Strategic Engineering Economic Decisions

Project ideas such as the EV1 can originate from many different levels in an organization. Since some ideas will be good, while others will not, we need to establish procedures for screening projects. Many large companies have a specialized project analysis division that actively searches for new ideas, projects, and ventures. Once project ideas are identified, they are typically classified as (1) service or quality improvement, (2) new products or product expansion, (3) equipment and process selection, (4) cost reduction, or (5) equipment replacement. This classification scheme allows management to address key questions such as the following: can the existing plant be used to achieve the new production levels? Does the firm have the capital to undertake this new investment? Does the new proposal warrant the recruitment of new technical personnel? The answers to these questions help firms screen out proposals that are not feasible given a company's resources.

The EV1 project represents a fairly complex engineering decision that required the approval of top executives and the board of directors. Virtually all big businesses at some time face investment decisions of this magnitude. In general, the larger the investment, the more detailed is the analysis required to support the expenditure. For example, expenditures to increase the output of existing products or to manufacture a new product would invariably require a very detailed economic justification. Final decisions on new products and marketing decisions are generally made at a high level within the company. On the other hand, a decision to repair damaged equipment can be made at a lower level within a company. In this section, we will provide many real examples to illustrate each class of engineering economic decisions. At this point, our intention is not to provide the solution to each example, but rather to describe the nature of decision problems that a typical engineer might face in the real world.

• How many more jeans would Levi need to sell to justify
the cost of additional robotic tailors?

A new computerized system being installed at some Original Levi's Stores allows women to order customized
blue jeans. Levi Strauss declined to have its factory photographed so here is an artist's conception of how
the process works

A sales clerk measures the customer using
instructions from a computer as an aid.

The clerk enters the measurements and
adjusts the data based on the customer's
reaction to the samples.

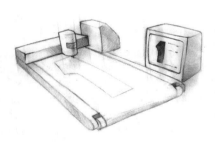

The final measurements are relayed to a
computerized fabric cutting machine at
the factory.

Bar codes are attached to the clothing to
track it as it is assembled, washed, and
prepared for shipment.

FIGURE 1.6 "From Data to Denim": Making customized blue jeans for
women, a new computerized system being installed at some Original Levi's
Stores allows women to order customized blue jeans

• **Service or Quality Improvement:** Investments in this category include any
activities to support the improvement of productivity, quality, and customer
satisfaction in the service sector, such as in the financial, healthcare, and retail industries. See Figure 1.6 for an example of a service improvement in retail. The service sector of the U.S. economy dominates both gross domestic
product (GDP) and total employment. It is also the fastest growing part of
the economy and the one offering the most fertile opportunities for productivity improvement. For example, service activities now approach 80% of

U.S. employment, far outstripping sectors like manufacturing (14%) and agriculture (2%). New service activities are continually emerging throughout the economy as forces such as globalization, e-commerce, and environmental reuse concerns produce ever more decentralization and outsourcing of operations and process.

- **New Products or Product Expansion:** Investments in this category are those that increase the revenues of a company if output is increased. There are two common types of expansion decision problems. The first type includes decisions about expenditures to increase the output of existing production or distribution facilities. In these situations, we are basically asking, "Shall we build or otherwise acquire a new facility?" The expected future cash inflows in this investment category are the revenues from the goods and services produced in the new facility.

  The second type of decision problem includes the consideration of expenditures necessary to produce a new product (e.g., see Figure 1.7) or to expand into a new geographic area. These projects normally require large sums of money over long periods. For example, after ten years of research and development (R & D), Gillette introduced its SensorExcel twin-blade shaving system in 1990. With blades mounted on springs that allowed the razor to adjust to a man's face as he shaved, Sensor raised the shaving bar to new heights. Soon after the introduction of the revolutionary twin-blade system, scientists at Gillette's research lab in Reading, Great Britain, were already trying to figure out how to create the world's first triple-blade shaving system. The MACH3 group worked for five full years in concert with R & D to produce and orchestrate the introduction of the new product. It took seven years and a whopping $750 million, but Gillette finally did it, introducing the MACH3 razor in 1998. Although the MACH3 was priced about 35% higher than SensorExcel, in the United States alone, MACH3 razors outsold Sensor four to one compared with Sensor's first six months on the market. MACH3 generated $68 million in sales in its first six months; Sensor brought in just $20 million in its first six months.

- R&D investment: $750 million
- Product promotion through advertising: $300 million
- Priced to sell at 35% higher than SensorExcel (about $1.50 extra per shave)
- Question 1: Would consumers pay $1.50 extra for a shave with greater smoothness and less irritation?
- Question 2: What would happen if blade consumption dropped more than 10% due to the longer blade life of the new razor?

Gillette's MACH3 Project

FIGURE 1.7 Launching a new product: Gillette's MACH3 project

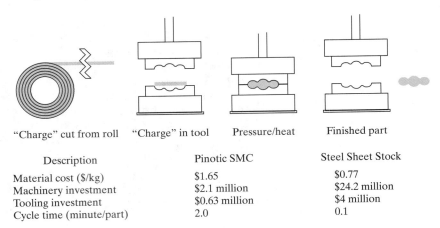

Description	Pinotic SMC	Steel Sheet Stock
Material cost ($/kg)	$1.65	$0.77
Machinery investment	$2.1 million	$24.2 million
Tooling investment	$0.63 million	$4 million
Cycle time (minute/part)	2.0	0.1

FIGURE 1.8 Sheet-molding compound process: material selection for an automotive exterior body (Courtesy of Dow Plastics, a business group of the Dow Chemical Company)

- **Equipment and Process Selection:** This class of engineering decision problem involves selecting the best course of action when there are several ways to meet a project's requirements. Which of several proposed items of equipment shall we purchase for a given purpose? The choice often hinges on which item is expected to generate the largest savings (or return on the investment). The choice of material will dictate the manufacturing process involved. (See Figure 1.8 for material selection for an automotive exterior body.[9]) Many factors will affect the ultimate choice of the material, and engineers should consider all major cost elements, such as machinery and equipment, tooling, labor, and material. Other factors may include press and assembly, production and engineered scrap, the number of dies and tools, and the cycle times for various processes.

- **Cost Reduction:** A cost-reduction project is a project that attempts to lower a firm's operating costs. Typically, we need to consider whether a company should buy equipment to perform an operation now done manually or spend money now in order to save more money later. The expected future cash inflows on this investment are savings resulting from lower operating costs. (See Figure 1.9.)

- **Equipment Replacement:** This category of investment decisions involves considering the expenditure necessary to replace worn-out or obsolete equipment. For example, a company may purchase 10 large presses with the expectation that they will produce stamped metal parts for 10 years. After five

---

[9]Since plastic is petroleum based, it is inherently more expensive than steel, and because the plastic-forming process involves a chemical reaction, it has a slower cycle time. However, both machinery and tool costs for plastic are lower than for steel because of relatively low-forming pressures, lack of tool abrasion, and single-stage pressing involved in handling. Thus, the plastic would require a lower investment, but would incur higher material costs.

• Should a company buy equipment to perform an operation now done manually?

• Should a company spend money now in order to save more money later?

FIGURE 1.9 **Cost-reduction decision**

• <u>Now</u> is the time to replace the old machine?

• If not, <u>when</u> is the right time to replace the old equipment?

FIGURE 1.10 **Replacement decision: Is it worth fixing or replacing?**

years, however, it may become necessary to produce the parts in plastic, which would require retiring the presses early and purchasing plastic-molding machines. Similarly, a company may find that, for competitive reasons, larger and more accurate parts are required, which will make the purchased machines obsolete earlier than expected. (See Figure 1.10.)

1-4 Fundamental Principles in Engineering Economics

This book is focused on the principles and procedures for making sound engineering economic decisions. To the first-time student of engineering economics, anything related to money matters may seem quite strange compared with other engineering subjects. However, the decision logic involved in the problem solving is quite similar to any other engineering subject matter; there are basic fundamental principles to follow in any engineering economic decision. These principles unite to form the concepts and techniques presented in the text, thereby allowing us to focus on the logic underlying the practice of engineering economics.

The four principles of engineering economics are as follows:

- **Principle 1: A nearby dollar is worth more than a distant dollar.** A fundamental concept in engineering economics is that money has a time value associated with it. Because we can earn interest on money received today, it is better to receive money earlier than later. This concept will be the basic foundation for all engineering project evaluation.

- **Principle 2: All that counts is the differences among alternatives.** An economic decision should be based on the differences among alternatives considered. All that is common is irrelevant to the decision. Certainly, any economic decision is no better than the alternatives being considered. Therefore, an economic decision should be based on the objective of making the best use of limited resources. Whenever a choice is made, something is given up. The opportunity cost of a choice is the value of the best alternative given up.

- **Principle 3: Marginal revenue must exceed marginal cost.** Any increased economic activity must be justified based on the following fundamental economic principle: marginal revenue must exceed marginal cost. Here, the marginal revenue is the additional revenue made possible by increasing the activity by one unit (or a small unit). Similarly, marginal cost is the additional cost incurred by the same increase in activity. Productive resources such as natural resources, human resources, and capital goods available to make goods and services are limited. Therefore, people cannot have all the goods and services they want; as a result, they must choose those things that produce the most.

- **Principle 4: Additional risk is not taken without the expected additional return.** For delaying consumption, investors demand a minimum return that must be greater than the anticipated rate of inflation or any perceived risk. If they didn't receive enough to compensate for anticipated inflation and perceived investment risk, investors would purchase whatever goods they desired ahead of time or invest in assets that would provide a sufficient return to compensate for any loss from inflation or potential risk.

These four principles are as much statements of common sense as they are theoretical statements. They provide the logic behind what is to follow in this text. We build on them and attempt to draw out their implications for decision making. As we continue, try to keep in mind that while the topics being treated may change from chapter to chapter, the logic driving our treatment of them is constant and rooted in these four principles.

Summary

- This chapter has provided an overview of a variety of engineering economic problems that commonly are found in the business world. We examined the place of engineers in a firm, and we saw that engineers have been playing an increasingly important role in companies, as evidenced in General Motors's development of an electrical vehicle known as the EV1. Commonly, engineers are called upon to participate in a variety of strategic business decisions ranging from product design to marketing.

- The term "engineering economic decision" refers to all investment decisions relating to engineering projects. The most interesting facet of an economic decision, from an engineer's point of view, is the evaluation of costs and benefits associated with making a capital investment.

- The five main types of engineering economic decisions are (1) service or quality improvement, (2) new products or product expansion, (3) equipment and process selection, (4) cost reduction, and (5) equipment replacement.

- The factors of time and uncertainty are the defining aspects of any investment project.

To Take or Not to Take the Offer[1]

Recently, a suburban Chicago couple won Powerball, a multistate lottery game. The game had rolled over for several weeks, so a huge jackpot was at stake. Ticket buyers had the choice between a single lump sum of $104 million or a total of $198 million paid out over 25 years (or $7.92 million per year) should they win the game. The winning couple opted for the lump sum. From a strictly economic standpoint, did the couple make the more lucrative choice?

[1]"It's official: Illinois couple wins $104 million Powerball prize," CNN.com, May 22, 1998—the couple's $104.3 million payout will be reduced by a 28 percent assessment in federal taxes and 6.87 percent in state taxes, leaving the couple with $67,940,000.

Time Value
of Money

I f you were the winner of the aforementioned jackpot, you might well
wonder why the value of the single lump-sum payment—$104 million,
paid immediately—is so much lower than the total value of the annuity
payments—$198 million, paid in 25 installments. Isn't receiving $198 million overall a lot better than receiving just $104 million now? The answer to
your question involves the principles we will discuss in this chapter, namely,
the operation of interest and the time value of money.

The question we just posed provides a good starting point for this chapter. Everyone knows that it is better to receive a dollar today than it is to receive a dollar in 10 years, but how do we quantify the difference? Our
Powerball example is a bit more involved. Instead of a choice between two
single payments, our lottery winners were faced with a decision between a
single payment now and an entire series of future payments. First, most people familiar with investments would tell you that receiving $104 million
today, like the Chicago couple did, is likely to prove a better deal than taking
$7.92 million a year for 25 years. In fact, based on the principles you will
learn in this chapter, the real present value of the 25-year payment series—
the value that you could receive today in the financial marketplace for the

promise of $7.92 million a year for the next 25 years—can be shown to be considerably less than $104 million. And that is even before we consider the effects of inflation! The reason for this surprising result is the **time value of money**; that is, the earlier a sum of money is received, the more it is worth, because over time money can earn more money, or interest.

In engineering economic analysis, the principles discussed in this chapter are regarded as the underpinnings of nearly all project investment analysis. These principles are so important because we always need to account for the effect of interest operating on sums of cash over time. Interest formulas allow us to place different cash flows received at different times in the same time frame and to compare them. As will become apparent, almost our entire study of engineering economic analysis is built on the principles introduced in this chapter.

2-1 Interest: The Cost of Money

Most of us are familiar in a general way with the concept of interest. We know that money left in a savings account earns interest so that the balance over time is greater than the sum of the deposits. We know that borrowing to buy a car means repaying an amount over time, that it includes interest, and that the amount paid is therefore greater than the amount borrowed. What may be unfamiliar to us is the idea that, in the financial world, money itself is a commodity, and like other goods that are bought and sold, money costs money.

The cost of money is established and measured by an **interest rate**, a percentage that is periodically applied and added to an amount (or varying amounts) of money over a specified length of time. When money is borrowed, the interest paid is the charge to the borrower for the use of the lender's property; when money is loaned or invested, the interest earned is the lender's gain for providing a good to another. **Interest**, then, may be defined as the cost of having money available for use. In this section, we examine how interest operates in a free-market economy and establish a basis for understanding the more complex interest relationships that are presented later on in the chapter.

2.1.1 The Time Value of Money

The time value of money seems like a sophisticated concept, yet it is one that you grapple with every day. Should you buy something today or save your money and buy it later? Here is a simple example of how your buying behavior can have varying results: Pretend you have $100 and you want to buy a $100 refrigerator for your dorm room. If you buy it now, you end up broke. But if you invest your money at 6% annual interest, then in a year you can still buy the refrigerator, and you will have $6 left over. However, if the price of the refrigerator increases at an annual rate of 8% due to inflation, then you will not have enough money (you will be $2 short) to buy the refrigerator a year from now. In that case, you probably are better off buying the refrigerator now (Case 1

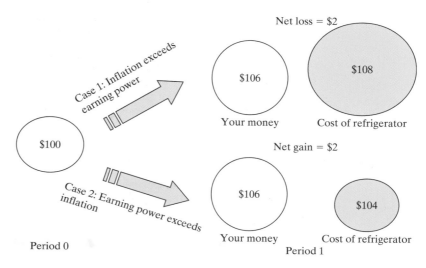

FIGURE 2.1 Gains achieved or losses incurred by delaying consumption

in Figure 2.1). If the inflation rate is running at only 4%, then you will have $2 left over if you buy the refrigerator a year from now (Case 2 in Figure 2.1).

Clearly, the rate at which you earn interest should be higher than the inflation rate in order to make any economic sense of the delayed purchase. In other words, in an inflationary economy, your purchasing power will continue to decrease as you further delay the purchase of the refrigerator. In order to make up this future loss in purchasing power, the rate at which you earn interest should be sufficiently larger than the anticipated inflation rate. After all, time, like money, is a finite resource. There are only 24 hours in a day, so time has to be budgeted, too. What this example illustrates is that we must connect *earning power* and *purchasing power* to the concept of time.

The way interest operates reflects the fact that money has a time value. This is why amounts of interest depend on lengths of time; interest rates, for example, are typically given in terms of a percentage per year. We may define the principle of the time value of money as follows: The economic value of a sum depends on when the sum is received. Because money has both **earning power** and **purchasing power** over time (i.e., it can be put to work, earning more money for its owner), a dollar received today has a greater value than a dollar received at some future time. When we deal with large amounts of money, long periods of time, or high interest rates, the change in the value of a sum of money over time becomes extremely significant. For example, at a current annual interest rate of 10%, $1 million will earn $100,000 in interest in a year; thus, to wait a year to receive $1 million clearly involves a significant sacrifice. When deciding among alternative proposals, we must take into account the operation of interest and the time value of money in order to make valid comparisons of different amounts at various times.

When financial institutions quote lending or borrowing interest rates in the marketplace, those interest rates reflect the desired earning rate, as well as any protection from loss in the future purchasing power of money because of inflation.

Interest rates, adjusted for inflation, rise and fall to balance the amount saved with the amount borrowed, which affects the allocation of scarce resources between present and future uses.

Unless stated otherwise, we will assume that the interest rates used in this book reflect the **market interest rate**, which considers the earning power of money as well as the effect of inflation perceived in the marketplace. We will also assume that all cash flow transactions are given in terms of **actual dollars**, where the effect of inflation, if any, is reflected in the amount.

2.1.2   Elements of Transactions Involving Interest

Many types of transactions involve interest—e.g., borrowing money, investing money, or purchasing machinery on credit—but certain elements are common to all of these types of transactions:

1. The initial amount of money invested or borrowed in transactions is called the **principal** (P).
2. The **interest rate** (i) measures the cost or price of money and is expressed as a percentage per period of time.
3. A period of time called the **interest period** (n) determines how frequently interest is calculated. (Note that, even though the length of time of an interest period can vary, interest rates are frequently quoted in terms of an annual percentage rate. We will discuss this potentially confusing aspect of interest in Chapter 3.)
4. A specified length of time marks the duration of the transaction and thereby establishes a certain **number of interest periods** (N).
5. **A plan for receipts or disbursements** (A_n) that yields a particular cash flow pattern over a specified length of time. (For example, we might have a series of equal monthly payments that repay a loan.)
6. A **future amount of money** (F) results from the cumulative effects of the interest rate over a number of interest periods.

Example of an Interest Transaction   As an example of how the elements we have just defined are used in a particular situation, let us suppose that an electronics manufacturing company borrows $20,000 from a bank at a 9% annual interest rate in order to buy a machine. In addition, the company pays a $200 loan origination fee[2] when the loan commences. The bank offers two repayment plans, one with equal payments made at the end of every year for the next five years and the other with a single payment made after the loan period of five years. These payment plans are summarized in Table 2.1.

- In Plan 1, the principal amount, P, is $20,000, and the interest rate, i, is 9%. The interest period, n, is one year, and the duration of the transaction is five years,

---

[2]The loan origination fee covers the administrative costs of processing the loan. It is often expressed in points. One point is 1% of the loan amount. For example, a $100,000 loan with a loan origination fee of one point would mean you pay $1,000. This is equivalent to financing $99,000, but your payments are based on a $100,000 loan.

End of Year	Receipts	Payments	
		Plan 1	Plan 2
Year 0	$20,000.00	$200.00	$200.00
Year 1		5,141.85	0
Year 2		5,141.85	0
Year 3		5,141.85	0
Year 4		5,141.85	0
Year 5		5,141.85	30,772.48

TABLE 2.1 Repayment Plans Offered by the Lender

Both payment plans are based on a rate of 9% interest.

which means that there are five interest periods ($N = 5$). It bears repeating that while one year is a common interest period, interest is frequently calculated at other intervals as well—monthly, quarterly, or semiannually, for instance. For this reason, we used the term **period** rather than **year** when we defined the preceding list of variables. The receipts and disbursements planned over the duration of this transaction yield a cash flow pattern of five equal payments A of $5,141.85 each, paid at year end during years one through five. (You'll have to accept these amounts on faith for now; the following section presents the formula used to arrive at the amount of these equal payments, given the other elements of the problem.)

- Plan 2 has most of the elements of Plan 1, except that instead of five equal repayments, we have a grace period followed by a single future repayment F of $30,772.48.

Cash Flow Diagrams   Problems involving the time value of money can be conveniently represented in graphic form with a **cash flow diagram** (Figure 2.2). Cash flow diagrams represent time by a horizontal line marked off with the number of interest periods specified. Arrows represent the cash flows over time at relevant periods: Upward arrows represent positive flows (receipts), and downward arrows represent negative flows (disbursements). Note, too, that the arrows actually represent **net cash flows**: Two or more receipts or disbursements made at the same time are summed and shown as a single arrow. For example, $20,000 received during the same period as a $200 payment would be recorded as an upward arrow of $19,800. The lengths of the arrows can also suggest the relative values of particular cash flows.

Cash flow diagrams function in a manner similar to free-body diagrams or circuit diagrams, which most engineers frequently use. Cash flow diagrams give a convenient summary of all the important elements of a problem as well as serve as a reference point for determining whether the elements of a problem have been converted into their appropriate parameters. This text frequently uses this graphic tool,

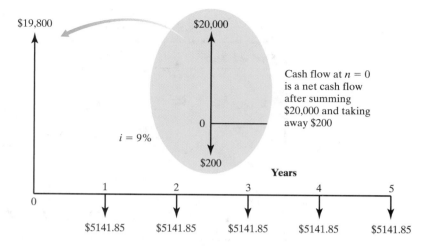

FIGURE 2.2 **A cash flow diagram for Plan 1 of the loan repayment example**

and you are strongly encouraged to develop the habit of using well-labeled cash flow diagrams as a means to identify and summarize pertinent information in a cash flow problem. Similarly, a table such as Table 2.1 can help you organize information in another summary format.

End-of-Period Convention   In practice, cash flows can occur at the *beginning* or in the *middle* of an interest period or at practically any point in time. One of the simplifying assumptions we make in engineering economic analysis is the **end-of-period convention**, which is the practice of placing all cash flow transactions at the *end* of an interest period. This assumption relieves us of the responsibility of dealing with the effects of interest within an interest period, which would greatly complicate our calculations. Like many of the simplifying assumptions and estimates we make in modeling engineering economic problems, the end-of-period convention inevitably leads to some discrepancies between our model and real-world results.

Suppose, for example, that $100,000 is deposited during the first month of the year in an account with an interest period of one year and an interest rate of 10% per year. In such a case, if the deposit is withdrawn of one month before the end of the year, the investor would experience an interest income loss of $10,000—all the interest! This is because, under the end-of-period convention, the $100,000 deposit made during the interest period is viewed as if it were made at the end of the year as opposed to 11 months earlier. This example gives you a sense of why financial institutions choose interest periods that are less than one year, even though they usually quote their rate in terms of *annual percentage*. Armed with an understanding of the basic elements involved in interest problems, we can now begin to look at the details of calculating interest.

2.1.3 **Methods of Calculating Interest**

Money can be loaned and repaid in many ways, and, equally, money can earn interest in many different ways. Usually, however, at the end of each interest period, the interest earned on the principal amount is calculated according to a specified interest rate. The two computational schemes for calculating this earned interest yield either **simple interest** or **compound interest**. *Engineering economic analysis uses the compound-interest scheme exclusively*, as it is most frequently practiced in the real world.

Simple Interest   The first scheme considers interest earned on only the principal amount during each interest period. In other words, under simple interest, the interest earned during each interest period does not earn additional interest in the remaining periods, *even though you do not withdraw it.*

In general, for a deposit of P dollars at a simple interest rate of i for N periods, the total earned interest I would be

$$I = (iP)N. \tag{2.1}$$

The total amount available at the end of N periods, F, thus would be

$$F = P + I = P(1 + iN). \tag{2.2}$$

Simple interest is commonly used with add-on loans or bonds.

Compound Interest   Under a compound-interest scheme, the interest earned in each period is calculated based on the total amount at the end of the previous period. This total amount includes the original principal plus the accumulated interest that has been left in the account. In this case, you are in effect increasing the deposit amount by the amount of interest earned. In general, if you deposited (invested) P dollars at an interest rate i, you would have $P + iP = P(1 + i)$ dollars at the end of one interest period. If the entire amount (principal and interest) were reinvested at the same rate i for another period, you would have at the end of the second period

$$P(1 + i) + i[P(1 + i)] = P(1 + i)(1 + i)$$
$$= P(1 + i)^2.$$

Continuing, we see that the balance after period three is

$$P(1 + i)^2 + i[P(1 + i)^2] = P(1 + i)^3.$$

This interest-earning process repeats, and after N periods, the total accumulated value (balance) F will grow to

$$F = P(1 + i)^N. \tag{2.3}$$

Example 2.1 Simple versus Compound Interest

Suppose you deposit $1,000 in a bank savings account that pays interest at a rate of 8% per year. Assume that you don't withdraw the interest earned at the end of each period (year), but instead let it accumulate. (a) How much would you have at the end of year three with simple interest? (b) How much would you have at the end of year three with compound interest?

SOLUTION

Given: $P = \$1,000$, $N = 3$ years, and $i = 8\%$ per year.

Find: F.

(a) Simple interest: We calculate F as

$$F = \$1,000[1 + (0.08)3] = \$1,240.$$

- P = Principal amount
- i = Interest rate
- N = Number of interest periods
- Example:
  $P = \$1,000$
  $i = 8\%$
  $N = 3$ years

End of Year	Beginning Balance	Interest Earned	Ending Balance
0			$1,000
1	$1,000	$80	$1,080
2	$1,080	$80	$1,160
3	$1,160	$80	$1,240

(b) Compound interest: Applying Eq. (2.3) to our three-year, 8% case, we obtain

$$F = \$1000(1 + 0.08)^3 = \$1259.71.$$

The total interest earned is $259.71, which is $19.71 more than was accumulated under the simple-interest method. We can keep track of the interest-accruing process more precisely as follows:

Period (n)	Amount at Beginning of Interest Period	Interest Earned for Period	Amount at End of Interest Period
1	$1,000.00	$80.00	$1,080.00
2	$1,080.00	$86.40	$1,166.40
3	$1,166.40	$93.31	$1,259.71

Comments: At the end of the first year, you would have $1,000 plus $80 in interest, or a total of $1,080. In effect, at the beginning of the second year, you would be depositing $1,080, rather than $1,000. Thus, at the end of the second year, the interest earned would be 0.08($1,080) = $86.40, and the balance would be $1,080 + $86.40 = $1,166.40. This is the amount you would be depositing at the beginning of the third year, and the interest earned for that period would be 0.08($1,166.40) = $93.31. With a beginning principal amount of $1,166.40 plus the $93.31 interest, the total balance would be $1,259.71 at the end of year three.

2-2   Economic Equivalence

The observation that money has a time value leads us to an important question: If receiving $100 today is not the same as receiving $100 at any future point, how do we measure and compare various cash flows? How do we know, for example, whether we should prefer to have $20,000 today, and $50,000 ten years from now or $8,000 each year for the next 10 years (Figure 2.3)? In this section, we will describe the basic analytical techniques for making these comparisons. Then, in Section 2.3, we will use these techniques to develop a series of formulas that can greatly simplify our calculations.

2.2.1   Definition and Simple Calculations

The central factor in deciding among alternative cash flows involves comparing their economic worth. This would be a simple matter if, in the comparison, we did not need to consider the time value of money: We could simply add the individual payments within a cash flow, treating receipts as positive cash flows and payments (disbursements) as negative cash flows. The fact that money has a time value makes our calculations more complicated. Calculations for determining the economic effects of one or more cash flows are based on the concept of economic equivalence.

Economic equivalence exists between cash flows that have the same economic effect and could therefore be traded for one another in the financial marketplace, which we assume to exist. Economic equivalence refers to the fact that a cash flow—whether a single payment or a series of payments—can be converted to an *equivalent* cash flow at any point in time. The important fact to remember about the present value of future cash flows is that the present sum is equivalent in value to the future cash flows. It is equivalent because if you had the present value today, you

- Economic equivalence exists between cash flows that have the *same economic effect* and could therefore be traded for one another

- Even though the amounts and timing of the cash flows may differ, the appropriate interest rate makes them equal

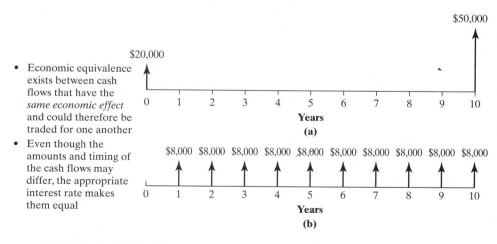

FIGURE 2.3 Which option would you prefer? (a) Two payments ($20,000 now and $50,000 at the end of 10 years) or (b) 10 equal annual receipts in the amount of $8,000 each

- If you deposit P dollars today for N periods at i, you will have F dollars at the end of period N
- F dollars at the end of period N is equal to a single sum of P dollars now if your earning power is measured in terms of the interest rate i

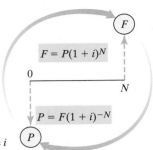

$$F = P(1 + i)^N$$

$$P = F(1 + i)^{-N}$$

FIGURE 2.4 Using compound interest to establish economic equivalence

could transform it into the future cash flows simply by investing it at the interest rate, also referred to as the **discount rate.**

The strict conception of equivalence may be extended to include the comparison of alternatives. For example, we could compare the value of two proposals by finding the equivalent value of each at any common point in time. If financial proposals that appear to be quite different turn out to have the same monetary value, then we can be *economically indifferent* to choosing between them. That is, in terms of economic effect, one would be an even exchange for the other, so no reason exists to prefer one over the other in terms of their economic value.

A way to see the concepts of equivalence and economic indifference at work in the real world is to note the variety of payment plans offered by lending institutions for consumer loans. Recall Table 2.1, where we showed two different repayment plans for a loan of $20,000 for five years at an annual interest rate of 9%. You will notice, perhaps to your surprise, that the two plans require significantly different repayment patterns and different total amounts of repayment. However, because money has time value, these plans are equivalent. In other words, economically, the bank is indifferent to the consumer's choice of plan. We will now discuss how such equivalence relationships are established.

Equivalence Calculations: A Simple Example   Equivalence calculations can be viewed as an application of the compound-interest relationships we developed in Section 2.1. Suppose, for example, that we invest $1,000 at 12% annual interest for five years. The formula developed for calculating compound interest, $F = P(1 + i)^N$ [Eq. (2.3)], expresses the equivalence between some present amount P and a future amount F for a given interest rate i and a number of interest periods, N. Therefore, at the end of the investment period, our sums grow to

$$\$1,000(1 + 0.12)^5 = \$1,762.34.$$

Thus, we can say that at 12% interest, $1,000 received now is equivalent to $1,762.34 received in five years, and thus we could trade $1,000 now for the promise of receiving $1,762.34 in five years. Example 2.2 further demonstrates the application of this basic technique.

Example 2.2 Equivalence

Suppose you are offered the alternative of receiving either $3,000 at the end of five years or P dollars today. There is no question that the $3,000 will be paid in full (i.e., no risk). Because you have no current need for the money, you would deposit the P dollars in an account that pays 8% interest. What value of P would make you indifferent to your choice between P dollars today and the promise of $3,000 at the end of five years?

Discussion: Our job is to determine the present amount that is economically equivalent to $3,000 in five years, given the investment potential of 8% per year. Note that the statement of the problem assumes that you would exercise the option of using the earning power of your money by depositing it. The "indifference" ascribed to you refers to economic indifference; that is, within a marketplace where 8% is the applicable interest rate, you could trade one cash flow for the other.

SOLUTION

Given: $F = \$3,000$, $N = 5$ years, $i = 8\%$ per year.
Find: P.
Equation: Eq. (2.3), $F = P(1 + i)^N$.

Rearranging Eq. (2.3) to solve for P, we obtain

$$P = \frac{F}{(1 + i)^N}.$$

Substituting the given quantities into the equation yields

$$P = \frac{\$3,000}{(1 + 0.08)^5} = \$2,042.$$

We can summarize the problem graphically as in Figure 2.5.

- Step 1: Determine the *base period*, say, year five
- Step 2: Identify the *interest rate* to use
- Step 3: Calculate *equivalence value*

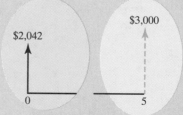

$i = 6\%, P = \$3,000/(1 + 0.06)^5 = \$2,242$
$i = 8\%, P = \$3,000/(1 + 0.08)^5 = \$2,042$
$i = 10\%, P = \$3,000/(1 + 0.10)^5 = \$1,863$

FIGURE 2.5 Equivalence calculations at varying interest rate

Comments: In this example, it is clear that if P is anything less than \$2,042, you would prefer the promise of \$3,000 in five years to P dollars today; if P is greater than \$2,042, you would prefer P. As you may have already guessed, at a lower interest rate, P must be higher in order to be equivalent to the future amount. For example, at $i = 4\%$, $P = \$2,466$.

2.2.2   Equivalence Calculations Require a Common Time Basis for Comparison

Just as we must convert fractions to common denominators in order to add them together, we must convert cash flows to a common basis in order to compare their value. One aspect of this basis is the choice of a single point in time at which to make our calculations. In Example 2.2, if we had been given the magnitude of each cash flow and had been asked to determine whether they were equivalent, we could have chosen any reference point and used the compound-interest formula to find the value of each cash flow at that point. As you can readily see, the choice of $n = 0$ or $n = 5$ would make our problem simpler, because we would need to make only one set of calculations: At 8% interest, either convert \$2,042 at time 0 to its equivalent value at time 5, or convert \$3,000 at time 5 to its equivalent value at time 0.

When selecting a point in time at which to compare the value of alternative cash flows, we commonly use either the present time, which yields what is called the **present worth** of the cash flows, or some point in the future, which yields their **future worth**. The choice of the point in time often depends on the circumstances surrounding a particular decision, or it may be chosen for convenience. For instance, if the present worth is known for the first two of three alternatives, simply calculating the present worth of the third will allow us to compare all three. For an illustration, consider Example 2.3.

Example 2.3 Equivalence Calculation

Consider the cash flow series given in Figure 2.6. Compute the equivalent lump-sum amount at $n = 3$ at 10% annual interest.

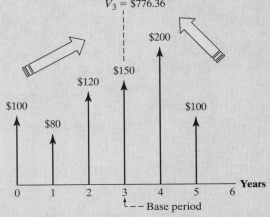

FIGURE 2.6 Equivalent-worth calculation at $n = 3$

SOLUTION

Given: The cash flows given in Figure 2.6, and $i = 10\%$.

Find: V_3 (or equivalent worth at $n = 3$).

We find the equivalent worth at $n = 3$ in two steps. First, we find the future worth of each cash flow at $n = 3$ for all cash flows that occur before $n = 3$. Second, we find the present worth of each cash flow at $n = 3$ for all cash flows that occur after $n = 3$.

(handwritten: what are we solving?)

- **Step 1:** Find the equivalent lump-sum payment of the first four payments at $n = 3$:

$$\$100(1 + 0.10)^3 + \$80(1 + 0.10)^2 + \$120(1 + 0.10)^1 + \$150 = \$511.90.$$

- **Step 2:** Find the equivalent lump-sum payment of the remaining two payments at $n = 3$:

$$\$200(1 + 0.10)^{-1} + \$100(1 + 0.10)^{-2} = \$264.46.$$

- **Step 3:** Find V_3, the total equivalent value:

$$V_3 = \$511.90 + \$264.46 = \$776.36.$$

2-3 Interest Formulas for Single Cash Flows

We begin our coverage of interest formulas by considering the simplest of cash flows: single cash flows.

2.3.1 Compound-Amount Factor

Given a present sum P invested for N interest periods at interest rate i, what sum will have accumulated at the end of the N periods? You probably noticed right away that this description matches the case we first encountered in describing compound interest. To solve for F (the future sum), we use Eq. (2.3):

$$F = P(1 + i)^N.$$

Because of its origin in the compound-interest calculation, the factor $(1 + i)^N$ is known as the **compound-amount factor**. Like the concept of equivalence, this factor is one of the foundations of engineering economic analysis. Given this factor, all other important interest formulas can be derived.

This process of finding F is often called the **compounding process**. The cash flow transaction is illustrated in Figure 2.7. (Note the time-scale convention: the first period begins at $n = 0$ and ends at $n = 1$.) If a calculator is handy, it is easy enough to calculate $(1 + i)^N$ directly.

Interest Tables Interest formulas such as the one developed in Eq. (2.3), $F = P(1 + i)^N$, allow us to substitute known values from a particular situation

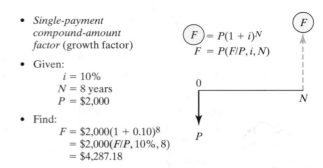

- *Single-payment compound-amount factor* (growth factor)
- Given:
  $i = 10\%$
  $N = 8$ years
  $P = \$2,000$
- Find:
  $F = \$2,000(1 + 0.10)^8$
  $= \$2,000(F/P, 10\%, 8)$
  $= \$4,287.18$

$\boxed{F} = P(1 + i)^N$
$F = P(F/P, i, N)$

FIGURE 2.7 Compounding process: Find *F*, given *P*, *i*, and *N*

into the equation and to solve for the unknown. Before the hand calculator was developed, solving these equations was very tedious. With a large value of N, for example, one might need to solve an equation such as $F = \$20,000(1 + 0.12)^{15}$. More complex formulas required even more involved calculations. To simplify the process, tables of compound-interest factors were developed. These tables allow us to find the appropriate factor for a given interest rate and the number of interest periods. Even though hand calculators are now readily available, it is still often convenient to use these tables, which are included in this text in Appendix B. Take some time now to become familiar with their arrangement. If you can, locate the compound-interest factor for the example just presented, in which we know P and, to find F, we need to know the factor by which to multiply $\$20,000$ when the interest rate i is 12% and the number of periods is 15:

$$F = \$20,000\underbrace{(1 + 0.12)^{15}}_{5.4736} = \$109,472.$$

Factor Notation   As we continue to develop interest formulas in the rest of this chapter, we will express the resulting compound-interest factors in a conventional notation that can be substituted into a formula to indicate precisely which table factor to use in solving an equation. In the preceding example, for instance, the formula derived as Eq. (2.3) is $F = P(1 + i)^N$. To specify how the interest tables are to be used, we may also express that factor in a functional notation as $(F/P, i, N)$, which is read as "Find F, given P, i, and N." This factor is known as the **single-payment compound-amount factor**. When we incorporate the table factor into the formula, the formula is expressed as follows:

$$F = P(1 + i)^N = P(F/P, i, N).$$

Thus, in the preceding example, where we had $F = \$20,000(1.12)^{15}$, we can now write $F = \$20,000(F/P, 12\%, 15)$. The table factor tells us to use the 12%-interest table and find the factor in the F/P column for $N = 15$. Because using the interest tables is often the easiest way to solve an equation, this factor notation is included for each of the formulas derived in the upcoming sections.

Example 2.4 Single Amounts: Find F, Given P, i, and N

If you had $2,000 now and invested it at 10% interest compounded annually, how much would it be worth in eight years (Figure 2.8)?

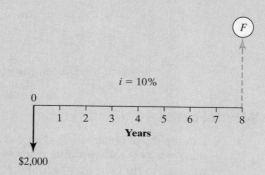

FIGURE 2.8 **Cash flow diagram**

SOLUTION

Given: $P = \$2,000$, $i = 10\%$ per year, and $N = 8$ years.

Find: F.

We can solve this problem in any of three ways:

1. Using a calculator: You can simply use a calculator to evaluate the $(1 + i)^N$ term (financial calculators are preprogrammed to solve most future-value problems):

$$F = \$2,000(1 + 0.10)^8$$
$$= \$4,287.18.$$

2. Using compound-interest tables: The interest tables can be used to locate the compound-amount factor for $i = 10\%$ and $N = 8$. The number you get can be substituted into the equation. Compound-interest tables are included in Appendix B of this book. Using this method, we obtain

$$F = \$2,000\ (F/P, 10\%, 8) = \$2,000(2.1436) = \$4,287.20.$$

This amount is essentially identical to the value obtained by direct evaluation of the single cash flow compound-amount factor. This slight deviation is due to rounding differences.

3. Using a computer: Many financial software programs for solving compound-interest problems are available for use with personal computers. As summarized in Appendix A, many spreadsheet programs such as Excel also provide financial functions to evaluate various interest formulas. With Excel, the future worth calculation looks like the following:

$$= \text{FV}(10\%, 8, 0, 2000, 0).$$

- *Single-payment present-worth factor (discount factor)*

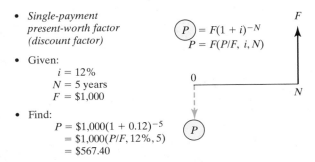

- Given:
  $i = 12\%$
  $N = 5$ years
  $F = \$1,000$

- Find:
  $P = \$1,000(1 + 0.12)^{-5}$
  $= \$1,000(P/F, 12\%, 5)$
  $= \$567.40$

FIGURE 2.9 Discounting process: Find *P*, given *F*, *i*, and *N*

2.3.2 Present-Worth Factor

Finding the present worth of a future sum is simply the reverse of compounding and is known as the **discounting process**. (See Figure 2.9.) In Eq. (2.3), we can see that if we need to find a present sum *P*, given a future sum *F*, we simply solve for *P*:

$$P = F\left[\frac{1}{(1 + i)^N}\right] = F(P/F, i, N). \tag{2.4}$$

The factor $1/(1 + i)^N$ is known as the **single-payment present-worth factor** and is designated $(P/F, i, N)$. Tables have been constructed for P/F factors and for various values of *i* and *N*. The interest rate *i* and the P/F factor are also referred to as the **discount rate** and the **discounting factor**, respectively.

Example 2.5 Single Amounts: Find *P*, Given *F*, *i*, and *N*

A zero-coupon bond[3] is a popular variation on the bond theme for some investors. What should be the price of an eight-year zero-coupon bond with a face value of $1,000 if similar, non-zero-coupon bonds are yielding 6% annual interest?

Discussion: As an investor of a zero-coupon bond, you do not receive any interest payments until the bond reaches maturity. When the bond matures, you will receive $1,000 (the face value). In lieu of getting interest payments, you can buy the bond at a discount. The question is what the price of the bond should be in order to realize a 6% return on your investment. (See Figure 2.10.)

---

[3]Bonds are loans that investors make to corporations and governments. In Example 2.5, the $1,000 of principal is the **face value** of the bond, the yearly interest payment is its **coupon**, and the length of the loan is bond's **maturity**.

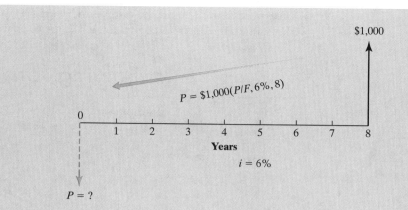

FIGURE 2.10 Cash flow diagram

SOLUTION

Given: $F = \$1,000$, $i = 6\%$ per year, and $N = 8$ years.
Find: P.

Using a calculator, we obtain

$$P = \$1,000(1 + 0.06)^{-8} = \$1,000(0.6274) = \$627.40.$$

Using a calculator may be the best way to make this simple calculation. It is equivalent to finding the present value of the $1,000 face value at 6% interest. We can also use the interest tables to find that

$$P = \$1,000 \overbrace{(P/F, 6\%, 8)}^{(0.6274)} = \$627.40.$$

Again, you could also use a financial calculator or computer to find the present worth. With Excel, the present-value calculation looks like the following:

$$= \text{PV}(6\%, 8, 0, 1000, 0).$$

2.3.3 Solving for Time and Interest Rates

At this point, you should realize that the compounding and discounting processes are reciprocals of one another and that we have been dealing with one equation in two forms:

$$\text{Future-value form: } F = P(1 + i)^N$$

and

$$\text{Present-value form: } P = F(1 + i)^{-N}.$$

There are four variables in these equations: P, F, N, and i. If you know the values of any three, you can find the value of the fourth. Thus far, we have always given you the interest rate (i) and the number of years (N), plus either P or F. In many situations, though, you will need to solve for i or N, as we discuss next.

Example 2.6 Solving for i

Suppose you buy a share of stock for $10 and sell it for $20; your profit is thus $10. If that happens within a year, your rate of return is an impressive 100% ($10/$10 = 1). If it takes five years, what would be the rate of return on your investment? (See Figure 2.11.)

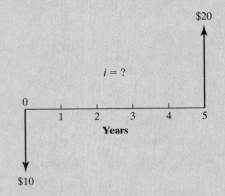

FIGURE 2.11 Cash flow diagram

SOLUTION

Given: $P = \$10$, $F = \$20$, and $N = 5$.
Find: i.

Here, we know P, F, and N, but we do not know i, the interest rate you will earn on your investment. This type of rate of return calculation is straightforward, since you make only a one-time lump-sum investment. We start with the following relationship:

$$F = P(1 + i)^N.$$

We then substitute in the given values:

$$\$20 = \$10(1 + i)^5.$$

Next, we solve for i by one of the following methods:

- **Method 1:** Go through a trial-and-error process in which you insert different values of i into the equation until you find a value that "works" in the sense that the right-hand side of the equation equals $20. The solution value is $i = 14.87\%$. The trial-and-error procedure is extremely tedious and inefficient for most problems, so it is not widely practiced in the real world.

- **Method 2:** You can solve the problem by using the interest tables in Appendix B. Start with the equation

$$\$20 = \$10(1 + i)^5,$$

which is equivalent to

$$2 = (1 + i)^5 = (F/P, i, 5).$$

Now look across the $N = 5$ row under the $(F/P, i, 5)$ column until you can locate the value of 2. This value is approximated in the 15% interest table at $(F/P, 15\%, 5) = 2.0114$, so the interest rate at which $10 grows to $20 over five years is very close to 15%. This procedure will be very tedious for fractional interest rates or when N is not a whole number, as you may have to approximate the solution by linear interpolation.

- **Method 3:** The most practical approach is to use either a financial calculator or an electronic spreadsheet such as Excel. A financial function such as RATE(N, 0, P, F) allows us to calculate an unknown interest rate. The precise command statement would be as follows:

$$= \text{RATE}(5,0, -10,20) = 14.87\%.$$

Note that we enter the present value (P) as a negative number in order to indicate a cash outflow in Excel.

Example 2.7 Single Amounts: Find N, Given P, F, and i

You have just purchased 100 shares of General Electric stock at $30 per share. You will sell the stock when its market price doubles. If you expect the stock price to increase 12% per year, how long do you expect to wait before selling the stock (Figure 2.12)?

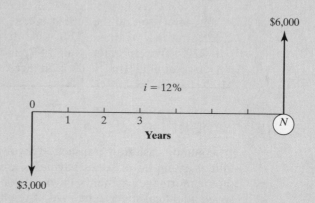

FIGURE 2.12 Cash flow diagram

SOLUTION

Given: $P = \$3{,}000$, $F = \$6{,}000$, and $i = 12\%$ per year.

Find: N (years).

Using the single-payment compound-amount factor, we write

$$F = P(1 + i)^N = P(F/P, i, N),$$

which in this case is

$$\$6{,}000 = \$3{,}000(1 + 0.12)^N = \$3{,}000(F/P, 12\%, N),$$

or

$$2 = (1.12)^N = (F/P, 12\%, N).$$

Again, we could use a calculator or a computer spreadsheet program to find N.

1. **Using a calculator:** We start with

$$\log 2 = N \log 1.12.$$

   Solving for N gives

$$N = \frac{\log 2}{\log 1.12}$$
$$= 6.11 \approx 6 \text{ years}.$$

2. **Using a spreadsheet program:** Within Excel, the financial function NPER(i, 0, P, F) computes the number of compounding periods it will take an investment P to grow to a future value F, earning a fixed interest rate i per compounding period. In our example, the Excel command would look like this:

$$= \text{NPER}(12\%, 0, -3000, 6000).$$

   The calculated result is 6.1163.

Comments: A very handy rule of thumb called the *Rule of 72* can determine approximately how long it will take for a sum of money to double. The rule states that, to find the time it takes for the present sum of money to grow by a factor of two, we divide 72 by the interest rate. For our example, the interest rate is 12%. Therefore, the Rule of 72 indicates that it will take $\frac{72}{12} = 6$ years for a sum to double. This result is, in fact, relatively close to our exact solution.

2-4   Uneven-Payment Series

A common cash flow transaction involves a series of disbursements or receipts. Familiar examples of series payments are payment of installments on car loans and home mortgage payments. Payments on car loans and home mortgages typically involve identical sums to be paid at regular intervals. When there is no clear pattern over the series, we call the transaction an uneven cash flow series.

We can find the present worth of any uneven stream of payments by calculating the present worth of each individual payment and summing the results. Once the present worth is found, we can make other equivalence calculations (e.g., future worth can be calculated by using the interest factors developed in the previous section).

Example 2.8 Present Values of an Uneven Series by Decomposition into Single Payments

Wilson Technology, a growing machine shop, wishes to set aside money now to invest over the next four years in automating its customer service department. The company can earn 10% on a lump sum deposited now, and it wishes to withdraw the money in the following increments:

- Year 1: $25,000 to purchase a computer and database software designed for customer service use;
- Year 2: $3,000 to purchase additional hardware to accommodate anticipated growth in use of the system;
- Year 3: No expenses; and
- Year 4: $5,000 to purchase software upgrades.

How much money must be deposited now in order to cover the anticipated payments over the next four years?

Discussion: This problem is equivalent to asking what value of P would make you indifferent in your choice between P dollars today and the future expense stream of ($25,000, $3,000, $0, $5,000). One way to deal with an uneven series of cash flows is to calculate the equivalent present value of each single cash flow and then to sum the present values to find P. In other words, the cash flow is broken into three parts as shown in Figure 2.13.

SOLUTION

Given: Uneven cash flow in Fig. 2.13; $i = 10\%$ per year.
Find: P.

We sum the individual present values as follows:

$$P = \$25,000(P/F, 10\%, 1) + \$3,000(P/F, 10\%, 2) + \$5,000(P/F, 10\%, 4)$$
$$= \$28,622.$$

Comments: To see if $28,622 is indeed a sufficient amount, let's calculate the balance at the end of each year. If you deposit $28,622 now, it will grow to (1.10)($28,622), or $31,484, at the end of year one. From this balance, you pay out $25,000. The remaining balance, $6,484, will again grow to (1.10)($6,484), or $7,132, at the end of year two. Now you make the second payment ($3,000) out of this balance, which will leave you with only $4,132 at the end of year two.

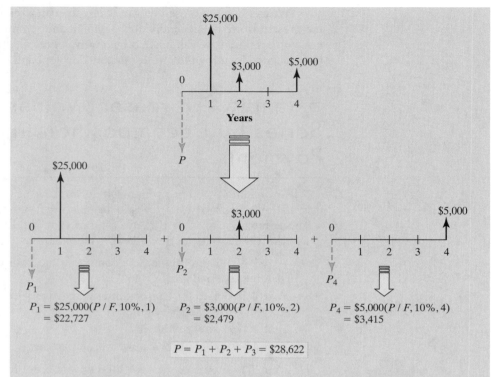

FIGURE 2.13 Decomposition of uneven cash flow series

Since no payment occurs in year three, the balance will grow to $\$(1.10)^2(\$4,132)$, or \$5,000, at the end of year four. The final withdrawal in the amount of \$5,000 will deplete the balance completely.

2-5 Equal-Payment Series

As we learned in Example 2.8, the present worth of a stream of future cash flows can always be found by summing the present worth of each individual cash flow. However, if cash flow regularities are present within the stream, the use of shortcuts, such as finding the present worth of a uniform series, may be possible. We often encounter transactions in which a uniform series of payments exists. Rental payments, bond interest payments, and commercial installment plans are based on uniform payment series. Our interest is to find the equivalent present worth (P) or future worth (F) of such a series, as illustrated in Figure 2.14.

2.5.1 Compound-Amount Factor: Find *F*, Given *A*, *i*, and *N*

Suppose we are interested in the future amount F of a fund to which we contribute A dollars each period and on which we earn interest at a rate of i per period. The contributions are made at the end of each of the N periods. These transactions are graphically illustrated in Figure 2.15. Looking at this diagram, we see that, if an amount A is invested at the end of each period for N periods, the total amount F

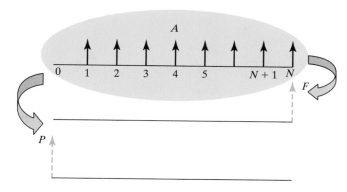

FIGURE 2.14 Equal payment series: Find equivalent P or F

that can be withdrawn at the end of N periods will be the sum of the compound amounts of the individual deposits.

As shown in Figure 2.15, the A dollars we put into the fund at the end of the first period will be worth $A(1 + i)^{N-1}$ at the end of N periods. The A dollars we put into the fund at the end of the second period will be worth $A(1 + i)^{N-2}$, and so forth. Finally, the last A dollars that we contribute at the end of the Nth period will be worth exactly A dollars at that time. This means we have a series in the form

$$F = A(1 + i)^{N-1} + A(1 + i)^{N-2} + \cdots + A(1 + i) + A,$$

or, expressed alternatively,

$$F = A + A(1 + i) + A(1 + i)^2 + \cdots + A(1 + i)^{N-1}. \tag{2.5}$$

Multiplying Eq. (2.5) by $(1 + i)$ results in

$$(1 + i)F = A(1 + i) + A(1 + i)^2 + \cdots + A(1 + i)^N. \tag{2.6}$$

Subtracting Eq. (2.5) from Eq. (2.6) to eliminate common terms gives us

$$F(1 + i) - F = -A + A(1 + i)^N.$$

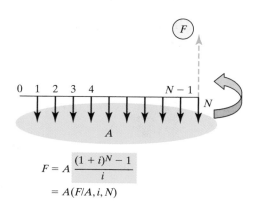

FIGURE 2.15 Cash flow diagram of the relationship between A and F

Solving for F yields

$$F = A\left[\frac{(1 + i)^N - 1}{i}\right] = A(F/A, i, N).\qquad(2.7)$$

The bracketed term in Eq. (2.7) is called the **equal-payment-series compound-amount factor**, or the **uniform-series compound-amount factor**; its factor notation is $(F/A, i, N)$. This interest factor has been calculated for various combinations of i and N in the tables in Appendix B.

Example 2.9 Equal-Payment Series: Find F, Given i, A, and N

Suppose you make an annual contribution of \$5,000 to your savings account at the end of each year for five years. If your savings account earns 6% interest annually, how much can be withdrawn at the end of five years (Figure 2.16)?

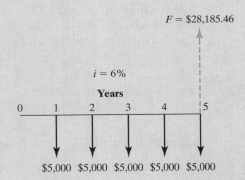

FIGURE 2.16 Cash flow diagram

SOLUTION

Given: $A = \$5,000$, $N = 5$ years, and $i = 6\%$ per year.

Find: F.

Using the equal-payment-series compound-amount factor, we obtain

$$\begin{aligned}F &= \$5,000(F/A, 6\%, 5)\\&= \$5,000(5.6371)\\&= \$28,185.46.\end{aligned}$$

To obtain the future value of the annuity on Excel, we may use the following financial command:

$$= \text{FV}(6\%,5,5000,,0).$$

We may be able to keep track of how the periodic balances grow in the savings account as follows:

End of Year	1	2	3	4	5
Beginning Balance	0	$5,000.00	$10,300.00	$15,918.00	$21,873.08
Interest Earned (6%)	0	300.00	618.00	955.08	1,312.38
Deposit Made	5,000.00	5,000.00	5,000.00	5,000.00	5,000.00
Ending Balance	$5,000.00	$10,300.00	$15,918.00	$21,873.08	$28,185.46

Example 2.10 Handling Time Shifts in a Uniform Series

In Example 2.9, the first deposit of the five-deposit series was made at the *end* of period one, and the remaining four deposits were made at the end of each following period. Suppose that all deposits were made at the *beginning* of each period instead. How would you compute the balance at the end of period five?

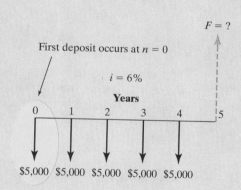

FIGURE 2.17 Cash flow diagram

SOLUTION

Given: Cash flow diagram in Figure 2.17; $i = 6\%$ per year.
Find: F_5.

Compare Figure 2.17 with Figure 2.16: Each payment in Figure 2.17 has been shifted one year earlier; thus, each payment is compounded for one extra year. Note that with the end-of-year deposit, the ending balance F was $28,185.46. With the beginning-of-year deposit, the same balance accumulates by the end of period four. This balance can earn interest for one additional year. Therefore, we can easily calculate the resulting balance as

$$F_5 = \$28{,}185.46(1.06) = \$29{,}876.59.$$

Annuity due can be easily evaluated using the following financial command available on Excel:

$$= \text{FV}(6\%,5,5000,,1).$$

Comments: Another way to determine the ending balance is to compare the two cash flow patterns. By adding the $5,000 deposit at period zero to the original cash flow and subtracting the $5,000 deposit at the end of period five, we obtain the second cash flow. Therefore, the ending balance can be found by making an adjustment to the $28,185.46:

$$F_5 = \$28{,}185.46 + \$5{,}000(F/P, 6\%, 5) - \$5{,}000 = \$29{,}876.59.$$

2.5.2 Sinking-Fund Factor: Find *A*, Given *F*, *i*, and *N*

If we solve Eq. (2.7) for A, we obtain

$$A = F\left[\frac{i}{(1 + i)^N - 1}\right] = F(A/F, i, N). \tag{2.8}$$

The term within the brackets is called the **equal-payment-series sinking-fund factor**, or just **sinking-fund factor**, and is referred to with the notation $(A/F, i, N)$. A sinking fund is an interest-bearing account into which a fixed sum is deposited each interest period; it is commonly established for the purpose of replacing fixed assets.

Example 2.11 College Savings Plan: Find *A*, Given *F*, *N*, and *i*

You want to set up a college savings plan for your daughter. She is currently 10 years old and will go to college at age 18. You assume that when she starts college, she will need at least $100,000 in the bank. How much do you need to save each year in order to have the necessary funds if the current rate of interest is 7%? Assume that end-of-year payments are made.

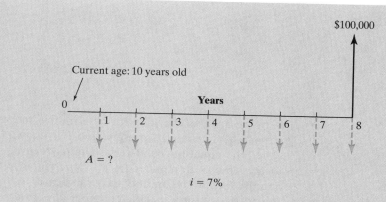

FIGURE 2.18 Cash flow diagram

SOLUTION

Given: Cash flow diagram in Figure 2.18, $i = 7\%$ per year, and $N = 8$ years.
Find: A.

Using the sinking-fund factors, we obtain

$$A = \$100,000(A/F, 7\%, 8)$$
$$= \$9,746.78.$$

2.5.3 Capital-Recovery Factor (Annuity Factor): Find **A**, Given **P**, **i**, and **N**

We can determine the amount of a periodic payment, A, if we know P, i, and N. Figure 2.19 illustrates this situation. To relate P to A, recall the relationship between P and F in Eq. (2.3): $F = P(1 + i)^N$. By replacing F in Eq. (2.8) by $P(1 + i)^N$, we get

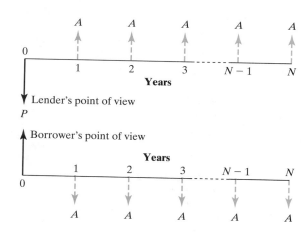

FIGURE 2.19 Cash flow diagram of the relationship between *P* and *A*

$$A = P(1 + i)^N \left[\frac{i}{(1 + i)^N - 1} \right],$$

or

$$A = P \left[\frac{i(1 + i)^N}{(1 + i)^N - 1} \right] = P(A/P, i, N). \tag{2.9}$$

Now we have an equation for determining the value of the series of end-of-period payments, A, when the present sum P is known. The portion within the brackets is called the **equal-payment-series capital-recovery factor**, or simply **capital-recovery factor**, which is designated $(A/P, i, N)$. In finance, this A/P factor is referred to as the **annuity factor**. The annuity factor indicates a series of payments of a fixed, or constant, amount for a specified number of periods.

Example 2.12 Paying Off an Educational Loan: Find A, Given P, i, and N

You borrowed $21,061.82 to finance the educational expenses for your senior year of college. The loan will be paid off over five years. The loan carries an interest rate of 6% per year and is to be repaid in equal annual installments over the next five years. Assume that the money was borrowed at the beginning of your senior year and that the first installment will be due a year later. Compute the amount of the annual installments (Figure 2.20).

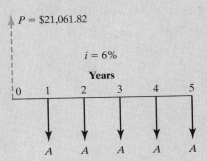

$P = \$21,061.82$

$i = 6\%$

Years

FIGURE 2.20 **Cash flow diagram**

SOLUTION

Given: $P = \$21,061.82$, $i = 6\%$ per year, and $N = 5$ years.
Find: A.
Using the capital-recovery factor, we obtain

$$A = \$21,061.82(A/P, 6\%, 5)$$
$$= \$21,061.82(0.2374)$$
$$= \$5,000.$$

The following table illustrates how the $5,000 annual repayment plan would re-tire the debt in five years:

End of Year	1	2	3	4	5
Beginning Balance	$21,061.82	$17,325.53	$13,365.06	$9,166.96	$4,716.98
Interest Charged (6%)	1,263.71	1,039.53	801.90	550.02	283.02
Payment Made	−5,000.00	−5,000.00	−5,000.00	−5,000.00	−5,000.00
Ending Balance	**$17,325.53**	**$13,365.06**	**$9,166.96**	**$4,716.98**	**$0.00**

The Excel solution using annuity function commands is as follows:

$$= \text{PMT}(i,N,P)$$
$$= \text{PMT}(6\%, 5, 21061.82).$$

The result of this formula is $5,000.

Example 2.13 Deferred Loan Repayment

Suppose in Example 2.12 that you had wanted to negotiate with the bank to defer the first loan installment until the end of year two (but still desire to make five equal installments at 6% interest). If the bank wishes to earn the same profit as in Example 2.12, what should be the annual installment (Figure 2.21)?

SOLUTION

Given: $P = \$21,061.82$, $i = 6\%$ per year, and $N = 5$ years, but the first payment occurs at the end of year two.
Find: A.

In deferring one year, the bank will add the interest accrued during the first year to the principal. In other words, we need to find the equivalent worth of $21,061.82 at the end of year 1, P':

$$P' = \$21,061.82(F/P, 6\%, 1)$$
$$= \$22,325.53.$$

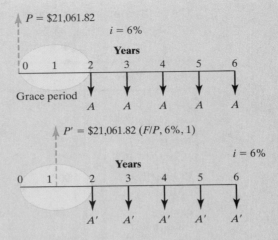

FIGURE 2.21 A deferred-loan cash flow diagram

Thus, you are borrowing $22,325.53 for five years. To retire the loan with five equal installments, the deferred equal annual payment, A', will be

$$A' = \$22,325.53(A/P, 6\%, 5)$$
$$= \$5,300.$$

By deferring the first payment for one year, you need to make an additional $300 in payments each year.

2.5.4 Present-Worth Factor: Find *P*, Given *A*, *i*, and *N*

What would you have to invest now in order to withdraw A dollars at the end of each of the next N periods? We now face just the opposite of the equal-payment capital-recovery factor situation: A is known, but P has to be determined. With the capital-recovery factor given in Eq. (2.9), solving for P gives us

$$P = A\left[\frac{(1 + i)^N - 1}{i(1 + i)^N}\right] = A(P/A, i, N). \qquad (2.10)$$

The bracketed term is referred to as the **equal-payment-series present-worth factor** and is designated $(P/A, i, N)$.

Example 2.14 Uniform Series: Find *P*, Given *A*, *i*, and *N*

Let us revisit the lottery problem introduced in the chapter opening. Recall that the Chicago couple gave up the installment plan of $7.92 million a year for 25 years to receive a cash lump sum of $104 million. If the couple could invest its money at 8% annual interest, did it make the right decision? What is the lump-sum amount that would make the couple indifferent to each payment plan? (See Figure 2.22.)

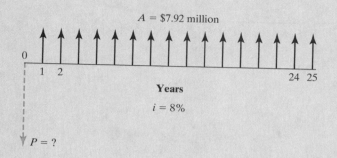

FIGURE 2.22 Cash flow diagram

SOLUTION

Given: i = 8% per year, A = \$7.92 million, and N = 25 years.

Find: P.

- **Tabular Solution:**

$$P = \$7.92(P/A, 8\%, 25) = \$7.92(10.6748)$$
$$= \$84.54 \text{ million.}$$

- **Excel solution:**

$$= PV(8\%,25,7.92,,0) = \$84.54 \text{ million.}$$

Comments: Clearly, we can tell the couple that giving up \$7.92 million a year for 25 years to receive \$104 million today is a winning proposition if they can earn an 8% return on its investment. At this point, we may be interested in knowing the minimum rate of return at which accepting the \$104 million lump sum would make sense. Since we know that P = \$104 million, N = 25, and A = \$7.92 million, we solve for i.

   If you know the cash flows and the present value (or future value) of a cash flow stream, you can determine the interest rate. In this case, we are looking for the interest rate that causes the P/A factor to equal $(P/A, i, 25) = (\$104/\$7.92) = 13.1313$. Since we are dealing with an annuity, we could proceed as follows:

- With a financial calculator, enter N = 25, P = -104, and A = 7.92, and then press the i key to find that i = 5.7195%. For a typical financial calculator, the symbols such as PV, PMT, and FV are commonly adopted on its key pad. These symbols correspond to PV = P, PMT = A, and FV = F, respectively. Note that just like Excel either P or A must be entered as a negative number to evaluate i correctly.

- If you want to use interest tables, first recognize that $104 = \$7.92$ $(P/A, i, 25)$, or $(P/A, i, 25) = 13.1313$. Look up 13.1313 or a close value in Appendix B. In the P/A column with $N = 25$ in Table B.23, you will find that $(P/A, 6\%, 25) = 12.7834$ and $(P/A, 5\%, 25) = 14.0939$, indicating that the interest rate should be rather close to 6%. If a factor does not appear in the table, the interest rate is not a whole number. In such cases, you cannot use this procedure to find the exact rate. In practice, this is not a problem, because in business people use financial calculators to find interest rates.

- In Excel, simply evaluate the following command to solve the unknown-interest-rate problem for an annuity:

$$= \text{RATE}(N, A, P, F, type, guess)$$
$$= \text{RATE}(25, 7.92, 104, 0, 0, 5\%)$$
$$= 5.7195\%.$$

It is likely the Chicago couple will find a financial investment that provides a rate of return higher than 5.7195%. Therefore, their decision to go with the lump-sum payment option appears to be a good deal.

Example 2.15 Start Saving Money as Soon as Possible: Composite Series That Requires Both (F/P, i, N) and (F/A, i, N) Factors

Consider the following two savings plans that you consider starting at the age of 21:

- Option 1: Save $2,000 a year for 10 years. At the end of 10 years, make no further investments, but invest the amount accumulated at the end of 10 years until you reach the age of 65. (Assume that the first deposit will be made when you are 22.)
- Option 2: Do nothing for the first 10 years. Start saving $2,000 a year every year thereafter until you reach the age of 65. (Assume that the first deposit will be made when you turn 32.)

If you were able to invest your money at 8% over the planning horizon, which plan would result in more money saved by the time you are 65? (See Figure 2.23.)

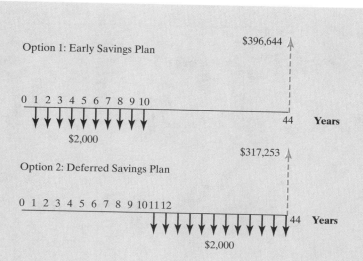

FIGURE 2.23 Cash flow diagrams for two different savings plans

SOLUTION

Given: $i = 8\%$, deposit scenarios shown in Figure 2.23.
Find: F when you are 65.

- Option 1: Compute the final balance in two steps. First, compute the accumulated balance at the end of 10 years (when you are 31). Call this amount F_{31}.

$$F_{31} = \$2,000(F/A, 8\%, 10) = \$28,973.$$

Then use this amount to compute the result of reinvesting the entire balance for another 34 years. Call this final result F_{65}.

$$F_{65} = \$28,973(F/P, 8\%, 34) = \$396,645.$$

- Option 2: Since you have only 34 years to invest, the resulting balance will be

$$F_{65} = \$2,000(F/A, 8\%, 34) = \$317,253.$$

With the early savings plan, you will be able to save $79,391 more.

Comments: In this example, the assumed interest rate was 8%. Certainly, we would be interested in knowing at what interest rate these two options would be equivalent. We can use Excel's **Goal Seek**[4] function to answer this question. As

[4]Goal Seek is part of a suite of commands sometimes called *what-if analysis* tools in Excel. When you know the desired result of a single *formula* but not the input value the formula needs to determine the result, you can use the Goal Seek feature by clicking **Goal Seek** on the **Tools** menu. When *goal seeking*, Microsoft Excel varies the value in one specific cell until a formula that is dependent on that cell returns the result you want.

TABLE 2.2 Using the Goal Seek to Find the Break-Even Interest Rate to Make Two Options Equivalent

	A	B	C	D	E	F
1						
2	Year	Option 1	Option 2			
3	0					
4	1	$ (2,000)				
5	2	$ (2,000)			Interest rate	0.08
6	3	$ (2,000)				
7	4	$ (2,000)			FV of Option 1	$396,645.95
8	5	$ (2,000)				
9	6	$ (2,000)			FV of Option 2	$317,253.34
10	7	$ (2,000)				
11	8	$ (2,000)			Target cell	$79,392.61
12	9	$ (2,000)				
13	10	$ (2,000)				
14	11		$ (2,000)			
15	12		$ (2,000)			
16	13		$ (2,000)			
17	14		$ (2,000)			
18	15		$ (2,000)			
19	16		$ (2,000)			
20	17		$ (2,000)			
21	18		$ (2,000)			
22	19		$ (2,000)			
40	37		$ (2,000)			
41	38		$ (2,000)			
42	39		$ (2,000)			
43	40		$ (2,000)			
44	41		$ (2,000)			
45	42		$ (2,000)			
46	43		$ (2,000)			
47	44		$ (2,000)			

=FV(F5,10, -2000)*(1+F5)^(34)

=FV(F5,34,-2000)

shown in Table 2.2, we enter the amount of deposits over 44 years in the second and third columns. Cells F7 and F9 respectively display the future value of each option. Cell F11 contains the difference between the future values of the two options, or " = F7 − F9". To begin using the Goal Seek function, first define Cell F11 as your *set cell*. Specify "set value" as "0", and set the "*By changing cell*" to be F5. Use the Goal Seek function to change the interest rate in Cell F5 incrementally until the value in Cell F11 equals "0." The break-even interest rate is 6.538%.

	E	**F**
5	Interest	6.5383607%
6		
7	FV of Option 1	$232,902.09
8		
9	FV of Option 2	$232,902.09
10		
11	Option 1 - Option 2	$0

"By changing cell" ← (row 5)

"Set cell" ← (row 11)

2.5.5 Present Value of Perpetuities

A perpetuity is a stream of cash flows that continues forever. A good example is a share of preferred stock that pays a fixed cash dividend each period (usually a quarter-year) and never matures. An interesting feature of any perpetual annuity is that you cannot compute the future value of its cash flows, because it is infinite. However, it has a well-defined present value. It appears counterintuitive that a series of cash flows that lasts forever can have a finite value today.

What is the value of a perpetuity? We know how to calculate the present value for an equal-payment series with the finite stream as shown in Eq. (2.10). If we take a limit on this equation by letting $N \to \infty$, we can find the closed-form solution as follows:

$$P = \frac{A}{i}. \tag{2.11}$$

To illustrate, consider a perpetual stream of $1,000 per year, as depicted in Figure 2.24. If the interest rate is 10% per year, how much is this perpetuity worth today? The answer is $10,000. To see why, consider how much money you would have to put into a bank account offering interest of 10% per year in order to be able to take out $1,000 every year forever. Certainly, if you put in $10,000, then at the end of the first year you

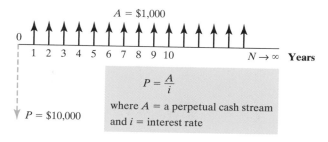

FIGURE 2.24 Present value of perpetual cash streams

would have $11,000 in the account. You take out $1,000, leaving $10,000 for the next year. Clearly, if the interest rate stays at 10% per year, you will not eat into the principal, so you could continue to take out $1,000 every year forever.

2-6   Dealing with Gradient Series

Engineers frequently encounter situations involving periodic payments that in- crease or decrease by a constant amount G or constant percentage (growth rate) from period to period. We can easily develop a series of interest formulas for this sit- uation, but Excel will be a more practical tool to calculate equivalent values for these types of cash flows.

2.6.1   Handling Linear Gradient Series

Sometimes cash flows will vary linearly, that is, they increase or decrease by a set amount, G, the gradient amount. This type of series is known as a **strict gradient series**, as seen in Figure 2.25. Note that each payment is $A_n = (n - 1)G$. Note also that the series begins with a zero cash flow at the end of period zero. If $G > 0$, the series is referred to as an *increasing* gradient series. If $G < 0$, it is referred to as a *decreasing* gradient series.

Linear Gradient Series as Composite Series   Unfortunately, the strict form of the increasing or decreasing gradient series does not correspond to the form that most engineering economic problems take. A typical problem involving a linear gra- dient series includes an initial payment during period one that increases by G during some number of interest periods, a situation illustrated in Figure 2.26. This configu- ration contrasts with the strict form illustrated in Figure 2.25, in which no payment is made during period one and the gradient is added to the previous payment begin- ning in period two.

In order to use the strict gradient series to solve typical problems, we must view cash flows as shown in Figure 2.26 as a **composite series**, or a set of two cash flows, each corresponding to a form that we can recognize and easily solve. Figure 2.26

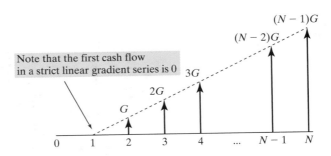

FIGURE 2.25 Cash flow diagram of a strict gradient series

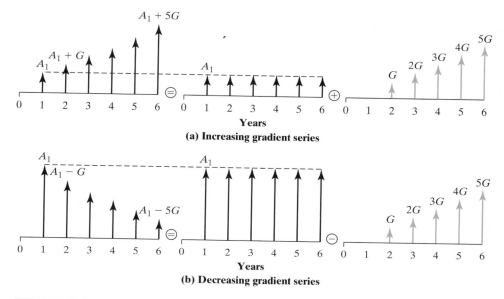

FIGURE 2.26 Two types of linear gradient series as composites of a uniform series of N payments of A_1 and a gradient series of increments of a constant amount G

illustrates that the form in which we find a typical cash flow can be separated into two components: a uniform series of N payments of amount A_1 and a gradient series of increments of a constant amount G. The need to view cash flows that involve linear gradient series as composites of two series is very important for the solution of problems, as we shall now see.

Present-Worth Factor: Linear Gradient: Find P, Given G, N, and i How much would you have to deposit now in order to withdraw the gradient amounts specified in Figure 2.27? To find an expression for the present amount P, we apply the single-payment present-worth factor to each term of the series, obtaining

$$P = 0 + \frac{G}{(1+i)^2} + \frac{2G}{(1+i)^3} + \cdots + \frac{(N-1)G}{(1+i)^N},$$

or

$$P = \sum_{n=1}^{N}(n-1)G(1+i)^{-n}. \tag{2.12}$$

After some algebraic operations, we obtain

$$P = G\left[\frac{(1+i)^N - iN - 1}{i^2(1+i)^N}\right] = G(P/G, i, N). \tag{2.13}$$

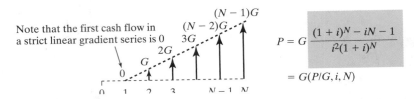

Note that the first cash flow in a strict linear gradient series is 0

$$P = G\,\frac{(1+i)^N - iN - 1}{i^2(1+i)^N}$$

$$= G(P/G, i, N)$$

FIGURE 2.27 Cash flow diagram of a strict gradient series

The resulting factor in brackets is called the **gradient-series present-worth factor**[5] and is designated by the notation $(P/G, i, N)$.

Example 2.16 Using Excel to Find Present Worth for a Linear Gradient Series

So, what could be better than winning a SuperLotto Plus jackpot? Choosing how to receive your winnings! Before playing a SuperLotto Plus jackpot, you have a choice between getting the entire jackpot in 26 annual graduated payments or receiving one lump sum that will be less than the announced jackpot. (See Figure 2.28.) What would these choices come out to for an announced jackpot of $7 million?

- **Lump-sum cash-value option:** The winner would receive the present cash value of the announced jackpot in one lump sum. In this case, the winner would receive about 49.14%, or $3.44 million, in one lump sum (less tax withholdings). This cash value is based on average market costs determined by U.S. Treasury zero-coupon bonds with 5.3383% annual yield.

- **Annual-payments option:** The winner would receive the jackpot in 26 graduated annual payments. In this case, the winner would receive $175,000 as the first payment (2.5% of the total jackpot amount). The second payment would be $189,000. Over the course of the next 25 years,

---

[5]We can obtain an equal payment series equivalent to the gradient series by multiplying Eq. (2.13) by Eq. (2.9). The resulting factor is referred to as the **gradient-to-equal payment series conversion factor** with designation of $(A/G, i, N)$, or

$$A = G\left[\frac{(1+i)^N - iN - 1}{i[(1+i)^N - 1]}\right].$$

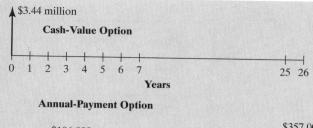

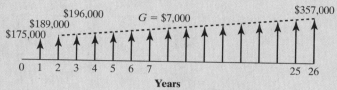

FIGURE 2.28 Cash flow diagram

these payments would gradually increase each year by $7,000 to a final payment of $357,000.

If the U.S. Treasury zero-coupon rate is reduced to 4.5% (instead of 5.338%) at the time of winning, what would be the equivalent cash value of the lottery?

SOLUTION

Given: $A_1 = \$175,000$, $A_2 = \$189,000$, $G = \$7,000$ (from payment periods 3 to 26), $i = 4.5\%$ per year, and $N = 26$ years, as shown in Figure 2.28.
Find: P.

This problem is equivalent to asking what the equivalent present worth for this annual-payment series is at 4.5% interest. Since the linear gradient series starts at period 2 for this example (i.e., is delayed by one period), we can calculate the present value in two steps: First compute the value at $N = 1$ and then extend it to $N = 0$. This method yields the following:

$$P = [\$175,000 + \$189,000(P/A, 4.5\%, 25)$$
$$+ \$7,000(P/G, 4.5\%, 25)](P/F, 4.5\%, 1)$$
$$= \$3,818,363.$$

Or, using Excel, we could reproduce the same result, as in Table 2.3. In obtaining column C of the spreadsheet in Table 2.3, the annual cash flow amount in each year is obtained by adding $7,000 to the amount in the previous period. For example, Cell C12 is obtained by = C11 + $C$6. Then Cell C13 is obtained by = C12 + $C$6, and so on.

The cash value now has increased from $3.44 million to $3.818 million. In other words, if you mark the "Cash Value" box on your lottery ticket and you win, you will receive the present cash value of the announced jackpot in one lump sum in the amount of $3.818 million.

TABLE 2.3 Excel Spreadsheet Presentation of the Linear Gradient Series

	A	B	C	D	E
1	Example 2.16 Cash Value Calculation Where the Payment				
2		Schedule Follows a Gradient Series			
3		Winning Jackpot	$ 7,000,000		
4		Interest Rate (%)	4.5%		
5		Base Amount	$ 189,000		
6		Gradient Amount	$ 7,000		
7					
8	Payment	Payment Schedule	Annual Payment	Discounting	Present
9	Number	as % of Jackpot	before Taxes	Factor (4.5%)	Cash Value
10	1	2.5%	$ 175,000	0.9569	$ 167,464
11	2	2.7%	$ 189,000	0.9157	$ 173,073
12	3	2.8%	$ 196,000	0.8763	$ 171,754
13	4	2.9%	$ 203,000	0.8386	$ 170,228
14	5	3.0%	$ 210,000	0.8025	$ 168,515
15	6	3.1%	$ 217,000	0.7679	$ 166,633
16	7	3.2%	$ 224,000	0.7348	$ 164,602
17	8	3.3%	$ 231,000	0.7032	$ 162,436
18	9	3.4%	$ 238,000	0.6729	$ 160,151
19	10	3.5%	$ 245,000	0.6439	$ 157,762
20	11	3.6%	$ 252,000	0.6162	$ 155,282
21	12	3.7%	$ 259,000	0.5897	$ 152,723
22	13	3.8%	$ 266,000	0.5643	$ 150,096
23	14	3.9%	$ 273,000	0.5400	$ 147,413
24	15	4.0%	$ 280,000	0.5167	$ 144,682
25	16	4.1%	$ 287,000	0.4945	$ 141,913
26	17	4.2%	$ 294,000	0.4732	$ 139,114
27	18	4.3%	$ 301,000	0.4528	$ 136,293
28	19	4.4%	$ 308,000	0.4333	$ 133,457
29	20	4.5%	$ 315,000	0.4146	$ 130,613
30	21	4.6%	$ 322,000	0.3968	$ 127,766
31	22	4.7%	$ 329,000	0.3797	$ 124,922
32	23	4.8%	$ 336,000	0.3634	$ 122,086
33	24	4.9%	$ 343,000	0.3477	$ 119,262
34	25	5.0%	$ 350,000	0.3327	$ 116,456
35	26	5.1%	$ 357,000	0.3184	$ 113,670
37	Total	100.0%	$ 7,000,000		$ 3,818,363

2.6.2 Handling Geometric Gradient Series

Another kind of gradient series is formed when the series in a cash flow is determined not by some fixed amount like $500, but by some *fixed rate* expressed as a percentage. Many engineering economic problems, particularly those relating to construction costs or maintenance costs, involve cash flows that increase or decrease over time by a constant percentage (**geometric**), a process that is called **compound growth**. Price changes caused by inflation are a good example of such a geometric series.

If we use g to designate the percentage change in a payment from one period to the next, the magnitude of the nth payment, A_n, is related to the first payment A_1 as follows:

$$A_n = A_1(1 + g)^{n-1}, n = 1, 2, \ldots, N. \tag{2.14}$$

The g can take either a positive or a negative sign, depending on the type of cash flow. If $g > 0$, the series will increase; if $g < 0$, the series will decrease. Figure 2.29 illustrates the cash flow diagram for this situation.

Present-Worth Factor: Find P, Given A_1, g, i, and N   Notice that the present worth P_n of any cash flow A_n at an interest rate i is

$$P_n = A_n(1 + i)^{-n} = A_1(1 + g)^{n-1}(1 + i)^{-n}.$$

To find an expression for the present amount for the entire series, P, we apply the **single-payment present-worth factor** to each term of the series:

$$P = \sum_{n=1}^{N} A_1(1 + g)^{n-1}(1 + i)^{-n}. \tag{2.15}$$

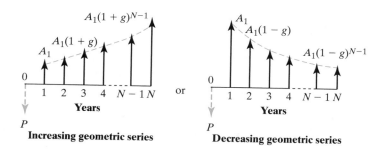

$$P = \begin{cases} A_1 \dfrac{1 - (1 + g)^N (1 + i)^{-N}}{i - g}, & \text{if } i \neq g; \\[2ex] A_1 \dfrac{N}{(1 + i)}, & \text{if } i = g. \end{cases}$$

FIGURE 2.29 **A geometrically increasing or decreasing gradient series**

The expression in Eq. (2.15) has the following closed expression.

$$P = \begin{cases} A_1 \left[\dfrac{1 - (1 + g)^N (1 + i)^{-N}}{i - g} \right] & \text{if } i \neq g, \\[3mm] A_1 \left(\dfrac{N}{1 + i} \right) & \text{if } i = g. \end{cases} \tag{2.16}$$

Or we can write

$$P = A_1 (P/A_1, g, i, N).$$

The factor within brackets is called the **geometric-gradient-series present-worth factor** and is designated $(P/A_1, g, i, N)$. In the special case where $i = g$, Eq. (2.16) becomes $P = [A_1/(1 + i)]N$.

There is an alternative way to derive the geometric-gradient-series present-worth factor. Bringing the constant term $A_1(1 + g)^{-1}$ in Eq. (2.15) outside the summation yields

$$P = \frac{A_1}{(1 + g)} \sum_{n=1}^{N} \left[\frac{1 + g}{1 + i} \right]^n = \frac{A_1}{(1 + g)} \sum_{n=1}^{N} \frac{1}{\left[1 + \dfrac{i - g}{1 + g} \right]^n}. \tag{2.17}$$

If we define

$$g' = \frac{i - g}{1 + g}$$

then we can rewrite P as follows:

$$P = \frac{A_1}{(1 + g)} \sum_{n=1}^{N} (1 + g')^{-n} = \frac{A_1}{(1 + g)} (P/A, g', N). \tag{2.18}$$

We don't need another interest-factor table for this **geometric-gradient-series present-worth factor**, as we can evaluate the factor with $(P/A, g', N)$. In the special case where $i = g$, Eq. (2.18) becomes $P = [A_1/(1 + i)]N$, as $g' = 0$.

Example 2.17 Required Cost-of-Living Adjustment Calculation

Suppose that your retirement benefits during your first year of retirement are $50,000. Assume that this amount is just enough to meet your cost of living during the first year. However, your cost of living is expected to increase at an annual rate of 5%, due to inflation. Suppose you do not expect to receive any cost-of-living adjustment in your retirement pension. Then, some of your future cost of living has to come from your savings other than retirement pension. If your savings account earns 7% interest a year, how much should you set aside in order to meet this future increase in cost of living over 25 years?

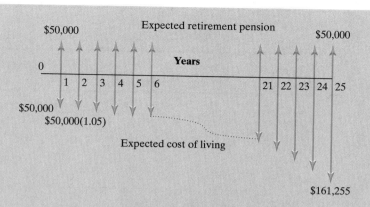

FIGURE 2.30 **Cash flow diagram**

SOLUTION

Given: $A_1 = \$50,000$, $g = 5\%$, $i = 7\%$, and $N = 25$ years, as shown in Figure 2.30.
Find: P.

- Find the equivalent amount of total benefits paid over 25 years:

$$P = \$50,000(P/A, 7\%, 25)$$
$$= \$582,679.$$

- Find the equivalent amount of total cost of living with inflation. To use Eq. (2.17), we need to find the value of g':

$$g' = \frac{0.07 - 0.05}{1 + 0.05} = 0.019, \text{ or } 1.9\%.$$

Then, using Eq. (2.17), we find P to be

$$P = \frac{\$50,000}{1 + 0.05}(P/A, 1.9\%, 25)$$
$$= \$940,696.$$

- The required additional savings to meet the future increase in cost of living will be

$$\Delta P = \$940,696 - \$582,679$$
$$= \$358,017.$$

Table 2.4 summarizes the interest formulas developed in this section and the cash flow situations in which they should be used. Recall that all the interest formulas developed in this section are applicable only to situations *where the interest (compounding) period is the same as the payment period* (e.g., annual compounding with annual payment). We also present some useful Excel's financial commands in this table.

TABLE 2.4 Formula Summary Table

Flow Type	Factor Notation	Formula	Excel Command	Cash Flow Diagram
S I N G L E	Compound amount $(F/P, i, N)$	$F = P(1 + i)^N$	$= \text{FV}(i, N, P,, 0)$	
	Present worth $(P/F, i, N)$	$P = F(1 + i)^{-N}$	$= \text{PV}(i, N, F,, 0)$	
E Q U A L P A Y M E N T S E R I E S	Compound amount $(F/A, i, N)$	$F = A\left[\dfrac{(1 + i)^N - 1}{i}\right]$	$= \text{PV}(i, N, A,, 0)$	
	Sinking fund $(A/F, i, N)$	$A = F\left[\dfrac{i}{(1 + i)^N - 1}\right]$	$= \text{PMT}(i, N, P, F, 0)$	
	Present worth $(P/A, i, N)$	$P = A\left[\dfrac{(1 + i)^N - 1}{i(1 + i)^N}\right]$	$= \text{PV}(i, N, A,, 0)$	
	Capital recovery $(A/P, i, N)$	$A = P\left[\dfrac{i(1 + i)^N}{(1 + i)^N - 1}\right]$	$= \text{PMT}(i, N,, P)$	
G R A D I E N T S E R I E S	Linear gradient Present worth $(P/G, i, N)$ Conversion factor $(A/G, i, N)$	$P = G\left[\dfrac{(1 + i)^N - iN - 1}{i^2(1 + i)^N}\right]$ $A = G\left[\dfrac{(1 + i)^N - iN - 1}{i[(1 + i)^N - 1]}\right]$		
S E R I E S	Geometric gradient Present worth $(P/A_1, g, i, N)$	$P = \left[\begin{array}{l} A_1\left[\dfrac{1 - (1 + g)^N(1 + i)^{-N}}{i - g}\right] \\ A_1\left(\dfrac{N}{1 + i}\right) \text{ (if } i = g) \end{array}\right.$		

2-7 Composite Cash Flows

Although many financial decisions involve constant or systematic changes in cash flows, many investment projects contain several components of cash flows that do not exhibit an overall pattern. Consequently, it is necessary to expand our analysis to deal with these mixed types of cash flows.

To illustrate, consider the cash flow stream shown in Figure 2.31. We want to compute the equivalent present worth for this mixed-payment series at an interest rate of 15%. Two different methods are presented:

1. **Method 1:** A "brute force" approach is to multiply each payment by the appropriate $(P/F, 10\%, n)$ factors and then to sum these products to obtain the present worth of the cash flows, $543.72 in this case. Recall that this is exactly the same procedure we used to solve the category of problems called the uneven-payment series, which were described in Section 2.4. Figure 2.31 illustrates this computational method. Excel is the best tool for this type of calculation.

2. **Method 2:** We may group the cash flow components according to the type of cash flow pattern that they fit, such as the single payment, equal-payment series, and so forth, as shown in Figure 2.32. Then the solution procedure involves the following steps:

   - Group 1: Find the present worth of $50 due in year one:

     $$\$50(P/F, 15\%, 1) = \$43.48.$$

   - Group 2: Find the equivalent worth of a $100 equal-payment series at year one (V_1), and then bring in this equivalent worth at year zero:

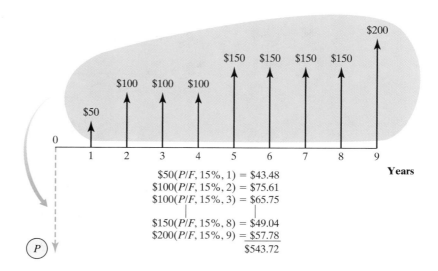

$\$50(P/F, 15\%, 1) = \$43.48$
$\$100(P/F, 15\%, 2) = \$75.61$
$\$100(P/F, 15\%, 3) = \$65.75$

$\$150(P/F, 15\%, 8) = \$49.04$
$\$200(P/F, 15\%, 9) = \underline{\$57.78}$
$\$543.72$

FIGURE 2.31 **Method 1: brute-force approach using** P/F **factors**

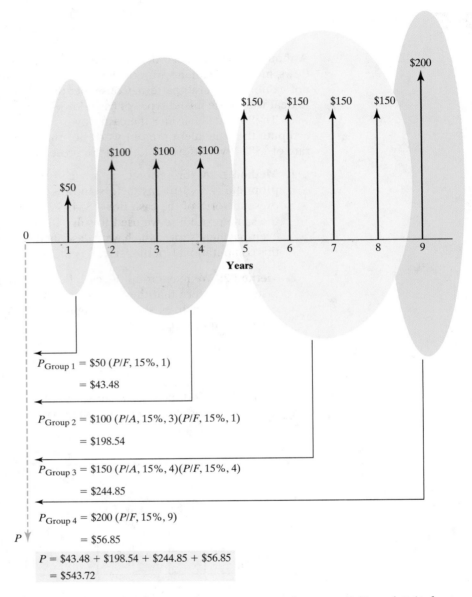

FIGURE 2.32 **Method 2: grouping approach using** *P/F* **and** *P/A* **factors**

$$\underbrace{\$100(P/A, 15\%, 3)}_{V_1}(P/F, 15\%, 1) = \$198.54.$$

- Group 3: Find the equivalent worth of a $150 equal-payment series at year four (V_4), and then bring in this equivalent worth at year zero.

$$\underbrace{\$150(P/A, 15\%, 4)}_{V_4}(P/F, 15\%, 4) = \$244.85.$$

- Group 4: Find the equivalent present worth of the $200 due in year nine:

$$\$200(P/F, 15\%, 9) = \$56.85$$

- For the group total, sum the components:

$$P = \$43.48 + \$198.54 + \$244.85 + \$56.85$$
$$= \$543.72$$

A pictoral view of this computational process is given in Figure 2.32.

Either the brute-force method in Figure 2.31 or the method using both the $(P/A, i, n)$ and $(P/F, i, n)$ factors in Figure 2.32 can be used to solve problems of this type. Method 2 is much easier if the annuity component runs for many years, however. For example, this solution would be clearly superior for finding the equivalent present worth of a payment stream consisting of $50 in year one, $200 in years two through 19, and $500 in year 20.

Summary

- Money has a time value because it can earn more money over time. A number of terms involving the time value of money were introduced in this chapter:

  Interest *is the cost of money. More specifically, it is a cost to the borrower and an earning to the lender above and beyond the initial sum borrowed or loaned.*

  Interest rate *is a percentage periodically applied to a sum of money to determine the amount of interest to be added to that sum.*

  Simple interest *is the practice of charging an interest rate only to an initial sum.*

  Compound interest *is the practice of charging an interest rate to an initial sum and to any previously accumulated interest that has not been withdrawn from the initial sum. Compound interest is by far the most commonly used system in the real world.*

  Economic equivalence *exists between individual cash flows or patterns of cash flows that have the same value. Even though the amounts and timing of the cash flows may differ, the appropriate interest rate makes them equal.*

- The following compound-interest formula is perhaps the single most important equation in this text:

$$F = P(1 + i)^{N}.$$

In this formula, P is a present sum, i is the interest rate, N is the number of periods for which interest is compounded, and F is the resulting future sum. All other important interest formulas are derived from this one.

- **Cash flow diagrams** are visual representations of cash inflows and outflows along a timeline. They are particularly useful for helping us detect which of the five patterns of cash flow is represented by a particular problem.
- The five patterns of cash flow are as follows:
  1. Single payment: A single present or future cash flow.
  2. Uniform series: A series of flows of equal amounts at regular intervals.

3. Linear gradient series: A series of flows increasing or decreasing by a fixed amount at regular intervals. Excel is one of the most convenient tools to solve this type of cash flow series.

4. Geometric gradient series: A series of flows increasing or decreasing by a fixed percentage at regular intervals. Once again, this type of cash flow series is a good candidate for solution by Excel.

5. Uneven series: A series of flows exhibiting no overall pattern. However, patterns might be detected for portions of the series.

• **Cash flow patterns** are significant because they allow us to develop **interest formulas**, which streamline the solution of equivalence problems. Table 2.4 summarizes the important interest formulas that form the foundation for all other analyses you will conduct in engineering economic analysis.

Problems

Methods of Calculating Interest

2.1 What is the amount of interest earned on $2,000 for five years at 10% simple interest per year?

2.2 You deposit $3,000 in a savings account that earns 9% simple interest per year. How many years will it take to double your balance? If instead you deposit the $3,000 in another savings account that earns 8% interest compounded yearly, how many years will it take to double your balance?

2.3 Compare the interest earned on $1,000 for 10 years at 7% simple interest with the amount of interest earned if interest were compounded annually.

2.4 You are considering investing $1,000 at an interest rate of 6% compounded annually for five years or investing the $1,000 at 7% per year simple interest for five years. Which option is better?

2.5 You are about to borrow $3,000 from a bank at an interest rate of 9% compounded annually. You are required to make three equal annual repayments in the amount of $1,185.16 per year, with the first repayment occurring at the end of year one. For each year, show the interest payment and principal payment.

The Concept of Equivalence

2.6 Suppose you have the alternative of receiving either $10,000 at the end of five years or P dollars today. Currently, you have no need for the money, so you deposit the P dollars into a bank account that pays 6% interest compounded annually. What value of P would make you indifferent in your choice between P dollars today and the promise of $10,000 at the end of five years?

2.7 Suppose that, to cover some of your college expenses, you are obtaining a personal loan from your uncle in the amount of $10,000 (now) to be repaid in two years. If your uncle always earns 10% interest (compounded annually) on his money invested in various sources, what minimum lump-sum payment two years from now would make your uncle happy economically?

2.8 Which of the following alternatives would you rather receive, assuming an interest rate of 8% compounded annually?

Alternative 1: Receive $100 today;

Alternative 2: Receive $120 two years from now.

Single Payments (Use of *F/P* or *P/F* Factors)

2.9 What will be the amount accumulated by each of the following present investments?

(a) $7,000 in 8 years at 9% compounded annually.

(b) $1,250 in 12 years at 4% compounded annually.

(c) $5,000 in 31 years at 7% compounded annually.

(d) $20,000 in 7 years at 6% compounded annually.

2.10 What is the present worth of the following future payments?

(a) $4,500 6 years from now at 7% compounded annually.

(b) $6,000 15 years from now at 8% compounded annually.

(c) $20,000 5 years from now at 9% compounded annually.

(d) $12,000 8 years from now at 10% compounded annually.

2.11 Assuming an interest rate of 8% compounded annually, answer the following questions:

(a) How much money can be loaned now if $6,000 is to be repaid at the end of five years?

(b) How much money will be required in four years in order to repay a $15,000 loan borrowed now?

2.12 How many years will it take an investment to triple itself if the interest rate is 9% compounded annually?

2.13 You bought 200 shares of Motorola stock at $3,800 on December 31, 2000. Your intention is to keep the stock until it doubles in value. If you expect 15% annual growth for Motorola stock, how many years do you expect to hold onto the stock? Compare your answer with the solution obtained by the Rule of 72 (discussed in Example 2.7).

2.14 If you want to withdraw $35,000 at the end of four years, how much should you deposit now in an account that pays 11% interest compounded annually? See the accompanying cash flow diagram.

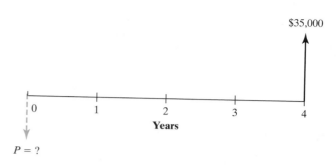

2.15 John and Susan just opened savings accounts at two different banks. They each deposited $1,000. John's bank pays simple interest at an annual rate of 10%, whereas Susan's bank pays compound interest at an annual rate of 9.5%. No interest will be taken out of the accounts for a period of three years. At the end of three years, whose balance will be greater and by how much (to the nearest dollar)?

Uneven–Payment Series

2.16 If you desire to withdraw the following amounts over the next five years from a savings account that earns 7% interest compounded annually, how much do you need to deposit now?

Year	Amount
2	$2,000
3	$3,000
4	$6,000
5	$8,000

2.17 If $1,000 is invested now, $1,500 two years from now, and $2,000 four years from now at an interest rate of 6% compounded annually, what will be the total amount in 10 years?

2.18 A local newspaper headline blared, "Bo Smith Signs for $30 Million." The article revealed that, on April 1, 2002, Bo Smith, the former record-breaking running back from Football University, signed a $30 million package with the Nebraska Lions. The terms of the contract were $3 million immediately, $2.4 million per year for the first five years (with the first payment after one year), and $3 million per year for the next five years (with the first payment at the end of year six). If the interest rate is 8% compounded annually, what is Bo's contract worth at the time of contract signing?

2.19 How much invested now at an interest rate of 6% compounded annually would be just sufficient to provide three payments as follows: the first payment in the amount of $3,000 occurring two years from now, the second payment in the amount of $4,000 five years thereafter, and the third payment in the amount of $5,000 seven years thereafter?

2.20 You deposit $100 today, $200 one year from now, and $300 three years from now. How much money will you have at the end of year three if there are different annual compound-interest rates per period according to the following diagram?

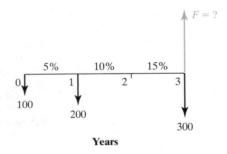

Years

2.21 A company borrowed $120,000 at an interest rate of 9% compounded annually over six years. The loan will be repaid in installments at the end of each year according to the accompanying repayment schedule. What will be the size of the last payment (X) that will pay off the loan?

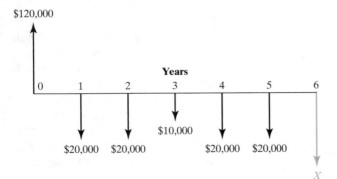

2.22 You are preparing to buy a vacation home eight years from now. The home will cost $50,000 at that time. You plan on saving three deposits at an interest rate of 10%:

> Deposit 1: Deposit $10,000 today.
> Deposit 2: Deposit $12,000 two years from now.
> Deposit 3: Deposit X five years from now.

How much do you need to invest in year five to ensure that you have the necessary funds to buy the vacation home at the end of year eight?

2.23 The accompanying diagram shows the anticipated cash dividends for Delta Electronics over the next four years. John is interested in buying some shares of this stock for a total of $80 and will hold them for four years. If John's interest rate is known to be 8% compounded annually, what would be the desired (minimum) total selling price for the set of shares at the end of the fourth year?

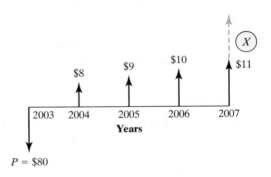

Years

$P = \$80$

Equal-Payment Series

2.24 What is the future worth of a series of equal year-end deposits of $5,000 for 10 years in a savings account that earns 8% annual compound interest if

(a) all deposits are made at the *end* of each year?

(b) all deposits are made at the *beginning* of each year?

2.25 What is the future worth of the following series of payments?

(a) $4,000 at the end of each year for six years at 7% compounded annually.

(b) $6,000 at the end of each year for nine years at 8.25% compounded annually.

(c) $3,000 at the end of each year for 22 years at 9% compounded annually.

(d) $9,000 at the end of each year for 30 years at 10.75% compounded annually.

2.26 What equal annual series of payments must be paid into a sinking fund in order to accumulate the following amounts?

(a) $12,000 in 13 years at 4% compounded annually.

(b) $25,000 in eight years at 7% compounded annually.

(c) $15,000 in 25 years at 9% compounded annually.

(d) $8,000 in eight years at 8.85% compounded annually.

2.27 Part of the income that a machine generates is put into a sinking fund to pay for replacement of the machine when it wears out. If $2,000 is deposited annually at 7% interest compounded annually, how many years must the machine be kept before a new machine costing $30,000 can be purchased?

2.28 A no-load (commission-free) mutual fund has grown at a rate of 9% compounded annually since its beginning. If it is anticipated that it will continue to grow at this rate, how much must be invested every year so that $10,000 will be accumulated at the end of five years?

2.29 You open a bank account, making a deposit of $200 now and deposits of $200 every other year. What is the total balance at the end of 10 years from now if your deposits earn 10% interest compounded annually?

2.30 What equal-annual-payment series is required in order to repay the following present amounts?

(a) $25,000 in five years at 8% interest compounded annually.

(b) $2,500 in four years at 9.5% interest compounded annually.

(c) $9,000 in three years at 11% interest compounded annually.

(d) $23,000 in 20 years at 7% interest compounded annually.

2.31 You have borrowed $20,000 at an interest rate of 12% compounded annually. Equal payments will be made over a three-year period, with each payment made at the end of the corresponding year. What is the amount of the annual payment? What is the interest payment for the second year?

2.32 What is the present worth of the following series of payments?

(a) $1,000 at the end of each year for eight years at 6% compounded annually.

(b) $1,500 at the end of each year for 10 years at 9% compounded annually.

(c) $2,500 at the end of each year for six years at 7.25% compounded annually.

(d) $5,000 at the end of each year for 30 years at 8.75% compounded annually.

2.33 From the interest tables in Appendix B, determine the value of the following factors by interpolation, and compare the results with those obtained from evaluating the A/P and P/A interest formulas:

(a) The capital-recovery factor for 36 periods at 6.25% compound interest.

(b) The equal-payment-series present-worth factor for 125 periods at 9.25% compound interest.

2.34 If $400 is deposited in a savings account at the beginning of each year for 15 years and the account earns 9% interest compounded annually, what will be the balance on the account the end of the 15 years (F)?

Linear Gradient Series

2.35 Kim deposits her annual bonus into a savings account that pays 6% interest compounded annually. The size of the bonus increases by $1,000 each year, and the initial bonus amount is $3,000. Determine how much will be in the account immediately after the fifth deposit.

2.36 Five annual deposits in the amounts of $1,200, $1,000, $800, $600, and $400 are made into a fund that pays interest at a rate of 9% compounded annually. Determine the amount in the fund immediately after the fifth deposit.

2.37 Compute the value of P for the accompanying cash flow diagram. Assume $i = 8\%$.

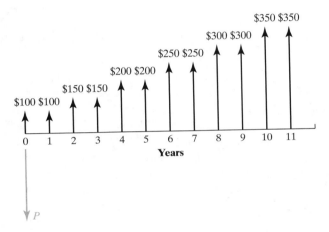

Years

2.38 What is the equal-payment series for 10 years that is equivalent to a payment series starting with $12,000 at the end of the first year and decreasing by $1,000 each year over 10 years? Interest is 8% compounded annually.

2.39 The maintenance expense on a machine is expected to be $800 during the first year and to increase $150 each year for the following seven years. What present sum of money should be set aside now to pay for the required maintenance expenses over the eight-year period? (Assume 9% compound interest per year.)

2.40 Consider the cash flow series given in the accompanying table. Which of the following values of C makes the deposit series equivalent to the withdrawal series at an interest rate of 12% compounded annually?

    (a) $C = \$200.00$.
    (b) $C = \$282.70$.

    (c) $C = \$394.65$.
    (d) $C = \$458.90$.

End of Period	Deposit	Withdrawal
0	$1,000	
1	800	
2	600	
3	400	
4	200	
5		
6		C
7		$2C$
8		$3C$
9		$4C$
10		$5C$

Geometric Gradient Series

2.41 Suppose that an oil well is expected to produce 100,000 barrels of oil during its first production year. However, its subsequent production (yield) is expected to decrease by 10% over the previous year's production. The oil well has a proven reserve of 1,000,000 barrels.

    (a) Suppose that the price of oil is expected to be $30 per barrel for the next several years. What would be the present worth of the anticipated revenue stream at an interest rate of 12% compounded annually over the next seven years?

    (b) Suppose that the price of oil is expected to start at $30 per barrel during the first year, but to increase at the rate of 5% over the previous year's price. What would be the present worth of the anticipated revenue stream at an interest rate of 12% compounded annually over the next seven years?

    (c) Assume the conditions of part (b). After three years of production, you decide to sell the oil well. What would be the fair price for the oil well?

2.42 A city engineer has estimated the annual toll revenues from a newly proposed highway construction over 20 years as follows:

$$A_n = (\$2{,}000{,}000)(n)(1.06)^{n-1},$$
$$n = 1, 2, \ldots, 20.$$

To determine the amount of debt financing through bonds, the engineer was asked to present the estimated total present value of toll revenue at an interest rate of 6%. Assuming annual compounding, find the present value of the estimated toll revenue.

2.43 What is the amount of 10 equal annual deposits that can provide five annual withdrawals, where a first withdrawal of $3,000 is made at the end of year 11 and subsequent withdrawals increase at the rate of 6% per year over the previous year's, if

(a) the interest rate is 8% compounded annually?
(b) the interest rate is 6% compounded annually?

Equivalence Calculations

2.44 Find the present worth of the cash receipts in the accompanying diagram if $i = 10\%$ compounded annually, with only *four interest factors*.

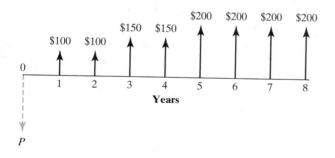

2.45 In computing the equivalent present worth of the following cash flow series at period zero, which of the following expressions is *incorrect*?

(a) $P = \$100(P/A, i, 4)(P/F, i, 4).$

(b) $P = \$100(F/A, i, 4)(P/F, i, 7).$

(c) $P = \$100(P/A, i, 7) - \$100(P/A, i, 3).$

(d) $P = \$100[(P/F, i, 4) + (P/F, i, 5)$
$$+ (P/F, i, 6) + (P/F, i, 7)].$$

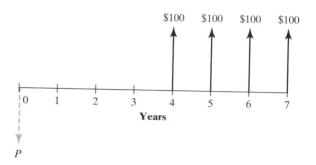

2.46 Find the equivalent present worth of the cash receipts in the accompanying diagram, where $i = 10\%$ compounded annually. In other words, how much do you have to deposit now (with the second deposit in the amount of $200 at the end of the first year) so that you will be able to withdraw $200 at the end of second year, $120 at the end of third year, and so forth, where the bank pays you 10% annual interest on your balance?

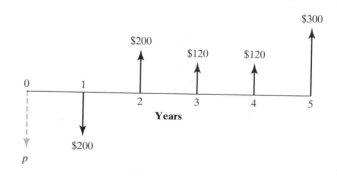

2.47 What value of A makes the two annual cash flows shown in the accompanying diagram equivalent at 10% interest compounded annually?

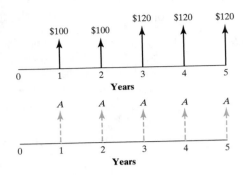

2.48 The two cash flow transactions shown in the accompanying cash flow diagram are said to be equivalent at 10% interest compounded annually. Find the unknown X value that satisfies the equivalence.

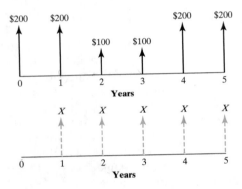

2.49 Solve for the present worth of the cash flow shown in the accompanying diagram, using at most three interest factors at 10% interest compounded annually.

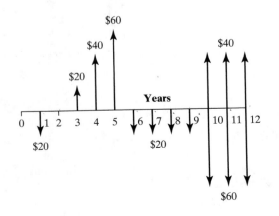

2.50 From the accompanying cash flow diagram, find the value of C that will establish economic equivalence between the deposit series and the withdrawal series at an interest rate of 8% compounded annually.

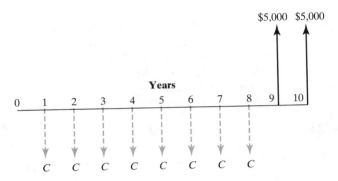

2.51 The following equation describes the conversion of a cash flow into an equivalent equal-payment series with $N = 10$:

$$A = [800 + 20(A/G, 6\%, 7)]$$
$$\times (P/A, 6\%, 7)(A/P, 6\%, 10)$$
$$+ [300(F/A, 6\%, 3) - 500](A/F, 6\%, 10).$$

Given the equation, reconstruct the original cash flow diagram.

2.52 Consider the accompanying cash flow diagram. What value of C makes the inflow series equivalent to the outflow series at an interest rate of 12% compounded annually?

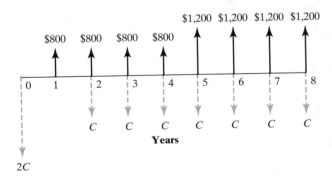

2.53 Find the value of X so that the two cash flows in the accompanying figure are equivalent for an interest rate of 10% compounded annually.

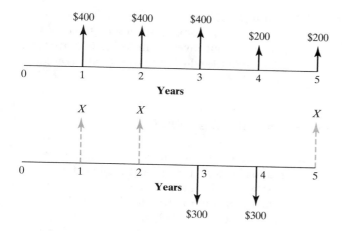

2.54 What single amount at the end of five years is equivalent to a uniform annual series of $3,000 per year for 10 years if the interest rate is 6% compounded annually?

2.55 On the day his baby was born, a father decided to establish a savings account for his child's college education. Any money that is put into the account will earn an interest rate of 8% compounded annually. The father will make a series of annual deposits in equal amounts on each of his child's birthdays from the 1st birthday through the 18th birthday, so that the child can make four annual withdrawals from the account in the amount of $20,000 on each of his 18th, 19th, 20th, and 21st birthdays. Assuming that the first withdrawal will be made on the child's 18th birthday, which of the following statements are correct to calculate the required annual deposit A?

(a) $A = (\$20,000 \times 4)/18$.

(b) $A = \$20,000(F/A, 8\%, 4)$
$\times (P/F, 8\%, 21)(A/P, 8\%, 18)$.

(c) $A = \$20,000(P/A, 8\%, 18)$
$\times (F/P, 8\%, 21)(A/F, 8\%, 4)$.

(d) $A = [\$20,000(P/A, 8\%, 3)$
$+ \$20,000](A/F, 8\%, 18)$.

(e) $A = \$20,000[(P/F, 8\%, 18)$
$+ (P/F, 8\%, 19) + (P/F, 8\%, 20)$
$+ (P/F, 8\%, 21)](A/P, 8\%, 18)$.

2.56 Find the equivalent equal-payment series C, using an A/G factor, such that the two accompanying cash flow diagrams are equivalent at 10% compounded annually.

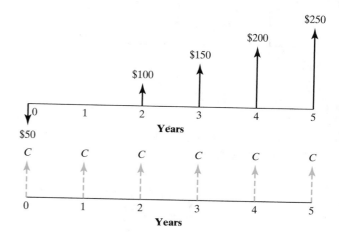

2.57 Consider the following cash flow:

Year End	Payment
0	$500
1–5	$1,000

In computing F at the end of year 5 at an interest rate of 12% compounded annually, which of the following statements is incorrect?

(a) $F = \$1,000(F/A, 12\%, 5)$
$- \$500(F/P, 12\%, 5)$.

(b) $F = \$500(F/A, 12\%, 6)$
$+ \$500(F/A, 12\%, 5)$.

(c) $F = [\$500 + \$1,000(P/A, 12\%, 5)]$
$\times (F/P, 12\%, 5)$.

(d) $F = [\$500(A/P, 12\%, 5)$
$+ \$1,000] \times (F/A, 12\%, 5)$.

2.58 Consider the cash flow series given in the accompanying diagram. In computing the equivalent worth at $n = 4$, which of the following statements is incorrect?

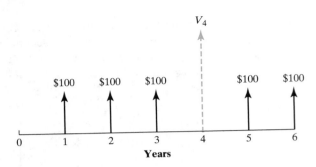

(a) $V_4 = [\$100(P/A, i, 6)$
$\qquad - \$100(P/F, i, 4)](F/P, i, 4).$

(b) $V_4 = \$100(F/A, i, 3) + \$100(P/A, i, 2).$

(c) $V_4 = \$100(F/A, i, 4)$
$\qquad - \$100 + \$100(P/A, i, 2).$

(d) $V_4 = [\$100(F/A, i, 6)$
$\qquad - \$100(F/P, i, 2)](P/F, i, 2).$

2.59 Henry Cisco is planning to make two deposits, $25,000 now and $30,000 at the end of six years. He wants to withdraw C each year for the first six years and $(C + \$1,000)$ each year for the next six years. Determine the value of C if the deposits earn 10% interest compounded annually.

Solving for Unknown Interest Rate

2.60 At what rate of interest compounded annually will an investment double in five years? Find the answers by using (1) the exact formula and (2) the Rule of 72.

2.61 Determine the interest rate i that makes the pairs of cash flows shown in the accompanying diagrams economically equivalent.

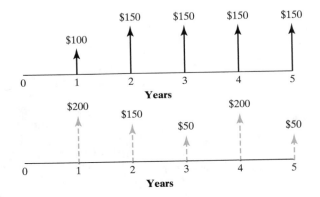

2.62 You have $10,000 available for investment in stock. You are looking for a growth stock that can grow your investment to $35,000 over five years. What kind of growth rate are you looking for?

Short Case Studies with Excel

2.63 The state of Florida sold a total of $36.1 million lottery tickets at $1 each during the first week of January 2003. As prize money, a total of $82 million will be distributed over the next 21 years ($1,952,381 at the *beginning* of each year). The distribution of the first-year prize money occurs now, and the remaining lottery proceeds are put into the state's educational reserve funds, which earn interest at the rate of 6% compounded annually. After the last prize distribution has been made (at the beginning of year 21), how much will be left over in the reserve account?

2.64 A newspaper headline reads "Millionaire Babies: How to Save Our Social Security System." It sounds a little wild, but the concept expressed in the title of this case study is probably the point of an economic plan proposed by a member of Congress. Senator Bob Kerrey, D–Nebraska, has proposed giving every newborn baby a $1,000 government savings account at birth, followed by five annual contributions of $500 each. If the funds are left untouched in an investment account, Kerrey says, then by the time each baby reaches age 65, his or her $3,500 contribution will have grown to $600,000 over the years, even at medium returns for a thrift-savings plan. At about 9.4% compounded annually, the balance would grow to be $1,005,132. (How would you calculate this number?) With about 4 million babies born each year, the proposal would cost the federal government $4 billion annually. Kerrey offered this idea in a speech devoted to tackling Social Security reform. About 90% of the total annual Social Security tax collections of more than $300 billion is used to pay current beneficiaries, which is the largest federal program. The remaining 10% is invested in interest-bearing government bonds that finance the day-to-day expenses of the federal government. Discuss the economics of Senator Bob Kerrey's Social Security savings plan.

2.65 Kevin Jones, Texas Tigers's quarterback, agreed to an eight-year, $50 million contract that at the time made him the highest-paid player in professional football history. The contract included a signing bonus of $11 million and called for annual salaries of $2.5 million in 2003, $1.75 million in 2004, $4.15 million in 2005, $4.90 million in 2006, $5.25 million in 2007, $6.2 million in 2008, $6.75 million in 2009, and $7.5 million in 2010. The $11-million signing bonus was prorated over the course of the contract so that an additional $1.375 million was paid each year over the eight-year contract period. With the salary paid at the beginning of each season, what is the worth of his contract at an interest rate of 6%?

.3

Investors are always looking for a sure thing, a risk-free way to get a big return for their money. Yet they often ignore one of the best and safest ways to make the most of their money: paying off credit card bills. Most college students carry one or two different credit cards so that they may purchase items that are necessary while they are pursuing college degrees. Here is an example of the consequences of making only the required minimum payments on your credit card balances: Say that you

Pay the Minimum, Pay for Years[1]

owe $2,705 on a credit card that charges an 18.38% annual percentage rate, and you make only the minimum 2% payment every month. With this payment schedule, it would take more than 27 years to pay off the debt!

Pay the minimum, pay for years

Making minimum payments on your credit cards can cost you a bundle over a lot of years. Here's what would happen if you paid the minimum—or more—every month on a $2,705 card balance, with a 18.38% interest rate.

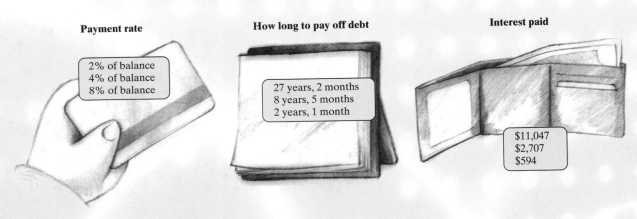

Payment rate

2% of balance
4% of balance
8% of balance

How long to pay off debt

27 years, 2 months
8 years, 5 months
2 years, 1 month

Interest paid

$11,047
$2,707
$594

[1]Source: Elys A. McLean, "Pay the Minimum. Pay for Years," *USA Today*, Page C2, 1998.

Understanding Money Management

What is the meaning of the annual percentage rate (APR) of 18.38% quoted by the credit card company? And how does the credit card company calculate the interest payment? In this chapter, we will consider several concepts crucial to managing money. In Chapter 2, we examined how time affects the value of money, and we developed various interest formulas for that purpose. Using these basic formulas, we will now extend the concept of equivalence to determine interest rates implicit in many financial contracts. To this end, we will introduce several examples in the area of loan transactions. For example, many commercial loans require that interest compound more frequently than once a year—for instance, monthly or daily. To consider the effect of more frequent compounding, we must begin with an understanding of the concepts of nominal and effective interest.

3-1   Market Interest Rates

As briefly mentioned in Chapter 2, the market interest rate is defined as the interest rate quoted by the financial market, such as by banks and financial institutions. This interest rate is supposed to consider any anticipated changes in earning power as well as purchasing power in the economy. In this section, we will review the nature of this interest rate in more detail.

3.1.1   Nominal Interest Rates

Take a closer look at the billing statement from any credit card. Or, if you financed a new car recently, examine the loan contract. You should be able to find the interest that the bank charges on your unpaid balance. Even if a financial institution uses a unit of time other than a year—for example, a month or a quarter—when calculating interest payments and in other matters, the institution usually quotes the interest rate on an *annual basis*. Many banks, for example, state the interest arrangement for credit cards in the following manner:

> "18% compounded monthly."

This statement means simply that each month the bank will charge 1.5% interest (12 months per year × 1.5% per month = 18% per year) on the unpaid balance. As shown in Figure 3.1, we say that 18% is the **nominal interest rate** or **annual percentage rate** (APR) and that the compounding frequency is monthly (12 times per year).

Although the APR is commonly used by financial institutions and is familiar to many customers, it does not explain precisely the amount of interest that will accumulate in a year. To explain the true effect of more frequent compounding on annual interest amounts, we will introduce the term *effective interest rate*, commonly known as *annual effective yield*, or annual percentage yield (APY).

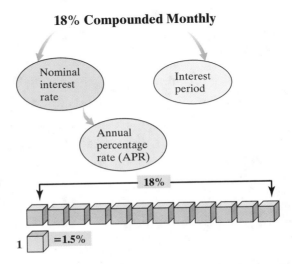

FIGURE 3.1 Relationship between APR and interest period

3.1.2 Annual Effective Yields

The **annual effective yield** (or **effective annual interest rate**) is the one rate that truly represents the interest earned in a year. On a yearly basis, you are looking for a cumulative rate—1.5% each month for 12 times. This cumulative rate predicts the actual interest payment on your outstanding credit card balance.

We could calculate the total annual interest payment for a credit card debt of $1,000 by using the formula given in Eq. (2.3). If $P = \$1,000$, $i = 1.5\%$, and $N = 12$, we obtain

$$
\begin{aligned}
F &= P(1 + i)^N \\
&= \$1,000(1 + 0.015)^{12} \\
&= \$1,195.62.
\end{aligned}
$$

Clearly, the bank is earning more than 18% on your original credit card debt. In fact, you are paying $195.62. The implication is that, for each dollar owed, you are paying an equivalent annual interest of 19.56 cents. In terms of an effective annual interest rate (i_a), the interest payment can be rewritten as a percentage of the principal amount:

$$
i_a = \$195.62/\$1,000 = 0.19562 \text{ or } 19.562\%.
$$

In other words, paying 1.5% interest per month for 12 months is equivalent to paying 19.56% interest just one time each year. This relationship is depicted in Figure 3.2.

Table 3.1 shows effective interest rates at various compounding intervals for 4%–12% APRs. As you can see, depending on the frequency of compounding, the effective interest earned or paid by the borrower can differ significantly from the APR. Therefore, truth-in-lending laws require that financial institutions quote both nominal and effective interest rates when you deposit or borrow money.

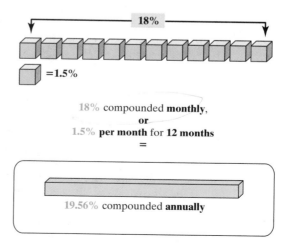

FIGURE 3.2 Relationship between nominal and effective interest rates

TABLE 3.1 Annual Effective Yields at Various Compounding Intervals

Nominal Rate	Annual Effective Yields Compounding Frequency				
	Annually	Semiannually	Quarterly	Monthly	Daily
4%	4.00%	4.04%	4.06%	4.07%	4.08%
5%	5.00%	5.06%	5.09%	5.12%	5.13%
6%	6.00%	6.09%	6.14%	6.17%	6.18%
7%	7.00%	7.12%	7.19%	7.23%	7.25%
8%	8.00%	8.16%	8.24%	8.30%	8.33%
9%	9.00%	9.20%	9.31%	9.38%	9.42%
10%	10.00%	10.25%	10.38%	10.47%	10.52%
11%	11.00%	11.30%	11.46%	11.57%	11.62%
12%	12.00%	12.36%	12.55%	12.68%	12.74%

Certainly, more frequent compounding increases the amount of interest paid over a year at the same nominal interest rate. Assuming that the nominal interest rate is r and that M compounding periods occur during the year, the annual effective yield i_a can be calculated as follows:

$$i_a = \left(1 + \frac{r}{M}\right)^M - 1. \tag{3.1}$$

When $M = 1$, we have the special case of annual compounding. Substituting $M = 1$ into Eq. (3.1) reduces it to $i_a = r$. That is, when compounding takes place once annually, effective interest is equal to nominal interest. Thus, in most of our examples in Chapter 2, where only annual interest was considered, we were, by definition, using annual effective yields.

Example 3.1 Determining a Compounding Period

Consider the following bank advertisement that appeared in a local newspaper: "Open a Liberty Bank Certificate of Deposit (CD) and get a guaranteed rate of return on as little as $500. It's a smart way to manage your money for months."

In this advertisement, no mention is made of specific interest compounding frequencies. Find the compounding period for each CD.

Type of Certificate	Interest Rate (APR)	Annual Percentage Yield (APY)	Minimum Required to Open
1-Year Certificate	2.23%	2.25%	$500
2-Year Certificate	3.06%	3.10%	$500
3-Year Certificate	3.35%	3.40%	$500
4-Year Certificate	3.45%	3.50%	$500
5- to 10-Year Certificates	4.41%	4.50%	$500

SOLUTION

Given: $r = 2.23\%$ per year, i_a.
Find: M.

First, we will consider the one-year CD. The nominal interest rate is 2.23% per year, and the effective annual interest rate (or APY) is 2.25%. Using Eq. (3.1), we obtain the expression

$$0.0225 = (1 + 0.0223/M)^M - 1,$$

or

$$1.0225 = (1 + 0.0223/M)^M.$$

By trial and error, we find that $M = 100$, which indicates daily compounding. Thus, the one-year CD earns 2.23% interest compounded daily. Similarly, we can find that the interest periods for the other CDs are daily as well.

3–2 Calculating Effective Interest Rates Based on Payment Periods

We can generalize the result of Eq. (3.1) to compute the effective interest rate for *any time duration*. As you will see later, the effective interest rate is usually computed based on the payment (transaction) period. We will look into two types of compounding situations: (1) discrete compounding and (2) continuous compounding.

3.2.1 Discrete Compounding

If cash flow transactions occur *quarterly*, but interest is compounded *monthly*, we may wish to calculate the effective interest rate on a *quarterly basis*. To consider this situation, we may redefine Eq. (3.1) as

$$i = (1 + r/M)^C - 1$$
$$= (1 + r/CK)^C - 1, \tag{3.2}$$

where

M = the number of interest periods per year,

C = the number of interest periods per payment period, and

K = the number of payment periods per year.

Note that $M = CK$ in Eq. (3.2).

Example 3.2 Effective Rate per Payment Period

Suppose that you make quarterly deposits into a savings account that earns 8% interest compounded monthly. Compute the effective interest rate per quarter.

SOLUTION

Given: $r = 8\%$, $C = 3$ interest periods per quarter, $K = 4$ quarterly payments per year, and $M = 12$ interest periods per year.

Find: i.

Using Eq. (3.2), we compute the effective interest rate per quarter as

$$i = (1 + 0.08/12)^3 - 1$$
$$= 2.013\%.$$

Comments: The annual effective interest rate i_a is $(1 + 0.02013)^4 = 8.24\%$. For the special case of annual payments with annual compounding, we obtain $i = i_a$, with $C = M$ and $K = 1$. Figure 3.3 illustrates the relationship between the nominal and effective interest rates per payment period.

Case 1: 8% compounded monthly

Payment Period = Quarterly
Interest Period = Monthly

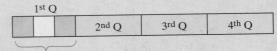

1st Q | 2nd Q | 3rd Q | 4th Q

3 interest periods

Given: $r = 8\%$;
$K = 4$ quarterly payments per year;
$C = 3$ interest periods per quarter;
$M = 12$ interest periods per year.

$i = [1 + r/CK]^C - 1$
$= [1 + 0.08/(3)(4)]^3 - 1$
$= 2.013\%$ per quarter.

FIGURE 3.3 Computing the effective interest rate per quarter

3.2.2 Continuous Compounding

To be competitive on the financial market, or to entice potential depositors, some financial institutions offer more frequent compounding. As the number of compounding periods (M) becomes very large, the interest rate per compounding period (r/M) becomes very small. As M approaches infinity and r/M approaches zero, we approximate the situation of **continuous compounding**.

By taking limits on the right side of Eq. (3.2), we obtain the effective interest rate per payment period as

$$i = \lim_{CK \to \infty} \left[(1 + r/CK)^C - 1 \right]$$
$$= \lim_{CK \to \infty} (1 + r/CK)^C - 1$$
$$= (e^r)^{1/K} - 1.$$

In sum, the effective interest rate per payment period is

$$i = e^{r/k} - 1. \tag{3.3}$$

To calculate the effective *annual* interest rate for continuous compounding, we set K equal to 1, resulting in

$$i_a = e^r - 1. \tag{3.4}$$

As an example, the effective annual interest rate for a nominal interest rate of 12% compounded continuously is $i_a = e^{0.12} - 1 = 12.7497\%$.

Example 3.3 Calculating an Effective Interest Rate

Find the effective interest rate per quarter at a nominal rate of 8% compounded (a) weekly, (b) daily, and (c) continuously.

SOLUTION

Given: $r = 8\%$, $K = 4$ payments per year.
Find: i per quarter.

(a) Weekly compounding:

With $r = 8\%$, $M = 52$, and $C = 13$ interest periods per quarter, we have

$$i = (1 + 0.08/52)^{13} - 1$$
$$= 2.0186\% \text{ per week.}$$

Figure 3.4 illustrates this result.

(b) Daily compounding:

With $r = 8\%$, $M = 365$, and $C = 91.25$ days per quarter, we have

$$i = (1 + 0.08/365)^{91.25} - 1$$
$$= 2.0199\% \text{ per quarter.}$$

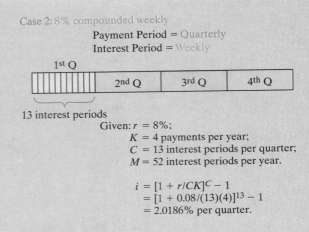

Case 2: 8% compounded weekly
Payment Period = Quarterly
Interest Period = Weekly

Given: $r = 8\%$;
$K = 4$ payments per year;
$C = 13$ interest periods per quarter;
$M = 52$ interest periods per year.

$$i = [1 + r/CK]^C - 1$$
$$= [1 + 0.08/(13)(4)]^{13} - 1$$
$$= 2.0186\% \text{ per quarter.}$$

FIGURE 3.4 Effective interest rate per payment period: quarterly payments with weekly compounding

(c) Continuous compounding:

With $r = 8\%$, $M \to \infty$, $C \to \infty$, and $K = 4$ and using Eq. (3.3), we obtain

$$i = e^{0.08/4} - 1 = 2.0201\% \text{ per quarter.}$$

Comments: Note that the difference between daily compounding and continuous compounding is often negligible. Many banks offer continuous compounding to entice deposit customers, but the extra benefits are small. Figure 3.5 summarizes the varying effective interest rates per payment period (quarterly in this case) under various compounding frequencies.

	Base	Case 1	Case 2	Case 3	Case 4
Interest Rate	8% compounded quarterly	8% compounded monthly	8% compounded weekly	8% compounded daily	8% compounded continuously
Payment Period	Payments occur quarterly	Payments occur quarterly	Payments occur quarterly	Payments occur quarterly	Payments occur quarterly
Effective Interest Rate per Payment Period	2.000% per quarter	2.013% per quarter	2.0186% per quarter	2.0199% per quarter	2.0201% per quarter

FIGURE 3.5 Effective interest rates per payment period

3-3 Equivalence Calculations with Effective Interest Rates

When calculating equivalent values, we need to identify both the interest period and the payment period. If the time interval for compounding is different from the time interval for cash transaction (or payment), we need to find *the effective interest rate that covers the payment period*. We illustrate this concept with specific examples.

3.3.1 Compounding Period Equal to Payment Period

All the examples in Chapter 2 assumed annual payments and annual compounding. Whenever a situation occurs where the compounding and payment periods are equal ($M = K$), no matter whether the interest is compounded annually or at some other interval, the following solution method can be used:

1. Identify the number of compounding (or payment) periods ($M = K$) per year.
2. Compute the effective interest rate per payment period, i.e,

$$i = r/M.$$

3. Determine the number of payment periods:

$$N = M \times (\text{number of years}).$$

Example 3.4 Calculating Auto Loan Payments

Suppose you want to buy a car. You have surveyed the dealers' newspaper advertisements, and the one in Figure 3.6 has caught your attention.

- 8.5% Annual Percentage Rate! 48-month financing on all Mustangs in stock. 60 to choose from
- ALL READY FOR DELIVERY! Prices starting as low as $21,599
- You just add sales tax and 1% for dealer's freight. We will pay the tag, title, and license
- Add 4% sales tax = $863.96
- Add 1% dealer's freight = $215.99
- Total purchase price = $22,678.95

FIGURE 3.6 Financing an automobile

You can afford to make a down payment of $2,678.95, so the net amount to be financed is $20,000.

(a) What would the monthly payment be?

(b) After the 25th payment, you want to pay off the remaining loan in a lump-sum amount. What is the required amount of this lump sum?

SOLUTION

(a) The advertisement does not specify a compounding period, but in automobile financing, the interest and the payment periods are almost always monthly. Thus, the 8.5% APR means 8.5% compounded monthly.

Given: $P = \$20,000$, $r = 8.5\%$ per year, $K = 12$ payments per year, $N = 48$ months, and $M = 12$ interest periods per year.

Find: (a) A

In this situation, we can easily compute the monthly payment by using Eq. (2.9). Since

$$i = 8.5\%/12 = 0.7083\% \text{ per month}$$

and

$$N = (12)(4) = 48 \text{ months,}$$

we have

$$A = \$20,000(A/P, 0.7083\%, 48) = \$492.97.$$

Figure 3.7 shows the cash flow diagram for this part of the example.

(b) In this part of the example, we need to calculate the remaining balance after the 25th payment. We can compute the amount you owe after you make the 25th payment by calculating the equivalent worth of the remaining 23 payments at the end of the 25th month, with the time scale shifted by 25 months:

Given: $A = \$492.97$, $i = 0.7083\%$ per month, and $N = 23$ months.

Find: Remaining balance after 25 months (B_{25}).

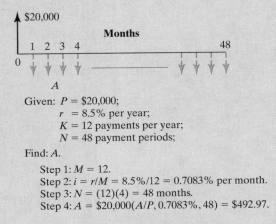

Given: $P = \$20,000$;
      $r = 8.5\%$ per year;
      $K = 12$ payments per year;
      $N = 48$ payment periods;
Find: A.
    Step 1: $M = 12$.
    Step 2: $i = r/M = 8.5\%/12 = 0.7083\%$ per month.
    Step 3: $N = (12)(4) = 48$ months.
    Step 4: $A = \$20,000(A/P, 0.7083\%, 48) = \$492.97$.

FIGURE 3.7 Cash flow diagram for part (a)

The balance is calculated as follows:

$$B_{25} = \$492.97(P/A, 0.7083\%, 23) = \$10,428.96.$$

So, if you desire to pay off the remainder of the loan at the end of the 25th payment, you must come up with $10,428.96 in addition to the payment for that month of $492.97. (See Figure 3.8.)

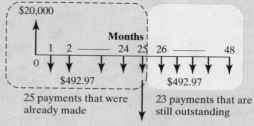

Suppose you want to pay off the remaining loan in a lump sum right after making the 25th payment. How much would this payoff amount be?

25 payments that were already made

23 payments that are still outstanding

$P = \$492.97 \ (P/A, 0.7083\%, 23)$
$= \$10,428.96.$

FIGURE 3.8 Process of calculating the remaining balance of the auto loan

3.3.2 Compounding Occurs at a Different Rate Than That at Which Payments Are Made

The computational procedure for dealing with compounding periods and payment periods that cannot be compared is as follows:

1. Identify the number of compounding periods per year (M), the number of payment periods per year (K), and the number of interest periods per payment period (C).

2. Compute the effective interest rate per payment period:

   - For discrete compounding, compute

     $$i = (1 + r/M)^C - 1.$$

   - For continuous compounding, compute

     $$i = e^{r/K} - 1.$$

3. Find the total number of payment periods:

   $$N = K \times (\text{number of years}).$$

4. Use i and N in the appropriate formulas in Table 2.4.

Example 3.5 Compounding Occurs More Frequently than Payments Are Made

Suppose you make equal quarterly deposits of $1,000 into a fund that pays interest at a rate of 12% compounded monthly. Find the balance at the end of year three.

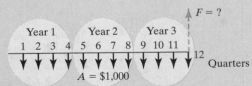

Step 1: $M = 12$ compounding periods/year;
$K = 4$ payment periods/year;
$C = 3$ interest periods per quarter.
Step 2: $i = [1 + 0.12/(3)(4)]^3 - 1$
$= 3.030\%.$
Step 3: $N = 4(3) = 12.$
Step 4: $F = \$1,000\,(F/A, 3.030\%, 12)$
$= \$14,216.24.$

FIGURE 3.9 Cash flow diagram

SOLUTION

Given: $A = \$1,000$ per quarter, $r = 12\%$ per year, $M = 12$ compounding periods per year, $N = 12$ quarters, and the cash flow diagram in Figure 3.9.
Find: F.

We follow the procedure for noncomparable compounding and payment periods as described previously:

1. Identify the parameter values for M, K, and C, where

$M = 12$ compounding periods per year,
$K = $ four payment periods per year, and
$C = $ three interest periods per payment period.

2. Use Eq. (3.1) to compute effective interest:

$$i = (1 + 0.12/12)^3 - 1$$
$$= 3.030\% \text{ per quarter.}$$

3. Find the total number of payment periods, N, where

$$N = K(\text{number of years}) = 4(3) = 12 \text{ quarters.}$$

4. Use i and N in the appropriate equivalence formulas:

$$F = \$1,000(F/A, 3.030\%, 12) = \$14,216.24.$$

Comment: Appendix B does not provide interest factors for $i = 3.030\%$, but the interest factor can still be evaluated by $F = \$1,000(A/F, 1\%, 3)$ $(F/A, 1\%, 36)$, where the first interest factor finds its equivalent monthly payment and the second interest factor converts the monthly payment series to an equivalent lump-sum future payment. If continuous compounding is assumed, the accumulated balance would be $14,228.37, which is about $12 more than the balance for the monthly compounding situation. (See Figure 3.10.)

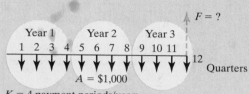

Step 1: $K = 4$ payment periods/year;
$C = \infty$ interest periods per quarter.
Step 2: $i = e^{0.12/4} - 1$
$= 3.045\%$ per quarter.
Step 3: $N = 4(3) = 12$.
Step 4: $F = \$1,000 \, (F/A, 3.045\%, 12)$
$= \$14,228.37.$

FIGURE 3.10 Equivalence calculation for continuous compounding

3-4   Debt Management

Credit card debt and commercial loans are among the most significant financial transactions involving interest. Many types of loans are available, but here we will focus on those most frequently used by individuals and in business.

3.4.1   Borrowing with Credit Cards

When credit cards were introduced in 1959, people were able to handle their personal finances in a dramatically different way. From a consumer's perspective, one's ability to use credit cards means that one does not have to wait for a paycheck to reach the bank before making a purchase. Most credit cards operate as *revolving credit*. With revolving credit, you have a line of borrowing that you can tap into at will and pay back as quickly or slowly as you want—as long as you pay the minimum required each month.

Your monthly bill is an excellent source of information about what your card really costs. Four things affect your card-based credit costs: annual fees, finance charges, the grace period, and the method of calculating interest. In fact, there are three different ways to compute interest charges, as summarized in Table 3.2. The average-daily-balance approach is the most common.

	TABLE 3.2 Methods of Calculating Interests on Your Credit Cards	
Method	**Description**	**Example of the Interest You Owe, Given a Beginning Balance of $3,000 at 18%**
Adjusted Balance	The bank subtracts the amount of your payment from the beginning balance and charges you interest on the remainder. This method costs you the least.	With a $1,000 payment, your new balance will be $2,000. You pay 1.5% interest for the month on this new balance, which comes out to (1.5%) ($2,000) = $30.
Average Daily Balance	The bank charges you interest on the average of the amount you owe each day during the period. So the larger the payment you make, the lower the interest you pay.	With a $1,000 payment on the 15th day, your balance reduced to $2,000. Therefore the interest on your average daily balance for the month will be (1.5%)($3,000 + $2,000)/2 = $37.50.
Previous Balance	The bank does not subtract any payments you make from your previous balance. You pay interest on the total amount you owe at the beginning of the period. This method costs you the most.	The annual interest rate is 18% compounded monthly. Regardless of your payment size, the bank will charge 1.5% on your beginning balance of $3,000. Therefore, your interest for the month is (1.5%)($3,000) = $45.

Example 3.6 Paying Off Cards Saves a Bundle

Suppose that you owe $2,000 on a credit card that charges 18% APR, and you make either the minimum 10% payment or $20, whichever is larger, every month. How long will it take to pay off debt? Assume that the bank uses the adjusted-balance method to calculate your interest, meaning that the bank subtracts the amount of your payment from the beginning balance and charges you interest on the remainder.

SOLUTION

Given: APR = 18% (or 1.5% per month), beginning balance = $2,000, and monthly payment = 10% of outstanding balance or $20, whichever is larger.

Find: Number of months to pay off the loan, assuming that no new purchases are made during this payment period.

With the initial balance of $2,000 ($n = 0$), the interest for the first month will be $30 (= $2,000(0.015)), so you will be billed $2,030. Then you make a $203 payment (10% of the outstanding balance), so the remaining balance will be $1,827. At the end of the second month, the billing statement will show that you owe the bank in the amount of $1,854.41, of which $27.41 is interest. With a $185.44 payment, the balance is reduced to $1,668.96. This process repeats

Period	Beg. Bal	Interest	Payment	End. Bal.
0				$2,000.00
1	$2,000.00	$30.00	$203.00	$1,827.00
2	$1,827.00	$27.41	$185.44	$1,668.96
3	$1,668.96	$25.03	$169.40	$1,524.60
4	$1,524.60	$22.87	$154.75	$1,392.72
5	$1,392.72	$20.89	$141.36	$1,272.25
6	$1,272.25	$19.08	$129.13	$1,162.20
7	$1,162.20	$17.43	$117.96	$1,061.67
8	$1,061.67	$15.93	$107.76	$969.84
9	$969.84	$14.55	$98.44	$885.95
10	$885.95	$13.29	$89.92	$809.31
11	$809.31	$12.14	$82.15	$739.31
12	$739.31	$11.09	$75.04	$675.36
13	$675.36	$10.13	$68.55	$616.94
14	$616.94	$9.25	$62.62	$563.57
15	$563.57	$8.45	$57.20	$514.82
16	$514.82	$7.72	$52.25	$470.29
17	$470.29	$7.05	$47.73	$429.61
18	$429.61	$6.44	$43.61	$392.45
19	$392.45	$5.89	$39.83	$358.50
20	$358.50	$5.38	$36.39	$327.49
21	$327.49	$4.91	$33.24	$299.16
22	$299.16	$4.49	$30.37	$273.29
23	$273.29	$4.10	$27.74	$249.65
24	$249.65	$3.74	$25.34	$228.05
25	$228.05	$3.42	$23.15	$208.33
26	$208.33	$3.12	$21.15	$190.31
27	$190.31	$2.85	$20.00	$173.16
28	$173.16	$2.60	$20.00	$155.76
29	$155.76	$2.34	$20.00	$138.09
30	$138.09	$2.07	$20.00	$120.17
31	$120.17	$1.80	$20.00	$101.97
32	$101.97	$1.53	$20.00	$83.50
33	$83.50	$1.25	$20.00	$64.75
34	$64.75	$0.97	$20.00	$45.72
35	$45.72	$0.69	$20.00	$26.41
36	$26.41	$0.40	$20.00	$6.80
37	$6.80	$0.10	$6.91	0
	Total:	$330.42	$2,330.42	

TABLE 3.3 Creating a Loan Repayment Schedule

until the 26th payment. For the 27th payment and all those thereafter, 10% of the outstanding balance is less than $20, so you pay $20. As shown in Table 3.3, it would take 37 months to pay off the $2,000 debt, with a total interest payment of $330.42.

Comments: If the bank uses the average-daily-balance method, meaning that the bank charges you interest on the average of the amount you owe each day during the period, it would take a little longer to pay off the debt.

3.4.2 Commercial Loans—Calculating Principal and Interest Payments

One of the most important applications of compound interest involves loans that are paid off in **installments** over time. If a loan is to be repaid in equal periodic amounts (e.g., weekly, monthly, quarterly, or annually), it is said to be an **amortized loan**. Examples include automobile loans, loans for appliances, home mortgage loans, and most business debts other than very short-term loans. Most commercial loans have interest that is compounded monthly. With a car loan, for example, a local bank or a dealer advances you the money to pay for the car, and you repay the principal plus interest in monthly installments, usually over a period of three to five years. The car is your collateral. If you don't keep up with your payments, the lender can repossess, or take back, the car and keep all the payments you have made.

Two factors determine what borrowing will cost you: the finance charge and the length of the loan. The cheapest loan is not necessarily the loan with the lowest payments, or even the loan with the lowest interest rate. Instead, you have to look at the total cost of borrowing, which depends on the interest rate, fees, and the term (i.e., the length of time it takes you to repay the loan). While you probably cannot influence the rate and fees, you may be able to arrange for a shorter term.

So far, we have considered many instances of amortized loans in which we calculated present or future values of the loans or the amounts of the installment payments. An additional aspect of amortized loans, which will be of great interest to us, is calculating the amount of interest contained in each installment versus the portion of the principal that is paid off in each installment. As we shall explore more fully in Chapter 9, the interest paid on a loan is an important element in calculating taxable income. We will further show how we may calculate the interest and principal paid at any point in the life of a loan, using Excel's financial commands. As illustrated in Example 3.7, the amount of interest owed for a specified period is calculated based on the *remaining balance* of the loan at the beginning of the period.

Example 3.7 Using Excel to Determine a Loan's Balance, Principal, and Interest

Suppose you secure a home improvement loan in the amount of $5,000 from a local bank. The loan officer gives you the following loan terms:

- Contract amount = $5,000;
- Contract period = 24 months;
- Annual percentage rate = 12%;
- Monthly installment = $235.37.

Figure 3.11 shows the cash flow diagram for this loan. Construct the loan payment schedule by showing the remaining balance, interest payment, and principal payment at the end of each period over the life of the loan.

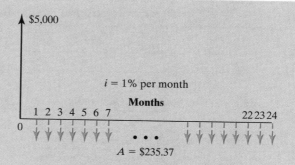

FIGURE 3.11 Cash flow diagram

SOLUTION

Given: $P = \$5,000$, $A = \$235.37$ per month, $r = 12\%$ per year, $M = 12$ compounding periods per year, and $N = 24$ months.

Find: B_n and I_n for $n = 1$ to 24.

(a) Tabular Approach:

We can easily see how the bank calculated the monthly payment of $235.37. Since the effective interest rate per payment period on this loan transaction is 1% per month, we establish the following equivalence relationship:

$$\$235.37(P/A, 1\%, 24) = \$235.37(21.2431) = \$5,000.$$

The loan payment schedule can be constructed as in Table 3.4. The interest due at $n = 1$ is $50.00, 1% of the $5,000 outstanding during the first month. The $185.37 left over is applied to the principal, reducing the amount outstanding in the second month to $4,814.63. The interest due in the second month is 1% of $4,814.63, or $48.15, leaving $187.22 for repayment of the principal. At $n = 24$, the last $235.37 payment is just sufficient to pay the interest on the unpaid loan principal and to repay the remaining principal.

Sample cell formulas:

    B8: = PMT($B$7/1200,$B$6,$B$5,0)
    C11: = $B$8
    D11: = PPMT($B$7/1200,B11,$B$6,$B$5,0)
    E11: = IPMT($B$7/1200,B11,$B$6,$B$5,0)
    F11: = $B$5 + D11

(b) Using Excel's Financial Commands:

Table 3.4 was constructed using some of Excel's financial commands. When you need to compute monthly payments, interest payments, and

TABLE 3.4 Loan Repayment Schedule Generated by Excel

	A	B	C	D	E	F
1						
2						
3	**Example 3.7 Loan Repayment Schedule**					
4						
5	Contract Amount	$ 5,000.00		Total payment		$ 5,648.82
6	Contract Period	24		Total interest		$648.82
7	APR (%)	12				
8	Monthly Payment	($235.37)				
9						
10		Payment Number	Payment Size	Principal Payment	Interest Payment	Loan Balance
11		1	($235.37)	($185.37)	($50.00)	$4,814.63
12		2	($235.37)	($187.22)	($48.15)	$4,627.41
13		3	($235.37)	($189.09)	($46.27)	$4,438.32
14		4	($235.37)	($190.98)	($44.38)	$4,247.33
15		5	($235.37)	($192.89)	($42.47)	$4,054.44
16		6	($235.37)	($194.82)	($40.54)	$3,859.62
17		7	($235.37)	($196.77)	($38.60)	$3,662.85
18		8	($235.37)	($198.74)	($36.63)	$3,464.11
19		9	($235.37)	($200.73)	($34.64)	$3,263.38
20		10	($235.37)	($202.73)	($32.63)	$3,060.65
21		11	($235.37)	($204.76)	($30.61)	$2,855.89
22		12	($235.37)	($206.81)	($28.56)	$2,649.08
23		13	($235.37)	($208.88)	($26.49)	$2,440.20
24		14	($235.37)	($210.97)	($24.40)	$2,229.24
25		15	($235.37)	($213.08)	($22.29)	$2,016.16
26		16	($235.37)	($215.21)	($20.16)	$1,800.96
27		17	($235.37)	($217.36)	($18.01)	$1,583.60
28		18	($235.37)	($219.53)	($15.84)	$1,364.07
29		19	($235.37)	($221.73)	($13.64)	$1,142.34
30		20	($235.37)	($223.94)	($11.42)	$918.40
31		21	($235.37)	($226.18)	($9.18)	$692.21
32		22	($235.37)	($228.45)	($6.92)	$463.77
33		23	($235.37)	($230.73)	($4.64)	$233.04
34		24	($235.37)	($233.04)	($2.33)	$0.00
35						

principal payments, several commands are available to facilitate a typical loan analysis:

Function Description	Excel Command
1. Calculating the periodic loan payment size (A)	$= \text{PMT}(i, N, P, type)$
2. Calculating the portion of loan interest payment for a given period n	$= \text{IPMT}(i, n, N, P, F, type)$
3. Calculating the cumulative interest payment between two periods	$= \text{CUMIPMT}(i, N, P, startperiod, endperiod, type)$
4. Calculating the portion of loan principal payment for a given period n	$= \text{PPMT}(i, n, N, P, F, type)$
5. Calculating the cumulative principal payment between two periods	$= \text{CUMPRINC}(i, N, P, startperiod, endperiod, type)$

i is the interest rate. N is the total number of payment periods. P is the present value. F is the future value. *start period* is the first period in the calculation. Payment periods are numbered beginning with 1. *end period* is the last period in the calculation. *type* is the timing of the payment: *type* is 0 if payments are at the end of the period and 1 if payments are at the beginning of the period. The default value is 0.

As an example, to compute the interest and principal payments for $n = 20$, we may use the following Excel commands:

- Interest payment ($n = 20$): $= \text{IPMT}(1\%, 20, 24, 5000,, 0) = \$11.42$.

- Principal payment ($n = 20$): $= \text{PPMT}(1\%, 20, 24, 5000,, 0) = \$223.94$.

- Total interest payments between $n = 1$ and $n = 20$: $= \text{CUMIPMT}(1\%, 24, 5000, 1, 20, 0) = \$625.73$.

Comments: Certainly, generation of a loan repayment schedule such as that in Table 3.4 can be a tedious and time-consuming process unless a computer is used. At this book's website,[2] you can download an Excel file that creates the loan repayment schedule. You can make any adjustment to this file in order solve a typical loan problem of your choice.

---

[2]http://www.prenhall.com/park.

3.4.3 Comparing Different Financing Options

When you choose a car, you also choose how to pay for it. If you do not have the cash on hand to buy a new car outright—and most of us don't—you can consider taking out a loan or leasing the car in order to spread out the payments over time. Your decision to pay cash, take out a loan, or sign a lease depends on a number of personal as well as economic factors. Leasing is an option that lets you pay for the portion of a vehicle you expect to use over a specified term, plus rent charge, taxes, and fees. For example, you might want a $20,000 vehicle. Assume that vehicle might be worth about $9,000 (its residual value) at the end of a three-year lease. Your options are as follows:

- If you have enough money to buy the car, you could purchase the car in cash. If you pay cash, however, you will lose the opportunity to earn interest on the amount you spend. That could be substantial if you have access to investments paying good returns.
- If you purchase the vehicle via debt financing, your monthly payments will be based on the entire $20,000 value of the vehicle. You will own the vehicle at the end of your financing term, but the interest you will pay on the loan will drive up the real cost of the car.
- If you lease the vehicle, your monthly payments will be based on the amount of the vehicle you expect to "use up" over the lease term. This value ($11,000 in our example) is the difference between the original cost ($20,000) and the estimated value at lease end ($9,000). With leasing, the length of your lease agreement, the monthly payments, and the yearly mileage allowance can be tailored to your driving needs. The greatest financial appeal for leasing is its low initial outlay costs: usually, you pay only a leasing administrative fee, one month's lease payment, and a refundable security deposit. The terms of your lease will include a specific mileage allowance; if you put additional miles on your car, you will have to pay more for each extra mile.

Which Interest Rate Do We Use in Comparing Different Financing Options?

The dealer's (bank's) interest rate is supposed to reflect the time value of money of the dealer (or the bank) and is factored into the required payments. However, the correct interest rate to use when comparing financing options is the interest rate that reflects *your* earning opportunity. For most individuals, this interest rate might be equivalent to the savings rate from their deposits. To illustrate, we provide two examples. Example 3.8 compares two different financing options for an automobile. Example 3.9 examines a lease-versus-buying decision on an automobile.

Example 3.8 Buying a Car: Paying in Cash versus Taking a Loan

Consider the following two options proposed by an auto dealer:

- **Option A:** Purchase the vehicle at the normal price of $26,200, and pay for the vehicle over 36 months with equal monthly payments at 1.9% APR financing.

- **Option B:** Purchase the vehicle at a discounted price of $24,048 to be paid immediately. The funds that would be used to purchase the vehicle are presently earning 5% annual interest compounded monthly.

Which option is more economically sound?

Discussion: In calculating the net cost of financing the car, we need to decide which interest rate to use in discounting the loan repayment series. Note that the 1.9% APR represents the dealer's interest rate to calculate the loan payments. With the 1.9% interest, your monthly payments will be $A = \$26,200(A/P, 1.9\%/12, 36) = \$749.29$. On the other hand, the 5% APR represents your earning opportunity rate. In other words, if you do not buy the car, your money continues to earn 5% APR. Therefore, this 5% rate represents your opportunity cost of purchasing the car. Which interest rate should we use in this analysis? Since we wish to calculate each option's present worth to you, given your money and financial situation, we must use your 5% interest rate to value these cash flows.

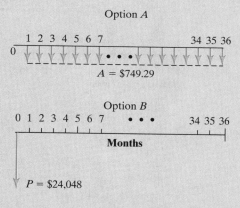

FIGURE 3.12 Cash flow diagram

SOLUTION

Given: The loan payment series shown in Figure 3.12, $r = 5\%$, payment period = monthly, and compounding period = monthly.

Find: The most economical financing option.

For each option, we will calculate the net equivalent cost (present worth) at $n = 0$. Since the loan payments occur monthly, we need to determine the effective interest rate per month, which is 5%/12.

- Option A (conventional financing):

    The equivalent present cost of the total loan repayments is calculated as

$$P_A = \$749.29(P/A, 5\%/12, 36)$$
$$= \$25,000.$$

● Option B (cash payment):

Since the cash payment is a lump sum to be paid presently, its equivalent present cost is equal to its value:

$$P_B = \$24,048.$$

Thus, there would be $952 of savings in present value with the cash payment option.

Example 3.9 Buying versus Leasing a Car

Two types of financing options are offered for an automobile at a local dealer, as shown in the below Table.

Buying versus Leasing		
	Option 1 **Debt Financing**	**Option 2** **Lease Financing**
Price	$14,695	$14,695
Down payment	$2,000	$0
APR (%)	3.6%	
Monthly payment	$372.55	$236.45
Length	36 months	36 months
Fees		$495
Cash due at lease end		$300
Purchase option at lease end		$8,673.10
Cash due at signing	$2,000	$731.45

The calculations are based on special financing programs available at participating dealers for a limited time. For each option, license, title, registration fees, taxes, and insurance are extra. For the lease option, the lessee must come up with $731.45 at signing. This cash due at signing includes the first month's lease payment of $236.45 and a $495 administrative fee. No security deposit is required. However, a $300 disposition fee is due at lease end. The lessee has the option to purchase the vehicle at lease end for $8,673.10. The lessee is also responsible for excessive wear and use. If your earning interest rate is 6% compounded monthly, which financing option is a better choice?

Discussion: With a lease payment, you pay for the portion of the vehicle you expect to use. At the end of the lease, you simply return the vehicle to the dealer and pay the agreed-upon disposal fee. With traditional financing, your monthly payment is based on the entire $14,695 value of the vehicle, and you will own the vehicle at the end of your financing terms. Since you are comparing the options over three years, you must explicitly consider the unused portion (resale value) of the vehicle at the end of the term. In other words, you must consider the resale value of the vehicle in order to figure out the net cost of owning the vehicle. You could use the $8,673.10 quoted by the dealer in the lease option as the resale value. Then you have to ask yourself if you can get that kind of resale value after three years of ownership. (See Figure 3.13.)

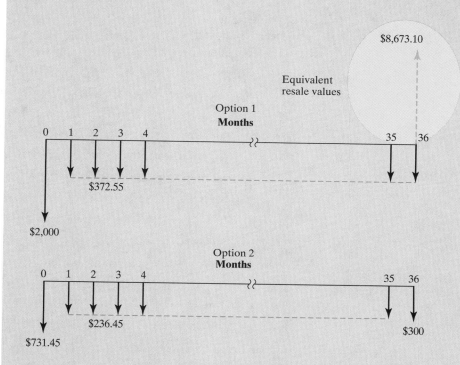

FIGURE 3.13 Cash flow diagrams for buying and leasing the car

SOLUTION

Given: The lease payment series shown in Figure 3.13, $r = 6\%$, payment period = monthly, and compounding period = monthly.

Find: The most economical financing option, assuming that you will be able to sell the vehicle for $8,673.10 at the end of three years.

For each option, we will calculate the net equivalent total cost at $n = 0$. Since the loan payments occur monthly, we need to determine the effective interest rate per month, which is 0.5%.

● **Conventional financing:**
The equivalent present cost of the total loan payments is calculated as

$$P_1 = \$2,000 + \$372.55(P/A, 0.5\%, 36)$$
$$= \$14,246.10.$$

The equivalent present worth of the resale value is calculated as

$$P_2 = \$8,673.10(P/F, 0.5\%, 36) = \$7,247.63.$$

The equivalent present net financing cost is therefore

$$P = P_1 + P_2 = \$14,246.10 - \$7,247.63$$
$$= \$6,998.47.$$

• **Lease financing:**

The equivalent present cost of the total lease payments is calculated as

$$P_1 = \$731.45 + \$236.45(P/A, 0.5\%, 35)$$
$$= \$731.45 + \$7,574.76$$
$$= \$8,306.21.$$

The equivalent present cost of the disposal fee is calculated as

$$P_2 = \$300(P/F, 0.5\%, 36) = \$250.69.$$

The equivalent present net lease cost is therefore

$$P = P_1 + P_2 = \$8,306.21 + \$250.69$$
$$= \$8,556.90.$$

It appears that the traditional financing program to purchase the car is more economical at 6% interest compounded monthly.

Comments: By varying the resale value S, we can find the break-even resale value that makes traditional financing equivalent to lease financing for this case:

$$\$8,556.90 = \$14,246.10 - S(P/F, 0.5\%, 36).$$

Thus, the break-even resale value is

$$S = (\$14,246.10 - \$8,556.90)/0.8356$$
$$= \$6,808.15.$$

So, at a resale value greater than $6,808.15, the conventional financing plan would be the more economical choice.

Summary

- Interest is most frequently quoted by financial institutions as an **APR**, or **annual percentage rate**. However, compounding often occurs more frequently. Simply multiplying the APR with the amount of debt does not account for the effect of this more frequent compounding. This situation leads to the distinction between nominal and effective interest.
  - **Nominal interest** is a stated rate of interest for a given period (usually a year).
  - **Effective interest** is the actual rate of interest, which accounts for the interest amount accumulated over a given period. The **effective rate** is related to the APR by

$$i = (1 + r/M)^M - 1,$$

where r = the APR, M = the number of compounding periods, and i = the effective interest rate.

- In any equivalence problem, the interest rate to use is the effective interest rate per payment period, which is expressed as

$$i = [1 + r/(CK)]^C - 1,$$

where C = the number of interest periods per payment period, K = the number of payment periods per year, and r/K = the nominal interest rate per payment period.

- The equation for determining the effective interest of continuous compounding is as follows:

$$i = e^{r/K} - 1.$$

- The difference in accumulated interest between continuous compounding and daily compounding is relatively small.
- Whenever payment and compounding periods differ with each other, it is recommended to compute the effective interest rate per payment period. The reason is that, to proceed with equivalency analysis, the compounding and payment periods must be the same.
- The cost of a loan will depend on many factors, such as loan amount, loan term, payment frequency, fees, and interest rate.
- In comparing different financing options, the interest rate to use is the one that reflects the decision maker's time value of money, not the interest rate quoted by the financial institution(s) lending the money.

Problems

Market Interest Rates (Nominal versus Effective Interest Rates)

3.1 A loan company offers money at 1.5% per month compounded monthly.

(a) What is the nominal interest rate?

(b) What is the effective annual interest rate?

3.2 A department store has offered you a credit card that charges interest at 0.95% per month compounded monthly. What is the nominal interest (annual percentage) rate for this credit card? What is the effective annual interest rate?

3.3 A California bank, Berkeley Savings and Loan, advertised the following information: interest 7.55% and effective annual yield 7.842%. No mention is made of the interest period in the advertisement. Can you figure out the compounding scheme used by the bank?

3.4 American Eagle Financial Sources, which makes small loans to college students, offers to lend a student $400. The borrower is required to pay $26.61 at the end of each week for 16 weeks. Find the interest rate per week. What is the nominal interest rate per year? What is the effective interest rate per year?

3.5 A financial institution is willing to lend you $40. However, you must repay $45 at the end of one week.

(a) What is the nominal interest rate?

(b) What is the effective annual interest rate?

3.6 A loan of $12,000 is to be financed to assist a person's purchase of an automobile. Based upon monthly compounding for 30 months, the end-of-the-month equal payment is quoted as $435. What nominal interest rate is being charged?

3.7 You are purchasing a $9,000 used automobile, which is to be paid for in 36 monthly installments of $288.72. What nominal interest rate are you paying on this financing arrangement?

3.8 You obtained a loan of $20,000 to finance your purchase of an automobile. Based on monthly compounding over 24 months, the end-of-the-month equal payment was figured to be $922.90. What is the APR used for this loan?

Calculating an Effective Interest Rate Based on a Payment Period

3.9 James Hogan is purchasing a $24,000 automobile, which is to be paid for in 48 monthly installments of $583.66. What is the effective annual interest rate for this financing arrangement?

3.10 Find the effective interest rate per payment period for an interest rate of 9% compounded monthly if the payment period is

   (a) monthly
   (b) quarterly
   (c) semiannual
   (d) annual

3.11 What is the effective interest rate per quarter if the interest rate is 6% compounded monthly?

3.12 What is the effective interest rate per month if the interest rate is 8% compounded continuously?

Equivalence Calculations Using Effective Interest Rates

3.13 What will be the amount accumulated by each of the following present investments?

   (a) $4,638 in 10 years at 6% compounded semiannually.
   (b) $6,500 in 15 years at 8% compounded quarterly.
   (c) $28,300 in seven years at 9% compounded monthly.

3.14 What is the future worth of the following series of payments?

   (a) $3,000 at the end of each six-month period for 10 years at 6% compounded semiannually.
   (b) $4,000 at the end of each quarter for six years at 8% compounded quarterly.
   (c) $7,000 at the end of each month for 14 years at 9% compounded monthly.

3.15 What equal series of payments must be paid into a sinking fund in order to accumulate the following amounts?

   (a) $12,000 in 10 years at 6% compounded semiannually when payments are semiannual.
   (b) $7,000 in 15 years at 9% compounded quarterly when payments are quarterly.
   (c) $34,000 in five years at 7.55% compounded monthly when payments are monthly.

3.16 What is the present worth of the following series of payments?

   (a) $500 at the end of each six-month period for 10 years at 8% compounded semiannually.
   (b) $2,000 at the end of each quarter for five years at 8% compounded quarterly.
   (c) $3,000 at the end of each month for eight years at 9% compounded monthly.

3.17 What is the amount C of quarterly deposits such that you will be able to withdraw the amounts shown in the accompanying cash flow diagram if the interest rate is 8% compounded quarterly?

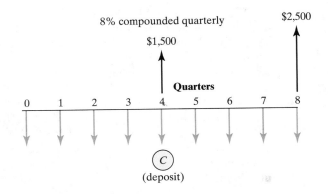

3.18 A series of equal quarterly deposits of $1,000 extends over a period of three years. It is desired to compute the future worth of this quarterly deposit series at 12% compounded monthly. Which of the following equations is correct for this operation?

(a) $F = 4(\$1,000)(F/A, 12\%, 3)$.
(b) $F = \$1,000(F/A, 3\%, 12)$.
(c) $F = \$1,000(F/A, 1\%, 12)$.
(d) $F = \$1,000(F/A, 3.03\%, 12)$.

3.19 Suppose you deposit $500 at the end of each quarter for five years at an interest rate of 8% compounded monthly. Which of the following formulas will determine the equal annual end-of-year deposit over five years that would accumulate the same amount under the same interest compounding?

(a) $A = [\$500(F/A, 2\%, 20)] \times (A/F, 8\%, 5)$.
(b) $A = \$500(F/A, 2.013\%, 4)$.
(c) $A = \$500(F/A, {}^{8\%}\!/_{12}, 20) \times (A/F, 8\%, 5)$.
(d) None of the above.

3.20 Suppose a young newlywed couple is planning to buy a home two years from now. To save the down payment required at the time of purchasing a home worth $220,000 (let's assume this down payment is 10% of the sales price, or $22,000), the couple has decided to set aside some money from their salaries at the end of each month. If the couple can earn 6% interest (compounded monthly) on their savings, determine the equal amount the couple must deposit each month so that they may buy the home at the end of two years.

3.21 Georgi Rostov deposits $5,000 in a savings account that pays 6% interest compounded monthly. Three years later, he deposits $4,000. Two years after the $4,000 deposit, he makes another deposit in the amount of $2,500. Four years after the $2,500 deposit, half of the accumulated funds is transferred to a fund that pays 8% interest compounded quarterly. How much money will be in each account six years after the transfer?

3.22 A man is planning to retire in 25 years. He wishes to deposit a regular amount every three months until he retires so that, beginning one year following his retirement, he will receive annual payments of $32,000 for the next 10 years. How much must he deposit if the interest rate is 8% compounded quarterly?

3.23 A building is priced at $125,000. If a buyer makes a down payment of $25,000 and a payment of $1,000 every month thereafter, how many months will it take for the buyer to completely pay for the building? Interest is charged at a rate of 9% compounded monthly.

3.24 A couple is planning to finance its three-year-old son's college education. The couple can deposit money at 6% compounded quarterly. What quarterly deposit must be made from the son's 3rd birthday to his 18th birthday in order to provide $50,000 on each birthday from the 18th to the 21st? (Note that the last deposit is made on the date of the first withdrawal.)

3.25 Sam Salvetti is planning to retire in 15 years. He can deposit money at 8% compounded quarterly. What deposit must he make at the end of each quarter until he retires so that he can make a withdrawal of $25,000 semiannually over the five years after his retirement? Assume that his first withdrawal occurs at the end of six months after his retirement.

3.26 Emily Lacy received $500,000 from an insurance company after her husband's death. She wants to deposit this amount in a savings account that earns interest at a rate of 6% compounded monthly. Then she would like to make 60 equal monthly withdrawals over five years such that, when she makes the last withdrawal, the savings account will have a balance of zero. How much should she withdraw each month?

3.27 Anita Tahani, who owns a travel agency, bought an old house to use as her business office. She found that the ceiling was poorly insulated and that the heat loss could be cut significantly if six inches of foam insulation were installed. She estimated that, with the insulation, she could cut the heating bill by $40 per month and the air-conditioning cost by $25 per month. Assuming that the summer season is three months (June, July, and August) of the year and that the winter season is another three months (December, January, and February) of the year, what's the most that Anita can spend on insulation that would make installation worthwhile, given that she expects to keep the property for five years? Assume that neither heating nor air conditioning would be required during the fall and spring seasons. If she decides to install the insulation, it will be done at the beginning of May. Anita's interest rate is 9% compounded monthly.

3.28 You want to open a savings plan for your future retirement. You are considering the following two options:

- Option 1: You deposit $1,000 at the end of each quarter for the first 10 years. At the end of 10 years, you make no further deposits, but you leave the amount accumulated at the end of 10 years for the next 15 years.
- Option 2: You do nothing for the first 10 years. Then you deposit $6,000 at the end of each year for the next 15 years.

If your deposits or investments earn an interest rate of 6% compounded quarterly and you choose Option 2 over Option 1, then at the end of 25 years from now, you will have accumulated

(a) $7,067 more.
(b) $8,523 more.
(c) $14,757 less.
(d) $13,302 less.

3.29 Don Harrison's current salary is $60,000 per year, and he is planning to retire 25 years from now. He anticipates that his annual salary will increase by $3,000 each year (i.e., in the first year he will earn $60,000, in the second year $63,000, in the third year $66,000, and so forth),

and he plans to deposit 5% of his yearly salary into a retirement fund that earns 7% interest compounded daily. What will be the amount accumulated at the time of his retirement?

Equivalence Calculations with Continuous Compounding

3.30 How many years will it take an investment to triple if the interest rate is 8% compounded

(a) quarterly?
(b) monthly?
(c) continuously?

3.31 A series of equal quarterly payments of $5,000 for 12 years is equivalent to what present amount at an interest rate of 9% compounded

(a) quarterly?
(b) monthly?
(c) continuously?

3.32 What is the future worth of an equal-payment series of $5,000 per year for five years if the interest rate is 8% compounded continuously?

3.33 Suppose that $1,000 is placed in a bank account at the end of each quarter over the next 20 years. What is the account's future worth at the end of 20 years when the interest rate is 8% compounded.

(a) quarterly?
(b) monthly?
(c) continuously?

3.34 If the interest rate is 7.5% compounded continuously, what is the required quarterly payment to repay a loan of $10,000 in four years?

3.35 What is the future worth of a series of equal monthly payments of $3,000 if the series extends over a period of six years at 12% interest compounded

(a) quarterly?
(b) monthly?
(c) continuously?

3.36 What is the required quarterly payment to repay a loan of $20,000 in five years if the interest rate is 8% compounded continuously?

3.37 A series of equal quarterly payments of $1,000 extends over a period of five years. What is the present worth of this quarterly-payment series at 9.75% interest compounded continuously?

3.38 A series of equal quarterly payments of $2,000 for 15 years is equivalent to what future lump-sum amount at the end of 10 years at an interest rate of 8% compounded continuously?

Borrowing with Credit Cards

3.39 You have just received credit card applications from two banks, A and B. The interest terms on your unpaid balance are stated as follows:

1. Bank A: 15% compounded quarterly.
2. Bank B: 14.8% compounded daily.

Which of the following statements is incorrect?

(a) The effective annual interest rate for Bank A is 15.865%.
(b) The nominal annual interest rate for Bank B is 14.8%.
(c) Bank B's term is a better deal, because you will pay less interest on your unpaid balance.
(d) Bank A's term is a better deal, because you will pay less interest on your unpaid balance.

3.40 Jim Norton, an engineering major in his junior year, has received in the mail two guaranteed-line-of-credit applications from two different banks. Each bank offers a different annual fee and finance charge. Jim expects his average monthly balance after payment to be $300 and plans to keep the card he chooses for only 24 months. (After graduation, he will apply for a new card.) Jim's interest rate (on his savings account) is 6% compounded daily.

Terms	Bank A	Bank B
Annual fee	$20	$30
Finance charge	1.55% monthly interest rate	16.5% annual percentage rate

(a) Compute the effective annual interest rate for each card.
(b) Which bank's credit card should Jim choose?

Commercial Loans

3.41 An automobile loan of $15,000 at a nominal rate of 9% compounded monthly for 48 months requires equal end-of-month payments of $373.28. Complete the following table for the first six payments as you would expect a bank to calculate the values:

End of Month (n)	Interest Payment	Repayment of Principal	Remaining Loan Balance
1			$14,739.22
2			
3		$264.70	
4	$106.59		
5	$104.59		
6			$13,405.71

3.42 You borrow $120,000 with a 30-year payback term and a variable APR that starts at 9% and can be changed every five years.

(a) What is the initial monthly payment?
(b) If, at the end of five years, the lender's interest rate changes to 9.75% (APR), what will the new monthly payment be?

3.43 Mr. Smith wants to buy a new car that will cost $18,000. He will make a down payment in the amount of $4,000. He would like to borrow the remainder from a bank at an interest rate of 9% compounded monthly. He agrees to make monthly payments for a period of two years in order to pay off the loan. Select the correct answer for each of the following questions.

(a) What is the amount of the monthly payment (A)?

  1. $A = \$14,000(A/P, 0.75\%, 24)$.
  2. $A = \$14,000(A/P, 9\%, 2)/12$.
  3. $A = \$14,000(A/F, 0.75\%, 24)$.
  4. $A = \$14,000(A/F, 9\%, 2)/12$.

(b) Mr. Smith has made 12 payments and wants to figure out the remaining balance immediately after the 12th payment. What is that remaining balance?

  1. $B_{12} = 12A$.
  2. $B_{12} = A(P/A, 9\%, 1)/12$.
  3. $B_{12} = A(P/A, 0.75\%, 12)$.
  4. $B_{12} = 10,000 - 12A$.

3.44 Talhi Hafid is considering the purchase of a used automobile. The price, including the title and taxes, is $8,260. Talhi is able to make a $2,260 down payment. The balance, $6,000, will be borrowed from his credit union at an interest rate of 9.25% compounded daily. The loan should be paid in 48 equal monthly payments. Compute the monthly payment. What is the total amount of interest Talhi has to pay over the life of the loan?

3.45 Bob Pearson borrowed $20,000 from a bank at an interest rate of 12% compounded monthly. This loan is to be repaid in 36 equal monthly installments over three years. Immediately after his 20th payment, Bob desires to pay the remainder of the loan in a single payment. Compute the total amount he must pay at that time.

3.46 You are buying a home for $190,000. If you make a down payment of $40,000 and take out a mortgage on the rest at 8.5% compounded monthly, what will be your monthly payment if the mortgage is to be paid off in 15 years?

3.47 For a $250,000 home mortgage loan with a 20-year term at 9% APR compounded monthly, compute the total payments on principal and interest over the first five years of ownership.

3.48 A lender requires that monthly mortgage payments be no more than 25% of gross monthly income, with a maximum term of 30 years. If you can make only a 15% down payment, what is the minimum monthly income needed in order to purchase a $200,000 house when the interest rate is 9% compounded monthly?

3.49 To buy a $150,000 house, you put down $30,000 and take out a mortgage for $120,000 at an APR of 9% compounded monthly. Five years later, you sell the house for $185,000 (after all selling expenses are factored in). What equity (the amount that you can keep before any taxes are taken out) would you realize with a 30-year repayment term? (Assure that the loan is paid off when the house is sold in lump sum.)

3.50 Just before the 15th payment,

  • Family A had a balance of $80,000 on a 9%, 30-year mortgage;
  • Family B had a balance of $80,000 on a 9%, 15-year mortgage; and
  • Family C had a balance of $80,000 on a 9%, 20-year mortgage.

All of the APRs are compounded monthly. How much interest did each family pay on its 15th payment?

3.51 Home mortgage lenders often charge points on a loan in order to avoid exceeding a legal limit on interest rates or to make their rates appear competitive with those of other lenders. As an example, with a two-point loan, the lender would loan only $98 for each $100 borrowed. The borrower would receive only $98, but would have to make payments just as if he or she had received $100. In this way, the lender can make more money while keeping his or her interest rate lower. Suppose that you receive a loan of $130,000 payable at the end each month for 30 years with an interest rate of 9% compounded monthly, but you have been charged three points. What is the effective interest rate on this home-mortgage loan?

3.52 A restaurant is considering purchasing the lot adjacent to its business to provide adequate parking space for its customers. The restaurant needs to borrow $35,000 to secure the lot. A deal has been made between a local bank and the restaurant such that the restaurant would pay the loan back over a five-year period with the following payment terms: 15%, 20%, 25%, 30%, and 35% of the initial loan at the end of the first, second, third, fourth, and fifth years, respectively.

(a) What rate of interest is the bank earning from this loan transaction?

(b) What would be the total interest paid by the restaurant over the five-year period?

3.53 Alice Harzem wanted to purchase a new car for $18,400. A dealer offered her financing through a local bank at an interest rate of 13.5% compounded monthly. The dealer's financing required a 10% down payment and 48 equal monthly payments. Because the interest rate was rather high, Alice checked with her credit union for other possible financing options. The loan officer at the credit union quoted her 10.5% interest for a new-car loan and 12.25% for a used-car loan. But to be eligible for the loan, Paula had to have been a member of the credit union for at least six months. Since she joined the credit union two months ago, she has to wait four more months to apply for the loan. Alice decides to go ahead with the dealer's financing and, four months later, refinances the balance through the credit union at an interest rate of 12.25% (because the car is no longer new).

(a) Compute the monthly payment to the dealer.

(b) Compute the monthly payment to the credit union.

(c) What is the total interest payment for each loan transaction?

3.54 David Kapamagian borrowed money from a bank to finance a small fishing boat. The bank's loan terms allowed him to defer payments (including interest) for six months and then to make 36 equal end-of-month payments thereafter. The original bank note was for $4,800, with an interest rate of 12% compounded monthly. After 16 monthly payments, David found himself in a financial bind and went to a loan company for assistance in lowering his monthly payments. Fortunately, the loan company offered to pay his debts in one lump sum, provided that he pays the company $104 per month for the next 36 months. What monthly rate of interest is the loan company charging on this transaction?

3.55 A loan of $10,000 is to be financed over a period of 24 months. The agency quotes a nominal interest rate of 8% for the first 12 months and a nominal interest rate of 9% for any remaining unpaid balance after 12 months, with both rates compounded monthly. Based on these rates, what equal end-of-the-month payment for 24 months would be required in order to repay the loan?

3.56 Robert Carré financed his office furniture through the furniture dealer from which he bought it. The dealer's terms allowed him to defer payments (including interest) for six months and then to make 36 equal end-of-month payments thereafter. The original note was for $12,000, with interest at 12% compounded monthly. After 26 monthly payments, Robert found himself in a financial bind and he went to a loan company for assistance. The loan company offered to pay his debts in one lump sum provided that he will pay the company $204 per month for the next 30 months.

(a) Determine the original monthly payment made to the furniture store.

(b) Determine the lump-sum payoff amount the loan company will make.

(c) What monthly rate of interest is the loan company charging on this loan?

Comparing Different Financing Options

3.57 Suppose you are in the market for a new car worth $18,000. You are offered a deal to make a $1,800 down payment now and to pay the balance in equal end-of-month payments of $421.85 over a 48-month period. Consider the following situations:

(a) Instead of going through the dealer's financing, you want to make a down payment of $1,800 and take out an auto loan from a bank at 11.75% compounded monthly. What would be your monthly payment to pay off the loan in four years?

(b) If you were to accept the dealer's offer, what would be the effective rate of interest per month charged by the dealer on your financing?

3.58 A local dealer is advertising a 24-month lease of a sport utility vehicle for $520 payable at the beginning of each month. The lease requires a $2,500 down payment, plus a $500 refundable security deposit. As an alternative, the company offers a 24-month lease with a single up-front payment of $12,780, plus a $500 refundable security deposit. The security deposit will be refunded at the end of the 24-month lease. Assuming you have access to a deposit account that pays an interest rate of 6% compounded monthly, which lease is more favorable?

3.59 You want to purchase a house for $85,000, and you have $17,000 cash available for a down payment. You are considering the following two financing options:

- Option 1: get a new standard mortgage with 10% (APR) interest compounded monthly and a 30-year term.

- Option 2: assume the seller's old mortgage that has an interest rate of 8.5% (APR) compounded monthly, a remaining term of 25 years (from an original term of 30 years), a remaining balance of $35,394, and payments of $285 per month. You can obtain a second mortgage for the remaining balance, $32,606, from your credit union at 12% (APR) compounded monthly, with a 10-year repayment period.

  (a) What is the effective interest rate for Option 2?

  (b) Compute the monthly payments for each option over the life of the mortgage.

  (c) Compute the total interest payment for each option.

(d) What homeowner's interest rate (home owner's time value of money) makes the two financing options equivalent?

Short Case Studies with Excel

3.60 You are considering buying a new car worth $15,000. You can finance the car either by withdrawing cash from your savings account, which earns 8% interest compounded monthly, or by borrowing $15,000 from your dealer for four years at 11% interest compounded monthly. You could earn $5,635 in interest from your savings account in four years if you leave the money in the account. If you borrow $15,000 from your dealer, you only pay $3,609 in interest over four years, so it makes sense to borrow for your new car and keep your cash in your savings account. Do you agree or disagree with the foregoing statement? Justify your reasoning with a numerical calculation.

3.61 Suppose you are going to buy a home worth $110,000 and make a down payment in the amount of $50,000. The balance will be borrowed from the Capital Savings and Loan Bank. The loan officer offers the following two financing plans for the property:

- Option 1: a conventional fixed loan at an interest rate of 13% compounded monthly over 30 years with 360 equal monthly payments.

- Option 2: a graduated payment schedule (FHA 235 plan) at 11.5% interest compounded monthly with the following monthly payment schedule:

Year (n)	Monthly Payment	Monthly Mortgage Insurance
1	$497.76	$25.19
2	$522.65	$25.56
3	$548.78	$25.84
4	$576.22	$26.01
5	$605.03	$26.06
6–30	$635.28	$25.96

For the FHA 235 plan, mortgage insurance is a must.

(a) Compute the monthly payment for Option 1.

(b) What is the effective annual interest rate you are paying for Option 2?

(c) Compute the outstanding balance for each option at the end of five years.

(d) Compute the total interest payment for each option.

(e) Assuming that your only investment alternative is a savings account that earns an interest rate of 6% compounded monthly, which option is a better deal?

3.62 Ms. Kennedy borrowed $4,909 from a bank to finance a car at an add-on interest rate[3] of 6.105%. The bank calculated the monthly payments as follows:

- Contract amount = $4,909 and contract period = 42 months (or 3.5 years). Thus add-on interest is = $4,909(0.06105)(3.5) = $1,048.90.
- Acquisition fee = $25, thus total loan charge = $1,048.90 + $25 = $1,073.90.
- Total of payments = $4,909 + 1,073.90 = $5,982.90, and monthly installment = $5,982.90/42 = $142.45.

After making the seventh payment, Ms. Kennedy wants to pay off the remaining balance. The following is the letter from the bank explaining the net balance Ms. Kennedy owes:

Dear Ms. Kennedy,

The following is an explanation of how we arrived at the payoff amount on your loan account:

Original note amount	$5,982.90
Less 7 payments @ $142.45 each	997.15
	4,985.75
Loan charge (interest)	1,073.90
Less acquisition fee	25.00
	$1,048.90

Rebate factor from Rule of 78[th]s chart is 0.6589 (loan ran 8 months on a 42-month term).

$1,048.90 multiplied by 0.6589 = $691.12. $691.12 represents the unearned interest rebate.

Therefore, your payoff amount is computed as follows:

Balance	$4,985.75
Less unearned interest rebate	691.12
Payoff amount	$4,294.63

If you have any further questions concerning these matters, please contact us.

Sincerely,

S. Govia

Vice President

Hint: The Rule of 78[th]s is used by some financial institutions to determine the outstanding loan balance. According to the Rule of 78[th]s, the interest charged during a given month is figured out by applying a changing fraction to the total interest over the loan period. For example, in the case of a one-year loan, the fraction used in determining the interest charge for the first month would be 12/78, 12 being the number of remaining months of the loan and 78 being the sum of $1 + 2 + \ldots + 11 + 12$. For the second month, the fraction would be 11/78, and so on. In the case of a two-year loan, the fraction during the first month is 24/300, because there are 24 remaining payment periods and the sum of the loan periods is $300 = 1 + 2 + \ldots + 24$.

(a) Compute the effective annual interest rate for this loan.

(b) Compute the annual percentage rate (APR) for this loan.

(c) Show how would you derive the rebate factor (0.6589).

(d) Verify the payoff amount by using the Rule of 78[th]s formula.

(e) Compute the payoff amount by using the interest factor $(P/A, i, N)$.

---

[3]The add-on loan is totally different from the popular amortized loan. In this type of loan, the total interest to be paid is precalculated and added to the principal. The principal plus the precalculated interest amount is then paid in equal installments. In such a case, the interest rate quoted is not the effective interest rate, but what is known as add-on interest.

All-Time Top-10 Movies at the Box Office, Adjusted for Inflation (2001, North America)[1]

The accompanying table contains a list of the top-10 movies of all time in North America as of 2001, based on their domestic box-office gross, adjusted for inflation. Admittedly, there are factors that make this list not as accurate as it should be (examples are given momentarily), but it gives a general idea of what the top movies are of all time. In compiling the list, all total gross figures were expressed in terms of year-2001 dollars, assuming 2.54% inflation. Note that the list can't be 100% accurate, since box-office results aren't always counted accurately, and on top of that, studios are not always 100% accurate with their reports. The figures given here include only theater-generated domestic grosses. They do not include proceeds from sources like foreign sales, video rentals, and sales, sales of TV rights, and merchandising. Also, if we use a higher inflation rate to find the adjusted gross figures in year-2001 dollars, the rankings could also change. Our interest in this case is how we may incorporate the loss of purchasing power into our dollar comparison from Chapters 2 and 3.

#	Top Ten Adjusted Films	Distributor	Release	Top Producer	Admissions	Total Gross	Adjusted Total
1	Gone with the Wind	MGM	12/15/39	David O. Seiznick	283,100,000	$198,700,000	$1,599,500,000
2	Snow White & the Seven Dwarfs	Disney	02/04/37	Walt Disney	225,300,000	$185,000,000	$1,271,200,000
3	Star Wars	Fox	05/25/77	Gary Kurtz	176,900,000	$461,000,000	$999,500,000
4	E.T. The Extra-Terrestrial	Universal	06/11/82	Kathleen Kennedy	164,000,000	$435,000,000	$926,600,000
5	101 Dalmatians (1961)	Disney	01/25/61	Walt Disney	143,100,000	$153,000,000	$808,500,000
6	Bambi	Disney	08/13/42	Walt Disney	140,800,000	$102,900,000	$795,500,000
7	Titanic	Paramount	12/19/97	James Cameron	130,900,000	$600,800,000	$739,600,000
8	Jaws	Universal	06/20/75	Richard D. Zanuck	128,600,000	$260,000,000	$726,600,000
9	The Sound of Music	Fox	05/18/65	Robert Wise	119,300,000	$163,000,000	$674,000,000
10	The Ten Commandments	Paramount	10/05/56	Cecil B. De Mille	117,800,000	$85,400,000	$665,600,000

[1] *Best Digest*, at http://www.bestdigest.com/movies/bo/alltime_inflationadjusted.htm.

Equivalence Calculations under Inflation

U p to this point, we have demonstrated how to compute equivalence values under constant conditions in the general economy. In other words, we have assumed that prices remain relatively unchanged over long periods. As you know from personal experience, this is not a realistic assumption. In this chapter, we define and quantify the loss of purchasing power, or **inflation**, and then go on to apply it in several equivalence analyses.

4-1   Measure of Inflation

Historically, the general economy has usually fluctuated in such a way as to experience **inflation**, a loss in the purchasing power of money over time. Inflation means that the cost of an item tends to increase over time, or, to put it another way, the same dollar amount buys less of an item over time. **Deflation** is the opposite of inflation, in that prices decrease over time, and hence a specified dollar amount gains purchasing power. Inflation is far more common than deflation in the real world, so our consideration in this chapter will be restricted to accounting for inflation in economic analyses.

4.1.1   Consumer Price Index

Before we can introduce inflation into an equivalence calculation, we need a means of isolating and measuring its effect. Consumers usually have a relative, if not a precise, sense of how their purchasing power is declining. This sense is based on their experience of shopping for food, clothing, transportation, and housing over the years. Economists have developed a measure called the **consumer price index** (CPI), which is based on a typical **market basket** of goods and services required by the average consumer. This market basket normally consists of items from eight major groups: (1) food and alcoholic beverages, (2) housing, (3) apparel, (4) transportation, (5) medical care, (6) entertainment, (7) personal care, and (8) other goods and services.

The CPI compares the cost of the typical market basket of goods and services in a current month with its cost at a previous time, such as 1 month ago, 1 year ago, or 10 years ago. The point in the past to which current prices are compared is called the **base period**. The index value for this base period is set at $100. The original base period used by the Bureau of Labor Statistics (BLS) of the U.S. Department of Labor for the CPI index is 1967.

For example, let us say that, in 1967, the prescribed market basket could have been purchased for $100. Suppose the same combination of goods and services costs $538.80 in 2002. We can then compute the CPI for 2002 by multiplying the ratio of the current price to the base-period price by 100. In our example, the price index is ($538.80/$100)100 = 538.80, which means that the 2002 price of the contents of the market basket is 538.80% of its base-period price. (See Figure 4.1.)

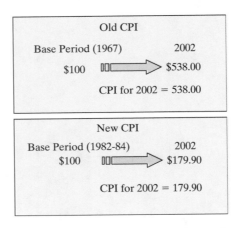

FIGURE 4.1 Measuring inflation based on CPI

The revised CPI introduced by the BLS in 1987 includes indices for two populations: (1) urban wage earners and clerical workers (CW) and (2) all urban consumers (CU). This change reflected the fact that differing populations had differing needs and thus differing "market baskets." Both the CW and the CU use updated expenditure weights based upon data tabulated from the three years of the Consumer Expenditure Survey (1982, 1983, and 1984) and incorporate a number of technical improvements. This method of assessing inflation does not imply, however, that consumers actually purchase the same goods and services year after year. Consumers tend to adjust their shopping practices to changes in relative prices and to substitute other items for those whose prices have greatly increased in relative terms. We must understand that the CPI does not take into account this sort of consumer behavior, because it is predicated on the purchase of a fixed market basket of the same goods and services, in the same proportions, month after month. For this reason, the CPI is called a **price index** rather than a **cost-of-living index**, although the general public often refers to it as a cost-of-living index.

4.1.2 Producer Price Index

The consumer price index is a good measure of the general price increase of consumer products. However, it is not a good measure of industrial price increases. When performing engineering economic analysis, the appropriate price indices must be selected to accurately estimate the price increases of raw materials, finished products, and operating costs. The *Survey of Current Business*, a monthly publication prepared by the BLS, provides the industrial-product price index for various industrial goods. Table 4.1 lists the CPI together with several price indexes over a number of years.[2]

TABLE 4.1 Selected Price Indexes between 1993 and 2002

Year (Base Period)	New CPI (1982–84)	Old CPI (1967)	Gasoline (1982)	Steel (1982)	Passenger Car (1982)
1993	144.0	461.2	63.9	116.0	129.8
1994	147.4	441.4	61.7	122.0	133.3
1995	152.2	455.0	63.7	128.8	134.0
1996	156.6	468.2	72.8	125.8	135.2
1997	160.2	479.7	71.9	126.5	135.2
1998	162.5	487.1	53.4	122.5	132.2
1999	166.2	497.8	64.7	114.0	121.4
2000	171.2	512.9	94.6	116.0	133.4
2001	176.6	530.4	90.5	109.7	138.9
2002	178.9	535.8	126.6	117.4	128.8

[2]CPI data are now available on the Internet at http://stats.bls.gov.

From Table 4.1, we can easily calculate the price index (or inflation rate) of gasoline from 2000 to 2001 as follows:

$$\frac{90.5 - 94.6}{94.6} = -0.04334 = -4.334\%.$$

Since the price index calculated is negative, the price of gasoline actually decreased at an annual rate of 4.334% over the year 2000, which was one of the better years for consumers who drive. However, in 2002, the price of gasoline increased at an annual rate of 39.88% over the price in 2001.

4.1.3 Average Inflation Rate (f)

To account for the effect of varying yearly inflation rates over a period of several years, we can compute a single rate that represents an **average inflation rate**. Since each year's inflation rate is based on the previous year's rate, these rates have a compounding effect. As an example, suppose we want to calculate the average inflation rate for a two-year period. The first year's inflation rate is 4%, and the second year's rate is 8%, with a base price of $100. To calculate the average inflation rate for the two years, we employ the following procedure:

- **Step 1:** To find the price at the end of the second year, we use the process of compounding:

$$\underbrace{\overbrace{\$100(1 + 0.04)}^{\text{First year}}(1 + 0.08)}_{\text{Second year}} = \$112.32.$$

- **Step 2:** To find the average inflation rate f, we establish the following equivalence equation:

$$\$100(1 + f)^2 = \$112.32, \text{ or } \$100(F/P, f, 2) = \$112.32.$$

Solving for f yields

$$f = 5.98\%.$$

Thus, we can say that the price increases in the last two years are equivalent to an average annual percentage rate of 5.98% per year. Note that the average is a geometric average, not an arithmetic average, over a several-year period. Using a single average rate such as this, rather than a different rate for each year's cash flows, simplifies our economic analysis, as you will see later.

Example 4.1 Average Inflation Rate

Consider the price increases for the 13 items in the following table over the last three years:

Category	2003 Price	2000 Price	Average Inflation Rate
Postage	$0.37	$0.33	3.89%
Homeowners insurance (per year)	$603.00	$500.00	6.44%
Auto insurance (per year)	$855.00	$687.00	7.56%
Private college tuition and fees	$18,273.00	$15,518.00	5.60%
Gasoline (per gallon)	$1.65	$1.56	1.89%
Haircut	$12.00	$10.50	4.55%
Car (Toyota Camry)	$22,000.00	$21,000.00	1.56%
Natural gas (per million BTUs)	$5.67	$3.17	21.38%
Baseball tickets (family of four)	$148.66	$132.44	3.92%
Cable TV (per month)	$47.97	$36.97	9.07%
Movies (average ticket)	$5.80	$5.39	2.47%
Movies (concessions)	$2.17	$1.98	3.10%
Health care (per year)	$2,088.00	$1,656.00	8.03%
Consumer price index (CPI) Base period: 1982 − 84 = 100	**184.20**	**171.20**	**2.47%**

Explain how the average inflation rates are calculated in the table.

SOLUTION

Let's take the fourth item, the cost of private college tuition, for a sample calculation. Since we know the prices during both 2000 and 2003, we can use the appropriate equivalence formula (single-payment compound amount factor or growth formula).

Given: $P = \$15,518$, $F = \$18,273$, and $N = 2003 - 2000 = 3$.
Find: f.

To solve this problem, we use the equation $F = P(1 + f)^N$:

$$\$18,273 = \$15,518(1 + f)^3.$$

Solving for f yields

$$f = \sqrt[3]{1.1775} - 1$$
$$= 0.0560 = 5.60\%.$$

In a similar fashion, we can obtain the average inflation rates for the remaining items as shown in the table. Clearly, the cost of natural gas increased the most among the items listed in the table.

4.1.4 General Inflation Rate ($\bar{f}$) versus Specific Inflation Rate (f_j)

When we use the CPI as a base to determine the average inflation rate, we obtain the **general inflation rate**. We need to distinguish carefully between the general inflation rate and the average inflation rate for specific goods:

- **General inflation rate ($\bar{f}$):** This average inflation rate is calculated based on the CPI for all items in the market basket. The market interest rate is expected to respond to this general inflation rate.

- **Specific inflation rate (f_j):** This rate is based on an index (or the CPI) specific to segment j of the economy. For example, we must often estimate the future cost for an item such as labor, material, housing, or gasoline. (When we refer to the average inflation rate for just one item, we will drop the subscript j, for simplicity.) In terms of CPI, we define the general inflation rate as

$$\text{CPI}_n = \text{CPI}_0(1 + \bar{f})^n, \tag{4.1}$$

or

$$\bar{f} = \left[\frac{\text{CPI}_n}{\text{CPI}_0}\right]^{1/n} - 1, \tag{4.2}$$

where $\bar{f}$ = the general inflation rate,

$\quad \text{CPI}_n$ = the consumer price index at the end period n, and

$\quad \text{CPI}_0$ = the consumer price index for the base period.

If we know the CPI values for two consecutive years, we can calculate the annual general inflation rate as

$$\bar{f}_n = \frac{\text{CPI}_n - \text{CPI}_{n-1}}{\text{CPI}_{n-1}}, \tag{4.3}$$

where $\bar{f}_n$ = the general inflation rate for period n.

As an example, let us calculate the general inflation rate for the year 2002, where $\text{CPI}_{2001} = 530.4$ and $\text{CPI}_{2002} = 538.8$:

$$\frac{538.8 - 530.4}{530.4} = 0.0158 = 1.58\%.$$

This calculation demonstrates that 2002 was an unusually good year for the U.S. economy, as its 1.58% general inflation rate is far lower than the average general inflation rate of 4.94% over the last 35 years.[3]

---

[3]To calculate the average general inflation rate from the base period (1967) to 2002, we perform the following calculation:

$$f = \left[\frac{538.8}{100}\right]^{1/35} - 1 = 4.94\%.$$

Example 4.2 Yearly and Average Inflation Rates

The accompanying table shows a utility company's cost to supply a fixed amount of power to a new housing development; the indices are specific to the utilities industry. Assume that year 0 is the base period.

   Determine the inflation rate for each period, and calculate the average inflation rate over the three years.

Year	Cost
0	$504,000
1	$538,400
2	$577,000
3	$629,500

SOLUTION

Given: History of utility cost.

Find: The yearly inflation rate (f_j) and the average inflation rate over the three-year time period (f).

The inflation rate during year 1 (f_1) is

$$(\$538,400 - \$504,000)/\$504,000 = 6.83\%.$$

The inflation rate during year 2 (f_2) is

$$(\$577,000 - \$538,400)/\$538,400 = 7.17\%.$$

The inflation rate during year 3 (f_3) is

$$(\$629,500 - \$577,000)/\$577,000 = 9.10\%.$$

The average inflation rate over the three years is

$$f = \left(\frac{\$629,500}{\$504,000}\right)^{1/3} - 1 = 0.0769 = 7.69\%.$$

Note that, although the average inflation rate is 7.69% for the period taken as a whole, none of the years within the period had this rate.[4]

4-2 Actual versus Constant Dollars

To introduce the effect of inflation into our economic analysis, we need to define two inflation-related terms:[5]

- **Actual (current) dollars (A_n):** Actual dollars are estimates of future cash flows for year n that take into account any anticipated changes in amount caused by inflationary or deflationary effects. Actual dollars are the amount of dollars that will be paid or received irrespective of how much these dollars are worth. Usually, these amounts are determined by applying an inflation rate to base-year dollar estimates.

---

[4]Since we obtained this average rate based on costs that are specific to the utilities industry, this rate is not the general inflation rate. It is a specific inflation rate for this utility.

[5]Based on the ANSI Z94 Standard Committee on Industrial Engineering Terminology, *The Engineering Economist*, 1988:33(2), 145–171.

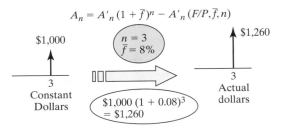

FIGURE 4.2 Conversion from constant to actual dollars

- **Constant (real) dollars (A'_n):** Constant dollars reflect constant purchasing power independent of the passage of time. Constant dollars are a measure of worth not an indicator of the number of dollars paid or received. In situations where inflationary effects were assumed when cash flows were estimated, these estimates can be converted to constant dollars (base-year dollars) by adjustment, using some readily accepted **general inflation rate**. We will assume that the base year is always time zero unless we specify otherwise.

4.2.1 Conversion from Constant to Actual Dollars

Since constant dollars represent dollar amounts expressed in terms of the purchasing power of the base year (see Figure 4.2), we may find the equivalent dollars in year n by using the general inflation rate $\bar{f}$ in the equation

$$A_n = A'_n(1 + \bar{f})^n = A'_n(\overline{F}/P, \bar{f}, n), \tag{4.4}$$

where A'_n = the constant-dollar expression for the cash flow occurring at the end of year n and

A_n = the actual-dollar expression for the cash flow occurring at the end of year n.

If the future price of a specific cost element (j) is not expected to follow the general inflation rate, we will need to use the appropriate average inflation rate applicable to this cost element, f_j, instead of $\bar{f}$.

Example 4.3 Conversion from Constant to Actual Dollars

Transco Company is considering making and supplying computer-controlled traffic-signal switching boxes to be used throughout Arizona. Transco has estimated the market for its boxes by examining data on new road construction and on deterioration and replacement of existing units. The current price per unit is $550; the before-tax manufacturing cost is $450. The start-up investment cost is $250,000. The projected sales and net before-tax cash flows in constant dollars are as below:

Assume that the price per unit and the manufacturing cost keep up with the general inflation rate, which is projected to be 5% annually. Convert the project's before-tax cash flows into the equivalent actual dollars.

Period	Unit Sales	Net Cash Flows in Constant $
0		−$250,000
1	1,000	$100,000
2	1,100	$110,000
3	1,200	$120,000
4	1,300	$130,000
5	1,200	$120,000

SOLUTION

Given: Net cash flows in Constant $, $\overline{f} = 5\%$.
Find: Net cash flows in actual $.

We first convert the constant dollars into actual dollars. Using Eq. (4.4), we obtain the following (note that the cash flow in period 0 is not affected by inflation):

Period	Net Cash Flow in Constant $	Conversion Factor	Cash Flow in Actual $
0	−$250,000	$(1 + 0.05)^0$	−$250,000
1	$100,000	$(1 + 0.05)^1$	$105,000
2	$110,000	$(1 + 0.05)^2$	$121,275
3	$120,000	$(1 + 0.05)^3$	$138,915
4	$130,000	$(1 + 0.05)^4$	$158,016
5	$120,000	$(1 + 0.05)^5$	$153,154

Figure 4.3 illustrates this conversion process graphically.

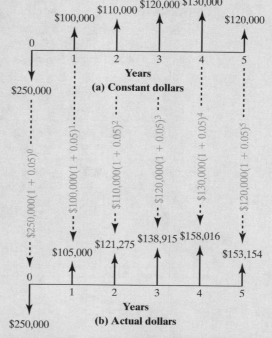

FIGURE 4.3 Cash flows for the project as expressed in constant dollars and actual dollars

4.2.2   Conversion from Actual to Constant Dollars

This process is the reverse of converting from constant to actual dollars. Instead of using the compounding formula, we use a discounting formula (single-payment present-worth factor):

$$A'_n = \frac{A_n}{(1 + \overline{f})^n} = A_n(P/F, \overline{f}, n). \tag{4.5}$$

Once again, we may substitute f_j for $\overline{f}$ if future prices are not expected to follow the general inflation rate.

Example 4.4 Conversion from Actual to Constant Dollars

Jagura Creek Fish Company, an aquacultural production firm, has negotiated a five-year lease on 20 acres of land, which will be used for fish ponds. The annual cost stated in the lease is $20,000, to be paid at the beginning of each of the five years. The general inflation rate $\overline{f}$ is 5%. Find the equivalent cost in constant dollars in each period.

Discussion: Although the $20,000 annual payments are *uniform*, they are not expressed in constant dollars. Unless an inflation clause is built into a contract, any stated amounts refer to *actual dollars*.

SOLUTION

Given: Five-year lease contract in actual $, $\overline{f} = 5\%$.
Find: Equivalent lease contract in constant $.

Using Eq. (4.5), we determine the equivalent lease payments in constant dollars as follows:

Note that, under the inflationary environment, the lease payment the lender receives in year 5 is worth only 82.27% of the first lease payment.

End of Period	Cash Flow in Actual $	Conversion at f	Cash Flow in Constant $	Loss in Purchasing Power
0	$20,000	$(1 + 0.05)^0$	$20,000	0%
1	$20,000	$(1 + 0.05)^{-1}$	$19,048	4.76%
2	$20,000	$(1 + 0.05)^{-2}$	$18,141	9.30%
3	$20,000	$(1 + 0.05)^{-3}$	$17,277	13.62%
4	$20,000	$(1 + 0.05)^{-4}$	$16,454	17.73%

4-3   Equivalence Calculations under Inflation

In previous chapters, our equivalence analyses took into consideration changes in the **earning power** of money, i.e., interest effects. To factor in changes in **purchasing power** as well—that is, inflation—we may use either (1) constant-dollar analysis or

(2) actual-dollar analysis. Either method produces the same solution; however, each method requires use of a different interest rate and procedure. Before presenting the two procedures for integrating interest and inflation, we will give a precise definition of the two interest rates used in them.

4.3.1 Market and Inflation-Free Interest Rates

Two types of interest rates are used in equivalence calculations: (1) the market interest rate and (2) the inflation-free interest rate. The difference between the two is analogous to the relationship between actual and constant dollars

- **Market interest rate (i):** this rate, commonly known as the **nominal interest rate**, takes into account the combined effects of the earning value of capital (earning power) and any anticipated inflation or deflation (purchasing power). Virtually all interest rates stated by financial institutions for loans and savings accounts are market interest rates. Most firms use a market interest rate (also known as **inflation-adjusted required rate of return**) in evaluating their investment projects.

- **Inflation-free interest rate (i'):** this rate is an estimate of the true earning power of money when the effects of inflation have been removed. This rate is commonly known as the **real interest rate**, and it can be computed if the market interest rate and the inflation rate are known. In fact, all the interest rates mentioned in previous chapters are inflation-free interest rates. As you will see later in this chapter, in the absence of inflation, the market interest rate is the same as the inflation-free interest rate.

In calculating any cash flow equivalence, we need to identify the nature of the cash flows. The three common cases are as follows:

Case 1: All cash flow elements are estimated in constant dollars.

Case 2: All cash flow elements are estimated in actual dollars.

Case 3: Some of the cash flow elements are estimated in constant dollars, and others are estimated in actual dollars.

For Case 3, we simply convert all cash flow elements into one type—either constant or actual dollars. Then we proceed with either constant-dollar analysis as for Case 1 or actual-dollar analysis as for Case 2.

4.3.2 Constant-Dollar Analysis

Suppose that all cash flow elements are already given in constant dollars and that you want to compute the equivalent present worth of the constant dollars (A'_n) in year n. In the absence of an inflationary effect, we should use i' to account for only the earning power of the money. To find the present-worth equivalent of this constant-dollar amount at i', we use

$$P_n = \frac{A'_n}{(1 + i')^n} \tag{4.6}$$

Constant-dollar analysis is common in the evaluation of many long-term public projects, because governments do not pay income taxes. Typically, income taxes are levied based on taxable incomes in actual dollars.

Example 4.5 Equivalence Calculations When Cash Flows Are Stated in Constant Dollars

Consider the constant-dollar flows given in Example 4.3. If Transco managers want the company to earn a 12% inflation-free rate of return (i') before tax on any investment, what would be the present worth of this project?

SOLUTION

Given: Cash flows stated in constant $\$$, $i' = 12\%$.
Find: Equivalent present worth of the cash flow series.

Since all values are in constant dollars, we can use the inflation-free interest rate. We simply discount the dollar inflows at 12% to obtain the following:

$$P = -\$250{,}000 + \$100{,}000(P/A, 12\%, 5)$$

$$+ \$10{,}000(P/G, 12\%, 4) + \$20{,}000(P/F, 12\%, 5)$$

$$= \$163{,}099 \text{ (in year zero dollars).}$$

4.3.3 Actual-Dollar Analysis

Now let us assume that all cash flow elements are estimated in actual dollars. To find the equivalent present worth of this actual dollar amount (A_n) in year n, we may use either the **deflation method** or the **adjusted-discount method**.

Actual-Dollar Analysis

- **Method 1: Deflation Method**

    Step 1: Bring all cash flows to a common purchasing power.

    Step 2: Consider the earning power.

- **Method 2: Adjusted-Discount Method**

    Combine Steps 1 and 2 into one step.

Deflation Method   The deflation method requires two steps to convert actual dollars into equivalent present-worth dollars. First, we convert actual dollars into equivalent constant dollars by discounting by the general inflation rate, a step that removes the inflationary effect. Now we can use i' to find the equivalent present worth.

Example 4.6 Equivalence Calculation When Cash Flows Are in Actual Dollars: Deflation Method

Applied Instrumentation, a small manufacturer of custom electronics, is contemplating an investment to produce sensors and control systems that have been requested by a fruit-drying company. The work would be done under a proprietary contract that would terminate in five years. The project is expected to generate the following cash flows in actual dollars:

n	Net Cash Flow in Actual Dollars
0	−$75,000
1	$32,000
2	$35,700
3	$32,800
4	$29,000
5	$58,000

(a) What are the equivalent year-zero dollars (constant dollars) if the general inflation rate $(\overline{f})$ is 5% per year?

(b) Compute the present worth of these cash flows in constant dollars at $i' = 10\%$.

SOLUTION

Given: Cash flows stated in actual $, $\overline{f} = 5\%$, and $i' = 10\%$.

Find: Equivalent present worth using the deflation method.

The net cash flows in actual dollars can be converted to constant dollars by deflating them, again assuming a 5% yearly deflation factor. The deflated (constant-dollar) cash flows can then be used to determine the present worth at i'. Figure 4.4 illustrates how the deflation method works in graphical form.

(a) We convert the actual dollars into constant dollars as follows:

Step 1: Convert Actual Dollars to Constant Dollars		
n	Cash Flows in Actual Dollars	Multiplied by Deflation Factor	Cash Flows in Constant Dollars
0	−$75,000	1	−$75,000
1	$32,000	$(1 + 0.05)^{-1}$	$30,476
2	$35,700	$(1 + 0.05)^{-2}$	$32,381
3	$32,800	$(1 + 0.05)^{-3}$	$28,334
4	$29,000	$(1 + 0.05)^{-4}$	$23,858
5	$58,000	$(1 + 0.05)^{-5}$	$45,445

(b) We use $i' = 10\%$ to compute the equivalent present worth of constant dollars:

	Step 2: Convert Constant Dollars to Equivalent Present Worth		
n	**Cash Flows in Constant Dollars**	**Multiplied by Discounting Factor**	**Equivalent Present Worth**
0	−$75,000	1	−$75,000
1	$30,476	$(1 + 0.10)^{-1}$	$27,706
2	$32,381	$(1 + 0.10)^{-2}$	$26,761
3	$28,334	$(1 + 0.10)^{-3}$	$21,288
4	$23,858	$(1 + 0.10)^{-4}$	$16,295
5	$45,445	$(1 + 0.10)^{-5}$	$28,218
			$\overline{\$45,268}$

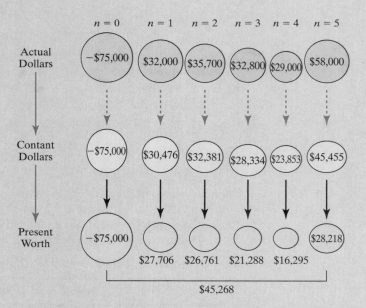

FIGURE 4.4 Deflation method. Converting actual dollars to *constant dollars* and then to equivalent *present worth*

Adjusted-Discount Method   The two-step process shown in Example 4.6 can be greatly streamlined by the efficiency of the **adjusted-discount method**, which performs deflation and discounting in one step. Mathematically, the two steps can be expressed as:

$$P_n = \frac{\dfrac{A_n}{(1 + \bar{f})^n}}{(1 + i')^n}$$

$$= \frac{A_n}{(1 + \bar{f})^n(1 + i')^n}$$

$$= \frac{A_n}{[(1 + \bar{f})(1 + i')]^n}. \qquad (4.7)$$

Since the market interest rate i reflects both the earning power and the purchasing power, we have the following relationship:

$$P_n = \frac{A_n}{(1 + i)^n}. \qquad (4.8)$$

The equivalent present-worth values in Eqs. (4.7) and (4.8) must be equal. Therefore,

$$\frac{A_n}{(1 + i)^n} = \frac{A_n}{[(1 + \bar{f})(1 + i')]^n}.$$

This equation leads to the following relationship among $\bar{f}$, i', and i:

$$(1 + i) = (1 + \bar{f})(1 + i').$$

Simplifying the terms yields

$$i = i' + \bar{f} + i'\bar{f}. \qquad (4.9)$$

This equation implies that the market interest rate is a function of two terms, i' and $\bar{f}$. See Figure 4.5 for a summary of the foregoing equations.

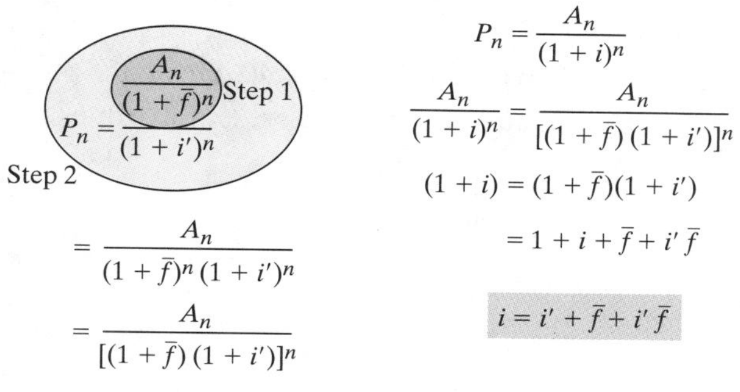

FIGURE 4.5 Derivation of the adjusted-discount method, showing how the market interest rate is related to the average inflation rate and the inflation-free interest rate

Note that, without an inflationary effect, the two interest rates are the same (if $\bar{f} = 0$, then $i = i'$). As either i' or $\bar{f}$ increases, i also increases. When prices increase due to inflation, bond rates climb, because promises of future payments from debtors are worth relatively less to lenders (i.e., banks, bondholders, money-market investors, CD holders, etc.). Thus, they demand and set higher interest rates. Similarly, if inflation were to remain at 3%, you might be satisfied with an interest rate of 7% on a bond, because your return would more than beat inflation. If inflation were running at 10%, however, you would not buy a 7% bond; you might insist instead on a return of at least 14%. On the other hand, when prices are coming down, or at least are stable, lenders do not fear the loss of purchasing power with the loans they make, so they are satisfied to lend at lower interest rates.

In practice, we often approximate the market interest rate i by simply adding the inflation rate $\bar{f}$ to the real interest rate i' and ignoring the product term $(i'\bar{f})$ if interest is not compounded continuously. *This practice is OK as long as either i' or $\bar{f}$ is relatively small.* With continuous compounding, the relationship among i, i', and $\bar{f}$ becomes

$$i' = i - \bar{f}. \tag{4.10}$$

So, if we assume a nominal APR (market interest rate) of 6% per year compounded continuously and an inflation rate of 4% per year compounded continuously, the inflation-free interest rate is exactly 2% per year compounded continuously.

Example 4.7 Equivalence Calculation When Flows Are in Actual Dollars: Adjusted-Discounted Method

Consider the cash flows in actual dollars in Example 4.6. Compute the equivalent present worth of these cash flows, using the adjusted-discount method.

SOLUTION

Given: Cash flows stated in actual $, $\bar{f} = 5\%$, and $i' = 10\%$.
Find: Equivalent present worth using the adjusted-discounted method.

First, we need to determine the market interest rate i. With $\bar{f} = 5\%$ and $i' = 10\%$, we obtain

$$i = i' + \bar{f} + i'\bar{f}$$
$$= 0.10 + 0.5 + (0.10)(0.05)$$
$$= 15.5\%.$$

The conversion process is shown in Figure 4.6. Note that the equivalent present worth that we obtain using the adjusted-discount method ($i = 15.5\%$) is exactly the same as the result we obtained in Example 4.6.

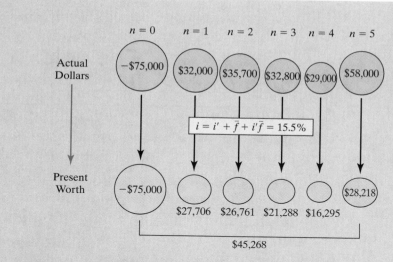

FIGURE 4.6 Conversion process from actual dollars to present-worth dollars: adjusted-discount method converting actual dollars to present-worth dollars by applying the market interest rate

Adjusted-Discount Method			
$i = i' + \bar{f} + i'\bar{f} = 0.10 + 0.05 + (0.10)(0.05) = 15.5\%$			
n	**Cash Flows in Actual Dollars**	**Multiplied by**	**Equivalent Present Worth**
0	$-\$75,000$	1	$-\$75,000$
1	$\$32,000$	$(1 + 0.155)^{-1}$	$\$27,706$
2	$\$35,700$	$(1 + 0.155)^{-2}$	$\$26,761$
3	$\$32,800$	$(1 + 0.155)^{-3}$	$\$21,288$
4	$\$29,000$	$(1 + 0.155)^{-4}$	$\$16,296$
5	$\$58,000$	$(1 + 0.155)^{-5}$	$\$28,217$
			$\overline{\$45,268}$

4.3.4 **Mixed-Dollar Analysis**

We now consider another situation in which some cash flow elements are expressed in constant (or today's) dollars and other elements in actual dollars. In this situation, we convert all cash flow elements into the same dollar units (either constant or actual). If the cash flow elements are all converted into actual dollars, the market interest rate i should be used in calculating the equivalence value. If the cash flow elements are all converted into constant dollars, the inflation-free interest rate i' should be used. Example 4.8 illustrates this situation.

Example 4.8 Equivalence Calculation with Composite Cash Flow Elements

A couple wishes to establish a college fund at a bank for their five-year-old child. The college fund will earn 8% interest compounded quarterly. Assuming that the child enters college at age 18, the couple estimates that an amount of $30,000 per year, in terms of today's dollars, will be required to support the child's college expenses for four years. College expenses are estimated to increase at an annual rate of 6%. Determine the equal quarterly deposits the couple must make until they send their child to college. Assume that the first deposit will be made at the end of the first quarter and that deposits will continue until the child reaches age 17. The child will enter college at age 18, and the annual college expense will be paid at the beginning of each college year. In other words, the first withdrawal will be made when the child is 18.

Discussion: In this problem, future college expenses are expressed in terms of today's dollars, whereas the quarterly deposits are in actual dollars. Since the interest rate quoted for the college fund is a market interest rate, we may convert the future college expenses into actual dollars:

Equivalence Calculation with Composite Cash Flow Elements		
Age	**College Expenses (in Today's Dollars)**	**College Expenses (in Actual Dollars)**
18 (freshman)	$30,000	$30,000(F/P, 6\%, 13) = \$63,988$
19 (sophomore)	$30,000	$30,000(F/P, 6\%, 14) = \$67,827$
20 (junior)	$30,000	$30,000(F/P, 6\%, 15) = \$71,897$
21 (senior)	$30,000	$30,000(F/P, 6\%, 16) = \$76,211$

SOLUTION

Given: A college savings plan, $i = 2\%$ per quarter, $N = 12$ years.

Find: Amount of quarterly deposit in actual $.

Approach: Convert any *cash flow elements in constant dollars* into *actual dollars*. Then use the *market interest rate* to find the equivalent present value.

The college expenses as well as the quarterly deposit series in actual dollars are shown in Figure 4.7. We first select $n = 12$, or age 17, as the base period for our equivalence calculation. (Note: inflation is compounded annually, thus the n we use here differs from the quarterly n we use next.) Then we calculate the accumulated total amount at the base period at 2% interest per quarter (8% APR/4 = 2% per quarter). Since the deposit period is 12 years and the first deposit is made at the end of the first quarter, we have a 48-quarter

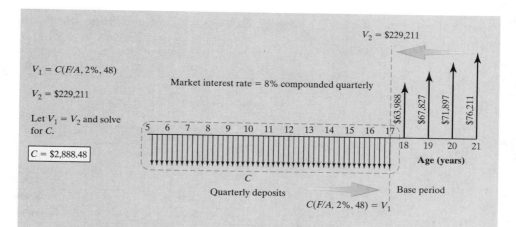

FIGURE 4.7 Establishing a college fund under an inflationary economy for a five-year-old child by making 48 quarterly deposits

deposit period. Therefore, the total balance of the deposits when the child is 17 would be

$$V_1 = C(F/A, 2\%, 48)$$
$$= 79.3535C.$$

The equivalent lump-sum worth of the total college expenditure at the base period would be

$$V_2 = \$63{,}988(P/F, 2\%, 4) + \$67{,}827(P/F, 2\%, 8)$$
$$+ \$71{,}897(P/F, 2\%, 12) + \$76{,}211(P/F, 2\%, 16)$$
$$= \$229{,}211.$$

By setting $V_1 = V_2$ and solving for C, we obtain $C = \$2{,}888.48$.

Comments: This is a good example for which Excel would help us understand the effects of variations in the circumstances of the situation. For example, how does changing the inflation rate affect the required quarterly savings plan? First, we must set up an Excel spreadsheet and utilize the Solver function.[6] Table 4.2 is an example spreadsheet that shows the deposit and withdrawal schedule for this scenario. After specifying the interest rate in E49 and using Excel functions to calculate equivalent total deposits and withdrawals in present-worth terms, we

---

[6]**Solver** is part of a suite of commands sometimes called *what-if analysis* tools. To access the command, go to the **Tool** menu and click on **Solver**. With Solver, you can find an optimal value for a *formula* in one cell—called the target cell—on a worksheet. Solver works with a group of cells that are related, either directly or indirectly, to the formula in the target cell. Solver adjusts the values in the changing cells you specify—called the adjustable cells—to produce the result you specify from the target cell formula. You can apply *constraints* to restrict the values Solver can use in the model, and the constraints can refer to other cells that affect the target cell formula.

TABLE 4.2 Excel Solution for Finding the Required Quarterly Deposits (Example 4.8)

	A	B	C	D	E
1					
2	Quarter	Deposits	Withdrawals		
3	0			College expense/year	$30,000
4	1	$(1)††		Inflation rate	6%
5	2	$(1)			College Expense in Actual $
6	3	$(1)		18 (Freshman)	$63,988
7	4	$(1)		19 (Sophomore)	$67,827
8†	5	$(1)		20 (Junior)	$71,897
40	37	$(1)		21 (Senior)	$76,211
41	38	$(1)			
42	39	$(1)			
43	40	$(1)			
44	41	$(1)			
45	42	$(1)			
46	43	$(1)			
47	44	$(1)		Target cell (E51-E52)	$0.57
48	45	$(1)			
49	46	$(1)		Interest rate (%)	2%
50	47	$(1)		Quarterly deposit	$2,888.47
51	48	$(1)		Equ. total deposits	$229,210.59
52	49			Equ. total withdrawals	−$229,210.02
53	50				
54	51				
55	52		$63,988		
56	53				
57	54				
58	55				
59	56		$67,827		
60	57				
61	58				
62	59				
63	60		$71,897		
64	61				
65	62				
66	63				
67	64		$76,211		
68					

†Note: Row 9 through Row 39 are hidden.

††Note: Initial guess.

designate cell E47 as the difference between E51 and E52 (i.e., E51−E52). To ensure that the accumulated balance of deposits is exactly sufficient to meet projected withdrawals, we specify that this target cell be zero (i.e., E51−E52 = 0) and tell Solver to adjust the quarterly deposit (E50), which is linked to the schedule of deposits in Column B) accordingly. The Solver function finds the required quarterly deposit amount to be $2,888.47 in Cell E50.

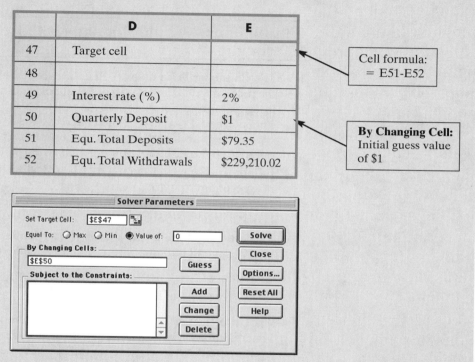

	D	E
47	Target cell	
48		
49	Interest rate (%)	2%
50	Quarterly Deposit	$1
51	Equ. Total Deposits	$79.35
52	Equ. Total Withdrawals	$229,210.02

Cell formula:
= E51-E52

By Changing Cell:
Initial guess value
of $1

Using this spreadsheet, we can then adjust the interest rate and see the change in quarterly deposits required. If we adjust annual inflation from 6% to 4%, we will find that the required quarterly deposit amount is $2,192.96, which is $695.51 less than in the 6% case.

Varying the annual inflation rate when the savings rate is fixed at 2% per quarter		Varying the quarterly savings rate when the inflation rate is fixed at 6% per year	
Annual Inflation Rate	Required Deposit Amount	Quarterly Savings Rate	Required Deposit Amount
2%	$1,657	1%	$4,131
4%	$2,193	1.5%	$3,459
6%	$2,888	2.0%	$2,888
8%	$3,786	2.5%	$2,404
10%	$4,941	3.0%	$1,995

Sample cells:

E6: = FV(E4,13,,-E3)
E47: = E51 − E52
E51: = FV(E49,48, −1*E50)
E52: = PV(E49,4,,C55) + PV(E49,8,,C59) + PV(E49,12,,C63)
        + PV(E49,16,,C67)

Summary

- The **Consumer Price Index** (**CPI**) is a statistical measure of change, over time, of the prices of goods and services in major expenditure groups—such as food, housing, apparel, transportation, and medical care—typically purchased by urban consumers. The CPI compares the cost of a sample "market basket" of goods and services in a specific period with the cost of the same market basket in an earlier reference period. This reference period is designated as the **base period**.

- **Inflation** is the term used to describe a **decline in purchasing power** evidenced in an economic environment of rising prices.

- **Deflation** is the opposite of inflation: it is an increase in purchasing power evidenced by falling prices.

- The **general inflation rate** $\bar{f}$ is an average inflation rate based on the CPI. An annual general inflation rate $\bar{f}_n$ can be calculated using the following equation:

$$\bar{f}_n = \frac{\mathrm{CPI}_n - \mathrm{CPI}_{n-1}}{\mathrm{CPI}_{n-1}}.$$

- The price changes of specific, individual commodities do not always reflect the general inflation rate. We can calculate an **average inflation rate** $\bar{f}_j$ for a specific commodity (*j*) if we have an index (that is, a record of historical costs) for that commodity.

- Project cash flows may be stated in one of two forms:

  1. **Actual dollars** (A_n): Dollar amounts that reflect the inflation or deflation rate.
  2. **Constant dollars** (A'_n): Dollar amounts that reflect the purchasing power of year zero dollars.

- Interest rates for the evaluation of projects may be stated in one of two forms:

  1. **Market interest rate** (*i*): A rate that combines the effects of interest and inflation; this rate is used with actual-dollar analysis. Unless otherwise mentioned, the interest rates used in the remainder of this text are the market interest rate.
  2. **Inflation-free interest rate** (*i'*): A rate from which the effects of inflation have been removed; this rate is used with constant-dollar analysis.

- To calculate the present worth of actual dollars, we can use either a two-step or a one-step process:

- **Deflation method—two steps:**

  1. Convert actual dollars to constant dollars by deflating with the general inflation rate of $\bar{f}$.

2. Calculate the present worth of constant dollars by discounting at i'.

- **Adjusted-discount method—one step (use the market interest rate):**

$$P_n = \frac{A_n}{[(1 + \overline{f})(1 + i')]^n}$$

$$= \frac{A_n}{(1 + i)^n},$$

where

$$i = i' + \overline{f} + i'\overline{f}.$$

Alternatively, just use the market interest rate to find the net present worth.

Problems

Note: In these problems, the term "market interest rate" represents the inflation-adjusted interest rate for equivalence calculations or the APR (annual percentage rate) quoted by a financial institution for commercial loans. Unless otherwise mentioned, all stated interest rates will be compounded annually.

Measure of Inflation

4.1 The median unleaded-gasoline price for California residents in 2003 was $1.82 per gallon. Assuming that the base period (price index = 100) is period 1996 and that the unleaded-gasoline price for that year was $1.10 per gallon, compute the average price index for the unleaded-gasoline price for the year 2003.

4.2 The following data indicate the price indices of lumber (price index for the base period (1982) = 100) during the last six years:

Period	Price Index
1997	144.5
1998	179.5
1999	188.2
2000	178.8
2001	171.6
2002	170.6
2003	?

(a) If the base period (price index = 100) is reset to the year 1997, compute the average price index for lumber between 1997 and 2002.

(b) If the past trend is expected to continue, how would you estimate the price of lumber in 2003?

4.3 Prices are increasing at an annual rate of 5% the first year and 8% the second year. Determine the average inflation rate $(\overline{f})$ over these two years.

4.4 Because of general price inflation in our economy, the purchasing power of the dollar shrinks with the passage of time. If the average general inflation rate is expected to be 7% per year for the foreseeable future, how many years will it take for the dollar's purchasing power to be one-half of what it is now?

Actual versus Constant Dollars

4.5 An annuity provides for 10 consecutive end-of-year payments of $4,500. The average general inflation rate is estimated to be 5% annually, and the market interest rate is 12% annually. What is the annuity worth in terms of a single equivalent amount of today's dollars?

4.6 A company is considering buying workstation computers to support its engineering staff. In today's dollars, it is estimated that the maintenance costs for the computers (paid at the end of each year) will be $25,000, $30,000, $32,000, $35,000, and $40,000 for years one through five, respectively. The general inflation rate $(\bar{f})$ is estimated to be 8% per year, and the company will receive 15% per year on its invested funds during the inflationary period. The company wants to pay for maintenance expenses in equivalent equal payments (in actual dollars) at the end of each of the five years. Find the amount of the company's annual payment.

4.7 Given the cash flows in actual dollars provided in the accompanying table, convert the cash flows to equivalent cash flows in constant dollars if the base year is time 0. Assume that the market interest rate is 16% and that the general inflation rate $(\bar{f})$ is estimated at 4% per year.

n	Cash Flow (in Actual $)
0	$1,500
4	$2,500
5	$3,500
7	$4,500

4.8 The purchase of a car requires a $25,000 loan to be repaid in monthly installments for four years at 12% interest compounded monthly. If the general inflation rate is 6% compounded monthly, find the actual and constant dollar value of the 20th payment of this loan.

4.9 A series of four annual constant-dollar payments beginning with $7,000 at the end of the first year is growing at the rate of 8% per year. Assume that the base year is the current year $(n = 0)$. If the market interest rate is 13% per year and the general inflation rate $(\bar{f})$ is 7% per year, find the present worth of this series of payments, based on

(a) constant-dollar analysis.

(b) actual-dollar analysis.

4.10 Consider the accompanying cash flow diagrams, where (a) the equal-payment cash flow in constant dollars is converted from (b) the equal-payment cash flow in actual dollars, at an annual general inflation rate of $\bar{f} = 3.8\%$. Also, $i = 9\%$. What is the amount A in actual dollars equivalent to $A' = \$1,000$ in constant dollars?

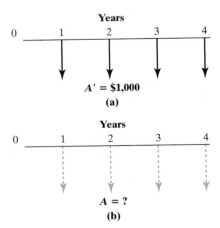

4.11 A 10-year $1,000 bond pays a nominal rate of 9% compounded semiannually. If the market interest rate is 12% compounded annually and the general inflation rate is 6% per year, find the actual and constant dollar amount (time = year zero dollars) of the 16th interest payment on the bond.

Equivalence Calculation under Inflation

4.12 Suppose that you borrow $20,000 at 12% compounded monthly over five years. Knowing that the 12% represents the market interest rate, you compute the monthly payment in actual dollars as $444.90. If the average monthly general inflation rate is expected to be 0.5%, determine the equivalent equal monthly payment series in constant dollars.

4.13 The annual fuel costs to operate a small solid-waste treatment plant are projected to be $1.5 million without considering any future inflation. The best estimates indicate that the annual inflation-free interest rate (i') will be 6% and the general inflation rate $\bar{f} = 5\%$. If the plant has a remaining useful life of five years, what is the present equivalent value of its fuel costs, using actual-dollar analysis?

4.14 Suppose that you just purchased a used car worth $6,000 in today's dollars. Assume also that, to finance the purchase, you borrowed $5,000 from a local bank at 9% compounded monthly over two years. The bank calculated your monthly payment at $228. Assume that average general inflation will run at 0.5% per month over the next two years,

(a) Determine the annual inflation-free interest rate (i') for the bank.

(b) What equal monthly payments, in terms of constant dollars over the next two years, are equivalent to the series of actual payments to be made over the life of the loan?

4.15 A man is planning to retire in 20 years. He can deposit money for his retirement at 6% compounded monthly. It is estimated that the future general inflation $(\bar{f})$ rate will be 5% compounded annually. What deposit must be made each month until the man retires so that he can make annual withdrawals of $40,000 in terms of today's dollars over the 15 years following his retirement? (Assume that his first withdrawal occurs at the end of the first six months after his retirement.)

4.16 On her 23rd birthday, an engineer decides to start saving toward building up a retirement fund that pays 8% interest compounded quarterly (market interest rate). She feels that $600,000 worth of purchasing power in today's dollars will be adequate to see her through her sunset years after her 63rd birthday. Assume a general inflation rate of 6% per year.

(a) If she plans to save by making 160 equal quarterly deposits, what should be the amount of her quarterly deposit in actual dollars?

(b) If she plans to save by making end-of-the-year deposits, increasing by $1,000 over

each subsequent year, how much would her first deposit be in actual dollars?

4.17 A father wants to save for his 8-year-old son's college expenses. The son will enter college 10 years from now. An annual amount of $40,000 in constant dollars will be required in order to support the son's college expenses for 4 years. Assume that these college payments will be made at the beginning of the school year. The future general inflation rate is estimated to be 6% per year, and the market interest rate on the savings account will average 8% compounded annually.

(a) What is the amount of the son's freshman-year expense in terms of actual dollars?

(b) What is the equivalent single-sum amount at the present time for these college expenses?

(c) What is the equal amount, in actual dollars, the father must save each year until his son goes to college?

4.18 Consider the following project's after-tax cash flow and the expected annual general inflation rate during the project period:

End of Year	Expected Cash Flow (in Actual $)	General Inflation Rate
0	−$45,000	
1	$26,000	6.5%
2	$26,000	7.7%
3	$26,000	8.1%

(a) Determine the average annual general inflation rate over the project period.

(b) Convert the cash flows in actual dollars into equivalent constant dollars, with year zero as the base year.

(c) If the annual inflation-free interest rate is 5%, what is the present worth of the cash flow?

Short Case Studies with Excel

4.19 You have $10,000 cash that you want to invest. Normally, you would deposit the money in a savings account that pays an annual interest rate of 6%. However, you are now considering the possibility of investing in a bond. Your alternatives are either a nontaxable municipal bond paying 9% or a taxable corporate bond paying 12%. Your marginal tax rate is 30% for both ordinary income and capital gains. (The marginal tax rate of 30% means that you will only keep 70% of your bond interest income.) You expect the general inflation rate to be 3% during the investment period. You can buy a high-grade municipal bond costing $10,000 that pays interest of 9% ($900) per year. This interest is not taxable. A comparable high-grade corporate bond for the same price is also available. This bond is just as safe as the municipal bond, but pays an interest rate of 12% ($1,200) per year. The interest for this bond is taxable as ordinary income. Both bonds mature at the end of year five.

(a) Determine the real (inflation-free) rate of return for each bond.

(b) Without knowing your earning-interest rate, can you make a choice between these two bonds?

4.20 *Business Week* magazine currently offers the following subscription options: one year for $39; two years for $72; three years for $106. These rates are expected to increase at the general inflation rate. You may vary the general inflation rate to be between 3% and 7%. What is your optimal strategy for subscribing at the lowest possible cost, assuming you intend to be a lifetime subscriber? What other assumptions are required in order to make your analysis complete?

4.21 The Michigan Legislature enacted the nation's first state-run program, the *Pay-Now, Learn-Later Plan*, to guarantee college tuition at public colleges in Michigan for students whose families invested in a special tax-free trust fund. The program is known as the Michigan Education Trust (MET), and there are three benefit plans from which to choose:

● **Full Benefits Plan:** It provides tuition and fees up to the number of credit hours required for a standard four-year undergraduate baccalaureate degree (usually up to 120 semester credit hours). Individuals may purchase 1, 2, 3, or 4 years of tuition under

this contract, and MET will pay for 30, 60, 90, or 120 credit hours, respectively.

● **Limited Benefits Plan:** It provides tuition and fees at Michigan public universities whose tuition *does not* exceed 105% of the weighted average tuition of all Michigan public four-year universities. For example, in 2002, if a student with a four-year limited benefits plan attends the relatively more expensive University of Michigan, MET will pay for 84 credit hours. If that student attends Michigan Technological University, MET will pay for 116 credit hours.

● **Community College Plan:** This plan provides in-district tuition and fees at Michigan public community colleges. Any individual may purchase 1 or 2 years under this contract.

The various contracts can be purchased in lump-sum or monthly payment plans. (See the price chart on the following page.)

For example, a lump sum four-year full benefits contact can be purchased at $24,252 for beneficiaries in grades eight and below. The lump sum contract cost is the same for eligible beneficiaries regardless of age or grade. However, for MET monthly purchase contracts, you can participate one of the three plans: 4-year purchase plan, 7-year purchase plan, and 10-year purchase plan.

The state government contends that the educational trust is a better deal than putting money into a certificate of deposit (CD) or a tuition prepayment plan at a bank because the state promises to stand behind the investment. Regardless of how high tuition goes, your trust funds will cover all of it. The disadvantage of a CD or a savings account is that you have to hope that tuition won't outpace the amount you save. The following chart also reveals how the tuition and mandatory fees at Michigan public four-year universities have increased between 1988 and 2002.

(a) Assuming that you are interested in the program for a newborn, would you join this program for a four-year full benefits contract?

(b) For a newborn, would you go with the lump sum payment plan or the 10-year monthly payment plan?

Price Chart: Met Lump-Sum Contracts				
	1 Year	2 Years	3 Years	4 Years
Full Benefits	$6,063	$12,126	$18,189	$24,252
Limited Benefits	$4,884	$9,768	$14,652	$19,536
Community College	$1,730	$3,460	NA	NA

A lump-sum full benefits contract can be purchased for beneficiaries in grades 8 and below. A lump sum limited benefits or community college contract can be purchased for beneficiaries in grades 10 and below. The lump-sum contract cost is the same for eligible beneficiaries regardless of age or grade.

Met Monthly Purchase Contracts

	4-Year Purchase Plan				**7-Year Purchase Plan**			
	1 Year	2 Years	3 Years	4 Years	1 Year	2 Years	3 Years	4 Years
Full Benefits	$148	$296	$444	$592	$95	$190	$285	$380
Limited Benefits	$119	$238	$357	$476	$77	$154	$231	$308
Community College	$42	$84	NA	NA	$27	$54	NA	NA

Monthly purchase contracts must be completely paid before the student is expected to enter college. Four-year monthly purchase plans can be purchased for beneficiaries in grades eight and below; 7-year monthly purchase plans can be purchased for beneficiaries in grades five and below; and 10-year monthly purchase plans can be purchased for beneficiaries in grades two and below. The monthly purchase cost is the same for eligible beneficiaries regardless of age or grade.

	10-Year Purchase Plan			
	1 Year	2 Years	3 Years	4 Years
Full Benefits	$74	$148	$222	$296
Limited Benefits	$60	$120	$180	$240
Community College	$21	$42	NA	NA

Source: Department of Education, State of Michigan, 2003 Contract Enrollment Booklet, Michigan's Section 529 Prepaid Tuition Program.

		1988–1989	2001–2002
CMU	Central Michigan University	$1,827	$4,247
EMU	Eastern Michigan University	$1,820	$4,753
FSU	Ferris State University	$1,947	$5,328
GVSU	Grand Valley State University	$1,794	$4,764
LSSU	Lake Superior State University	$1,767	$4,334
MSU	Michigan State University	$3,017	$6,118
MTU	Michigan Technological University	$2,193	$5,773
NMU	Northern Michigan University	$1,729	$4,172
OU	Oakland University	$2,065	$4,795
SVSU	Saginaw Valley State University	$1,959	$4,451
UM-AA	University of Michigan-Ann Arbor	$3,191	$7,560
UM-D	University of Michigan-Dearborn	$2,190	$5,281
UM-F	University of Michigan-Flint	$1,920	$4,391
WSU	Wayne State University	$2,289	$4,820
WMU	Western Michigan University	$2,104	$4,901

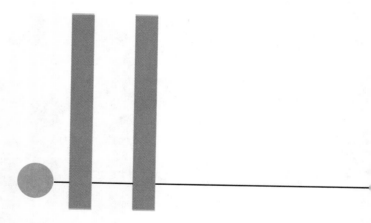

Evaluating Business
and Engineering Assets

FedEx to Roll Out PowerPads to Speed Delivery, Data Flow[1]

FedEx Corporation, in the latest move in a technology battle among the world's delivery carriers, said it will begin using PowerPads, a new generation of handheld, wireless computers that can capture detailed package information in its operation.

The Memphis, Tennessee, company plans to equip 40,000 U.S. couriers in its express-delivery unit with the devices over roughly the next two years. The computers, jointly developed with Motorola, Inc., are expected to save each courier more than 10 seconds per pickup stop, FedEx said. The company's current system, launched in 1986, requires drivers to return to their trucks in order to transmit information.

In the field, the PowerPads should be more intuitive than the existing truck-based system, according to FedEx, thus shrinking driver-training time and sharply reducing address errors on shipments. Drivers also will be able to immediately capture signatures electronically for the first time, matching an advantage claimed by rival United Parcel Service, Inc., since 1991. FedEx's current system requires paper-written signatures, which later can be scanned into the company's computers.

The devices will send information more quickly back into FedEx's delivery network. For example, if a big group of packages is being sent to Florida from New York—and the packages are running late—FedEx personnel in Memphis would quickly know to delay an outbound plane to Florida by an hour.

FedEx believes the PowerPad project will cost more than $150 million, including networking costs, and is projected to create savings of more than $20 million annually, just on picking up packages.

The new devices are the latest advance in a continuing effort among package carriers to harness the enormous flow of package-detail information that accompanies their daily loads of deliveries.

[1] Rick Brooks, "FedEx's New Hand-Held Devices Are Designed to Improve Service," *The Wall Street Journal*, November 27, 2002.

Present-Worth Analysis

The development and implementation of PowerPads is a cost-reduction project where the main question is, "Would there be enough annual savings from reducing pickup times to recoup the $150 million investment?" We can also pose the following questions: How long would it take to recover the initial investment? If a better information technology would make the PowerPad become obsolete in the near future, how would that affect the investment decision? These are the essential types of questions to address in evaluating any business investment decision. In business investment decisions, financial risk is by far the most critical element to consider.

What makes a company successful are the investments it makes today. Project ideas can originate from many different levels in an organization. Since some ideas will be good, while others will not, we need to establish procedures for screening projects to ensure that the right investments are made. Many large companies have a specialized project analysis division that actively searches for new ideas, projects, and ventures. The generation and evaluation of creative investment proposals is far too important a task to be left to just this project analysis group; instead, it is the ongoing responsibility of all managers and engineers throughout the organization. A key aspect of this process is the financial evaluation of investment proposals.

Our treatment of measures of investment worth is divided into three chapters. Chapter 5 begins with a consideration of the payback period, a project-screening tool that was the first formal method used to evaluate investment projects. Then it introduces two measures based on the basic cash flow equivalence technique of present-worth analysis. Because the annual-worth approach has many useful engineering applications related to estimating the unit cost, Chapter 6 is devoted to annual cash flow analysis. Chapter 7 presents measures of investment worth based on yield; these measures are known as rate-of-return analysis.

5-1   Loan versus Project Cash Flows

An investment made in a fixed asset is similar to an investment made by a bank when it lends money. The essential characteristic of both transactions is that funds are committed today in the expectation of their earning a return in the future. In the case of the bank loan, the future return takes the form of interest plus repayment of the principal. This return is known as the **loan cash flow**. In the case of the fixed asset, the future return takes the form of cash generated by productive use of the asset. The representation of these future earnings, along with the capital expenditures and annual expenses (such as wages, raw materials, operating costs, maintenance costs, and income taxes), is the **project cash flow**. This similarity between the loan cash flow and the project cash flow (see Figure 5.1) brings us to an important conclusion—that is, we can use the same equivalence techniques developed in earlier chapters to measure economic worth.

In Chapters 2 through 4, we presented the concept of the time value of money and developed techniques for establishing cash flow equivalence with compound-interest factors. This background provides a foundation for accepting or rejecting a capital investment—that is, for economically evaluating a project's desirability. The forthcoming coverage of investment worth in this chapter will allow us to go a step beyond accepting or rejecting an investment to making comparisons of alternative

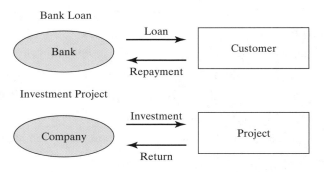

FIGURE 5.1 Loan versus project cash flows

investments. We will determine how to compare alternatives on an equal basis and select the wisest alternative from an economic standpoint.

We must also recognize that one of the most important parts of this capital-budgeting process is the estimation of relevant cash flows. For all examples in this chapter, and those in Chapters 6 and 7, net cash flows can be viewed as before-tax values or after-tax values for which tax effects have been recalculated. Since some organizations (e.g., governments and nonprofit organizations) are not subject to tax, the before-tax situation provides a valid base for this type of economic evaluation. Taking this view will allow us to focus on our main area of concern, the economic evaluation of investment projects. The procedures for determining after-tax net cash flows in taxable situations are developed in Part III. We will also assume that all cash flows are estimated in *actual dollars*, unless otherwise mentioned. Also, all interest rates used in project evaluation are assumed to be *market interest rates*.

5-2 Initial Project Screening Methods

Before studying the three measures of investment attractiveness, we will review a simple method that is commonly used to screen capital investments. One of the primary concerns of most businesspeople is whether, and when, the money invested in a project can be recovered. The **payback method** screens projects on the basis of how long it takes for net receipts to equal investment outlays:

- The payback period is determined by adding the expected cash flows for each year until the sum is equal to, or greater than, zero. The significance of this procedure can be easily explained. The cumulative cash flow equals zero at the point where cash inflows exactly match or pay back the cash outflows; thus, the project has reached the payback point. Once the cumulative cash flows exceed zero, cash inflows exceed cash outflows, and the project has begun to generate a profit, thus exceeding its payback point.

- This calculation can take one of two forms by either ignoring time-value-of-money considerations or including them. The former case is usually designated as the **conventional-payback method**, whereas the latter case is known as the **discounted-payback method**.

- A common standard used to determine whether to pursue a project is that a project does not merit consideration unless its payback period is shorter than some specified period of time. (This time limit is largely determined by management policy.)

For example, a high-tech firm, such as a computer-chip manufacturer, would set a short time limit for any new investment, because high-tech products rapidly become obsolete. If the payback period is within the acceptable range, a formal project evaluation (such as a present-worth analysis, which is the subject of this chapter) may begin. It is important to remember that **payback screening** is not an *end* in itself, but rather a method of screening out certain obviously unacceptable investment alternatives before progressing to an analysis of potentially acceptable ones.

Example 5.1 Conventional-Payback Period with Salvage Value

Ashland Company has just bought a new spindle machine at a cost of $105,000 to replace one that had a salvage value of $20,000. The projected annual after-tax savings due to improved efficiency, are as follows:

Period	Cash Flow	Cumulative Cash Flow
0	-$105,000 + $20,000	-$85,000
1	$15,000	-$70,000
2	$25,000	-$45,000
3	$35,000	-$10,000
4	$45,000	$35,000
5	$45,000	$80,000
6	$35,000	$115,000

SOLUTION

Given: Initial cost = $85,000; cash flow series as shown in Figure 5.2(a).

Find: Conventional-payback period.

The salvage value of retired equipment becomes a major consideration in most justification analyses. (In this example, the salvage value of the old machine should be taken into account, as the company already decided to replace the old machine.) When used, the salvage value of the retired equipment is subtracted from the purchase price of the new equipment, revealing a closer true cost of the investment. As we see from the cumulative cash flow series in Figure 5.2(b), the total investment is recovered at the end of year 4. If the firm's stated maximum payback period is three years, the project would not pass the initial screening stage.

Comments: In this example, we assumed that cash flows occur only in discrete lumps at the ends of years. If cash flows occur continuously throughout the year, our calculation of the payback period needs adjustment. A negative balance of $10,000 remains at the start of year 4. If the $45,000 is expected to be received as a more or less continuous flow during year 4, then the total investment will be recovered two-tenths ($10,000/$45,000) of the way through the fourth year. In this situation, the prorated payback period is thus 3.2 years.

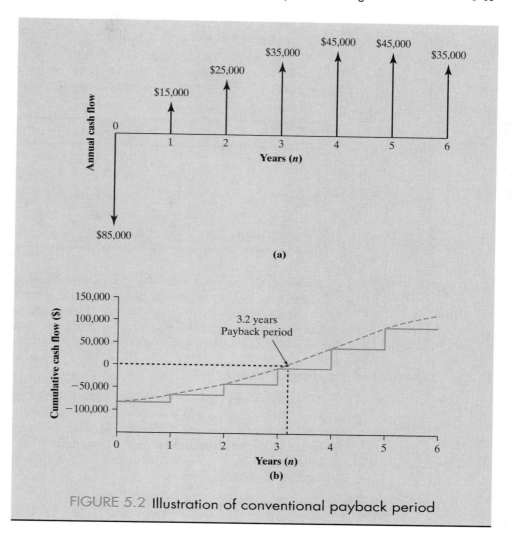

FIGURE 5.2 Illustration of conventional payback period

5.2.1   Benefits and Flaws of Payback Screening

The simplicity of the payback method is one of its most appealing qualities. Initial project screening by the payback method reduces the information search by focusing on that time at which the firm expects to recover the initial investment. The method may also eliminate some alternatives, thus reducing a firm's need to make further analysis efforts on those alternatives. But the much-used payback method of investment screening has a number of serious drawbacks as well:

- The principal objection to the payback method is that it fails to measure profitability; that is, it assumes that no profit is made during the payback period. Simply measuring how long it will take to recover the initial investment outlay contributes little to gauging the earning power of a project. (For instance, if you know that the money you borrowed for the drill press is costing you 12% per year, the payback method will not be able to tell you how much your invested money is contributing toward your interest expense.)

• Because payback-period analysis ignores differences in the timing of cash flows, it fails to recognize the difference between the present and future value of money. By way of illustration, consider two investment projects:

n	Project 1	Project 2
0	−$10,000	−$10,000
1	$1,000	$9,000
2	$9,000	$1,000
3	$1,000	$1,000
Payback period:	2 years	2 years

Although the payback period for both investments can be the same in terms of numbers of years, project 2 is better, because most investment being recovered at the end of year 1 is worth more than that to be gained later. Because payback screening also ignores all proceeds after the payback period, it does not allow for the possible advantages of a project with a longer economic life.

5.2.2   Discounted-Payback Period

To remedy one of the shortcomings of the payback period described previously, we may modify the procedure to consider the time value of money, i.e., the cost of funds (interest) used to support the project. This modified payback period is often referred to as the **discounted-payback period**. In other words, we may define the discounted-payback period as the number of years required to recover the investment from *discounted* cash flows.

Example 5.2 Discounted-Payback Periods

Consider the cash flow data given in Example 5.1. Assuming the firm's cost of funds to be 15%, compute the discounted-payback period.

SOLUTION

Given: $i = 15\%$; cash flow data in Example 5.1.
Find: Discounted-payback period.

To determine the period necessary to recover both the capital investment and the cost of funds required to support the investment, we may construct Table 5.1, which shows the cash flows and costs of funds to be recovered over the project's life. The cost of funds shown can be thought of as interest payments, if the initial investment is financed by loan, or as the opportunity cost of committing capital.

TABLE 5.1 Payback-Period Calculation Considering the Cost of Funds

Period	Cash Flow	Cost of Funds (15%)*	Cumulative Cash Flow
0	−$85,000	0	−$85,000
1	$15,000	−$85,000(0.15) = −$12,750	−$82,750
2	$25,000	−$82,750(0.15) = −$12,413	−$70,163
3	$35,000	−$70,163(0.15) = −$10,524	−$45,687
4	$45,000	−$45,687(0.15) = −$6,853	−$7,540
5	$45,000	−$7,540(0.15) = −$1,131	$36,329
6	$35,000	$36,329(0.15) = $5,449	$76,778

*Cost of funds = Unrecovered beginning balance × interest rate.

To illustrate, let's consider the cost of funds during the first year. With $85,000 committed at the beginning of the year, the interest in year one would be $12,750 ($85,000 × 0.15). Therefore, the total commitment grows to $97,750, but the $15,000 cash flow in year one leaves a net commitment of $82,750. The cost of funds during the second year would be $12,413 ($82,750 × 0.15). But with the $25,000 receipt from the project, the net commitment reduces to $70,163. When this process repeats for the remaining project years, we find that the net commitment to the project ends during year five. Depending on the cash flow assumption, the project must remain in use for about 4.2 years (continuous cash flows) or five years (year-end cash flows) in order for the company to cover its cost of capital and recover the funds invested in the project.

Comments: Inclusion of time-value-of-money effects has increased the payback period calculated for this example by a year. Certainly, this modified measure is an improved one, but, it, too, does not show the complete picture of the project's profitability.

5-3 Present-Worth Analysis

Until the 1950s, the payback method was widely used as a means of making investment decisions. As flaws in this method were recognized, however, businesspeople began to search for methods to improve project evaluations. The result was the development of **discounted cash flow techniques** (**DCFs**), which take into account the time value of money. One of the DCFs is the net-present-worth (or net-present-value) (PW or PV) method.

A capital-investment problem is essentially a matter of determining whether the anticipated cash inflows from a proposed project are sufficiently attractive to invest funds in the project. In developing the PW criterion, we will use the concept of cash flow equivalence discussed in Chapter 2. As we observed, the most convenient point at which to calculate the equivalent values is often at time zero. Under the PW criterion, the present worth of all cash inflows associated with an investment project is compared with the present worth of all cash outflows associated with that project. The difference between the present worths of these cash flows, referred to as the **net present worth** (PW), determines whether the project is or is not an acceptable investment. When two or more projects are under consideration, PW analysis further allows us to select the best project by comparing their PW figures directly.

5.3.1   Net-Present-Worth Criterion

We will first summarize the basic procedure for applying the net-present-worth criterion to a typical investment project, as well as for comparing alternative projects.

● Evaluation of a Single Project

Step 1: Determine the interest rate that the firm wishes to earn on its investments. The interest rate you determine represents the rate at which the firm can always invest the money in its *investment pool*. This interest rate is often referred to as either a *required rate of return* or a *minimum attractive rate of return* (MARR). Usually this selection is a policy decision made by top management. It is possible for the MARR to change over the life of a project, but for now we will use a single rate of interest when calculating PW.

Step 2: Estimate the service life of the project.

Step 3: Estimate the cash inflow for each period over the service life.

Step 4: Estimate the cash outflow for each period over the service life.

Step 5: Determine the net cash flows for each period (net cash flow = cash inflow − cash outflow).

Step 6: Find the present worth of each net cash flow at the MARR. Add up these present-worth figures; their sum is defined as the project's PW. That is,

$$\text{PW}(i) = \frac{A_0}{(1 + i)^0} + \frac{A_1}{(1 + i)^1} + \frac{A_2}{(1 + i)^2} + \cdots + \frac{A_N}{(1 + i)^N}$$

$$= \sum_{n=0}^{N} \frac{A_n}{(1 + i)^n}$$

$$= \sum_{n=0}^{N} A_n(P/F, i, n), \tag{5.1}$$

where
$$\text{PW}(i) = \text{PW calculated at } i,$$
$$A_n = \text{net cash flow at the end of period } n,$$
$$i = \text{MARR (or cost of capital), and}$$
$$n = \text{service life of the project.}$$

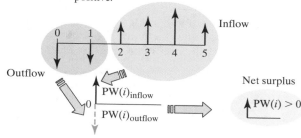

Principle: Compute the equivalent net surplus at $n = 0$ for a given interest rate of i.

Decision Rule: Accept the project if the net surplus is positive.

FIGURE 5.3 Illustration of how PW decision rules work

A_n will be positive if the corresponding period has a net cash inflow and negative if the period has a net cash outflow.

Step 7: In this context, a positive PW means that the equivalent worth of the inflows is greater than the equivalent worth of the outflows, so the project makes a profit. Therefore, if the PW(i) is positive for a single project, the project should be accepted; if it is negative, the project should be rejected. The process of applying the PW measure is shown in Figure 5.3 and is implemented with the following decision rule:

If PW(i) > 0, accept the investment.

If PW(i) = 0, remain indifferent.

If PW(i) < 0, reject the investment.

● **Comparing More Than One Alternative**

Note that the foregoing decision rule is for evaluation of a single project for which you can estimate the revenues as well as the costs.[2] The following guidelines should be used for evaluating and comparing more than one project:

1. If you need to select the best alternative, based on the net-present-worth criterion, select the one with the highest PW, as long as all the alternatives have the same service lives. Comparison of alternatives with unequal service lives requires special assumptions, as will be detailed in Section 5.4.

2. As you will find in Section 5.4, comparison of mutually exclusive alternatives with the same revenues is performed on a *cost-only basis*. In this situation, you should accept the project that results in the smallest PW of costs, or the least negative PW (because you are minimizing costs, rather than maximizing profits).

For now, we will focus on evaluating a single project. Techniques on how to compare multiple alternatives will be also detailed in Section 5.4.

---

[2]Some projects cannot be avoided—e.g., the installation of pollution-control equipment to comply with government regulations. In such a case, the project would be accepted even though its PW(i) < 0.

Example 5.3 Net Present Worth—Uneven Flows

Tiger Machine Tool Company is considering the acquisition of a new metal-cutting machine. The required initial investment of $75,000 and the projected cash benefits over a three-year project life are as follows:[3]

End of Year	Net Cash Flow
0	−$75,000
1	$24,400
2	$27,340
3	$55,760

You have been asked by the president of the company to evaluate the economic merit of the acquisition. The firm's MARR is known to be 15%.

SOLUTION

Given: Cash flows as tabulated; MARR = 15% per year.

Find: PW.

If we bring each flow to its equivalent at time zero as shown in Figure 5.4, we find that

$$PW(15\%) = -\$75,000 + \$24,000(P/F, 15\%, 1) + \$27,340(P/F, 15\%, 2)$$
$$+ \$55,760(P/F, 15\%, 3)$$
$$= \$3,553.$$

Since the project results in a surplus of $3,553, the project is acceptable. It is returning a profit greater than 15%.

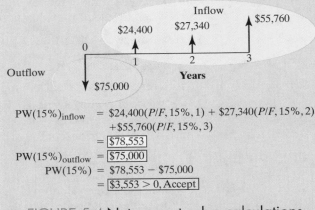

$$PW(15\%)_{inflow} = \$24,400(P/F, 15\%, 1) + \$27,340(P/F, 15\%, 2)$$
$$+\$55,760(P/F, 15\%, 3)$$
$$= \boxed{\$78,553}$$
$$PW(15\%)_{outflow} = \boxed{\$75,000}$$
$$PW(15\%) = \$78,553 - \$75,000$$
$$= \boxed{\$3,553 > 0, \text{Accept}}$$

FIGURE 5.4 Net-present-value calculations

[3]As we stated at the beginning of this chapter, we treat net cash flows in actual dollars as before-tax values or as having their tax effects precalculated. Explaining the process of obtaining cash flows requires an understanding of income taxes and the role of depreciation, which are discussed in Chapters 8 and 9.

TABLE 5.2 Present-Worth Amounts at Varying Interest Rates

i(%)	PW(i)	i(%)	PW(i)
0	$32,500	20	−$3,412
2	$27,743	22	−$5,924
4	$23,309	24	−$8,296
6	$19,169	26	−$10,539
8	$15,296	28	−$12,662
10	$11,670	30	−$14,673
12	$8,270	32	−$16,580
14	$5,077	34	−$18,360
16	$2,076	36	−$20,110
17.45*	$0	38	−$21,745
18	−$751	40	−$23,302

*Break-even interest rate.

In the above example, we computed the PW of the project at a fixed interest rate of 15%. If we compute the PW at varying interest rates, we obtain the data in Table 5.2. Plotting the PW as a function of interest rate gives the graph in Figure 5.5, the present-worth profile.

Figure 5.5 indicates that the investment project has a positive PW if the interest rate is below 17.45% and a negative PW if the interest rate is above 17.45%. As we will see in Chapter 7, this **break-even interest rate** is known as the **internal rate of return**. If the firm's MARR is 15%, the project has a PW of $3,553 and so may be accepted. The figure of $3,553 measures the equivalent immediate gain in present worth to the firm following the acceptance of the project. On the other hand, at $i = 20\%$, $\text{PW}(20\%) = -\$3,412$; the firm should reject the project in this case. Note that the decision to accept or reject an investment is influenced by the choice of a MARR, so it is crucial to estimate the MARR correctly. We will briefly describe the elements to consider for setting this interest rate for project evaluation.

5.3.2 Guidelines for Selecting a MARR

Return is what you get back in relation to the amount you invested. Return is one way to evaluate how your investments in financial assets or projects are doing in relation to each other and to the performance of investments in general. Let us look first at how we may derive rates of return. Conceptually, the rate of return that we realistically expect to earn on any investment is a function of three components:

- risk-free real return,
- inflation factor, and
- risk premium(s).

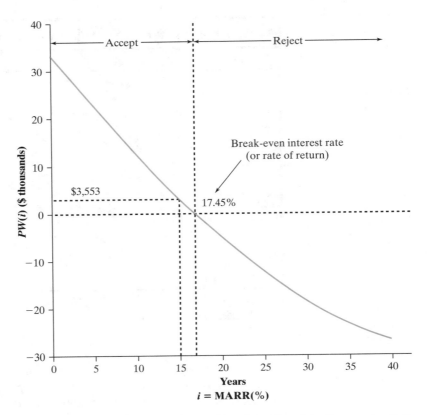

FIGURE 5.5 Present-worth profile described in Example 5.3

Suppose you want to invest in a stock. First, you would expect to be rewarded in some way for not being able to use your money while you hold the stock. Second, you would expect to be compensated for decreases in purchasing power between the time you invest and the time your investment is returned to you. Finally, you would demand additional rewards for having taken the risk of losing your money if the stock did poorly. If you did not expect your investment to compensate you for these factors, why would you tie up your money in this investment in the first place?

For example, if you were to invest $1,000 in risk-free U.S. Treasury bills for a year, you would expect a real rate of return of about 2%. Your risk premium would be also zero. You probably think that this does not sound like much. However, to that you have to add an allowance for inflation. If you expect inflation to be about 4% during that investment period, you should realistically expect to earn 6% during that interval (2% real return + 4% inflation factor + 0% for risk premium). This concept is illustrated in Figure 5.6.

How would it work out for a riskier investment, says an Internet stock? As you consider the stock to be a very volatile one, you would increase the risk premium to 20%. So you will not invest your money in the Internet stock unless you are reasonably confident of having it grow at an annual rate of 2% real return + 4% inflation factor + 20% risk premium = 26%. Again, the risk premium of 20% is a perceived value that can vary from one investor to another. We use the same concept in selecting the interest rate for project evaluation. In Chapter 9, we will

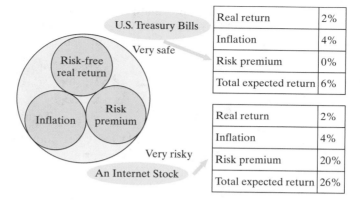

U.S. Treasury Bills	
Real return	2%
Inflation	4%
Risk premium	0%
Total expected return	6%

An Internet Stock	
Real return	2%
Inflation	4%
Risk premium	20%
Total expected return	26%

FIGURE 5.6 How to determine your expected return

consider this special issue in more detail. For now, we will assume that the firm has established a single interest rate for project evaluation, considering all relevant risk inherent in the project, and we will use it to measure the project's worth.

5.3.3 Meaning of Net Present Worth

In present-worth analysis, we assume that all the funds in a firm's treasury can be placed in investments that yield a return equal to the MARR. We may view these funds as an **investment pool**. Alternatively, if no funds are available for investment, we assume that the firm can borrow them at the MARR from the capital markets. In this section, we will examine these two views when explaining the meaning of MARR in PW calculations.

Investment-Pool Concept An investment pool is equivalent to a firm's treasury. It is where all fund transactions are administered and managed by the firm's comptroller. The firm may withdraw funds from this investment pool for other investment purposes, but if left in the pool, the funds will earn interest at the MARR. Thus, in investment analysis, net cash flows will be net cash flows relative to this investment pool. To illustrate the investment-pool concept, we consider again the project in Example 5.3, which required an investment of $75,000.

   If the firm did not invest in the project and instead left the $75,000 in the investment pool for three years, these funds would have grown as follows:

$$\$75,000(F/P, 15\%, 3) = \$114,066.$$

Suppose the company did decide to invest $75,000 in the project described in Example 5.3. Then the firm would receive a stream of cash inflows during the project life of three years in the following amounts:

Period (n)	Net Cash Flow (A_n)
1	$24,400
2	$27,340
3	$55,760

Since the funds that return to the investment pool earn interest at a rate of 15%, it would be of interest to see how much the firm would benefit from this investment. For this alternative, the returns after reinvestment are as follows:

$$\$24,400(F/P, 15\%, 2) = \$32,269$$
$$\$27,340(F/P, 15\%, 1) = \$31,441$$
$$\$55,760(F/P, 15\%, 0) = \underline{\$55,760}$$
$$\text{Total} \qquad \$119,470$$

These returns total $119,470. The additional cash accumulation at the end of three years from investing in the project is

$$\$119,470 - \$114,066 = \$5,404.$$

This $5,404 is also known as *net future worth of the project* at the project termination. If we compute the equivalent present worth of this net cash surplus at time zero, we obtain

$$\$5,404(P/F, 15\%, 3) = \$3,553,$$

which is exactly the same as the PW of the project as computed by Eq. (5.1). Clearly, on the basis of its positive PW, the alternative of purchasing a new machine should be preferred to that of simply leaving the funds in the investment pool at the MARR. Thus, in PW analysis, any investment is assumed to be returned at the MARR. If a surplus exists at the end of the project life, then we know that PW (MARR) > 0. Figure 5.7 summarizes the reinvestment concept as it relates to the firm's investment pool.

Borrowed-Funds Concept   Suppose that the firm does not have $75,000 at the outset. In fact, the firm doesn't have to maintain an investment pool at all. Let's further assume that the firm borrows all its capital from a bank at an interest rate of 15%, invests in the project, and uses the proceeds from the investment to pay off the

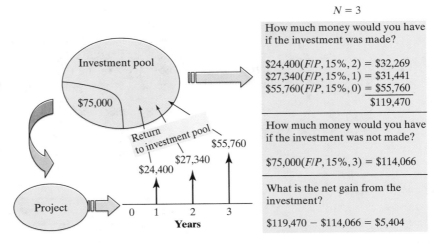

FIGURE 5.7 The concept of investment pool with the company as a lender and the project as a borrower

TABLE 5.3 Tabular Approach to Determining the Project Balances

N	0	1	2	3
Beginning Balance		−$75,000	−$61,850	−$43,788
Interest (15%)		−$11,250	−$9,278	−$6,568
Payment Received	−$75,000	+$24,400	+$27,340	+$55,760
Project Balance	−$75,000	−$61,850	−$43,788	$5,404

Net future worth, FW(15%)

$$PW(15\%) = \$5,404(P/F, 15\%, 3) = \$3,553$$

principal and interest on the bank loan. How much is left over for the firm at the end of the project period?

At the end of the first year, the interest on the bank loan would be $75,000(0.15) = $11,250. Therefore, the total loan balance grows to $75,000(1.15) = $86,250. Then the firm receives $24,400 from the project and applies the entire amount to repay the loan portion. This repayment leaves a balance due of

$$-\$75,000(1 + 0.15) + \$24,400 = -\$61,850.$$

As summarized in Table 5.3, this amount becomes the net amount the project is borrowing at the beginning of year two. This amount is also known as the **project balance**. At the end of year two, the debt to the bank grows to $61,850(1.15) = $71,128, but with the receipt of $27,340, the project balance reduces to

$$-\$61,850(1.15) + \$27,340 = -\$43,788.$$

Similarly, at the end of year three, the debt to the bank becomes $43,788(1.15) = $50,356, but with the receipt of $55,760 from the project, the firm should be able to pay off the remaining debt and come out with a surplus in the amount of $5,404

$$-\$43,788(1.15) + \$55,760 = \$5,404.$$

This terminal project balance is also known as the **net future worth** of the project. In other words, the firm fully repays its initial bank loan and interest at the end of year three, with a resulting profit of $5,404. Finally, if we compute the equivalent present worth of this net profit at time zero, we obtain

$$PW(15\%) = \$5,404(P/F, 15\%, 3) = \$3,553.$$

The result is identical to the case where we directly computed the PW of the project at $i = 15\%$, shown in Example 5.3. Figure 5.8 illustrates the project balance as a function of time. A negative project balance indicates the amount of remaining loan to be paid out or the amount of investment to be recovered.

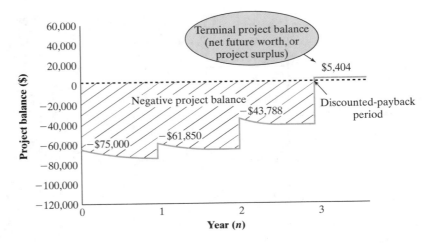

FIGURE 5.8 Project balance as a function of investment period

5.3.4 Capitalized-Equivalent Method

A special case of the PW criterion is useful when the life of a proposed project is **perpetual** or the planning horizon is extremely long (say, 40 years or more). Many public projects such as bridges, waterway constructions, irrigation systems, and hydroelectric dams are expected to generate benefits over an extended period of time (or forever). In this section, we examine the **capitalized-equivalent** [CE(i)] method for evaluating such projects.

Perpetual Service Life   Consider the cash flow series shown in Figure 5.9. How do we determine the PW for an infinite (or almost infinite) uniform series of cash flows or a repeated cycle of cash flows? The process of computing the PW for this infinite series is referred to as the **capitalization** of project cost. The cost is known as the **capitalized cost**. The capitalized cost represents the amount of money that must be invested today in order to yield a certain return A at the end of each and every period forever, assuming an interest rate of i. Observe the limit of the uniform-series

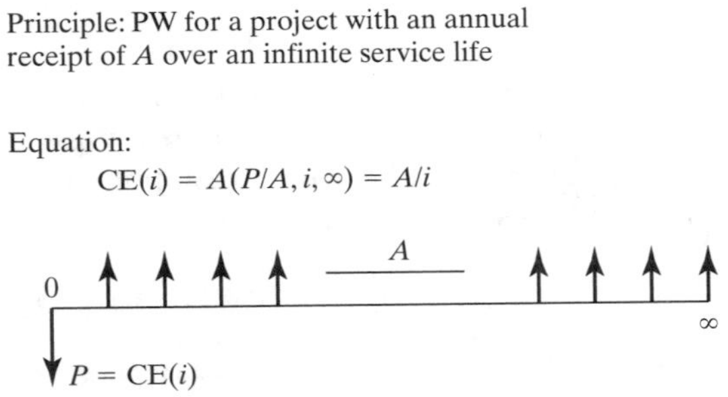

Principle: PW for a project with an annual
receipt of A over an infinite service life

Equation:
$$\text{CE}(i) = A(P/A, i, \infty) = A/i$$

FIGURE 5.9 Capitalized-equivalent worth—a project with a perpetual service life

present-worth factor as N approaches infinity:

$$\lim_{N \to \infty} (P/A, i, N) = \lim_{N \to \infty} \left[\frac{(1 + i)^N - 1}{i(1 + i)^N} \right] = \frac{1}{i}.$$

Thus, it follows that

$$PW(i) = A(P/A, i, N \to \infty) = \frac{A}{i}. \qquad (5.2)$$

This is the same result shown in Section 2.5.5. Another way of looking at this concept is to ask what constant income stream could be generated by $PW(i)$ dollars today in perpetuity. Clearly, the answer is $A = iPW(i)$. If withdrawals were greater than A, you would be eating into the principal, which would eventually reduce it to zero.

Example 5.4 Capitalized-Equivalent Cost

An engineering school has just completed a new engineering complex worth $50 million. A campaign targeting alumni is planned to raise funds for future maintenance costs, which are estimated at $2 million per year. Any unforeseen costs above $2 million per year would be obtained by raising tuition. Assuming that the school can create a trust fund that earns 8% interest annually, how much has to be raised now to cover the perpetual string of $2 million annual costs?

SOLUTION

Given: $A = $2 million, $i = 8\%$ per year, and $N = \infty$.
Find: CE(8%).

The capitalized-cost equation is

$$CE(i) = \frac{A}{i}.$$

Substituting in our given values, we obtain

$$CE(8\%) = \$2,000,000/0.08$$
$$= \$25,000,000.$$

Comments: It is easy to see that this lump-sum amount should be sufficient to pay maintenance expenses for the school forever. Suppose the school deposited $25 million at a bank that paid 8% interest annually. At the end of the first year, the $25 million would earn 8%($25 million) = $2 million interest. If this interest were withdrawn, the $25 million would remain in the account. At the end of the second year, the $25 million balance would again earn 8%($25 million) = $2 million. The annual withdrawal could be continued forever, and the endowment (gift funds) would always remain at $25 million.

5-4   Methods to Compare Mutually Exclusive Alternatives

Until now, we have considered situations involving only a single project or projects that were independent of each other. In both cases, we made the decision to reject or accept each project individually, based on whether it met the MARR requirements, evaluated using the PW.

In the real world of engineering practice, however, it is more typical for us to have two or more choices of projects for accomplishing a business objective. (As we shall see, even when it appears that we have only one project to consider, the implicit "do-nothing" alternative must be factored into the decision-making process.)

In this section, we extend our evaluation techniques to consider multiple projects that are mutually exclusive. Often, various projects or investments under consideration do not have the same duration or do not match the desired study period. Adjustments must be made when comparing multiple options in order to properly account for such differences. In this section, we explain the concepts of an analysis period and the process of accommodating for different lifetimes as important considerations for selecting among several alternatives. In the first few subsections of this section, all available options in a decision problem are assumed to have equal lifetimes. In Section 5.4.4, this restriction is relaxed.

When alternatives are **mutually exclusive**, any one of the alternatives will fulfill the same need, and the selection of one alternative implies that the others will be excluded. Take, for example, buying versus leasing an automobile for business use: When one alternative is accepted, the other is excluded. We use the terms **alternative** and **project** interchangeably to mean decision option. *One of the fundamental principles in comparing mutually exclusive alternatives is that they must be compared over an equal time span (or planning horizon).* In this section, we will present some of the fundamental principles that should be applied in comparing mutually exclusive investment alternatives. In doing so, we will consider two cases: (1) analysis period equals project lives and (2) analysis period differs from project lives. In each case, the required assumption for analysis can be varied. First, we will define some of the relevant terminology, such as "do-nothing alternative," "revenue project," and "service project."

5.4.1   Doing Nothing Is a Decision Option

When considering an investment, the project is one of two types: The project either is aimed at replacing an existing asset or system or is a new endeavor. In either case, a do-nothing alternative may exist. If a process or system already in place to accomplish our business objectives is still adequate, then we must determine which, if any, new proposals are economical replacements. If none are feasible, then we do nothing. On the other hand, if the existing system has terminally failed, the choice among proposed alternatives is mandatory (i.e., doing nothing is not an option).

New endeavors occur as alternatives to the do-nothing situation, which has zero revenues and zero costs. For most new endeavors, doing nothing is generally an alternative, as we won't proceed unless at least one of the proposed alternatives is economically sound. In fact, undertaking even a single project entails making a decision between two alternatives when the project is optional, because the do-nothing alternative is implicitly included. Occasionally, a new initiative *must* be undertaken, cost notwithstanding, and in this case the goal is to choose the most economical alternative, since doing nothing is not an option.

When the option of retaining an existing asset or system is available, there are two ways to incorporate it into the evaluation of the new proposals. One way is to treat the do-nothing option as a distinct alternative, but this approach will be covered primarily in Chapter 11, where methodologies specific to replacement analysis are presented. The second approach, used in this chapter, is to generate the cash flows of the new proposals relative to those of the do-nothing alternative. That is, for each new alternative, the **incremental costs** (and incremental savings or revenues if applicable) relative to those of the do-nothing alternative are used for the economic evaluation.

For a replacement-type problem, the incremental cash flow is calculated by subtracting the do-nothing cash flows from those of each new alternative. For new endeavors, the incremental cash flows are the same as the absolute amounts associated with each alternative, since the do-nothing values are all zero.

Because the main purpose of this section is to illustrate how to choose among mutually exclusive alternatives, most of the problems are structured so that one of the options presented must be selected. Therefore, unless otherwise stated, it is assumed that doing nothing is not an option and that the costs and revenues of the alternatives can be viewed as incremental to those of doing nothing.

5.4.2 Service Projects versus Revenue Projects

When comparing mutually exclusive alternatives, we need to classify investment projects as either service or revenue projects:

- **Service projects** are projects that generate revenues that do not depend on the choice of project, but *must produce the same amount of output (revenue).* In this situation, we certainly want to choose an alternative with the least input (or cost). For example, suppose an electric utility company is considering building a new power plant to meet the peak-load demand during either hot summer or cold winter days. Two alternative service projects could meet this peak-load demand: a combustion turbine plant or a fuel-cell power plant. No matter which type of plant is selected, the firm will generate the same amount of revenue from its customers. The only difference is how much it will cost to generate electricity from each plant. If we were to compare these service projects, we would be interested in knowing which plant could provide cheaper power (lower production cost). Further, if we were to use the PW criterion to compare these alternatives to minimize expenditures, *we would choose the alternative with the **lower present-value** production cost over the service life.*

- **Revenue projects** are projects that generate revenues that depend on the choice of alternative. For revenue projects, we are not limiting the amount of input to the project or the amount of output that the project would generate. Therefore, we want to select the alternative with the largest net gains (output − input). For example, a computer-monitor manufacturer is considering marketing two types of high-resolution monitors. With its present production capacity, the firm can market only one of them. Distinct production processes for the two models could incur very different manufacturing costs, and the revenues from each model would be expected to differ, due to divergent market prices and potentially different sales volumes. In this situation, if we were to use the PW criterion, *we would select the model that promises to bring in the* **largest net present worth**.

5.4.3 Analysis Period Equals Project Lives

Let's begin our analysis with the simplest situation, in which the project lives equal the analysis period. In this case, we compute the PW for each project and select the one with the highest PW. Example 5.5 illustrates this point.

Example 5.5 Comparing Two Mutually Exclusive Alternatives

Ansell, Inc., a medical-device manufacturer, uses compressed air in solenoids and pressure switches in its machines to control various mechanical movements. Over the years, the manufacturing floor has changed layouts numerous times. With each new layout, more piping was added to the compressed-air delivery system in order to accommodate new locations of manufacturing machines. None of the extra, unused old piping was capped or removed; thus, the current compressed-air delivery system is inefficient and fraught with leaks. Because of the leaks in the current system, the compressor is expected to run 70% of the time that the plant will be in operation during the upcoming year. This excessive usage will require 260 kWh of electricity at a rate of $0.05/kWh. (The plant runs 250 days a year, 24 hours per day.) Ansell may address this issue in one of two ways:

- Option 1—Continue current operation: If Ansell continues to operate the current air delivery system, the compressor's run time will increase by 7% per year for the next five years, because of ever-worsening leaks. (After five years, the current system will not be able to meet the plant's compressed-air requirement, so it will have to be replaced.)

- Option 2—Replace old piping now: If Ansell decides to replace all of the old piping now, it will cost $28,570. The compressor will still run for the same number of days; however, it will run 23% less (or will incur $70\%(1 - 0.23) = 53.9\%$ usage per day) because of the reduced air-pressure loss.

    If Ansell's interest rate is 12% compounded annually, is it worth fixing the air delivery system now?

SOLUTION

Given: Current power consumption, $g = 7\%$, $i = 12\%$, and $N = 5$ years.
Find: A_1 and P.

- **Step 1:** We need to calculate the cost of power consumption of the current piping system during the first year. The power consumption is determined as follows:

$$\begin{aligned}
\text{Power cost} &= \text{\% of day operating} \times \text{days operating per year} \\
&\quad \times \text{hours per day} \times \text{kWh} \times \text{\$/kWh} \\
&= (70\%) \times (250 \text{ days/year}) \times (24 \text{ hours/day}) \times (260 \text{ kWh}) \\
&\quad \times (\$0.05/\text{kWh}) \\
&= \$54{,}440.
\end{aligned}$$

● **Step 2:** Each year, if the current piping system is left in place, the annual power cost will increase at the rate of 7% over the previous year's cost. The anticipated power cost over the five-year period is summarized in Figure 5.10. The equivalent present lump-sum cost at 12% interest for this geometric gradient series is

$$\begin{aligned}
P_{\text{Option 1}} &= \$54{,}440(P/A_1, 7\%, 12\%, 5) \\
&= \$54{,}440\left[\frac{1 - (1 + 0.07)^5(1 + 0.12)^{-5}}{0.12 - 0.07}\right] \\
&= \$222{,}283.
\end{aligned}$$

● **Step 3:** If Ansell replaces the current compressed-air delivery system with the new one, the annual power cost will be 23% less during the first year and will remain at that level over the next five years. The equivalent present lump-sum cost at 12% interest is

$$\begin{aligned}
P_{\text{Option 2}} &= \$54{,}440(1 - 0.23)\,(P/A, 12\%, 5) \\
&= \$41{,}918.80(3.6048) \\
&= \$151{,}109.
\end{aligned}$$

• Option 1:
  $g = 7\%$
  $i = 12\%$
  $N = 5$ years
  $A_1 = \$54{,}440$

$$P_{\text{Option 1}} = \$54{,}440\,\frac{1 - (1 + 0.07)^5\,(1 + 0.12)^{-5}}{0.12 - 0.07}$$
$$= \$222{,}283$$

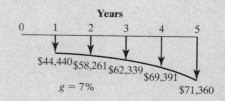

• Option 2:

$$\begin{aligned}
P_{\text{Option 2}} &= \$54{,}410(1 - 0.23)(P/A, 12\%, 5) \\
&= \$41{,}919(P/A, 12\%, 5) \\
&= \$151{,}109
\end{aligned}$$

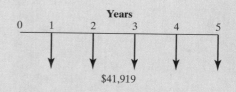

FIGURE 5.10 Comparing two mutually exclusive options

> **Step 4:** The net cost of not replacing the old system now is $71,174 (= $222,283 − $151,109). Since the new system costs only $28,570, the replacement should be made now.

5.4.4 Analysis Period Differs from Project Lives

In Example 5.5, we assumed the simplest scenario possible when analyzing mutually exclusive projects. The projects had useful lives equal to each other and to the required service period. In practice, this is seldom the case. Often, project lives do not match the required analysis period or do not match each other. For example, two machines may perform exactly the same function, but one lasts longer than the other, and both of them last longer than the analysis period for which they are being considered. In the upcoming sections and examples, we will develop some techniques for dealing with these complications.

Case I: Project's Life Is Longer Than Analysis Period   Consider the case of a firm that undertakes a five-year production project when all of the alternative equipment choices have useful lives of seven years. In such a case, we analyze each project only for as long as the required service period (in this case, five years). We are then left with some unused portion of the equipment (in this case, two years' worth), which we include as salvage value in our analysis. **Salvage value** is the amount of money for which the equipment could be sold after its service to the project has been rendered or the dollar measure of its remaining usefulness.

A common instance of project lives that are longer than the analysis period occurs in the construction industry, where a building project may have a relatively short completion time, but the equipment purchased (power tools, tractors, etc.) has a much longer useful life.

Example 5.6 Present-Worth Comparison: Project Lives Longer Than the Analysis Period

Waste Management Company (WMC) has won a contract that requires the firm to remove radioactive material from government-owned property and transport it to a designated dumping site. This task requires a specially made ripper–bulldozer to dig and load the material onto a transportation vehicle. Approximately 400,000 tons of waste must be moved in a period of two years. There are two possible models of ripper–bulldozer that WMC could purchase for this job:

- Model *A* costs $150,000 and has a life of 6,000 hours before it will require any major overhaul. Two units of model *A* will be required in order to remove the material within two years, and the operating cost for each unit will run $40,000 per year for 2,000 hours of operation. At this operational rate, each unit will be operable for three years, and at the end of that time, it is estimated that the salvage value will be $25,000 for each machine.

- The more efficient model B costs $240,000 each, has a life of 12,000 hours before requiring any major overhaul, and costs $22,500 to operate for 2,000 hours per year in order to complete the job within two years. The estimated salvage value of model B at the end of six years is $30,000. Once again, two units of model B will be required.

Since the lifetime of either model exceeds the required service period of two years, WMC has to assume some things about the used equipment at the end of that time. Therefore, the engineers at WMC estimate that, after two years, the model A units could be sold for $45,000 each and the model B units for $125,000 each. After considering all tax effects, WMC summarized the resulting cash flows (in thousands of dollars) for each project as follows:

Period	Model A ($ Thousands)	Model B ($ Thousands)
0	−$300	−$480
1	−$80	−$45
2	−$80 + $90	−$45 + $250
3	−$80 + $50	−$45
4		−$45
5		−$45
6		−$45 + $60

Here, the figures in the boxes represent the estimated salvage values at the end of the analysis period (end of year 2). Assuming that the firm's MARR is 15%, which option is acceptable?

SOLUTION

Given: Cash flows for the two alternatives as shown in Figure 5.11; $i = 15\%$ per year.

Find: PW for each alternative; the preferred alternative.

First, note that these projects are service projects, as we can assume the same revenues for both configurations. Since the firm explicitly estimated the market values of the assets at the end of the analysis period (two years), we can compare the two models directly. Since the benefits (removal of the wastes) are equal, we can concentrate on the costs:

$$PW(15\%)_A = -\$300 - \$80(P/A, 15\%, 2) + \$90(P/F, 15\%, 2)$$
$$= -\$362;$$
$$PW(15\%)_B = -\$480 - \$45(P/A, 15\%, 2) + \$250(P/F, 15\%, 2)$$
$$= -\$364.$$

Model A has the least negative PW costs and thus would be preferred.

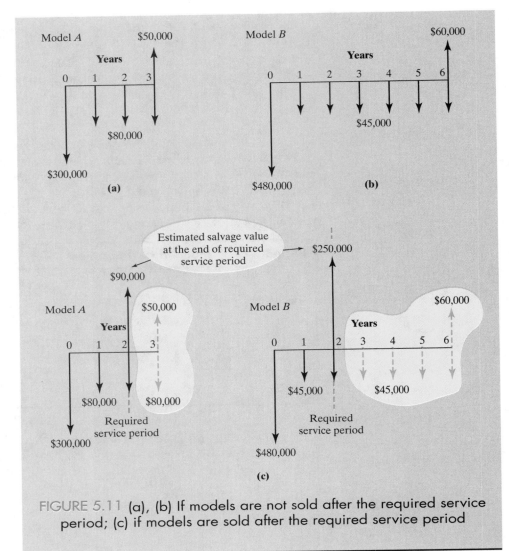

FIGURE 5.11 (a), (b) If models are not sold after the required service period; (c) if models are sold after the required service period

Case II: Project's Life Is Shorter Than Analysis Period   When project lives are shorter than the required service period, we must consider how, at the end of the project lives, we will satisfy the rest of the required service period. Replacement projects—additional projects to be implemented when the initial project has reached the limits of its useful life—are needed in such a case. Sufficient replacement projects must be analyzed to match or exceed the required service period.

To simplify our analysis, we could assume that the replacement project will be exactly the same as the initial project, with the same corresponding costs and benefits. In the case of an indefinitely ongoing service project, we typically select a finite analysis period by using the **lowest common multiple** of projects' lives. For example, if alternative A has a three-year useful life and alternative B has a four-year useful life, we may select 12 years as the analysis period. Because this assumption is rather unrealistic in most real-world problems, we will not advocate the method in this book. However, if such an analysis is warranted, we will demonstrate how the annual equivalent approach would simplify the mathematical aspect of the analysis in Example 6.7.

The assumption of an identical future replacement project is not necessary, however. For example, depending on our forecasting skills, we may decide that a different kind of technology—in the form of equipment, materials, or processes—will be a preferable and potential replacement. *Whether we select exactly the same alternative or a new technology as the replacement project, we are ultimately likely to have some unused portion of the equipment to consider as salvage value,* just as in the case when project lives are longer than the analysis period. On the other hand, we may decide to lease the necessary equipment or subcontract the remaining work for the duration of the analysis period. In this case, we can probably exactly match our analysis period and not worry about salvage values.

In any event, we must make some initial guess concerning the method of completing the analysis period at its outset. Later, when the initial project life is closer to its expiration, we may revise our analysis with a different replacement project. This approach is quite reasonable, since economic analysis is an ongoing activity in the life of a company and an investment project, and we should always use the most reliable, up-to-date data we can reasonably acquire.

Example 5.7 Present-Worth Comparison: Project Lives Shorter Than the Analysis Period

The Smith Novelty Company, a mail-order firm, wants to install an automatic mailing system to handle product announcements and invoices. The firm has a choice between two different types of machines. The two machines are designed differently, but have identical capacities and do exactly the same job. The $12,500 semiautomatic model A will last three years, while the fully automatic model B will cost $15,000 and last four years. The expected cash flows for the two machines, including maintenance costs, salvage values, and tax effects, are as follows:

n	Model A	Model B
0	−$12,500	−$15,000
1	−$5,000	−$4,000
2	−$5,500	−$4,500
3	−$6,000 + $2,000	−$5,000
4		−$5,500 + $1,500
5		

Once again, as business grows to a certain level, neither of the models may be able to handle the expanded volume at the end of year 5. If that happens, a fully computerized mail-order system will need to be installed to handle the increased business volume. With this scenario, which model should the firm select at MARR = 15%?

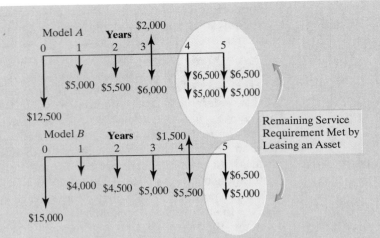

FIGURE 5.12 Comparison for service projects with unequal lives when the required service period is longer than the individual project life

SOLUTION

Given: Cash flows for the two alternatives as shown in Figure 5.12, analysis period of five years, and $i = 15\%$.

Find: PW for each alternative; the preferred alternative.

Since both models have a shorter life than the required service period (five years), we need to make an explicit assumption of how the service requirement is to be met. Suppose that the company considers leasing comparable equipment (Model A) that has an annual lease payment of $5,000 (after taxes), with an annual operating cost of $6,500 for the remaining required service period. The cash flow for this case is depicted in the top diagram in Figure 5.12. The anticipated cash flows for both models under this scenario are as follows:

n	Model A	Model B
0	−$12,500	−$15,000
1	−$5,000	−$4,000
2	−$5,500	−$4,500
3	−$6,000 + $2,000	−$5,000
4	−$6,500 − $5,000	−$5,500 + $1,500
5	−$6,500 − $5,000	−$6,500 − $5,000

Here, the boxed figures represent the annual lease payments. (It costs $5,000 to lease the equipment and $6,500 to operate it annually. Other maintenance costs will be paid by the leasing company.) Note that both alternatives now have the same required service period of five years. Therefore, we can use PW analysis:

$$PW(15\%)_A = -\$12,500 - \$5,000(P/F, 15\%, 1)$$
$$- \$5,500(P/F, 15\%, 2) - \$4,000(P/F, 15\%, 3)$$
$$- \$11,500(P/A, 15\%, 2)(P/F, 15\%, 3)$$
$$= -\$35,929;$$

$$PW(15\%)_B = -\$15,000 - \$4,000(P/F, 15\%, 1)$$
$$- \$4,500(P/F, 15\%, 2) - \$5,000(P/F, 15\%, 3)$$
$$- \$4,000(P/F, 15\%, 4) - \$11,500(P/F, 15\%, 5)$$
$$= -\$33,173.$$

Since these projects are service projects, model B is the better choice.

Summary

In this chapter, we presented the concept of present-worth analysis based on cash flow equivalence along with the payback period. We observed the following important results:

- Present worth is an equivalence method of analysis in which a project's cash flows are discounted to a single present value. It is perhaps the most efficient analysis method we can use for determining project acceptability on an economic basis. Other analysis methods, which we will study in Chapters 6 and 7, are built on a sound understanding of present worth.

- The MARR, or minimum attractive rate of return, is the interest rate at which a firm can always earn or borrow money. It is generally dictated by management and is the rate at which PW analysis should be conducted.

- Revenue projects are projects for which the income generated depends on the choice of project. Service projects are projects for which income remains the same, regardless of which project is selected.

- The term **mutually exclusive** as applied to a set of alternatives that meet the same need means that, when one of the alternatives is selected, the others will be rejected.

- When not specified by management or company policy, the analysis period to use in a comparison of mutually exclusive projects may be chosen by an individual analyst. Several efficiencies can be applied when selecting an analysis period. In general, the analysis period should be chosen to cover the required service period.

Problems

Note: Unless otherwise stated, all cash flows represent cash flows in *actual dollars* with tax effects considered. The interest rate (MARR) is also given on an after-tax basis, considering the effects of inflation in the economy. This interest rate is equivalent to the market interest rate. Also, all interest rates are assumed to be compounded annually.

Identifying Cash Inflows and Outflows

5.1 Camptown Togs, Inc., a children's clothing manufacturer, has always found payroll processing to be costly, because it must be done by a clerk so that the number of piece-goods coupons collected for each employee can be collected and the types of tasks performed by each employee can be calculated. Recently, however, an industrial engineer has designed a system that partially automates the process by means of a scanner that reads the piece-goods coupons. Management is enthusiastic about this system because it uses some personal computer systems that were purchased recently. It is expected that this new automated system will save $40,000 per year in labor. The new system will cost about $30,000 to build and test prior to operation. It is expected that operating costs, including income taxes, will be about $15,000 per year. The system will have a five-year useful life. The expected net salvage value of the system is estimated to be $3,000.

(a) Identify the cash inflows over the life of the project.
(b) Identify the cash outflows over the life of the project.
(c) Determine the net cash flows over the life of the project.

Payback Period

5.2 Refer to Problem 5.1 in answering the following questions:

(a) How long does it take to recover the investment?
(b) If the firm's interest rate is 15% after taxes, what would be the discounted-payback period for this project?

5.3 You are given the following financial data about a new system to be implemented at a company:

- investment cost at $n = 0$: $10,000;
- investment cost at $n = 1$: $15,000;
- useful life: 10 years;
- salvage value (at the end of 11 years): $5,000;
- annual revenues: $12,000 per year;
- annual expenses: $4,000 per year;
- MARR: 10%.

Note: The first revenues and expenses will occur at the end of year two.

(a) Determine the conventional-payback period.
(b) Determine the discounted-payback period.

5.4 Consider the following cash flows, for four different projects?

| n | Project's Cash Flow | | | |
	A	B	C	D
0	−$1,500	−$4,000	−$4,500	−$3,000
1	$300	$2,000	$2,000	$5,000
2	$300	$1,500	$2,000	$3,000
3	$300	$1,500	$2,000	−$2,000
4	$300	$500	$5,000	$1,000
5	$300	$500	$5,000	$1,000
6	$300	$1,500		$2,000
7	$300			$3,000
8	$300			

(a) Calculate the conventional payback period for each project.

(b) Determine whether it is meaningful to calculate a payback period for project D.

(c) Assuming $i = 10\%$, calculate the discounted-payback period for each project.

PW Criterion

5.5 Select the net present worth of the following cash flow series at an interest rate of 9% from the choices provided after the table:

End of Period	Cash Flow	End of Period	Cash Flow
0	-$100	5	-$300
1	-$150	6	-$250
2	-$200	7	-$200
3	-$250	8	-$150
4	-$300	9	-$100

(a) $P = -\$1,387$.

(b) $P = -\$1,246$.

(c) $P = -\$1,127$.

(d) $P = -\$1,027$.

5.6 Consider the following sets of investment projects, all of which have a three-year investment life:

	Project's Cash Flow			
n	A	B	C	D
0	-$1,000	-$1,000	-$1,000	-$1,000
1	$0	$600	-$1,200	$900
2	$0	$800	$800	$900
3	$3,000	$1,500	$1,500	$1,800

(a) Compute the net present worth of each project at $i = 10\%$.

(b) Plot the present worth as function of interest rate (from 0% to 30%) for project B.

5.7 You need to know if the building of a new warehouse is justified under the following conditions:

The proposal is for a warehouse costing $100,000. The warehouse has an expected useful life of 35 years and a net salvage value (net proceeds from sale after tax adjustments) of $25,000. Annual receipts of $17,000 are expected, annual maintenance and administrative costs will be $4,000, and annual income taxes are $2,000.

Given these data, which of the following statements is (are) correct?

(a) The proposal is justified for a MARR of 9%.

(b) The proposal has a net present worth of $62,730.50 when 6% is used as the interest rate.

(c) The proposal is acceptable as long as the MARR $\leq 10.77\%$.

(d) All of the above are correct.

5.8 Your firm is considering the purchase of an old office building with an estimated remaining service life of 25 years. The tenants have recently signed long-term leases, which leads you to believe that the current rental income of $150,000 per year will remain constant for the first five years. Then the rental income will increase by 10% for every five-year interval over the remaining asset life. For example, the annual rental income would be $165,000 for years six through 10, $181,500 for years 11 through 15, $199,650 for years 16 through 20, and $219,615 for years 21 through 25. You estimate that operating expenses, including income taxes, will be $45,000 for the first year and that they will increase by $3,000 each year thereafter. You estimate that razing the building and selling the lot on which it stands will realize a net amount of $50,000 at the end of the 25-year period. If you had the opportunity to invest your money elsewhere and thereby earn interest at the rate of 12% per annum, what would be the maximum amount you would be willing to pay for the building and lot at the present time?

5.9 Consider the following investment project:

n	A_n	i
0	−$2,000	10%
1	$2,400	12%
2	$3,400	14%
3	$2,500	15%
4	$2,500	13%
5	$3,000	10%

Suppose, as shown in the foregoing tables, that the company's reinvestment opportunities change over the life of the project (i.e., the firm's MARR changes over the life of the project). For example, the company can invest funds available now at 10% for the first year, 12% for the second year, and so forth. Calculate the net present worth of this investment, and determine the acceptability of the investment.

5.10 Cable television companies and their equipment suppliers are on the verge of installing new technology that will pack many more channels into cable networks, thereby creating a potential programming revolution with implications for broadcasters, telephone companies, and the consumer electronics industry. This technique, called digital compression, uses computer techniques to squeeze three to 10 programs into a single channel. A cable system fully using digital-compression technology would be able to offer well over 100 channels, compared with about 35 for the average cable television system now used. If the new technology is combined with the increased use of optical fibers, it might be possible to offer as many as 500 channels.

A cable company is considering installing this new technology in order to increase subscription sales and save on satellite time. The company estimates that the installation will take place over two years. The system is expected to have an eight-year service life and the following savings and expenditures:

Digital Compression	
Investment	
Now	$500,000
First year	$3,200,000
Second year	$4,000,000
Annual savings in	
satellite time	$2,000,000
Incremental annual revenues	
due to new subscriptions	$4,000,000
Incremental annual expenses	$1,500,000
Incremental annual	
income taxes	$1,300,000
Net salvage values	$1,200,000
Economic service life	8 years

Note that the project has a two-year investment period followed by an eight-year service life (for a total life of 10 years). This implies that the first annual savings will occur at the end of year three, and the last savings will occur at the end of year 10. If the firm's MARR is 15%, justify the economic worth of the project, based on the PW method.

5.11 A large food-processing corporation is considering using laser technology to speed up and eliminate waste in the potato-peeling process. To implement the system, the company anticipates needing $3 million to purchase the industrial-strength lasers. The system will save $1,200,000 per year in labor and materials. However, it will incur an additional operating and maintenance cost of $250,000 per year. Annual income taxes will also increase by $150,000. The system is expected to have a 10-year service life and a salvage value of about $200,000. If the company's MARR is 18%, justify the economics of the project, based on the PW method.

Future Worth and Project Balance

5.12 Consider the following sets of investment projects, all of which have a three-year investment life:

Period	Project's Cash Flow			
(n)	A	B	C	D
0	−$2,500	−$1,000	$2,500	−$3,000
1	$5,400	−$3,000	−$7,000	$1,500
2	$14,400	$1,000	$2,000	$5,500
3	$7,200	$3,000	$4,000	$6,500

Compute the net future worth of each project at $i = 13\%$.

5.13 Consider the following information for a typical investment project with a service life of five years:

n	Cash Flow	Project Balance
0	−$1,000	−$1,000
1	$200	−$900
2	$490	−$500
3	$550	$0
4	−$100	−$100
5	$200	$90

Which of the following interest rates is used in the project-balance calculation?

(a) 10%.
(b) 15%.
(c) 20%.
(d) 25%.

5.14 Consider the following project balances for a typical investment project with a service life of four years:

n	A_n	Project Balance
0	−$1,000	−$1,000
1	( )	−$1,100
2	( )	−$800
3	$460	−$500
4	( )	$0

(a) Construct the original cash flows of the project.

(b) Determine the interest rate used in computing the project balance.

(c) Would this project be acceptable at a MARR of 12%?

5.15 Consider the accompanying project-balance diagram for a typical investment project with a service life of five years. The numbers in the figure indicate the beginning project balances.

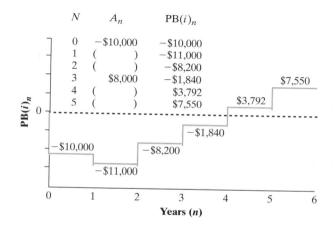

N	A_n	$PB(i)_n$
0	−$10,000	−$10,000
1	( )	−$11,000
2	( )	−$8,200
3	$8,000	−$1,840
4	( )	$3,792
5	( )	$7,550

(a) From the project-balance diagram, construct the project's original cash flows.

(b) What is the project's conventional-payback period (without interest)?

5.16 Consider the following project-balance profiles for proposed investment projects:

	Project Balances		
N	Project A	Project B	Project C
0	−$600	−$500	−$200
1	$200	$300	$0
2	$300	$650	$150
PW	?	$416	?
Rate used	15%	?	?

Now consider the following statements:

Statement 1: For project A, the cash flow at the end of year two is $100.

Statement 2: The future value of project C is $0.

Statement 3: The interest rate used in the project B balance calculations is 25%.

Which of the foregoing statements is (are) correct?

(a) Just statement 1.

(b) Just statement 2.

(c) Just statement 3.

(d) All of them.

5.17 Consider the following cash flows and present-worth profile:

Net Cash Flows

Year	Project 1	Project 2
0	−$100	−$100
1	$40	$30
2	$80	$Y
3	$X	$80

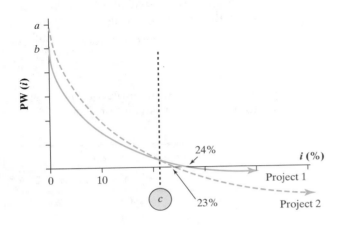

(a) Determine the values for X and Y.

(b) Calculate the terminal project balance of Project 1 at MARR = 24%.

(c) Find the values for a, b, and c in the PW plot.

5.18 Consider the following project balances for a typical investment project with a service life of five years:

n	A_n	Project Balance
0	−$1,000	−$1,000
1	( )	−$900
2	$490	−$500
3	( )	$0
4	( )	−$100
5	$200	( )

(a) Fill in the blanks by constructing the original cash flows of the project and determining the terminal balance.

(b) Determine the interest rate used in the project-balance calculation, and compute the present worth of this project at the computed interest rate.

5.19 Consider the following sets of investment projects:

n	**Project's Cash Flow**				
	A	**B**	**C**	**D**	**E**
0	−$1,500	−$5,000	−$3,500	−$3,000	−$5,500
1	$500	$2,000	$0	$500	$1,000
2	$900	−$3,000	$0	$2,000	$3,000
3	$1,000	$5,000	$3,000	$3,000	$2,000
4	$2,000	$5,000	$7,000	$4,000	
5	−$500	$3,500	$13,000	$1,250	

(a) Compute the future worth at the end of life for each project at i = 15%.

(b) Determine the acceptability of each project.

5.20 Perform the following tasks for the circumstances presented in Problem 5.19:

(a) Plot the future worth for each project as a function of interest rate (0%–50%).

(b) Compute the project balance for each project at i = 15%.

(c) Compare the terminal project balances calculated in (b) with the results obtained in

Problem 5.19(a). Without using the interest-factor tables, compute the future worth, based on the project-balance concept.

5.21 Consider the following set of independent investment projects:

Project Cash Flows

n	A	B	C
0	−$100	−$100	$100
1	$50	$40	−$40
2	$50	$40	−$40
3	$50	$40	−$40
4	−$100	$10	
5	$400	$10	
6	$400		

(a) For a MARR of 10%, compute the net present worth for each project, and determine the acceptability of each project.

(b) For a MARR of 10%, compute the net future worth of each project at the end of each project period, and determine the acceptability of each project.

(c) Compute the future worth of each project at the end of six years with variable MARRs as follows: 10% for $n = 0$ to $n = 3$ and 15% for $n = 4$ to $n = 6$.

5.22 Consider the following project-balance profiles for proposed investment projects:

Project Balances

n	A	B	C
0	−$1,000	−$1,000	−$1,000
1	−$1,000	−$650	−$1,200
2	−$900	−$348	−$1,440
3	−$690	−$100	−$1,328
4	−$359	$85	−$1,194
5	$105	$198	−$1,000
Interest rate used	10%	?	20%
PW	?	$79.57	?

Project-balance figures are rounded to the nearest dollar.

(a) Compute the net present worth of Projects A and C, respectively.

(b) Determine the cash flows for Project A.

(c) Identify the net future worth of Project C.

(d) What interest rate was used in the project-balance calculations for Project B?

5.23 Consider the following project-balance profiles for proposed investment projects, where the project-balance figures are rounded to the nearest dollar:

Project Balances

n	A	B	C
0	−$1,000	−$1,000	−$1,000
1	−$800	−$680	−$530
2	−$600	−$302	$X
3	−$400	−$57	−$211
4	−$200	$233	−$89
5	$0	$575	$0
Interest rate used	0%	18%	12%

(a) Compute the net present worth of each investment.

(b) Determine the project balance at the end of period two for Project C if $A_2 = 500.

(c) Determine the cash flows for each project.

(d) Identify the net future worth of each project.

Capitalized Equivalent Worth

5.24 Maintenance money for a new building at a college is being solicited from potential alumni donors. Mr. Kendall would like to make a donation to cover all future expected maintenance costs for the building. These maintenance costs are expected to be $40,000 each year for the first five years, $50,000 for each of years six through 10, and $60,000 each year after that. (The building has an indefinite service life.)

(a) If the money is placed in an account that will pay 13% interest compounded annually, how large should the gift be?

(b) What is the equivalent annual maintenance cost over the infinite service life?

5.25 Consider an investment project for which the cash flow pattern repeats itself every four years indefinitely as shown in the accompanying figure. At an interest rate of 12% compounded annually, compute the capitalized-equivalent amount for this project.

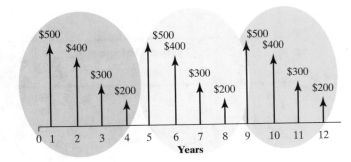

5.26 A group of concerned citizens has established a trust fund that pays 6% interest compounded monthly to preserve a historical building by providing annual maintenance funds of $12,000 forever. Compute the capitalized-equivalent amount for these building maintenance expenses.

5.27 A newly constructed bridge costs $5,000,000. The same bridge is estimated to need renovation every 15 years at a cost of $1,000,000. Annual repairs and maintenance are estimated to be $100,000 per year.

(a) If the interest rate is 5%, determine the capitalized equivalent cost of the bridge.

(b) Suppose that the bridge must be renovated every 20 years, not every 15 years. What is the capitalized cost of the bridge if the interest rate is the same as in (a)?

(c) Repeat (a) and (b) with an interest rate of 10%. What have you to say about the effect of interest on the results?

5.28 To decrease the costs of operating a lock in a large river, a new system of operation is proposed. The system will cost $650,000 to design and build. It is estimated that it will have to be reworked every 10 years at a cost of $100,000. In addition, an expenditure of $50,000 will have to be made at the end of the fifth year for a new type of gear that will not be available until then. Annual operating costs are expected to be $30,000 for the first 15 years and $35,000 a year thereafter. Compute the capitalized cost of perpetual service at $i = 8\%$.

Comparing Mutually Exclusive Alternatives

5.29 Consider the following two mutually exclusive projects:

End of Year	Net Cash Flow	
	Project A	Project B
0	−$1,000	−$2,000
1	$475	$915
2	$475	$915
3	$475	$915

At an interest rate of 12%, which project would you recommend choosing?

5.30 Consider the following cash flow data for two competing investment projects.

	Cash Flow Data (Thousands of $)	
n	Project A	Project B
0	−$800	−$2,635
1	−$1,500	−$565
2	−$435	$820
3	$775	$820
4	$775	$1,080
5	$1,275	$1,880
6	$1,275	$1,500

(*continued*)

n	Project A	Project B
7	$975	$980
8	$675	$580
9	$375	$380
10	$660	$840

At $i = 12\%$, which of the two projects would be a better choice?

5.31 Consider the following two mutually exclusive investment projects.

	Project's Cash Flow	
n	A	B
0	−$4,500	−$2,900
1	$2,610	$1,210
2	$2,930	$1,720
3	$2,300	$1,500

Assume that the MARR = 12%.

(a) Which alternative would be selected by using the PW criterion?

(b) Which alternative would be selected by using the net-future-worth criterion?

5.32 Consider the following two mutually exclusive investment projects:

	Project's Cash Flow	
n	A	B
0	−$3,000	−$8,000
1	$400	$11,500
2	$7,000	$400

Assume that the MARR = 15%.

(a) Using the PW criterion, which project would be selected?

(b) Sketch the PW(i) function for each alternative on the same chart for $i = 0\%$ and 50%. For what range of i would you prefer Project B?

5.33 Consider the cash flows for the following investment projects:

	Project's Cash Flow				
n	A	B	C	D	E
0	−$1,500	−$1,500	−$3,000	1,500	−$1,800
1	$1,350	$1,000	$1,000	−$450	$600
2	$800	$800	$X	−$450	$600
3	$200	$800	$1,500	−$450	$600
4	$100	$150	$X	−$450	$600

Assume that the MARR = 15%.

(a) Suppose that projects A and B are mutually exclusive. Which project would be selected, based on the PW criterion?

(b) Suppose that projects D and E are mutually exclusive. Which project would you select based on the FW criterion?

(c) Find the minimum value of X that makes project C acceptable.

(d) Would you accept project D at $i = 18\%$?

5.34 Consider the following two investment alternatives:

	Project's Cash Flow	
n	A	B
0	−$15,000	−$25,000
1	$9,500	$0
2	$12,500	$X
3	$7,500	$X
PW(15%)	?	$9,300

The firm's MARR is known to be 15%.

(a) Compute the PW (15%) for Project A.

(b) Compute the unknown cash flow X in years two and three for Project B.

(c) Compute the project balance (at 15%) for Project A at the end of year three.

(d) If these two projects are mutually exclusive alternatives, which project would you select?

5.35 Consider the following after-tax cash flows:

Project's Cash Flow

n	A	B	C	D
0	-$2,500	-$7,000	-$5,000	-$5,000
1	$650	-$2,500	-$2,000	-$500
2	$650	-$2,000	-$2,000	-$500
3	$650	-$1,500	-$2,000	$4,000
4	$600	-$1,500	-$2,000	$3,000
5	$600	-$1,500	-$2,000	$3,000
6	$600	-$1,500	-$2,000	$2,000
7	$300		-$2,000	$3,000
8	$300			

(a) Compute the project balances for projects A and D as a function of project year at $i = 10\%$.

(b) Compute the future worth values for Projects A and D at $i = 10\%$, at end of service life.

(c) Suppose that projects B and C are mutually exclusive. Assume also that the required service period is eight years and that the company is considering leasing comparable equipment that has an annual lease expense of $3,000 for the remaining years of the required service period. Which project is the better choice?

5.36 Consider the following two mutually exclusive investment projects:

Project's Cash Flow

n	A	B
0	-$10,000	-$22,000
1	$7,500	$15,500
2	$7,000	$18,000
3	$5,000	

Which project would be selected if you use the infinite planning horizon with project repeatability likely (same costs and benefits), based on the PW criterion? Assume that $i = 12\%$.

5.37 Consider the following two mutually exclusive investment projects, which have unequal service lives:

Project's Cash Flow

n	A	B
0	-$900	-$1,800
1	-$400	-$300
2	-$400	-$300
3	-$400 + $200	-$300
4		-$300
5		-$300
6		-$300
7		-$300
8		-$300 + $500

(a) What assumption(s) do you need in order to compare a set of mutually exclusive investments with unequal service lives?

(b) With the assumption(s) defined in (a) and using $i = 10\%$, determine which project should be selected.

(c) If your analysis period (study period) is just three years, what should be the salvage value of project B at the end of year three in order to make the two alternatives economically indifferent?

5.38 Consider the following two mutually exclusive investment projects:

	A		B	
n	Cash Flow	Salvage Value	Cash Flow	Salvage Value
0	-$12,000		-$10,000	
1	-$2,000	$6,000	-$2,100	$6,000
2	-$2,000	$4,000	-$2,100	$3,000
3	-$2,000	$3,000	-$2,100	$1,000
4	-$2,000	$2,000		
5	-$2,000	$2,000		

Salvage values represent the net proceeds (after tax) from disposal of the assets if they are sold at the end of the year listed. Both projects will be available (and can be repeated) with the same costs and salvage values for an indefinite period.

(a) With an infinite planning horizon, which project is a better choice at MARR = 12%?

(b) With a 10-year planning horizon, which project is a better choice at MARR = 12%?

5.39 Two methods of carrying away surface runoff water from a new subdivision are being evaluated:

- Method A: dig a ditch. The initial cost would be $30,000, and $10,000 of redigging and shaping would be required at five-year intervals forever.
- Method B: lay concrete pipe. The initial cost would be $75,000, and replacement pipe would be required at 50-year intervals at a net cost of $90,000 indefinitely.

At $i = 12\%$, which method is the better one? (*Hint*: Use the capitalized-equivalent-worth approach.)

5.40 A local car dealer is advertising a standard 24-month lease of $1,150 per month for its new XT 3000 series sports car. The standard lease requires a down payment of $4,500 plus a $1,000 refundable initial deposit now. The first lease payment is due at the end of month one. Alternatively, the dealer offers a 24-month lease plan that has a single up-front payment of $30,500 plus a refundable initial deposit of $1,000. Under both options, the initial deposit will be refunded at the end of month 24. Assume an interest rate of 6% compounded monthly. With the present-worth criterion, which option is preferred?

5.41 Two alternative machines are being considered for a manufacturing process. Machine A has an initial cost of $75,200, and its estimated salvage value at the end of its six years of service life is

$21,000. The operating costs of this machine are estimated to be $6,800 per year. Extra income taxes are estimated at $2,400 per year. Machine B has an initial cost of $44,000, and its estimated salvage value at the end of its six years of service life is estimated to be negligible. Its annual operating costs will be $11,500. Compare these two alternatives by the present-worth method at $i = 13\%$.

5.42 An electric motor is rated at 10 horsepower (HP) and costs $800. Its full-load efficiency is specified to be 85%. A newly designed, high-efficiency motor of the same size has an efficiency of 90%, but costs $1,200. It is estimated that the motors will operate at a rated 10-HP output for 1,500 hours a year, and the cost of energy will be $0.07 per kilowatt-hour. Each motor is expected to have a 15-year life. At the end of 15 years, the first motor will have a salvage value of $50, and the second motor will have a salvage value of $100. Consider the MARR to be 8%. (Note: 1 HP = 0.7457 kW.)

(a) Determine which motor should be installed, based on the PW criterion.

(b) What if the motors operated 2,500 hours a year instead of 1,500 hours a year? Would the motor selected in (a) still be the choice?

5.43 Consider the following cash flows for two types of models:

	Project's Cash Flow	
n	Model A	Model B
0	−$6,000	−$15,000
1	$3,500	$10,000
2	$3,500	$10,000
3	$3,500	

Both models will have no salvage value upon their disposal (at the end of their respective service lives). The firm's MARR is known to be 15%.

(a) Notice that both models have different service lives. However, model *A* will be available in the future, with the same cash flows. Model *B* is available now only. If you select Model *B* now, you will have to replace it with Model *A* at the end of year two. If your firm uses the present worth as a decision criterion, which model should be selected, assuming that your firm will need either model for an indefinite period?

(b) Suppose that your firm will need either model for only two years. Determine the salvage value of Model *A* at the end of year two that makes both models indifferent (equally likely).

5.44 An electric utility company is taking bids on the purchase, installation, and operation of microwave towers:

	Cost per Tower	
	Bid A	**Bid B**
Equipment cost	$65,000	$58,000
Installation cost	$15,000	$20,000
Annual maintenance and inspection fee	$1,000	$1,250
Annual extra income taxes		$500
Life	40 years	35 years
Salvage value	$0	$0

Which is the most economical bid, if the interest rate is considered to be 11%? Both towers will have no salvage value after 20 years of use.

5.45 A bi-level mall is under construction. Installation of only nine escalators is planned at the start, although the ultimate design calls for 16. The question arises of whether to provide necessary facilities that would permit the installation of the additional escalators (e.g., stair supports, wiring conduits, and motor foundations) at the mere cost of their purchase and installation now or to defer investment in these facilities until the escalators need to be installed. The two options are detailed as follows:

- Option 1: Provide these facilities now for all seven future escalators at $200,000.
- Option 2: Defer the investment as needed. Installation of two more escalators is planned in two years, three more in five years, and the last two in eight years. The installation of these facilities at the time they are required is estimated to cost $100,000 in year two, $160,000 in year five, and $140,000 in year 8. Additional annual expenses are estimated at $3,000 for each escalator facility installed.

At an interest rate of 12%, compare the net present worth of each option over eight years.

Short Case Studies with Excel

5.46 An electrical utility is experiencing sharp power demand, which continues to grow at a high rate in a certain local area. Two alternatives to address this situation are under consideration. Each alternative is designed to provide enough capacity during the next 25 years. Both alternatives will consume the same amount of fuel, so fuel cost is not considered in the analysis. The alternatives are detailed as follows:

- Alternative *A*: increase the generating capacity now so that the ultimate demand can be met with additional expenditures later. An initial investment of $30 million would be required, and it is estimated that this plant facility would be in service for 25 years and have a salvage value of $0.85 million. The annual operating and maintenance costs (including income taxes) would be $0.4 million.
- Alternative *B*: spend $10 million now, and follow this expenditure with additions during the 10th year and the 15th year. These additions would cost $18 million and $12 million, respectively. The facility would be sold 25 years from now with a salvage value of $1.5 million. The annual operating and maintenance costs (including income taxes)

initially will be $250,000, increasing to $350,000 after the second addition (from the 11th year to the 15th year) and to $450,000 during the final 10 years. (Assume that these costs begin one year subsequent to the actual addition.)

If the firm uses 15% as a MARR, which alternative should be undertaken, based on the present-worth criterion?

5.47 A large refinery and petrochemical complex is planning to manufacture caustic soda, which will use feed water of 10,000 gallons per day. Two types of feed-water storage installation are being considered over 40 years of useful life:

● Option 1: build a 20,000-gallon tank on a tower. The cost of installing the tank and tower is estimated to be $164,000. The salvage value is estimated to be negligible.
● Option 2: place a 20,000-gallon tank of equal capacity on a hill that is 150 yards away from the refinery. The cost of installing the tank on the hill, including the extra length of service lines, is estimated to be $120,000, with negligible salvage value. Because of its hill location, an additional investment of $12,000 in pumping equipment is required. The pumping equipment is expected to have a service life of 20 years, with a salvage value of $1,000 at the end of that time. The annual operating and maintenance cost (including any income-tax effects) for the pumping operation is estimated at $1,000.

If the firm's MARR is known to be 12%, which option is better, on the basis of the present-worth criterion?

5.48 Apex Corporation requires a chemical finishing process for a product under contract for a period of six years. Three options are available. Neither Option 1 nor Option 2 can be repeated after its process life. However, Option 3 will always be available from H&H Chemical Corporation at the same cost during the

contract period. The details of each option are as follows:

● Option 1: process device A, which costs $100,000, has annual operating and labor costs of $60,000 and a useful service life of four years, with an estimated salvage value of $10,000.
● Option 2: process device B, which costs $150,000, has annual operating and labor costs of $50,000 and a useful service life of six years, with an estimated salvage value of $30,000.
● Option 3: subcontract out the process at a cost of $100,000 per year.

According to the present-worth criterion, which option would you recommend at $i = 12\%$?

5.49 Tampa Electric Company, an investor-owned electric utility serving approximately 2,000 square miles in west central Florida, is faced with the job of providing electricity to a newly developed industrial-park complex. The distribution engineering department needs to develop guidelines for design of the distribution circuit. The main feeder, which is the backbone of each 13-kV distribution circuit, represents a substantial investment by the company.

Tampa Electric has four approved main-feeder construction configurations: (1) cross-arm, (2) vertical (horizontal line post), (3) vertical (stand-off pin), and (4) triangular. The width of the easement sought depends on the planned construction configuration. If cross-arm construction is planned, a 15-foot easement is sought. A 10-foot-wide easement is sought for vertical and triangular configurations. Once the required easements are obtained, the line-clearance department clears any foliage that would impede the construction of the line. The clearance cost is dictated by the typical tree densities along road rights of way. The average cost to trim one tree is estimated at $20, and the average tree density along the length of the service area is estimated to be 75 trees per mile. The costs of each construction type are as follows:

	Design Configurations			
Factors	**Cross-Arm**	**Triangular**	**Horizontal Line**	**Stand-off**
Easements	$487,000	$388,000	$388,000	$388,000
Line clearance	$613	$1,188	$1,188	$1,188
Line construction	$7,630	$7,625	$12,828	$8,812

Additional factors to consider in selecting the best main-feeder configuration are as follows:

- In certain sections of Tampa Electric's service territory, osprey often nest on transmission and distribution poles. These osprey nests reduce the structural and electrical integrity of the pole on which the nest is built. Cross-arm construction is most vulnerable to osprey nesting, since the cross-arm and braces provide a secure area for nest construction. Vertical and triangular construction do not provide such spaces. In areas where osprey are known to nest, vertical and triangular configuration have added advantages.

- The insulation strength of a construction configuration may favorably or adversely affect the reliability of the line for which the configuration is used. A common measure of line insulation strength is the critical flashover (CFO) voltage. The greater the value CFO, the less susceptible the line is to suffer from nuisance flashovers from lightning and other electrical phenomena. The average cost of each flashover repair would be $3,000.

- The existing inventory of cross-arms is used primarily for main-feeder construction and maintenance. Use of another configuration for main-feeder construction would result in a substantial reduction of cross-arm inventory.

- The line crews complain that line spacing on vertical and triangular construction is too restrictive for safe live-line work. Each accident would cost $65,000 in lost work and other medical expenses.

These factors and their associated costs are summarized as follows:

	Design Configurations			
Factors	**Cross-Arm**	**Triangular**	**Horizontal Line**	**Stand-Off**
Nesting	Severe	None	None	None
Insulation strength				
CFO (kV)	387	474	476	462
Annual flashover occurrence (n)	2	1	1	1
Annual inventory savings		$4,521	$4,521	$4,521
Safety	OK	Problem	Problem	Problem

All configurations would last about 20 years, with no salvage values. It appears that non-cross-arm designs are better, but engineers need to consider other design factors, such as safety, rather than just monetary factors when implementing the project. It is true that the line spacing on triangular construction is restrictive. However, with a better clearance design between phases for vertical construction, the safety issue would be minimized. In the utilities industry, the typical opposition to new construction types is caused by the confidence acquired from constructing lines in the cross-arm configuration for many years. As more vertical and triangular lines are built, the opposition to these configurations should decrease. Which of the four designs described in the table would you recommend to the management of Tampa Electric? Assume Tampa Electric's MARR to be 12%.

Here Come the Pint-Size Power Plants[1]

Capstone Turbine Corporation is the world's leading provider of micro-turbine based MicroCHP (combined heat and power) systems for clean, continuous, distributed-generation electricity. The MicroCHP unit is a compact turbine generator that delivers electricity on-site, or close to the point where it is needed. Designed to operate on a variety of gaseous and liquid fuels, this form of distributed-generation technology first debuted in 1998. The microturbine is designed to operate on demand or continously for up to a year between recommended maintenance (filter cleaning/replacement). The generator is cooled by airflow into the gas turbine, thus eliminating the need for liquid cooling. It can make electricity from a variety of fuels—natural gas, kerosene, diesel oil, and even waste gases from landfills, sewage plants, and oilfields.

Example of MicroCHP Economics

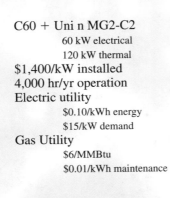

C60 + Uni n MG2-C2
- 60 kW electrical
- 120 kW thermal

$1,400/kW installed
4,000 hr/yr operation

Electric utility
- $0.10/kWh energy
- $15/kW demand

Gas Utility
- $6/MMBtu
- $0.01/kWh maintenance

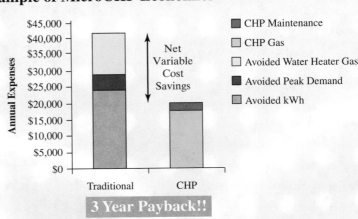

[1]Courtesy of Capstone Turbine Corporation (http://www.microturbine.com/index.cfm).

Annual Equivalence Analysis

Capstone's focus applications include combined heat and power, resource recovery of waste fuel from wellhead and biogas sites, power quality and reliability, and hybrid electric vehicles. And, unlike traditional backup power, this solution can support everyday energy needs and generate favorable payback. With the current design, which has a 60-KW rating, one of Capstone's generators would cost about $84,000. The expected annual expenses, including capital costs as well as operating costs, would run close to $19,000. These expenses yield an annual savings of close to $25,000 compared with the corresponding expenses for a conventional generator of the same size. The investment would pay for itself within three to four years.

One of the major questions among the Capstone executives is, "How low does the microturbine's production cost need to be for it to be a sensible option in some utility operations?" To answer this question, Capstone must first determine the cost per kilowatt of its generators.

How does Capstone come up with the capital cost of $1,400 per kilowatt? Suppose you plan to purchase the 60-KW microturbine and expect to operate it continuously for 10 years. How would you calculate the operating cost per kilowatt-hour? Similarly, suppose you are considering buying a new car. If you expect to drive 12,000 miles per year, can you figure out the per-mile price of the car? You would have good reason to want to know this cost if you were being reimbursed by your employer on a per-mile basis for the business use of your car. Or consider a real-estate developer who is planning to build a shopping center of 500,000 square feet. What would be the minimum annual rental fee per square foot required in order to recover the initial investment?

Annual cash flow analysis is the method by which these and other unit costs are calculated. Annual-equivalence analysis, along with present-worth analysis, is the second major equivalence technique for translating alternatives into a common basis of comparison. In this chapter, we develop the annual-equivalence criterion and demonstrate a number of situations in which annual-equivalence analysis is preferable to other methods of comparison.

6-1   Annual Equivalent Worth Criterion

The **annual equivalent worth (AE) criterion** provides a basis for measuring investment worth by determining equal payments on an annual basis. Knowing that any lump-sum cash amount can be converted into a series of equal annual payments, we may first find the net present worth of the original series and then multiply this amount by the capital-recovery factor:

$$AE(i) = PW(i)(A/P, i, N). \qquad (6.1)$$

We use this formula to evaluate the investment worth of projects as follows:

- **Evaluating a Single Project:** The accept–reject decision rule for a single *revenue* project is as follows:

    If $AE(i) > 0$, accept the investment.
    If $AE(i) = 0$, remain indifferent to the investment.
    If $AE(i) < 0$, reject the investment.

    Notice that the factor $(A/P, i, N)$ in Eq. (6.1) is positive for $-1 < i < \infty$, which indicates that the $AE(i)$ value will be positive if and only if $PW(i)$ is positive. In other words, accepting a project that has a positive $AE(i)$ value is equivalent to accepting a project that has a positive $PW(i)$ value. Therefore, the AE criterion provides a basis for evaluating a project that is consistent with the PW criterion.

- **Comparing Multiple Alternatives:** As with present-worth analysis, when you compare mutually exclusive *service* projects that have equivalent revenues, you may compare them on a *cost-only* basis. In this situation, the alternative with the least annual equivalent cost (or least negative annual equivalent worth) is selected.

Example 6.1 Finding Annual Equivalent Worth by Conversion from Present Worth (PW)

Singapore Airlines is planning to equip some of its Boeing 747 aircrafts with in-flight e-mail and Internet service on transoceanic flights. Passengers on these flights will be able to send and receive e-mail no matter where they are in the skies. As e-mail has become a communications staple, airlines have been under increasing pressure to offer access to it. But the rollout has been slow, because airlines are hesitant to invest in systems that could quickly become outdated. Hoping to provide added value to its passengers, Singapore Airlines has decided to offer the service through telephone modems for 10 Boeing 747s in 2004. As Boeing unveils a broadband e-mail and Internet system during 2004, Singapore Airlines will upgrade the systems on its planes similarly. If the project turns out to be a financial success, Singapore Airlines will introduce the service to the remaining 56 Boeing 747s in its fleet. The service will be free during the first year. After the promotional period, a nominal charge of about $10 will be instituted for each e-mail message sent or received. Singapore Airlines has estimated the projected cash flows (in millions of dollars) for the systems in the first 10 aircraft as follows:

n Year	A$_n$ (Unit: Million Dollars)
2004	−$15.0
2005	−$3.5
2005	$5.0
2006	$9.0
2007	$12.0
2008	$10.0
2009	$8.0

Determine whether this project can be justified at MARR = 15%, and calculate the annual benefit (or loss) that would be generated after installation of the systems.

Discussion: When a cash flow has no special pattern, it is easiest to find the AE in two steps: (1) find the PW of the flow and (2) find the AE of the PW. We use this method in the solution to this example. You might want to try another method with this type of cash flow in order to demonstrate how difficult using such a method can be.

SOLUTION

Given: The cash flow diagram in Figure 6.1; $i = 15\%$.
Find: The AE.

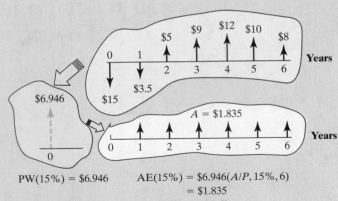

$$\text{PW}(15\%) = \$6.946 \qquad \text{AE}(15\%) = \$6.946(A/P, 15\%, 6)$$
$$= \$1.835$$

Note: All dollar values are in millions of dollars.

FIGURE 6.1 **Computing equivalent annual worth**

We first compute the PW at $i = 15\%$:

$$\text{PW}(15\%) = -\$15 - \$3.5(P/F, 15\%, 1) + \$5(P/F, 15\%, 2) + \cdots$$
$$+ \$10(P/F, 15\%, 5) + \$8(P/F, 15\%, 6)$$
$$= \$6.946 \text{ million.}$$

Since $\text{PW}(15\%) > 0$, the project would be acceptable under the PW analysis. Now, spreading the PW over the project life gives

$$\text{AE}(15\%) = \$6.946(A/P, 15\%, 6) = \$1.835 \text{ million.}$$

Since $\text{AE}(15\%) > 0$, the project is worth undertaking. The positive AE value indicates that the project is expected to bring in a net annual benefit of $1.835 million over the life of the project.

6.1.1 Benefits of AE Analysis

Example 6.1 should look familiar to you. It is exactly the situation we encountered in Chapter 2 when we converted an uneven cash flow series into a single present value and then into a series of equivalent cash flows. In the case of Example 6.1, you may wonder why we bother to convert PW to AE at all, since we already know from the PW analysis that the project is acceptable. In fact, the example was mainly an exercise to familiarize you with the AE calculation.

In the real world, a number of situations can occur in which AE analysis is preferred, or even demanded, over PW analysis. Consider, for example, that corporations issue annual reports and develop yearly budgets. For these purposes, a company may find it more useful to present the annual cost or benefit of an ongoing project rather than its overall cost or benefit. Some additional situations in which AE analysis is preferred include the following:

1. **When consistency of report formats is desired.** Financial managers more commonly work with annual rather than with overall costs in any number of internal

and external reports. Engineering managers may be required to submit project analyses on an annual basis for consistency and ease of use by other members of the corporation and stockholders.

2. **When there is a need to determine unit costs or profits.** In many situations, projects must be broken into unit costs (or profits) for ease of comparison with alternatives. Make-or-buy and reimbursement analyses are key examples of such situations and will be discussed in this chapter.

3. **When project lives are unequal.** As we saw in Chapter 5, comparison of projects with unequal service lives is complicated by the need to determine the common lifespan. For the special situation of an indefinite service period and replacement with identical projects, we can avoid this complication by use of AE analysis. This situation will also be discussed in more detail in this chapter.

6.1.2 Capital Costs versus Operating Costs

When only costs are involved, the AE method is sometimes called the **annual equivalent cost method**. In this case, revenues must cover two kinds of costs: **operating costs** and **capital costs**.

- *Operating costs* are incurred by the operation of physical plants or equipment needed to provide service; examples include the costs of items such as labor and raw materials.
- *Capital recovery costs* (or *ownership costs*) are incurred by purchasing assets to be used in production and service. Normally, capital costs are nonrecurring (i.e., one-time costs), whereas operating costs recur for as long as an asset is owned.

Because operating costs recur over the life of a project, they tend to be estimated on an annual basis, so for the purposes of annual equivalent cost analysis, no special calculation is required. However, because capital costs tend to be one-time costs, in conducting an annual equivalent cost analysis we must translate these one-time costs into its annual equivalent over the life of the project.

The annual equivalent of a capital cost is given a special name: **capital-recovery cost**, designated CR(i). (See Figure 6.2.) Two general monetary transactions are associated with the purchase and eventual retirement of a capital asset: the asset's initial

- Definition: The cost of owning a piece of equipment is associated with two amounts: (1) the equipment's initial cost (I) and (2) its salvage value (S).

- Capital costs: Taking these amounts into account, we calculate the capital costs as follows:

$$CR(i) = I(A/P, i, N) - S(A/F, i, N)$$
$$= (I - S)(A/P, i, N) + iS$$

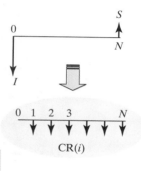

FIGURE 6.2 Calculation of capital recovery cost (with return)

cost (I) and its salvage value (S). Taking these amounts into account, we calculate the capital-recovery cost as follows:

$$CR(i) = I(A/P, i, N) - S(A/F, i, N). \qquad (6.2)$$

If we recall the algebraic relationships between factors shown in Table 2.4 and notice that the $(A/F, i, N)$ factor can be expressed as

$$(A/F, i, N) = (A/P, i, N) - i,$$

then we may rewrite the expression for $CR(i)$ as

$$\begin{aligned} CR(i) &= I(A/P, i, N) - S[(A/P, i, N) - i] \\ &= (I - S)(A/P, i, N) + iS. \end{aligned} \qquad (6.3)$$

We may interpret this result as follows. To obtain the machine, one borrows a total of I dollars, S dollars of which are returned at the end of the Nth year. The first term, $(I - S)(A/P, i, N)$, implies that the balance $(I - S)$ will be paid back in equal installments over the N-year period at a rate of i, and the second term implies that *simple interest* in the amount iS is paid on S until S is repaid. Many auto leases are based on this arrangement in that most require a guarantee of S dollars in salvage.

Table 6.1 shows the value that some popular vehicles are expected to hold after three years of ownership. For example, if you purchase a Mini Cooper at $19,800 and sell it at $12,078 after three years, your annual ownership cost (capital cost) would be calculated as follows, assuming an interest rate of 6% compounded annually:

$$\begin{aligned} CR(6\%) &= (\$19,800 - \$12,078)(A/P, 6\%, 3) + \$12,078(0.06) \\ &= \$3,613.55. \end{aligned}$$

The costs of owning the rest of vehicles are summarized in Table 6.1. Clearly, the Porsche 911 is the most expensive vehicle to own on an annual basis.

TABLE 6.1 Will Your Car Hold Its Value?				
Segment	Model	Asking Price	Price After Three Years	CR (6%)
Compact car	Mini Cooper	$19,800	$12,078	$3,614
Midsize car	Volkswagen Passat	$28,872	$15,013	$6,086
Sports car	Porsche 911	$87,500	$48,125	$17,618
Near-luxury car	BMW 3 Series	$39,257	$20,806	$8,151
Luxury car	Mercedes CLK	$51,275	$30,765	$9,519
Minivan	Honda Odyssey	$26,876	$15,051	$5,327
Subcompact SUV	Honda CR-V	$20,540	$10,681	$4,329
Compact SUV	Acura MDX	$37,500	$21,375	$7,315
Full-size SUV	Toyota Sequoia	$37,842	$18,921	$8,214
Compact truck	Toyota Tacoma	$21,200	$10,812	$4,535
Full-size truck	Toyota Tundra	$25,653	$13,083	$5,488

Source: "Will Your Car Hold Its Value? A New Study Does the Math," *The Wall Street Journal*, August 6, 2002, Karen Lundegaard, Page D1

From an industry viewpoint, $CR(i)$ is the annual cost to the firm of owning the asset. With this information, the amount of annual savings required in order to recover the capital and operating costs associated with a project can be determined. As an illustration, consider Example 6.2.

Example 6.2 Annual Equivalent Worth: Capital Recovery Cost

Consider a machine that costs $20,000 and has a five-year useful life. At the end of the five years, it can be sold for $4,000 after all tax adjustments have been factored in. If the firm could earn an after-tax revenue of $4,400 per year with this machine, should it be purchased at an interest rate of 10%? (All benefits and costs associated with the machine are accounted for in these figures.)

SOLUTION

Given: $I = \$20,000$, $S = \$4,000$, $A = \$4,400$, $N = 5$ years, and $i = 10\%$ per year.

Find: AE, and determine whether the firm should or should not purchase the machine.

We will compute the capital costs in two different ways:

Method 1: First compute the PW of the cash flows:

$$PW\,(10\%) = -\$20,000$$
$$+ \$4,400(P/A, 10\%, 5) + \$4,000\,(P/F, 10\%, 5)$$
$$= -\$20,000 + \$4,400(3.7908) + \$4,000(0.6209)$$
$$= -\$836.88.$$

Then compute the AE from the calculated PW:

$$AE(10\%) = -\$836.88(A/P, 10\%, 5) = -\$220.76.$$

This negative AE value indicates that the machine does not generate sufficient revenue to recover the original investment, so we may reject the project. In fact, there will be an equivalent loss of $220.76 per year over the machine's life.

Method 2: The second method is to separate cash flows associated with the asset acquisition and disposal from the normal operating cash flows. Since the operating cash flows—the $4,400 yearly income—are already given in equivalent annual flows $(AE(i)_2)$, we need only to convert the cash flows associated with asset acquisition and disposal into equivalent annual flows $(AE(i)_1)$. Equation (6.3) states that

$$CR(i) = (I - S)(A/P, i, N) + iS.$$

In this case,

$$AE(i)_1 = -CR(i)$$
$$= -[(\$20,000 - \$4,000)(A/P, 10\%, 5) + (0.10)\$4,000]$$
$$= -\$4,620.76.$$

Knowing that

$$AE(i)_2 = \$4,400,$$

we can calculate the total AE as follows:

$$
\begin{aligned}
AE(10\%) &= AE(i)_1 + AE(i)_2 \\
&= -\$4,620.76 + \$4,400 \\
&= -\$220.76.
\end{aligned}
$$

Comments: Obviously, Method 2 saves a calculation step, so we may prefer it over Method 1. We may interpret Method 2 as determining that the annual operating benefits must be at least $4,620.76 in order to recover the asset cost. However, the annual operating benefits actually amount to only $4,400, resulting in a loss of $220.76 per year. Therefore, the project is not worth undertaking. (See Figure 6.3.)

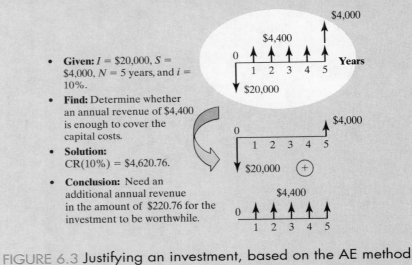

- **Given:** $I = \$20,000$, $S = \$4,000$, $N = 5$ years, and $i = 10\%$.
- **Find:** Determine whether an annual revenue of $4,400 is enough to cover the capital costs.
- **Solution:** CR(10%) = $4,620.76.
- **Conclusion:** Need an additional annual revenue in the amount of $220.76 for the investment to be worthwhile.

FIGURE 6.3 Justifying an investment, based on the AE method

6–2   Applying Annual-Worth Analysis

In general, most engineering economic analysis problems can be solved by the present-worth methods that were introduced in Chapter 5. However, some economic analysis problems can be solved more efficiently by annual-worth analysis. In this section, we introduce several applications that call for annual-worth analysis techniques.

6.2.1   Unit-Profit or Unit-Cost Calculation

In many situations, we need to know the *unit profit* (or *unit cost*) of operating an asset. To obtain a unit profit (or cost), we may proceed as follows:

- Determine the number of units to be produced (or serviced) each year over the life of the asset.
- Identify the cash flow series associated with production or service over the life of the asset.

- Calculate the present worth of the project's cash flow series at a given interest rate, and then determine the equivalent annual worth.
- Divide the equivalent annual worth by the number of units to be produced or serviced during each year. When the number of units varies each year, you may need to convert the units into equivalent annual units.

To illustrate the procedure, we will consider Example 6.3, in which the annual-equivalence concept is useful in estimating the savings per machine hour for a proposed machine acquisition.

Example 6.3 Unit Profit per Machine Hour When Annual Operating Hours Remain Constant

Consider the investment in the metal-cutting machine in Example 5.3. Recall that this three-year investment was expected to generate a PW of $3,553. Suppose that the machine will be operated for 2,000 hours per year. Compute the equivalent savings per machine hour at $i = 15\%$ compounded annually.

SOLUTION

Given: PW = $3,553, N = 3 years, i = 15% per year, and there are 2,000 machine-hours per year.

Find: Equivalent savings per machine-hour.

We first compute the annual equivalent savings from the use of the machine. Since we already know the PW of the project, we obtain the AE as follows:

$$AE(15\%) = \$3,553(A/P, 15\%, 3) = \$1,556.$$

With an annual usage of 2,000 hours, the equivalent savings per machine-hour would be calculated as follows:

Savings per machine hour = $1,556/2,000 hours = $0.78/hour.

See Figure 6.4.

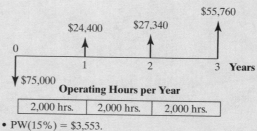

- PW(15%) = $3,553.
- AE(15%) = $3,553 ($A/P$, 15%, 3)
      = $1,556.
- Savings per machine–hour = $1,556/2,000 = $0.78/hour.

FIGURE 6.4 Computing equivalent savings per machine-hour

Comments: Note that we cannot simply divide the PW amount ($3,553) by the total number of machine-hours over the three-year period (6,000 hours), which would result in $0.59/hour. This $0.59/hour figure represents the instant savings in present worth for each hour of use of the equipment, but does not consider the time over which the savings occur. Once we have the annual equivalent worth, we can divide by the desired time unit if the compounding period is one year. If the compounding period is shorter, then the equivalent worth should be calculated for the compounding period.

Example 6.4 Unit Profit per Machine Hour When Annual Operating Hours Fluctuate

Reconsider Example 6.3, but suppose that the metal-cutting machine will be operated according to varying hours: 1,500 hours in the first year, 2,500 hours in the second year, and 2,000 hours in third year. The total number of operating hours is still 6,000 over three years. Compute the equivalent savings per machine-hour at $i = 15\%$ compounded annually.

SOLUTION

Given: PW = $3,553, $N = 3$ years, $i = 15\%$ compounded annually, operating hours of 1,500 hours in the first year, 2,500 hours in the second year, and 2,000 hours in the third year.

Find: Equivalent savings per machine-hour.

As calculated in Example 6.3, the annual equivalent savings are $1,556. Let C denote the equivalent annual savings per machine-hour that needs to be determined. Now, with varying annual usages of the machine, we can set up the equivalent annual savings as a function of C:

$$
\begin{aligned}
\text{Equivalent annual savings} = C[&(1,500)(P/F, 15\%, 1) \\
&+ C(2,500)(P/F, 15\%, 2) \\
&+ C(2,000)(P/F, 15\%, 3)](A/P, 15\%, 3) \\
= {}& 1,975.16C.
\end{aligned}
$$

We can equate this amount to $1,556 (from Example 6.3) and solve for C. This operation gives us

$$
C = \$1,556/1,975.16 = \$0.79/\text{hour},
$$

which is a penny more than in the situation in Example 6.3.

6.2.2 **Make-or-Buy Decision**

Make-or-buy problems are among the most common business decisions. At any given time, a firm may have the option of either buying an item or producing it. *If either the "make" or the "buy" alternative requires the acquisition of machinery or equipment besides the item itself, then the problem becomes an investment decision.* Since the cost of an outside service (the "buy" alternative) is usually quoted in terms of dollars per unit, it is easier to compare the two alternatives if the differential costs of the "make" alternative are also given in dollars per unit. This unit-cost comparison requires the use of annual-worth analysis. The specific procedure is as follows:

- **Step 1:** Determine the time span (planning horizon) for which the part (or product) will be needed.
- **Step 2:** Determine the annual quantity of the part (or product).
- **Step 3:** Obtain the unit cost of purchasing the part (or product) from the outside firm.
- **Step 4:** Determine the cost of the equipment, manpower, and all other resources required to make the part (or product).
- **Step 5:** Estimate the net cash flows associated with the "make" option over the planning horizon.
- **Step 6:** Compute the annual equivalent cost of producing the part (or product).
- **Step 7:** Compute the unit cost of making the part (or product) by dividing the annual equivalent cost by the required annual quantity.
- **Step 8:** Choose the option with the smallest unit cost.

Example 6.5 Unit Cost: Make or Buy

Ampex Corporation currently produces both videocassette cases (bodies) and metal-particle magnetic tape for commercial use. An increased demand for metal-particle videotapes is projected, and Ampex is deciding between increasing the internal production of both empty cassette cases and magnetic tape or purchasing empty cassette cases from an outside vendor. If Ampex purchases the cases from a vendor, the company must also buy specialized equipment to load the magnetic tape into the empty cases, since its current loading machine is not compatible with the cassette cases produced by the vendor under consideration. The projected production rate of cassettes is 79,815 units per week for 48 weeks of operation per year. The planning horizon is seven years. After considering the effects of income taxes, the accounting department has itemized the costs associated with each option as follows:

- "Make" Option:

Annual Costs:	
Labor	$1,445,633
Materials	$2,048,511
Incremental overhead	$1,088,110
Total annual cost	$4,582,254

- "Buy" Option:

Capital Expenditure:	
Acquisition of a new loading machine	$ 405,000
Salvage value at end of seven years	$ 45,000
Annual Operating Costs:	
Labor	$ 251,956
Purchase of empty cassette cases ($0.85/unit)	$3,256,452
Incremental overhead	$822,719
Total annual operating costs	$4,331,127

(Note the conventional assumption that cash flows occur in discrete lumps at the ends of years, as shown in Figure 6.5.) Assuming that Ampex's MARR is 14%, calculate the unit cost under each option.

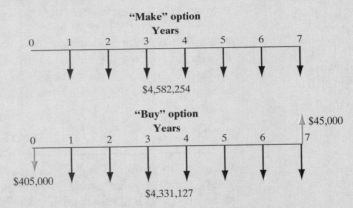

FIGURE 6.5 Make-or-buy analysis

SOLUTION

Given: Cash flows for both options; $i = 14\%$.

Find: Unit cost for each option and which option is preferred.

The required annual production volume is

79,815 units/week × 48 weeks = 3,831,120 units per year.

We now need to calculate the annual equivalent cost under each option:

- **Make Option:** Since the parameters for the make option are already given on an annual basis, we find that the annual equivalent cost is

$$AE(14\%)_{Make} = \$4,582,254.$$

- **Buy Option:** The two cost components are capital cost and operating cost.

We calculate the annual equivalent cost for each as follows:

Capital cost:

The capital-recovery cost is

$$CR(14\%) = (\$405{,}000 - \$45{,}000)(A/P, 14\%, 7) + (0.14)(\$45{,}000)$$
$$= \$90{,}249.$$

Therefore, the annual equivalent cost is

$$AE(14\%)_1 = CR(14\%) = \$90{,}249.$$

Operating cost:

The annual equivalent cost is

$$AE(14\%)_2 = \$4{,}331{,}127.$$

Total annual equivalent cost:

Therefore, the total annual equivalent cost is

$$AE(14\%)_{Buy} = AE(14\%)_1 + AE(14\%)_2 = \$4{,}421{,}376.$$

Obviously, this annual-equivalence calculation indicates that Ampex would be better off buying cassette cases from the outside vendor. However, Ampex wants to know the unit costs in order to set a price for the product. For this situation, we need to calculate the unit cost of producing the cassette tapes under each option. (Note that the negative sign indicates that the AE is a cash outflow, or cost. When comparing the options, we are interested in the magnitude of the costs and can thus ignore the negative sign.) We do this calculation by dividing the magnitude of the annual-equivalent cost for each option by the annual quantity required:

- **Make Option:**

$$\text{Unit cost} = \$4{,}582{,}254/3{,}831{,}120 = \$1.20/\text{unit.}$$

- **Buy Option:**

$$\text{Unit cost} = \$4{,}421{,}376/3{,}831{,}120 = \$1.15/\text{unit.}$$

Buying the empty cassette cases from the outside vendor and loading the tape in-house will save Ampex 5 cents per cassette before any tax consideration.

Comments: Two important noneconomic factors should also be considered. The first is the question of whether the quality of the supplier's component is better than (or equal to) or worse than that of the component the firm is presently manufacturing. The second is the reliability of the supplier in terms of providing the needed quantities of the cassette cases on a timely basis. A reduction in quality or reliability should virtually always rule out a switch from making to buying.

6-3 Comparing Mutually Exclusive Projects

In this section, we will consider a situation where two or more mutually exclusive alternatives need to be compared based on annual equivalent worth. In Section 5.4, we discussed the general principle that should be applied when mutually exclusive alternatives with unequal service lives are compared. The same general principle should be applied when comparing mutually exclusive alternatives based on annual equivalent worth—that is, mutually exclusive alternatives in equal time spans must be compared. Therefore, we must give a careful consideration of the time period covered by the analysis, the **analysis period**. We will consider two situations: (1) The analysis period equals the project lives, and (2) the analysis period differs from the project lives.

6.3.1 Analysis Period Equals Project Lives

Let's begin our comparison with a simple situation where the length of the projects' lives equals the length of the analysis period. In this situation, we compute the AE value for each project and select the project that has the least negative AE value (for service projects) or the largest AE value (for revenue projects).

In many situations, we need to compare a set of different design alternatives for which each design would produce the same number of units (constant revenues), but would require different amounts of investment and operating costs (because of different degrees of mechanization). Example 6.6 illustrates the use of the annual equivalent cost concept to compare the cost of operating a conventional electric motor with that of operating a premium-efficiency motor in a strip-processing mill.

Example 6.6 How Premium Efficiency Motors Can Cut Your Electric Costs

Birmingham Steel, Inc., is considering replacing 20 conventional 25-HP, 230-V, 60-Hz, 1800-rpm induction motors in its plant with modern premium-efficiency (PE) motors. Both types of motors have power outputs of 18.650 kW per motor (25 HP = 0.746 kW/HP). Conventional motors have a published efficiency of 89.5%, while the PE motors are 93% efficient. The initial cost of the conventional motors is $13,000, while the initial cost of the proposed PE motors is $15,600. The motors are operated 12 hours per day, 5 days per week, 52 weeks per year, with a local utility cost of $0.07 per kilowatt-hour (kWh). The motors are to be operated at 75% load, and the life cycle of both the conventional motor and the PE motor is 20 years, with no appreciable salvage value.

(a) At an interest rate of 13% compounded annually, what is the amount of savings per kWh gained by switching from the conventional motors to the PE motors?

(b) At what operating hours are the two types of motors equally economical?

Discussion: Whenever we compare machines with different efficiency ratings, we need to determine the input powers required to operate the machines. Since percent efficiency is equal to the ratio of output power to input power, we can determine the input power by dividing the output power by the motor's percent efficiency:

$$\text{Input power} = \frac{\text{output power}}{\text{percent efficiency}}.$$

For example, a 30-HP motor with 90% efficiency will require an input power calculated as follows:

$$\text{Input power} = \frac{(30 \text{ HP} \times 0.746 \text{ kW/HP})}{0.90}$$

$$= 24.87 \text{ kW}.$$

Once we determine the input-power requirement and the number of operating hours, we can convert this information into equivalent energy cost (power cost). Although the company needs 20 motors, we can compare the two types of motors based on a single unit.

SOLUTION

Given: Types of motors = (standard, PE), I = ($13,000, $15,000), S = (0, 0), N = (20 years, 20 years), rated power output = (18.65 kW, 18.65 kW), efficiency rating = (89.5%, 93%), i = 13%, utility rate = $0.07/kWh, operating hours = 3,120 hours per year, and number of motors required = 20.

Find: (a) The amount saved per kWh by operating the PE motor and (b) the break-even number of operating hours for the PE motor.

Mutually Exclusive Alternatives with Equal Project Lives		
	Standard Motor	**Premium-Efficiency Motor**
Size	25 HP	25 HP
Cost	$13,000	$15,600
Life	20 years	20 years
Salvage Value	$0	$0
Efficiency	89.5%	93%
Energy Cost	$0.07/kWh	$0.07/kWh
Operating Hours	3,120 hrs/yr	3,120 hrs/yr

(a) At i = 13%, determine the operating cost per kWh for each motor.

(b) At what number of operating hours are the two types of motor equivalent?

(a) Execute the following steps to determine the operating cost per kWh per unit:

- Determine total input power for both motor types:

Conventional motor:

$$\text{Input power} = \frac{18.650 \text{ kW}}{0.895} = 20.838 \text{ kW}.$$

PE motor:

$$\text{Input power} = \frac{18.650 \text{ kW}}{0.93} = 20.054 \text{ kW}.$$

Note that each PE motor requires 0.784 kW less input power (or 15.68 kW for 20 motors), which results in energy savings.

- Determine total kWh per year for each type of motor, assuming a total of 3,120 hours per year in motor operation:

Conventional motor:

$$3,120 \text{ hrs/year} \times 20.838 \text{ kW} = 65,018 \text{ kWh/year}.$$

PE motor:

$$3,120 \text{ hrs/year} \times 20.054 \text{ kW} = 62,568 \text{ kWh/year}.$$

- Determine annual energy costs for both motor types. Since the utility rate is $0.07/kWh, the annual energy cost for each type of motor is calculated as follows:

Conventional motor:

$$\$0.07/\text{kWh} \times 65,018 \text{ kWh/year} = \$4,551/\text{year}.$$

PE motor:

$$\$0.07/\text{kWh} \times 62,568 \text{ kWh/year} = \$4,380/\text{year}.$$

- Determine capital costs for both types of motors. Recall that we assumed that the useful life for both motor types is 20 years. To determine the annualized capital cost at 13% interest, we use the capital-recovery factor:

Conventional motor:

$$(\$13,000)(A/P, 13\%, 20) = \$1,851.$$

PE motor:

$$(\$15,600)(A/P, 13\%, 20) = \$2,221.$$

- Determine the total equivalent annual cost, which is equal to the capital cost plus the annual energy cost. Then calculate the unit cost per kWh based on output power. Note that the total output power is 58,188 kWh per year (25 HP $\times$ 0.746 kW/HP $\times$ 3120 hours/year). We execute these steps as follows:

Conventional motor:

We calculate that

$$AE(13\%) = \$4,551 + \$1,851 = \$6,402.$$

So,

$$\text{Cost per kWh} = \$6,402/58,188 \text{ kWh} = 11 \text{ cents/kWh.}$$

PE motor:

We calculate that

$$AE(13\%) = \$4,380 + \$2,221 = \$6,601.$$

So,

$$\text{Cost per kWh} = \$6,601/58,188 \text{ kWh} = 11.34 \text{ cents/kWh.}$$

Clearly, conventional motors are cheaper to operate if the motors are expected to run only 3,120 hours per year.

- Determine the savings (or loss) per operating hour obtained from switching from conventional to PE motors:

Additional capital cost required from switching from conventional to PE motors:

$$\text{Incremental capital cost} = \$2,221 - \$1,851 = \$370.$$

Additional energy-cost savings from switching from conventional to PE motors:

$$\text{Incremental energy savings} = \$4,551 - \$4,380 = \$171.$$

At 3,120 annual operating hours, it will cost the company an additional $370 to switch to PE motors, but the energy savings are only $171, which results in a $199 loss from each motor. In other words, for each operating hour, you lose about 6.38 cents.

(b) Execute the following steps to determine the break-even number of operating hours for the PE motors:

- Would that result in (a) change if the same motor were to operate 5,000 hours per year? If a motor is to run around the clock, the savings in kWh would result in a substantial annual savings in electricity bills, which are an operating cost. As we calculate the annual equivalent cost by varying the number of operating hours, we obtain the situation shown in Figure 6.6. Observe that if Birmingham Steel uses the PE motors for more than 6,742 hours annually, replacing the conventional motors with the PE motors would be justified.

	A	B	C	D	E	F	G
1	**Example 6.6 How Premium Efficiency Motors Can Cut Your Electric Costs**						
2					Operating Hours	Conventional Motor	PE Motor
3		Conventional Motor	Premium Efficiency Motor				
4							
5					0	$1,851	$2,221
6	Output power (hp)	25	25		500	$2,580	$2,923
7	Operating hours per year	6742	6742		1000	$3,309	$3,624
8	Efficiency (%)	89.5	93		1500	$4,039	$4,326
9					2000	$4,768	$5,028
10	Initial cost ($)	$13,000	$15,600		2500	$5,497	$5,730
11	Salvage value ($)	0	0		3000	$6,227	$6,432
12	Service life (year)	20	20		3500	$6,956	$7,134
13	Utility rate ($/kWh)	0.07	0.07		4000	$7,685	$7,836
14	Interest rate (%)	13	13		4500	$8,415	$8,538
15					5000	$9,144	$9,240
16					5500	$9,873	$9,941
17	Capital cost ($/year)	$1,850.59	$2,220.71		6000	$10,603	$10,643
18	Energy cost ($/year)	$9,834.28	$9,464.17		6500	$11,332	$11,345
19	Total equ. annual cost	$11,684.87	$11,684.89		7000	$12,061	$12,047
20	Cost per kWh	$0.0929	$0.0929		7500	$12,791	$12,749
21					8000	$13,520	$13,541
22					8500	$14,249	$14,153
23					8750	$14,614	$14,504
24							

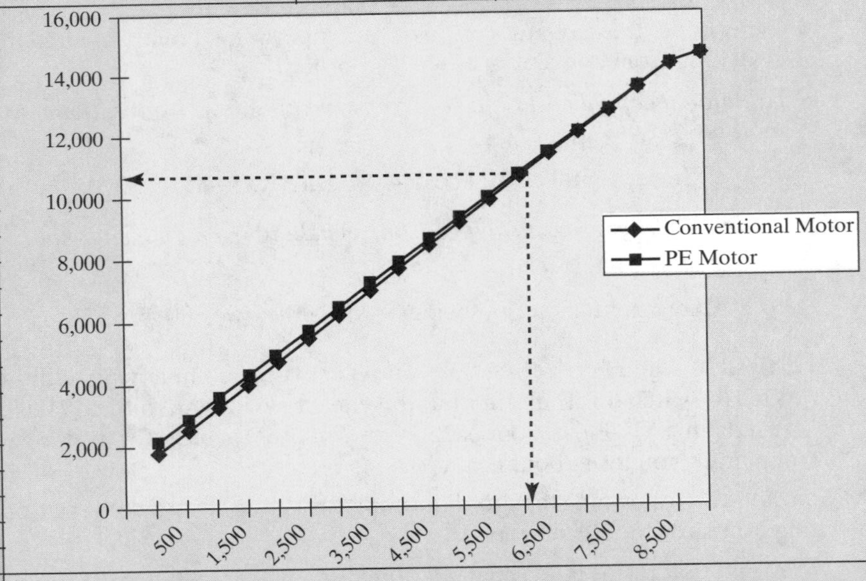

Cell B17: = PMT(B14/100,B12,B10-B11)+(B14/100)*B11
Cell B18: = (B13*(B6*0.746/(B8/100))*B7
Cell B19: = (B17+B18)
Cell B20: = B19/(B6*0.746*B7)

FIGURE 6.6 Break-even number of operating hours as calculated with a sensitivity graph in Excel

6.3.2　Analysis Period Differs from Projects' Lives

In Section 5.4, we learned that, in present-worth analysis, we must have a common analysis period when mutually exclusive alternatives are compared. Annual-worth analysis also requires establishing common analysis periods, but AE offers some computational advantages over present-worth analysis, provided the following criteria are met:

1. The service of the selected alternative is required on a continuous basis.
2. Each alternative will be replaced by an *identical* asset that has the same costs and performance.

When these two criteria are met, we may solve for the AE of each project based on its initial life span, rather than on the infinite streams of project cash flows.

Example 6.7 Annual Equivalent Cost Comparison—Unequal Project Lives

Consider the scenario in Example 5.7. Suppose that the current mode of operation is expected to continue for an indefinite period of time and not simply phased out at the end of five years. We also assume that these two models will be available in the future without significant changes in price and operating costs. At MARR = 15%, which model should the firm select? Apply the annual-equivalence approach to select the most economical equipment.

n	Model A	Model B
0	−$12,500	−$15,000
1	−$5,000	−$4,000
2	−$5,500	−$4,500
3	−$6,000 + $2,000	−$5,000
4		−$5,500 + $1,500

Discussion: A required service period of infinity may be assumed if we anticipate that an investment project will be ongoing at roughly the same level of production for some indefinite period. This assumption certainly is possible mathematically, though the analysis is likely to be complicated and tedious. Therefore, in the case of an indefinitely ongoing investment project, we typically select a finite analysis period by using the **lowest common multiple** of project lives (12 years). We would consider alternative A through four life cycles and alternative B through three life cycles; in each case, we would use the alternatives completely. We then accept the finite model's results as a good prediction of what will be the economically wisest course of action for the foreseeable future. This example is a case in which we conveniently use the lowest common multiple of project lives as our analysis period. (See Figure 6.7.)

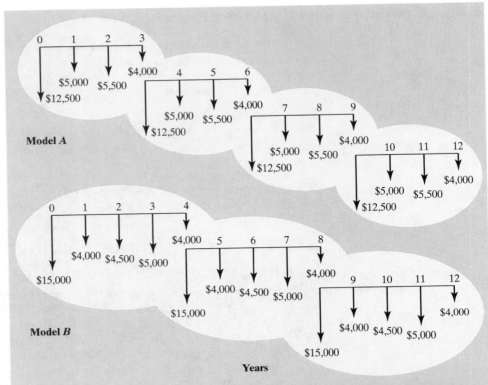

FIGURE 6.7 Comparing unequal-lived projects, over the lowest common multiple service period of 12 years

SOLUTION

Given: Cash flows for Model A and Model B and $i = 15\%$ compounded annually.

Find: AE cost, and which is the preferred alternative.

Our objective is to determine the AE cost of each model over the lowest common multiple period of 12 years. In doing so, we will compute the PW cost of the first cycle and then we convert it into its equivalent AE cost. We do the same for the entire cycle.

Model A: See Figure 6.8 for detailed calculations.

- For a three-year period (first cycle):

$$\text{PW}(15\%)_{\text{first cycle}} = -\$23,637$$
$$\text{AE}(15\%)_{\text{first cycle}} = -23,637(A/P, 15\%, 3)$$
$$= -\$10,352.$$

- For a 12-year period (four replacement cycles):

$$\text{PW}(15\%)_{\text{12-year period}} = -\$64,531,$$

$$\text{AE}(15\%)_{\text{12-year period}} = -64{,}531(A/P, 15\%, 12)$$
$$= -\$10{,}352.$$

Model A:

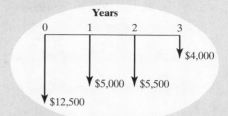

- First Cycle:
    PW(15%) = $-\$12{,}500 - \$5{,}000\ (P/F, 15\%, 1) - \$5{,}500\ (P/F, 15\%, 2)$
    $\qquad\quad -\$4{,}000\ (P/F, 15\%, 3)$
    $\qquad = -\$23{,}637.$
    AE(15%) = $-\$23{,}637\ (A/P, 15\%, 3) = \boxed{-\$10{,}352}.$

- Over four replacement cycles:
    PW(15%) = $-\$23{,}637\ [1 + (P/F, 15\%, 3)$
    $\qquad\qquad\qquad + (P/F, 15\%, 6) + (P/F, 15\%, 9)]$
    $\qquad = \qquad -\$64{,}531.$
    AE(15%) = $-\$64{,}531\ (A/P, 15\%, 12) = \boxed{-\$10{,}352}.$

FIGURE 6.8 **AE calculations for Model A**

Model B: See Figure 6.9 for detailed calculations.

- For a four-year life (first cycle)

$$\text{PW}(15\%)_{\text{first cycle}} = -\$27{,}456$$
$$\text{AE}(15\%)_{\text{first cycle}} = -\$27{,}456\ (A/P, 15\%, 34)$$
$$= -\$12{,}024.$$

- For a 12-year period (three replacement cycles):

$$\text{PW}(15\%)_{\text{12-year period}} = -\$74{,}954$$
$$\text{AE}(15\%)_{\text{12-year period}} = -\$74{,}954\ (A/P, 15\%, 12)$$
$$= -\$12{,}024.$$

We can see that the AE cost of Model B is much larger ($\$12{,}024 > \$10{,}352$), thus we select Model A, despite its shorter lifespans.

Comments: Notice that the AE costs that were calculated based on the lowest common multiple period are the same as those that were obtained over the initial lifespans. Thus, for alternatives with unequal lives, comparing the AE cost of each project over its first cycle is sufficient in determining the best alternative.

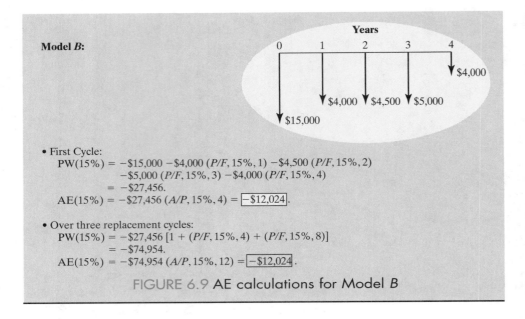

- First Cycle:
  PW(15%) = −$15,000 −$4,000 $(P/F, 15\%, 1)$ −$4,500 $(P/F, 15\%, 2)$
  −$5,000 $(P/F, 15\%, 3)$ −$4,000 $(P/F, 15\%, 4)$
  = −$27,456.
  AE(15%) = −$27,456 $(A/P, 15\%, 4)$ = $\boxed{-\$12,024}$.

- Over three replacement cycles:
  PW(15%) = −$27,456 $[1 + (P/F, 15\%, 4) + (P/F, 15\%, 8)]$
  = −$74,954.
  AE(15%) = −$74,954 $(A/P, 15\%, 12)$ = $\boxed{-\$12,024}$.

FIGURE 6.9 AE calculations for Model B

Summary

- Annual equivalent worth analysis, or AE, and present-worth analysis are the two main analysis techniques based on the concept of equivalence. The equation for AE is

$$AE(i) = PW(i)(A/P, i, N).$$

  AE analysis yields the same decision result as PW analysis.
- The capital-recovery cost factor, or CR(i), is one of the most important applications of AE analysis in that it allows managers to calculate an annual equivalent cost of capital for ease of itemization with annual operating costs. The equation for CR(i) is

$$CR(i) = (I - S)(A/P, i, N) + iS,$$

  where I = initial cost and S = salvage value.
- AE analysis is recommended over PW analysis in many key real-world situations for the following reasons:

  1. In many financial reports, an annual-equivalence value is preferred over a present-worth value, for ease of use and its relevance to annual results.
  2. Calculation of unit costs is often required in order to determine reasonable pricing for sale items.
  3. Calculation of cost per unit of use is required in order to reimburse employees for business use of personal cars.
  4. Make-or-buy decisions usually require the development of unit costs so that "make" costs can be compared to prices for "buying".

5. Comparisons of options with unequal service lives is facilitated by the AE method, assuming that the future replacements of the project have the same initial cost and operating costs. However, this method is not practical in general, as future replacement projects generally have quite different cost streams. It is recommended that you consider various future replacement options by estimating the cash flows associated with each of them.

Problems

Note 1: Unless otherwise stated, all cash flows given in the problems represent after-tax cash flows in *actual dollars*. The MARR also represents a market interest rate, which considers any inflationary effects in the cash flows.

Note 2: Unless otherwise stated, all interest rates presented in this set of problems assume annual compounding.

6.1 Consider the following cash flows and compute the equivalent annual worth at $i = 12\%$:

	A_n	
n	**Investment**	**Revenue**
0	−$10,000	
1		$2,000
2		$2,000
3		$3,000
4		$3,000
5		$1,000
6	$2,000	$500

6.2 The following investment has a net present value of zero at $i = 8\%$:

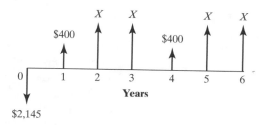

Which of the following is the net equivalent annual worth at 8% interest?

(a) $400

(b) $0

(c) $500

(d) $450

6.3 Consider the following sets of investment projects:

	Project's Cash Flow			
n	**A**	**B**	**C**	**D**
0	−$2,000	−$4,000	−$3,000	−$9,000
1	$400	$3,000	−$2,000	$2,000
2	$500	$2,000	$4,000	$4,000
3	$600	$1,000	$2,000	$8,000
4	$700	$500	$4,000	$8,000
5	$800	$500	$2,000	$4,000

Compute the equivalent annual worth of each project at $i = 10\%$, and determine the acceptability of each project.

6.4 What is the annual-equivalence amount for the following infinite series at $i = 12\%$?

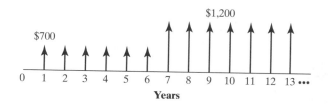

(a) $950

(b) $866

(c) $926

(d) None of the above

6.5 Consider the following sets of investment projects:

Period	Project's Cash Flow			
(n)	A	B	C	D
0	−$3,500	−$3,000	−$3,000	−$3,600
1	$0	$1,500	$3,000	$1,800
2	$0	$1,800	$2,000	$1,800
3	$5,500	$2,100	$1,000	$1,800

Compute the equivalent annual worth of each project at $i = 13\%$, and determine the acceptability of each project.

6.6 Consider an investment project with the following repeating cash flow pattern every four years for forever:

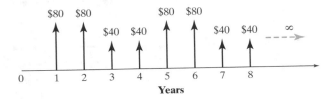

What is the annual-equivalence amount of this project at an interest rate of 12%?

6.7 The owner of a business is considering investing $55,000 in new equipment. He estimates that the net cash flows will be $5,000 during the first year and will increase by $2,500 per year each year thereafter. The equipment is estimated to have a 10-year service life and a net salvage value at the end of this time of $6,000. The firm's interest rate is 12%.

   (a) Determine the annual capital cost (ownership cost) for the equipment.

   (b) Determine the equivalent annual savings (revenues).

   (c) Determine whether this investment is wise.

Capital Recovery (Ownership) Cost

6.8 Susan wants to buy a car that she will keep for the next four years. She can buy a Honda Civic at $15,000 and then sell it for $8,000 after four years. If she bought this car, what would be her annual ownership cost (capital recovery cost)? Assume that her interest rate is 6%.

6.9 Nelson Electronics Company just purchased a soldering machine to be used in its assembly cell for flexible disk drives. This soldering machine costs $250,000. Because of the specialized function it performs, its useful life is estimated to be five years. At the end of that time, its salvage value is estimated to be $40,000. What is the capital cost for this investment if the firm's interest rate is 18%?

6.10 A construction firm is considering establishing an engineering computing center. This center will be equipped with three engineering workstations that each would cost $25,000 and have a service life of five years. The expected salvage value of each workstation is $2,000. The annual operating and maintenance cost would be $15,000 for each workstation. At a MARR of 15%, determine the equivalent annual cost of operating the engineering center.

6.11 Beginning next year, a foundation will support an annual seminar on campus by using the interest earnings on a $100,000 gift it received this year. It is determined that 8% interest will be realized for the first 10 years, but that plans should be made to anticipate an interest rate of only 6% after that time. What amount should be added to the foundation now in order to fund the seminar at a level of $10,000 per year into infinity?

Annual Equivalent Worth Criterion

6.12 The present price (year zero) of kerosene is $2.50 per gallon, and its cost is expected to increase by $.30 per year (e.g., kerosene at the end of year one will cost $2.80 per gallon). Mr. Garcia uses about 800 gallons of kerosene during a winter season for space heating. He has an opportunity to buy a storage tank for $700, and at the end of four years, he can sell the storage tank for $100. The tank has a capacity to supply four years of Mr. Garcia's heating needs, so he can buy four years of kerosene at its present price ($2.50). He can invest his money elsewhere at 8%. Should he purchase the storage tank? Assume that kerosene purchased on a pay-as-you-go basis is paid for at the end of the year. (However, kerosene purchased for the storage tank is purchased now.)

6.13 Consider the cash flows for the following investment projects:

	Project's Cash Flow	
n	A	B
0	−$4,000	$5,500
1	$1,000	−$1,400
2	$X	−$1,400
3	$1,000	−$1,400
4	$1,000	−$1,400

(a) For project A, find the value of X that makes the equivalent annual receipts equal the equivalent annual disbursement at $i = 13\%$.

(b) Would you accept project B at $i = 15\%$, based on the AE criterion?

6.14 An industrial firm can purchase a certain machine for $40,000. A down payment of $4,000 is required, and the balance can be paid in five equal year-end installments at 7% interest on the unpaid balance. As an alternative, the machine can be purchased for $36,000 in cash. If the firm's MARR is 10%, determine which alternative should be accepted, based on the annual-equivalence method.

6.15 An industrial firm is considering purchasing several programmable controllers and automating their manufacturing operations. It is estimated that the equipment will initially cost $100,000 and the labor to install it will cost $35,000. A service contract to maintain the equipment will cost $5,000 per year. Trained service personnel will have to be hired at an annual salary of $30,000. Also estimated is an approximate $10,000 annual income-tax savings

(cash inflow). How much will this investment in equipment and services have to increase the annual revenues after taxes in order to break even? The equipment is estimated to have an operating life of 10 years, with no salvage value, because of obsolescence. The firm's MARR is 10%.

6.16 A certain factory building has an old lighting system, with lighting this building costs, on average, of $20,000 a year. A lighting consultant tells the factory supervisor that the lighting bill can be reduced to $8,000 a year if $50,000 were invested in a new lighting system for the factory building. If the new lighting system is installed, an incremental maintenance cost of $3,000 per year must be considered. If the old lighting system has zero salvage value and the new lighting system is estimated to have a life of 20 years, what is the net annual benefit for this investment in new lighting? Consider the MARR to be 12%. Also consider that the new lighting system has zero salvage value at the end of its life.

Unit-Profit or Unit-Cost Calculation

6.17 Two 150-horsepower (HP) motors are being considered for installation at a municipal sewage-treatment plant. The first costs $4,500 and has an operating efficiency of 83%. The second costs $3,600 and has an operating efficiency of 80%. Both motors are projected to have zero salvage value after a life of 10 years. If all the annual charges, such as insurance and maintenance, amount to a total of 15% of the original cost of each motor, and if power costs are a flat 5 cents per kilowatt-hour, what is the minimum number of hours of full-load operation per year required in order to justify purchase of the more expensive motor at $i = 6\%$? (A conversion factor you might find useful is 1 HP = 746 watts = 0.746 kilowatts.)

6.18 Two 180-horsepower water pumps are being considered for installation in a municipal work. Financial data for these pumps are given as follows:

Item	Pump I	Pump II
Initial cost	$6,000	$4,000
Efficiency	86%	80%
Useful life	12 years	12 years
Annual operating cost	$500	$440
Salvage value	$0	$0

If power cost is a flat 6 cents per kWh over the study period, which of the following ranges includes the minimum number of hours of full-load operation per year required in order to justify purchase of the more expensive pump at an interest rate of 8% (1 HP = 746 watts = 0.746 kilowatts)?

(a) 340 hours/year < minimum number of operation hours/year ≤ 390 hours/year

(b) 390 hours/year < minimum number of operation hours/year ≤ 440 hours/year

(c) 440 hours/year < minimum number of operation hours/year ≤ 490 hours/year

(d) 490 hours/year < minimum number of operation hours/year ≤ 540 hours/year

6.19 You invest in a piece of equipment costing $20,000. The equipment will be used for two years, at the end of which time the salvage value of the machine is expected to be $10,000. The machine will be used for 6,000 hours during the first year and 8,000 hours during the second year. The expected annual net savings in operating costs will be $30,000 during the first year and $40,000 during the second year. If your interest rate is 10%, which of the following would be the equivalent net savings per machine hour?

(a) $4.00/hour
(b) $5.00/hour
(c) $6.00/hour
(d) $7.00/hour

6.20 A company is currently paying its employees $0.33 per mile to drive their own cars when on company business. However, the company is considering supplying employees with cars, which would involve the following cost components: car purchase at $22,000, with an estimated three-year life; a net salvage value of $5,000; taxes and insurance at a cost of $700 per year; and operating and maintenance expenses of $0.15 per mile. If the interest rate is 10% and the company anticipates an employee's annual travel to be 30,000 miles, what is the equivalent cost per mile (without considering income tax)?

6.21 Santa Fe Company, a farm-equipment manufacturer, currently produces 20,000 units of gas filters for use in its lawnmower production annually. The following costs are reported based on the previous year's production:

Item	Expense
Direct materials	$60,000
Direct labor	$180,000
Variable overhead (power and water)	$135,000
Fixed overhead light and heat	$70,000
Total cost	**$445,000**

It is anticipated that gas-filter production will last five years. If the company continues to produce the product in-house, annual direct-material costs will increase at a rate of 5%. (For example, the annual direct-material costs during the first production year will be $63,000.) In addition, direct-labor costs will increase at a rate of 6% per year, variable-overhead costs will increase at a rate of 3%, while fixed-overhead costs will remain at the current level over the next five years. Tompkins Company has offered to sell Santa Fe Company 20,000 units of gas filters for $25 per unit. If Santa Fe accepts the offer, some of the manufacturing facilities currently used to manufacture the gas filters could be rented to a third party at an annual rate of $35,000. In addition, $3.50 per unit of the fixed-overhead costs applied to gas-filter production would be eliminated. The firm's interest rate is known to be 15%. What is the unit cost of buying the gas filters from the outside source? Should Santa Fe accept Tompkins's offer? Why or why not?

6.22 An electric automobile can be purchased for $25,000. The automobile is estimated to have a life of 12 years, with annual travel of 20,000 miles. Every three years, a new set of batteries will have to be purchased at a cost of $3,000. Annual maintenance of the vehicle is estimated to cost $700 per year. The cost of recharging the batteries is estimated at $0.015 per mile. The salvage value of the batteries and the vehicle at the end of 12 years is estimated at $2,000. Consider the MARR to be 7%. What is the cost per mile to own and operate this vehicle, based on the foregoing estimates? The $3,000 cost of the batteries is a net value, with the old batteries traded in for the new ones.

6.23 A California utility firm is considering building a 50-megawatt geothermal plant that generates electricity from naturally occurring underground heat. The binary geothermal system will cost $85 million to build and $6 million (including any income-tax effect) to operate per year. (Unlike a conventional fossil-fuel plant, this system will require virtually no fuel costs.) The geothermal plant is to last for 25 years. At the end of that time, the expected salvage value will be about the same as the cost to remove the plant. The plant will be in operation for 70% (the plant-utilization factor) of the year (or 70% of 8,760 hours per year). If the firm's MARR is 14% per year, determine the cost of generating electricity per kilowatt-hour.

Break-Even Analysis

6.24 The estimated cost of a completely installed and ready-to-operate 40-kilowatt generator is $30,000. Its annual maintenance costs are estimated at $500. The energy that can be generated annually, at full load, is estimated to be 100,000 kilowatt-hours. If the value of the energy generated is considered to be $0.08 per kilowatt-hour, how long will it take before this machine becomes profitable? Consider the MARR to be 9% and the salvage value of the machine to be $2,000 at the end of its estimated life of 15 years.

6.25 A large state university, currently facing a severe parking shortage on its campus, is considering constructing parking decks off campus. A shuttle service composed of minibuses could pick up students at the off-campus parking deck and quickly transport them to various locations on campus. The university would charge a small fee for each shuttle ride, and the students could be quickly and economically transported to their classes. The funds raised by the shuttle would be used to pay for minibuses, which cost about $150,000 each. Each minibus has a 12-year service life, with an estimated salvage value of $3,000. To operate each minibus, the following additional expenses must be considered:

Item	Annual Expenses
Driver	$40,000
Maintenance	$7,000
Insurance	$2,000

If students pay 10 cents for each ride, determine the annual ridership (i.e., the number of shuttle rides per year) required to justify the shuttle project, assuming an interest rate of 6%.

6.26 Eradicator Food Prep, Inc., has invested $7 million to construct a food irradiation plant. This technology destroys organisms that cause spoilage and disease, thus extending the shelf life of fresh foods and the distances over which they can be shipped. The plant can handle about 200,000 pounds of produce in an hour, and it will be operated for 3,600 hours a year. The net expected operating and maintenance costs (considering any income-tax effects) would be $4 million per year. The plant is expected to have a useful life of 15 years, with a net salvage value of $700,000. The firm's interest rate is 15%.

(a) If investors in the company want to recover the plant investment within six years of operation (rather than 15 years), what would be the equivalent after-tax annual revenues that must be generated?

(b) To generate the annual revenues determined in part (a), what minimum processing fee per pound should the company charge to its customers?

6.27 A corporate executive jet with a seating capacity of 20 has the following cost factors:

Item	Cost
Initial cost	$12,000,000
Service life	15 years
Salvage value	$2,000,000
Crew costs per year	$225,000
Fuel cost per mile	$1.10
Landing fee	$250
Maintenance costs per year	$237,500
Insurance costs per year	$166,000
Catering cost per passenger trip	$75

The company flies three round-trips from Boston to London per week, a distance of 3,280 miles one way. How many passengers must be carried on an average trip in order to justify the use of the jet if the alternative commercial airline first-class round-trip fare is $3,400 per passenger? The firm's MARR is 15%. (Ignore income-tax consequences.)

6.28 The local government of Santa Catalina Island, off the coast of Long Beach, California, is completing plans to build a desalination plant to help ease a critical drought on the island. Both the drought and new construction on Catalina have left the island with an urgent need for a new water source. A modern desalination plant could produce freshwater from seawater for $1,000 an acre-foot. An acre-foot is 326,000 gallons, or enough to supply two households for 1 year. On Santa Catalina Island, the cost for using water from natural sources is about the same as for desalting. The $3 million plant, with a daily desalting capacity of 0.4 acre-foot, can produce 132,000 gallons of freshwater a day (enough to supply 295 households daily), more than a quarter of the island's total needs. The desalination plant has an estimated service life of 20 years, with no appreciable salvage value. The annual operating and maintenance costs would be about $250,000. Assuming an interest rate of 10%, what should be the average monthly water bill for each household?

Comparing Mutually Exclusive Alternatives by Using the AE Method

6.29 You are considering two types of electric motors for your paint shop. Financial information and operating characteristics are summarized as follows:

Summary Info and Characteristics	Brand X	Brand Y
Price	$4,500	$3,600
O&M cost per year	$300	$500
Salvage value	$250	$100
Capacity	150 HP	150 HP
Efficiency	83%	80%

If you plan to operate the motor for 2,000 hours annually, which of the given options represents the total cost savings per operating hour associated with the more efficient brand (Brand X) at an interest rate of 12%? The motor will be needed for 10 years. Assume that power costs are five cents per kilowatt-hour (1 HP = 0.746 kW).

O & M = Operating & Maintenance

(a) Less than 10 cents/hr
(b) Between 10 cents/hr and 20 cents/hr, inclusive
(c) Between 20.1 cents/hr and 30 cents/hr, inclusive
(d) Greater than or equal to than 30.1 cents/hr

6.30 The following cash flows represent the potential annual savings associated with two different types of production processes, each of which requires an investment of $12,000:

n	Process A	Process B
0	−$12,000	−$12,000
1	$9,120	$6,350
2	$6,840	$6,350
3	$4,560	$6,350
4	$2,280	$6,350

Assuming an interest rate of 15%, complete the following tasks:

(a) Determine the equivalent annual savings for each process.

(b) Determine the hourly savings for each process, assuming 2,000 hours of operation per year.

(c) Determine which process should be selected.

6.31 Travis Wenzel has $2,000 to invest. Usually, he would deposit the money in his savings account, which earns 6% interest, compounded monthly. However, he is considering three alternative investment opportunities:

- Option 1: Purchasing a bond for $2,000. The bond has a face value of $2,000 and pays $100 every six months for three years. The bond matures in three years.
- Option 2: Buying and holding a growth stock that grows 11% per year for three years.
- Option 3: Making a personal loan of $2,000 to a friend and receiving $150 interest per year for three years.

Determine the equivalent annual cash flows for each option, and select the best option.

6.32 A chemical company is considering two types of incinerators to burn solid waste generated by a chemical operation. Both incinerators have a burning capacity of 20 tons per day. The following data have been compiled for comparison of the two incinerators:

Item	Incinerator A	Incinerator B
Installed cost	$1,200,000	$750,000
Annual O&M costs	$50,000	$80,000
Service life	20 years	10 years
Salvage value	$60,000	$30,000
Income taxes	$40,000	$30,000

If the firm's MARR is known to be 13%, determine the processing cost per ton of solid waste for each incinerator. Assume that incinerator B will be available in the future at the same cost.

6.33 An airline is considering two types of engine systems for use in its planes:

- System A costs $100,000 and uses 40,000 gallons of fuel per 1,000 hours of operation at the average load encountered in passenger service.
- System B costs $200,000 and uses 32,000 gallons of fuel per 1,000 hours of operation at the average load encountered in passenger service.

Both engine systems have the same life and same maintenance and repair record, and both have a three-year life before any major overhaul is required. Each system has a salvage value of 100% of the initial investment. If jet fuel costs $1.80 a gallon currently and fuel consumption is expected to increase at the rate of 6% per year because of degrading engine efficiency, which engine system should the firm install? Assume 2,000 hours of operation per year and a MARR of 10%. Use the AE criterion. What is the equivalent operating cost per hour for each engine?

6.34 Norton Auto Parts, Inc., is considering two different forklift trucks for use in its assembly plant:

- Truck A costs $15,000 and requires $3,000 annually in operating expenses. It will have a $5,000 salvage value at the end of its 3-year service life.
- Truck B costs $20,000, but requires only $2,000 annually in operating expenses; its service life is four years, after which its expected salvage value is $8,000.

The firm's MARR is 12%. Assuming that the trucks are needed for 12 years and that no significant changes are expected in the future price and functional capacity of both trucks, select the most economical truck, based on AE analysis.

6.35 A small manufacturing firm is considering the purchase of a new machine to modernize one of its current production lines. Two types of machines are available on the market. The lives of Machine A and Machine B are four years and six years, respectively, but the firm does not expect to need the service of either machine for more than five years. The machines have the following expected receipts and disbursements:

Item	Machine A	Machine B
Initial cost	$6,500	$8,500
Service life	4 years	6 years
Estimated salvage value	$600	$1,000
Annual O&M costs	$800	$520
Cost to change oil filter every other year	$100	None
Engine overhaul	$200 (every 3 years)	$280 (every 4 years)

After four years of use, the salvage value for Machine B will be $1,000. The firm always has another option: to lease a machine at $3,000 per year, fully maintained by the leasing company. The lease payment will be made at the beginning of each year.

(a) How many decision alternatives are there?

(b) Which decision appears to be the best at $i = 10\%$?

6.36 A plastic-manufacturing company owns and operates a polypropylene-production facility that converts the propylene from one of its cracking facilities to polypropylene plastics for outside sale. The polypropylene-production facility is currently forced to operate at less than capacity, due to lack of enough propylene-production capacity in its hydrocarbon-cracking facility. The chemical engineers are considering alternatives for supplying additional propylene to the polypropylene-production facility. Some of the feasible alternatives are as follows:

- Option 1: Build a pipeline to the nearest outside supply source.
- Option 2: Provide additional propylene by truck from an outside source.

The engineers also gathered the following projected cost estimates:

- future costs for purchased propylene, excluding delivery: $0.215 per lb;
- cost of pipeline construction: $200,000 per pipeline mile;
- estimated length of pipeline: 180 miles;
- transportation costs by tank truck: $0.05 per lb, using a common carrier;
- pipeline operating costs: $0.005 per lb, excluding capital costs;
- projected additional propylene needs: 180 million lbs per year;
- projected project life: 20 years;
- estimated salvage value of the pipeline: 8% of the installed costs.

Determine the propylene cost per pound under each option, if the firm's MARR is 18%. Which option is more economical?

6.37 The city of Prattville is comparing two plans for supplying water to a newly developed subdivision:

- Plan A will take care of requirements for the next 15 years; at the end of that period, the initial cost of $400,000 will have to be doubled to meet the requirements of subsequent years. The facilities installed in years zero and 15 may be considered permanent; however, certain supporting equipment will have to be replaced every 30 years from the installation dates, at a cost of $75,000. Operating costs are $31,000 a year for the first 15 years and $62,000 thereafter. Beginning in the 21st year, they will increase by $1,000 a year.
- Plan B will supply all requirements for water indefinitely into the future, although it will operate at only half capacity for the first 15 years. Annual costs over this period will be $35,000 and will increase to $55,000 beginning in the 16th year. The initial cost of Plan B is $550,000; the facilities can be considered permanent, although it will be necessary to replace $150,000 of equipment every 30 years after the initial installation.

The city will charge the subdivision the use of water based on the equivalent annual cost. At an interest rate of 10%, determine the equivalent annual cost for each plan, and make a recommendation to the city as to the amount that should be charged to the subdivision.

6.38 Southern Environmental Consulting, Inc. (SEC) designs plans and specifications for asbestos abatement (removal) projects. These projects involve public, private, and governmental buildings. Currently, SEC must also conduct an air test before allowing the reoccupancy of a building from which asbestos has been removed. SEC subcontracts air-test samples to a laboratory for analysis by transmission electron microscopy (TEM). To offset the cost of TEM analysis, SEC charges its clients $100 more than the subcontractor's fee. The only expenses in this system are the costs of shipping the air-test samples to the subcontractor and the labor involved in this shipping. As business grows, SEC needs to consider either continuing to subcontract the TEM analysis to outside companies or developing its own TEM laboratory. Because of the passage of the Asbestos Hazard Emergency Response ACT (AHERA) by the U.S. Congress, SEC expects about 1,000 air-sample testings per year over eight years. The firm's MARR is known to be 15%. The details of each option for TEM analysis are as follows:

Subcontract Option: The client is charged $400 per sample, which is $100 above the subcontracting fee of $300. Labor expenses are $1,500 per year, and shipping expenses are estimated to be $2.50 per sample.

TEM-Laboratory Purchase Option: The purchase and installation cost for the TEM laboratory is $415,000. The equipment would last for eight years, after which it would have no salvage value. The design and renovation cost is estimated to be $9,500. The client is charged $300 per sample, based on the current market price. One full-time manager and two part-time technicians are needed to operate the laboratory. Their combined annual salaries will be $50,000. The material required to operate the

lab includes carbon rods, copper grids, filter equipment, and acetone. The costs of these materials are estimated at $6,000 per year. Utility costs, operating and maintenance costs, and indirect-labor costs needed to maintain the lab are estimated at $18,000 per year. The extra income-tax expenses would be $20,000.

(a) Determine the cost per air-sample test by the TEM laboratory (in-house).

(b) What is the number of air-sample tests per year that will make the two options equivalent?

Short Case Studies with Excel

6.39 Automotive engineers at Ford are considering the laser blank welding (LBW) technique to produce windshield frame-rail blanks. The engineers believe that the LBW technique as compared with the conventional process to manufacture sheet-metal blanks would result in significant savings because of the following factors:

1. scrap reduction through more efficient blank nesting on coil and

2. scrap reclamation (weld scrap offal made into a larger usable blank).

Based on an annual volume of 3,000 blanks, Ford engineers have estimated the following financial data:

Description	Blanking Method	
	Conventional	Laser Blank Welding
Weight per blank (lbs/part)	63.764	34.870
Steel cost per part	$14.98	$8.19
Transportation per part	$0.67	$0.42
Cost blanking per part	$0.50	$0.40
Die investment	$106,480.00	$83,000.00

The LBW technique appears to achieve significant savings, so Ford's engineers are leaning toward adopting it. However, since Ford engineers have had no prior experience with the LBW, they are not sure if producing the windshield frames in-house at this time is a good strategy. For these windshield frames, it may be cheaper to use the services of a supplier, that has both the experience and the machinery for laser blanking. Ford's lack of skill in laser blanking may require that it take six months to get up to the required production volume. On the other hand, if Ford relies on a supplier, it can only assume that supplier labor problems will not halt production of Ford's parts. The make-or-buy decision depends on two factors: the amount of new investment that is required in laser welding, and whether additional machinery will be required for future products. Assuming an analysis period of 10 years and an interest rate of 16%, recommend the best course of action. Assume also that the salvage value at the end of 10 years is estimated to be nonsignificant for either system. If Ford is considering the subcontracting option, what would be the acceptable range of contract bid (unit cost per part)?

6.40 A Veterans Administration (VA) hospital is to decide which type of boiler fuel system will most efficiently provide the required steam energy output for heating, laundry, and sterilization purposes. The present boilers were installed in the early 1950s and are now obsolete. Much of the auxiliary equipment is also old and in need of repair. Because of these general conditions, an engineering recommendation was made to replace the entire plant with a new boiler-plant building that would house modern equipment. The cost of demolishing the old boiler plant would be almost a complete loss, as the salvage value of the scrap steel and used brick is estimated to be only about $1,000. The VA hospital's engineer finally selected two alternative proposals as being worthy of more intensive analysis. The hospital's annual energy requirement, measured in terms of steam output, is approximately 145,000,000 pounds of steam. As a rule of thumb for analysis, one pound of steam is approximately 1,000 BTUs, and one cubic foot of natural gas is approximately 1,000 BTUs. The two alternatives are as follows:

● Proposal 1: Build a new, coal-fired boiler plant. This boiler plant would cost $1,770,300. To meet the requirements for particulate emission as set by the Environmental Protection Agency (EPA), this coal-fired boiler, even if it burned low-sulfur coal, would need an electrostatic precipitator, which would cost approximately $100,000. This plant would last for 20 years. One pound of dry coal yields about 14,300 BTUs. To convert the 145,000,000 pounds of steam energy to the common denominator of BTUs, it is necessary to multiply by 1,000. To find BTU input requirements, it is necessary to divide by the relative boiler efficiency for type of fuel. The boiler efficiency for coal is 0.75. The coal price is estimated to be $35.50 per ton.

● Proposal 2: Build a gas-fired boiler plant with No.-2 fuel oil as a standby. This system would cost $889,200, with an expected service life of 20 years. Since small household or commercial gas users entirely dependent on gas have priority, large plants must have oil switchover capabilities. It has been estimated that 6% of 145,000,000 pounds of steam energy (or 8,700,000 pounds) would come about as a result of the oil switch. The boiler efficiency would be 0.78 for gas and 0.81 for oil. The heat value of natural gas is approximately 1,000,000 BTU/MCF (million cubic feet), and for No.-2 fuel oil it is 139,400 BTU/gal. The estimated gas price is $2.50/MCF, and the No.-2 fuel-oil price is $0.82 per gallon.

(a) Calculate the annual fuel costs for each proposal.
(b) Determine the unit cost per steam pound for each proposal. Assume $i = 10\%$.
(c) Which proposal is more economical?

6.41 The following is a letter that I received from a local city engineer:

Dear Professor Park:

Thank you for taking the time to assist with this problem. I'm really embarrassed at not being able to solve it myself, since it seems rather straightforward. The situation is as follows:

A citizen of Opelika paid for concrete drainage pipe approximately 20 years ago to be installed on his property. (We have a policy that if drainage trouble exists on private property and the owner agrees to pay for the material, city crews will install it.) That was the case in this problem. Therefore, we are dealing with only material costs, disregarding labor.

However, this past year, we removed the pipe purchased by the citizen, due to a larger area drainage project. He thinks, and we agree, that he is due some refund for the salvage value of the pipe, due to its remaining life.

Problem:

- Known: 80′ of 48″ pipe purchased 20 years ago; original purchase price, current quoted price of

48″ pipe = $52.60/foot. ($52.60 per foot × 80 feet = $4,208 total cost in today's dollars.)

- Assumptions: 50-year life (therefore, assume 30 years of life remaining at removal after 20 years); a 4% price increase per year, on average, over 20 years.

Thus, we wish to calculate the cost of the pipe 20 years ago. We need to calculate the present salvage value after 20 years use, with 30 years of life remaining in today's dollars. Thank you again for your help. We look forward to your reply.

Charlie Thomas, P.E.

Director of Engineering

City of Opelika

Recommend a reasonable amount of compensation to the citizen for the replaced drainage pipe.

.7

It's the right time of year to take a fresh look at an old question: Why bother going to a college? With all the fees and debts they lumber you with these days, is it still worth it?

Investing in an Education Is Still a Good Deal—to a Degree[1]

According to a recent study, attending a full-time university costs the average student more than $17,000 a year. But here's the magic number that makes it all worthwhile. Over the years, university graduates earn an average of almost $10,000 a year more than high school graduates do—and that's an after-tax figure.

Putting it together, we find that a college degree will have an initial cost averaging about $52,000, but lead to increased after-tax earnings over the graduate's working life of about $433,000, thus yielding a net total lifetime gain of about $380,000.

If you view the acquisition of a degree as though it is a business investment, it yields an average rate of return of 14.5% a year.

Now, if you know of many investments that yield as much as 14.5%, please tell me. That's a good deal. This figure of 14.5% is the estimated return averaged over all four-year degrees. But you would expect some degrees to lead to more lucrative occupations than others.

What's the Degree Worth in 2003?[2]		
	Avg. Starting Salary	**Change from Yr. Before**
Chemical Engineering	$52,169	+1.8%
Electrical Engineering	$50,566	−0.4%
Computer Science	$46,536	−1.6%
Math/Statistics	$41,543	−6.2%
Management Information Systems	$41,543	−5%
Accounting	$41,360	+2.6%
Economics/Finance	$40,764	+1.8%
Civil Engineering	$41,067	+0.5%
Information Sciences	$39,800	−3.9%
Nursing	$37,803	+3.3%
Business	$36,515	+3.7%
History	$30,395	+0.7%
Biology	$29,554	−1.0%
Liberal Arts	$29,543	+3.1%
Political Science	$28,546	−12.6%
English	$28,438	−8.3%
Psychology	$26,738	−10.7%

[1] Ross Grittins, "Investing in an Education Is Still a Good Deal—to a Degree," *The Age*, January 29, 2003.
[2] Source: National Association of Colleges.

Rate-of-Return Analysis

For example, limited breakdown of the average confirms this suspicion: Arts and social science degrees, for instance, yield an average return of 11%, whereas economics and business degrees yield 18%.

But what about the case of the return on a five-year degree—or on the increasingly popular double degrees, which can take five years or longer? The study does not address this matter, but other research suggests that if you spend more than four years on one or more undergraduate degrees, you soon encounter diminishing returns.

It's probably right to expect that people who have done more years of study in their primary degrees will, on average, end up in better-paid jobs. But the extra income doesn't seem to be great enough to overcome the higher up-front cost caused by the delayed entry into the full-time workforce. And the study is clear on one thing: the rate of return on post-graduate degrees averages only 6.5%—that is, roughly half what you get on a primary degree.

What does the 14.5% rate-of-return figure for the college education really represent? How do we compute the figure from the projected cash flow series? And once we have computed this figure, how do we use it when evaluating an investment alternative? Our consideration of the concept of rate of return in this chapter will answer these and other questions.

Along with the PW and AE criteria, the third primary measure of investment worth is **rate of return**. As shown in Chapter 5, the PW measure is easy to calculate and apply. Nevertheless, many engineers and financial managers prefer rate-of-return analysis to the PW method, because they find it intuitively more appealing to analyze investments in terms of percentage rates of return rather than in dollars of PW. Consider the following statements regarding an investment's profitability:

- This project will bring in a 15% rate of return on the investment.
- This project will result in a net surplus of $10,000 in terms of PW.

Neither statement describes the nature of an investment project in any complete sense. However, the rate-of-return figure is somewhat easier to understand, because many of us are so familiar with savings and loan interest rates, which are in fact rates of return.

In this chapter, we will examine four aspects of rate-of-return analysis: (1) the concept of return on investment; (2) calculation of a rate of return; (3) development of an internal rate-of-return criterion; and (4) comparison of mutually exclusive alternatives, based on rate of return.

7–1 Rate of Return

Many different terms are used to refer to **rate of return**, including **yield** (e.g., yield to maturity, commonly used in bond valuation), **internal rate of return**, and **marginal efficiency of capital**. We will first review three common definitions of rate of return. Then we will use the definition of internal rate of return as a measure of profitability for a single investment project throughout the text.

7.1.1 Return on Investment

There are several ways of defining the concept of rate of return on investment. We will show two of them: The first is based on a typical loan transaction, and the second is based on the mathematical expression of the present-worth function.

Definition 1— *Rate of return is the interest earned on the unpaid balance of an amortized loan.*

Suppose that a bank lends $10,000, which is repaid in installments of $4,021 at the end of each year for three years. How would you determine the interest rate that the bank charges on this transaction? As we learned in Chapter 3, you would set up the following equivalence equation and solve for *i*:

$$\$10,000 = \$4,021(P/A, i, 3).$$

It turns out that $i = 10\%$. In this situation, the bank will earn a return of 10% on its investment of $10,000. The bank calculates the loan balances over the life of the loan as follows:

Year	Unpaid Balance at Beginning Year	Return on Unpaid Balance (10%)	Payment Received	Unpaid Balance at End of Year
0				−$10,000
1	−$10,000	−$1,000	$4,021	−$6,979
2	−$6,979	−$698	$4,021	−$3,656
3	−$3,656	−$366	$4,021	$0

A negative balance indicates an unpaid balance.

Observe that, for the repayment schedule shown, the 10% interest is calculated only for each year's outstanding balance. In this situation, only part of the $4,021 annual payment represents interest; the remainder goes toward repaying the principal. In other words, the three annual payments repay the loan itself and provide a return of 10% on the *amount still outstanding each year*.

Note also that, when the last payment is made, the outstanding principal is eventually reduced to zero.[3] If we calculate the PW of the loan transaction at its rate of return (10%), we see that

$$PW(10\%) = -\$10,000 + \$4,021(P/A, 10\%, 3) = 0,$$

which indicates that the bank can break even at a 10% rate of interest. In other words, the rate of return becomes the rate of interest that equates the present value of future cash repayments to the amount of the loan. This observation prompts the second definition of rate of return:

Definition 2— *Rate of return is the break-even interest rate i* at which the present worth of a project is zero or*

$$PW(i^*) = PW_{\text{Cash inflows}} - PW_{\text{Cash outflows}} = 0.$$

Note that the foregoing expression is equivalent to

$$PW(i^*) = \frac{A_0}{(1 + i^*)^0} + \frac{A_1}{(1 + i^*)^1} + \cdots + \frac{A_N}{(1 + i^*)^N} = 0. \quad (7.1)$$

Here, we know the value of A_n for each period, but not the value of i^*. Since it is the only unknown, we can solve for i^*. As we will discuss momentarily, this solution is not

[3]As we learned in Section 5.3.2, the terminal balance is equivalent to the net future worth of the investment. If the net future worth of the investment is zero, its PW should also zero.

always straightforward due to the nature of the PW function in Eq. (7.1), it is certainly possible to have more than one rate of return for certain types of cash flows.[4]

Note that the $i*$ formula in Eq. (7.1) is simply the PW formula in Eq. (5.1) solved for the particular interest rate ($i*$) at which PW(i) is equal to zero. By multiplying both sides of Eq. (7.1) by $(1 + i*)^N$, we obtain

$$PW(i*)(1 + i*)^N = FW(i*) = 0.$$

If we multiply both sides of Eq. (7.1) by the capital-recovery factor, ($A/P, i*, N$), we obtain the relationship AE($i*$) = 0. *Therefore, the $i*$ of a project may also be defined as the rate of interest that equates the present worth, future worth, and annual equivalent worth of the entire series of cash flows to zero.*

7.1.2 Return on Invested Capital

Investment projects can be viewed as analogous to bank loans. We will now introduce the concept of rate of return based on the return on invested capital in terms of a project investment. A project's return is referred to as the internal rate of return (IRR), or the **yield** promised by an **investment project** over its **useful life**.

Definition 3— *The internal rate of return is the interest rate charged on the unrecovered project balance of the investment such that, when the project terminates, the unrecovered project balance is zero.*

Suppose a company invests $10,000 in a computer with a three-year useful life and equivalent annual labor savings of $4,021. Here, we may view the investing firm as the lender and the project as the borrower. The cash flow transaction between them would be identical to the amortized loan transaction described under Definition 1:

Year	Beginning Project Balance	Return on Invested Capital (10%)	Cash Generated from Project	Project Balance at End of Year
0	−$10,000	$0	$0	−$10,000
1	−$10,000	−$1,000	$4,021	−$6,979
2	−$6,979	−$698	$4,021	−$3,656
3	−$3,656	−$366	$4,021	$0

In our project-balance calculation, we see that 10% is earned (or charged) on $10,000 during year one, 10% is earned on $6,979 during year two, and 10% is earned on $3,656 during year three. This information indicates that the firm earns a 10% rate of return on funds that remain *internally* invested in the project. Since it is a return *internal* to the project, we refer to it as the **internal rate of return**, or IRR. This means that the computer project under consideration brings in enough cash to

---

[4]You will always have N rates of return. The issue is whether these rates are real or imaginary numbers. If they are real, the question of whether they are in the (−100%, ∞) interval should be asked. A **negative rate of return** implies that you never recover your initial investment.

pay for itself in three years and to provide the firm with a return of 10% on its invested capital. To put it differently, if the computer is financed with funds costing 10% annually, the cash generated by the investment will be exactly sufficient to repay the principal and the annual interest charge on the fund in three years.

Notice also that only one cash outflow occurs in year 0, and the present worth of this outflow is simply $10,000. There are three equal receipts, and the present worth of these inflows is $4,021(P/A, 10\%, 3) = \$10,000$. Since $\text{PW} = \text{PW}_{\text{Inflow}} - \text{PW}_{\text{Outflow}} = \$10,000 - \$10,000 = 0$, 10% also satisfies Definition 2 for rate of return. Even though this simple example implies that i^* coincides with IRR, only Definitions 1 and 3 correctly describe the true meaning of internal rate of return. As we will see later, if the cash expenditures of an investment are not restricted to the initial period, several break-even interest rates (i^*s) may exist that satisfy Eq. (7.1). However, there may not be a rate of return *internal* to the project.

7-2 Methods for Finding Rate of Return

We may find i^* by several procedures, each of which has its advantages and disadvantages. To facilitate the process of finding the rate of return for an investment project, we will first classify the types of investment cash flow.

7.2.1 Simple versus Nonsimple Investments

We can classify an investment project by counting the number of sign changes in its net cash flow sequence. A change from either "−" to "+" or "+" to "−" is counted as one sign change. (We ignore a zero cash flow.) We can then establish the following categories:

- A **simple (or conventional) investment** is an investment in which the initial cash flows are negative and only one sign change occurs in the net cash flow series. If the initial flows are positive and only one sign change occurs in the subsequent net cash flows, the flows are referred to as **simple-borrowing** cash flows.

- A **nonsimple (or nonconventional) investment** is an investment in which more than one sign change occurs in the cash flow series.

Multiple i^*'s, as we will see later, occur only in nonsimple investments. If there is no sign change in the entire cash flow series, no rate of return exists. The different types of investment possibilities may be illustrated as follows:

Investment Type	Cash Flow Sign at Period						The Number of Sign Changes
	0	1	2	3	4	5	
Simple	−	+	+	+	+	+	1
Simple	−	−	+	+	0	+	1
Nonsimple	−	+	−	+	+	−	4
Nonsimple	−	+	+	−	0	+	2

Example 7.1 Investment Classification

Classify the following three cash flow series as either simple or nonsimple investments:

Period	Net Cash Flow		
n	Project A	Project B	Project C
0	−$1,000	−$1,000	$1,000
1	−$500	$3,900	−$450
2	$800	−$5,030	−$450
3	$1,500	$2,145	−$450
4	$2,000		

SOLUTION

Given: Cash flow sequences provided in the foregoing table.

Find: Classify the sequences as either simple or nonsimple investments.

- Project *A* represents many common simple investments. This type of investment reveals the PW profile shown in Figure 7.1(a). The curve crosses the *i*-axis only once.

Period (N)	Project A	Project B	Project C
0	−$1,000	−$1,000	+$1,000
1	−$500	$3,900	−$450
2	$800	−$5,030	−$450
3	$1,500	$2,145	−$450
4	$2,000		

Project *A* is a simple investment.
Project *B* is a nonsimple investment.
Project *C* is a simple-borrowing cash flow.

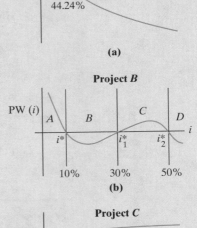

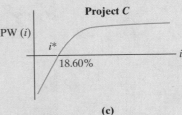

FIGURE 7.1 Classification of investments

- Project B represents a nonsimple investment. The PW profile for this investment has the shape shown in Figure 7.1(b). The i-axis is crossed at 10%, 30%, and 50%.
- Project C represents neither a simple nor a nonsimple investment, even though only one sign change occurs in the cash flow sequence. Since the first cash flow is positive, this flow is a **simple-borrowing** cash flow, not an investment flow. The PW profile for this type of investment looks like the one in Figure 7.1(c).

7.2.2 Computational Methods

Once we identify the type of an investment cash flow, there are several ways to determine its rate of return. We will discuss some of the most practical methods here. They are as follows:

- direct-solution method,
- trial-and-error method, and
- Excel method.

Direct-Solution Method   For the very special case of a project with only a two-flow transaction (an investment followed by a single future payment) or a project with a service life of two years of return, we can seek a direct analytical solution for determining the rate of return. These two cases are examined in Example 7.2.

Example 7.2 Finding $i*$ by Direct Solution: Two Flows and Two Periods

Consider two investment projects with the following cash flow transactions: Compute the rate of return for each project.

n	Project 1	Project 2
0	−$1,000	−$2,000
1	$0	$1,300
2	$0	$1,500
3	$0	
4	$1,500	

SOLUTION

Given: Cash flows for two projects.
Find: $i*$ for each project.

Project 1: Solving for $i*$ in PW$(i*) = 0$ is identical to solving for $i*$ in FW $(i*) = 0$, because FW equals PW times a constant. We could use either method here, but we choose FW $(i*) = 0$. Using the single-payment future-worth relationship, we obtain

$$FW(i) = -\$1,000(F/P, i, 4) + \$1,500 = 0.$$

Setting FW $(i) = 0$, we obtain

$$\$1,500 = \$1,000(F/P, i, 4) = \$1,000(1 + i)^4,$$

or

$$1.5 = (1 + i)^4.$$

Solving for i yields

$$i* = \sqrt[4]{1.5} - 1 = 0.1067, \text{ or } 10.67\%.$$

Project 2: We may write the PW expression for this project as follows:

$$PW(i) = -\$2,000 + \frac{\$1,300}{(1 + i)} + \frac{\$1,500}{(1 + i)^2} = 0.$$

Let

$$X = \frac{1}{(1 + i)}.$$

We may then rewrite the PW(i) expression as a function of X and set it equal to zero as follows:

$$PW(i) = -\$2,000 + \$1,300X + \$1,500X^2 = 0.$$

This expression is a quadratic equation that has the following solution:[5]

$$X = \frac{-1,300 \pm \sqrt{1,300^2 - 4(1,500)(-2,000)}}{2(1,500)}$$

$$= \frac{-1,300 \pm 3,700}{3,000}$$

$$= 0.8 \text{ or } -1.667.$$

Replacing the X values and solving for i gives us

$$0.8 = \frac{1}{(1 + i)} \rightarrow i = 25\%$$

and

$$-1.667 = \frac{1}{(1 + i)} \rightarrow i = -160\%.$$

Since an interest rate less than -100% has no economic significance, we find that the project's $i*$ is 25%.

Comments: In both projects, one sign change occurred in the net cash flow series, so we expected a unique $i*$. Also, these projects had very simple cash flows. When cash flows are more complex, generally we must use a trial-and-error method or a computer program to find $i*$.

---

[5]Given $aX^2 + bX + c = 0$, the solution of the quadratic equation is $X = \dfrac{-b \pm \sqrt{b^2 - 4ac}}{2a}$.

Trial-and-Error Method   The first step in the trial-and-error method is to make an estimated guess at the value of i^*.[6] For a simple investment, we use the guessed interest rate to compute the present worth of net cash flows and observe whether the result is positive, negative, or zero:

- **Case 1:** $PW(i) < 0$. Since we are aiming for a value of i that makes $PW(i) = 0$, we must raise the present worth of the cash flow. To do this, we lower the interest rate and repeat the process.
- **Case 2:** $PW(i) > 0$. We raise the interest rate in order to lower $PW(i)$. The process is continued until $PW(i)$ is approximately equal to zero.

Whenever we reach the point where $PW(i)$ is bounded by one negative value and one positive value, we use **linear interpolation** to approximate i^*. This process is somewhat tedious and inefficient. (The trial-and-error method does not work for nonsimple investments in which the PW function is not, in general, a monotonically decreasing function of interest rate.)

Example 7.3 Finding i^* by Trial and Error

Agdist Corporation distributes agricultural equipment. The board of directors is considering a proposal to establish a facility to manufacture an electronically controlled "intelligent" crop sprayer invented by a professor at a local university. This crop-sprayer project would require an investment of $10 million in assets and would produce an annual after-tax net benefit of $1.8 million over a service life of eight years. All costs and benefits are included in these figures. When the project terminates, the net proceeds from the sale of the assets would be $1 million (Figure 7.2). Compute the rate of return of this project.

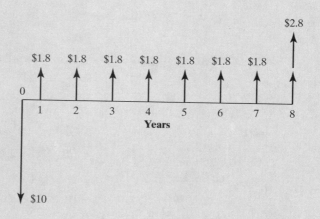

FIGURE 7.2 Cash flow diagram for a simple investment. All dollar amounts are in millions of dollars

---

[6]As we will see later in this chapter, the ultimate objective of finding i^* is to compare it with the MARR. Therefore, it is a good idea to use the MARR as the initial guess value.

SOLUTION

Given: $I = \$10$ million, $A = \$1.8$ million, $S = \$1$ million, and $N = 8$ years.
Find: i^*.

We start with a guessed interest rate of 8%. The present worth of the cash flows in millions of dollars is

$$PW(8\%) = -\$10 + \$1.8(P/A, 8\%, 8) + \$1(P/F, 8\%, 8) = \$0.88.$$

Since this present worth is positive, we must raise the interest rate in order to bring this value toward zero. When we use an interest rate of 12%, we find that

$$PW(12\%) = -\$10 + \$1.8(P/A, 12\%, 8) + \$1(P/F, 12\%, 8) = -\$0.65.$$

We have bracketed the solution. $PW(i)$ will be zero at i somewhere between 8% and 12%. Using straight-line interpolation, we approximate that

$$i^* \cong 8\% + (12\% - 8\%)\left[\frac{0.88 - 0}{0.88 - (-0.65)}\right]$$

$$= 8\% + 4\%(0.5752)$$

$$= 10.30\%.$$

Now we will check to see how close this value is to the precise value of i^*. If we compute the present worth at this interpolated value, we obtain

$$PW(10.30\%) = -\$10 + \$1.8(P/A, 10.30\%, 8) + \$1(P/A, 10.30\%, 8)$$
$$= -\$0.045.$$

As this result is not zero, we may recompute i^* at a lower interest rate, say 10%:

$$PW(10\%) = -\$10 + \$1.8(P/A, 10\%, 8) + \$1(P/A, 10\%, 8)$$
$$= \$0.069.$$

With another round of linear interpolation, we approximate that

$$i^* \cong 10\% + (10.30\% - 10\%)\left[\frac{0.069 - 0}{0.069 - (-0.045)}\right]$$

$$= 10\% + 0.30\%(0.6053)$$

$$= 10.18\%.$$

At this interest rate,

$$PW(10.18\%) = -\$10 + \$1.8(P/A, 10.18\%, 8) + \$1(P/F, 10.18\%, 8)$$
$$= \$0.0007,$$

which is practically zero, so we may stop here. In fact, there is no need to be more precise about these interpolations, because the final result can be no more accurate than the basic data, which ordinarily are only rough estimates. Incidentally, computing the i^* for this problem on a computer gives us 10.1819%.

Rate-of-Return Calculation using Excel Fortunately, we don't need to do laborious manual calculations to find i^*. Many financial calculators have built-in functions for calculating i^*. It is worth noting that many spreadsheet packages have i^* functions that solve Eq. (7.1) very rapidly. These functions are usually used by entering the cash flows via a computer keyboard or by reading a cash flow data file. For example, Microsoft Excel has an IRR financial function that analyzes investment cash flows, namely, =IRR(range, guess). We will demonstrate the IRR function with an example involving an investment in a corporate bond.

Example 7.4 Yield to Maturity

Consider buying a $1,000-denomination Atlanta Gas & Light Company (AGLC) bond at the market price of $996.25. The interest will be paid semiannually at an interest rate per payment period of 4.8125%. Twenty interest payments over 10 years are required. We show the resulting cash flow to the investor in Figure 7.3. Find the return on this bond investment (or yield to maturity).

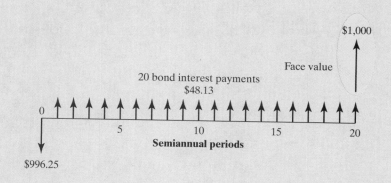

FIGURE 7.3 A typical cash flow transaction associated with an investment in the AGLC corporate bond

Discussion: You can trade bonds on the market just like stocks. Once you have purchased a bond, you can keep the bond until it reaches maturity or sell at any interest period before selling it. You can purchase or sell bonds at prices other than face value, depending on the economic environment, as bond prices change over time because of the risk of nonpayment of interest or face value, supply and demand, and the outlook for economic conditions. These factors affect the **yield to maturity** (or **return on investment**). The **yield to maturity** represents the actual interest earned from a bond over the holding period. In other words, the yield to maturity on a bond is the interest rate that establishes the equivalence between all future interest and face-value receipts and the market price of the bond. Some specific bond terms to understand are summarized as follows:

- **Face value:** The AGLC bond has a face value of $1,000.
- **Maturity date:** The AGLC bonds, which were issued on January 30, 1997, will mature on January 31, 2007; thus, they have a 10-year maturity at time of issue.
- **Coupon rate:** The AGLC bond's coupon rate is 9.625%, and interest is payable semiannually. For example, the AGLC bonds have a $1,000 face value, and they pay $96.25 in simple interest ($9\frac{5}{8}$%) each year (or $48.13 every six months).
- **Discount bond:** The AGLC bonds are offered at less than the par value at 99.625%, or 0.375% discount. For example, an AGLC bond with a face value of $1,000 can be purchased for just $996.25, which is known as the **market price** (or value) of the bond.

SOLUTION

Given: Initial purchase price = $996.25, coupon rate = 9.625% per year paid semiannually, and 10-year maturity with a face value of $1,000.

Find: Yield to maturity.

We find the yield to maturity by determining the interest rate that makes the present worth of the receipts equal to the market price of the bond:

$$\$996.25 = \$48.13(P/A, i, 20) + \$1000(P/F, i, 20).$$

The value of i that makes the present worth of the receipts equal to $996.25 lies between 4.5% and 5%. Solving for i by interpolation yields $i = 4.84$%.

Using the IRR function in Excel, we may easily calculate the yield to maturity as shown in Table 7.1. The initial guess value used in this calculation is 4%, with the cell range of B10:B30.

Comments: Note that this result is a 4.84% yield to maturity per semiannual period. The nominal (annual) yield is 2(4.84) = 9.68%, compounded semiannually. When compared with the coupon rate of $9\frac{5}{8}$% (or 9.625%), purchasing the bond with the price discounted at 0.375% brings about an additional 0.055% yield. The effective annual interest rate is then

$$i_a = (1 + 0.0484)^2 - 1 = 9.91\%.$$

This 9.91% represents the **effective annual yield** to maturity on the bond. Notice that when you purchase a bond at face value and sell at face value, the yield to maturity will be the *same* as the coupon rate of the bond.

TABLE 7.1 Yield-to-Maturity Calculation for the AGLC Bond

	A	B	C
1	**Example 7.4: Yield-to-Maturity Calculation**		
2			
3	Market price		$ 996.25
4	Maturity (semiannual periods)		20
5	Face value		$ 1,000
6	Coupon rate (per semiannual period)		4.8125%
7	Yield to maturity		4.8422%
8			
9	Period	Cash Flow	
10	0	−996.25	
11	1	48.125	
12	2	48.125	
13	3	48.125	
14	4	48.125	
15	5	48.125	
16	6	48.125	
17	7	48.125	
18	8	48.125	
19	9	48.125	
20	10	48.125	
21	11	48.125	
22	12	48.125	
23	13	48.125	
24	14	48.125	
25	15	48.125	
26	16	48.125	
27	17	48.125	
28	18	48.125	
29	18	48.125	
30	20	1048.125	

Row 7 annotation: ← `=IRR(B10:B30,4%)`

Cell range:
B10–B30

7-3   Internal-Rate-of-Return Criterion

Now that we have classified investment projects and learned methods to determine the i^* value for a given project's cash flows, our objective is to develop an accept-or-reject decision rule that gives results consistent with those obtained from PW analysis.

7.3.1   Relationship to the PW Analysis

As we already observed in Chapter 5, PW analysis is dependent on the rate of interest used for the PW computation. A different rate may change a project from being considered acceptable to being considered unacceptable, or it may change the ranking of several projects.

Consider again the PW profile as drawn for the simple project in Figure 7.1(a). For interest rates below i^*, this project should be accepted, as PW > 0; for interest rates above i^*, it should be rejected.

On the other hand, for certain nonsimple projects, the PW may look like the one shown in Figure 7.1(b). Use of PW analysis would lead you to accept the projects in regions A and C, but reject those in regions B and D. Of course, this result goes against intuition: a higher interest rate would change an unacceptable project into an acceptable one. The situation graphed in Figure 7.1(b) is one of the cases of multiple i^*'s mentioned regarding Definition 2.

Therefore, for the simple investment situation in Figure 7.1(a), the i^* can serve as an appropriate index for either accepting or rejecting the investment. However, for the nonsimple investment in Figure 7.1(b), it is not clear which i^* to use in order to make an accept-or-reject decision. Therefore, the i^* value fails to provide an appropriate measure of profitability for an investment project with multiple rates of return.

7.3.2   Decision Rule for Simple Investments

Suppose we have a simple investment. Why are we interested in finding the particular interest rate that equates a project's cost with the present worth of its receipts? Again, we may easily answer this question by examining Figure 7.1(a). In this figure, we notice two important characteristics of the PW profile. First, as we compute the project's PW(i) at a varying interest rate i, we see that the PW is positive for $i < i^*$, indicating that the project would be acceptable under the PW analysis for those values of i. Second, the PW is negative for $i > i^*$, indicating that the project is unacceptable for those values of i. Therefore, the i^* serves as a **benchmark** interest rate. By knowing this benchmark rate, we will be able to make an accept-or-reject decision consistent with the PW analysis:

- **Evaluating a Single Project:** Note that, for a simple investment, i^* is indeed the IRR of the investment. (See Section 7.1.2.) Merely knowing i^* is not enough to apply this method, however. Because firms typically wish to do better than break even (recall that at PW $= 0$, we were indifferent to the project), a minimum acceptable rate of return (MARR) is indicated by company policy, management, or the project decision maker. If the IRR exceeds this MARR, we are assured that the company will more than break even. Thus, the IRR becomes a useful gauge against which to judge project acceptability, and the decision rule for a simple project is as follows:

If IRR > MARR, accept the project.
If IRR = MARR, remain indifferent.
If IRR < MARR, reject the project.

- **Evaluating Mutually Exclusive Projects:** Note that the foregoing decision rule is designed to be applied for a single-project evaluation. When we have to compare mutually exclusive investment projects, we need to apply the **incremental analysis approach**, as we shall see in Section 7.4.2. For now, we will consider the single-project evaluation.

Example 7.5 Investment Decision for a Simple Investment

Merco, Inc., a machinery builder in Louisville, Kentucky, is considering making an investment of $1,250,000 in a complete structural beam-fabrication system. The increased productivity resulting from the installation of the drilling system is central to the project's justification. Merco estimates the following figures as a basis for calculating productivity:

- increased fabricated-steel production: 2,000 tons/year;
- average sales price per ton of fabricated steel: $2,566.50;
- labor rate: $10.50/hour;
- tons of steel produced in a year: 15,000;
- cost of steel per ton (2,205 lb): $1,950;
- number of workers on layout, hole making, sawing, and material handling: 17;
- additional maintenance cost: $128,500/year.

With the cost of steel at $1,950 per ton and the direct-labor cost of fabricating 1 lb at 10 cents, the cost of producing a ton of fabricated steel is about $2,170.50. With a selling price of $2,566.50 per ton, the resulting contribution to overhead and profit becomes $396 per ton. Assuming that Merco will be able to sustain an increased production of 2,000 tons per year by purchasing the system, the projected additional contribution has been estimated to be 2,000 tons ×$396 = $792,000.

Since the drilling system has the capacity to fabricate the full range of structural steel, two workers can run the system, one on the saw and the other on the drill. A third operator is required to operate a crane for loading and unloading materials. Merco estimates that to do the equivalent work of these three workers with a conventional manufacturing system would require, on average, an additional 14 people for center punching, hole making with a radial or magnetic drill, and material handling. This factor translates into a labor savings in the amount of $294,000 per year (14 × $10.50 × 40 hours/week × 50 weeks/year). The system can last for 15 years, with an estimated after-tax salvage value of $80,000. However, after an annual deduction of $226,000 in corporate income taxes, the net investment cost as well as savings are as follows:

- project investment cost: $1,250,000;
- projected annual net savings:

$$($792,000 + $294,000) - $128,500 - $226,000 = $731,500;$$

- projected after-tax salvage value at the end of year 15: $80,000.

(a) What is the projected IRR on this investment?
(b) If Merco's MARR is known to be 18%, is this investment justifiable?

SOLUTION

Given: Projected cash flows as shown in Figure 7.4; MARR = 18%.
Find: (a) IRR and (b) whether to accept or reject the investment.

(a) Since only one sign change occurs in the net cash flow series, the project is a simple investment. This factor indicates that there will be a unique rate of return that is internal to the project:

$$PW(i) = -$1,250,000 + $731,500(P/A, i, 15) + $80,000(P/F, i, 15) = 0.$$

We could use the trial-and-error approach outlined in Section 7.2.2 (or IRR function in Excel) to find the IRR, but an on-line financial calculator such as the Cash Flow Analyzer would be a more convenient way to calculate the internal rate of return. Figure 7.5 shows that the rate of return for the project is 58.46%, which far exceeds the MARR of 18%.

(b) The IRR figure far exceeds Merco's MARR, indicating that the project is an economically attractive one and should be accepted. Merco's management believes that, over a broad base of structural products, there is no doubt that the installation of the fabrication system would result in a significant savings, even after considering some potential deviations from the estimates used in the foregoing analysis.

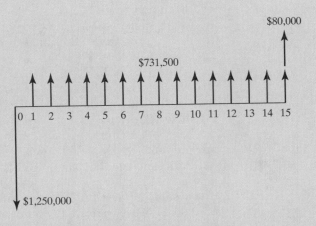

FIGURE 7.4 Cash flow diagram

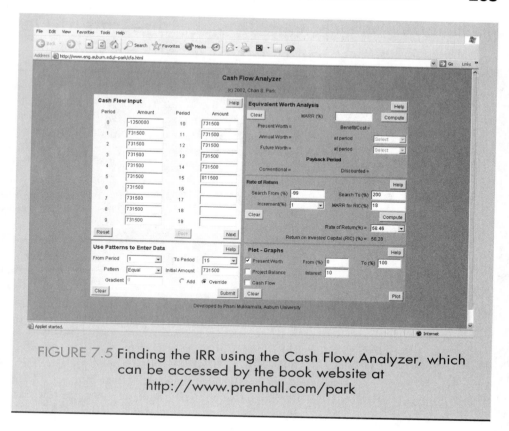

FIGURE 7.5 Finding the IRR using the Cash Flow Analyzer, which can be accessed by the book website at http://www.prenhall.com/park

7.3.3 Decision Rule for Nonsimple Investments

When applied to simple projects, the $i*$ provides an unambiguous criterion for measuring profitability. However, when multiple rates of return occur, none of them is an accurate portrayal of project acceptability or profitability. Clearly, then, we should place a high priority on discovering this situation early in our analysis of a project's cash flows. The quickest way to predict the existence of multiple $i*$'s is to generate a PW profile on the computer and check if it crosses the horizontal axis more than once.

In addition to the PW profile, there are good—although somewhat more complex—analytical methods for predicting the existence of multiple $i*$'s. Perhaps more importantly, there is a good method, which uses an **external interest rate**, for refining our analysis when we do discover that there are multiple $i*$'s. An external rate of return allows us to calculate a single true rate of return—it is discussed in Chapter 7A.

If you choose to avoid these more complex applications of rate-of-return techniques, you must be able to predict the existence of multiple $i*$'s via the PW profile and, when they occur, select an alternative method such as PW or AE analysis for determining project acceptability.

Example 7.6 Investment Decision for a Nonsimple Project

By outbidding its competitors, Trane Image Processing (TIP), a defense contractor, has received a contract worth $7,300,000 to build navy flight simulators for U.S. Navy pilot training over two years. For some defense contracts, the U.S. government makes an advance payment when the contract is signed, but in this case, the government will make two progressive payments: $4,300,000 at the end of the first year and the $3,000,000 balance at the end of the second year. The expected cash outflows required in order to produce these simulators are estimated to be $1,000,000 now, $2,000,000 during the first year, and $4,320,000 during the second year. The expected net cash flows from this project are summarized as follows:

Year	Cash Inflow	Cash Outflow	Net Cash Flow
0		$1,000,000	−$1,000,000
1	$4,300,000	$2,000,000	$2,300,000
2	$3,000,000	$4,320,000	−$1,320,000

In normal situations, TIP would not even consider a marginal project such as this one in the first place. However, hoping that TIP can establish itself as a technology leader in the field, management felt that it was worth outbidding its competitors by providing the lowest bid. Financially, what is the economic worth of outbidding the competitors for this project?

(a) Compute the values of the i^*'s for this project.

(b) Make an accept-or-reject decision, based on the results of part (a). Assume that the contractor's MARR is 15%.

SOLUTION

Given: Cash flow shown in the foregoing table; MARR = 15%.
Find: (a) i^* and (b) determine whether to accept the project.

(a) Since this project has a two-year life, we may solve the PW equation directly via the quadratic-formula method:

$$- \$1,000,000 + \frac{\$2,300,000}{(1 + i^*)} - \frac{\$1,320,000}{(1 + i^*)^2} = 0$$

If we let $X = 1/(1 + i^*)$, we can rewrite the expression as

$$-1,000,000 + 2,300,000X - 1,320,000X^2 = 0.$$

Solving for X gives $X = \frac{10}{11}$ and $\frac{10}{12}$, or $i^* = 10\%$ and 20%, respectively. As shown in Figure 7.6, the PW profile intersects the horizontal axis twice, once at 10% and again at 20%. The investment is obviously not a simple one, and thus neither 10% nor 20% represents the true internal rate of return of this government project.

(b) Since the project is a nonsimple project, we may abandon the IRR criterion for practical purposes and use the PW criterion. If we use the present-worth method at MARR = 15%, we obtain

$$PW(15\%) = -\$1,000,000 + \$2,300,000(P/F, 15\%, 1)$$
$$-\$1,320,000(P/F, 15\%, 2)$$
$$= \$1,890$$

which verifies that the project is marginally acceptable, and it is thus not as bad as initially believed.

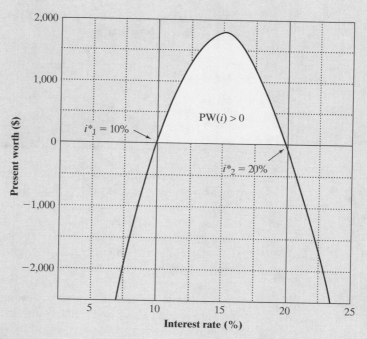

FIGURE 7.6 NPW plot for a nonsimple investment with multiple rates of return

7–4   Incremental Analysis for Comparing Mutually Exclusive Alternatives

In this section, we present the decision procedures that should be used when comparing two or more mutually exclusive projects, based on the rate-of-return measure. We will consider two situations: (1) alternatives that have the same economic service life and (2) alternatives that have unequal service lives.

7.4.1   Flaws in Project Ranking by IRR

Under PW or AE analysis, the mutually exclusive project with the highest worth figure was preferred. (This approach is known as the "total investment approach.") Unfortunately, the analogy does not carry over to IRR analysis. The project with the highest IRR may *not* be the preferred alternative. To illustrate the flaws of comparing IRRs in order to choose from mutually exclusive projects, suppose you have two mutually exclusive alternatives, each with a one-year service life. One requires an investment of $1,000 with a return of $2,000, and the other requires $5,000 with a return of $7,000. You already obtained the IRRs and PWs at MARR = 10% as follows:

Comparing Mutually Exclusive Alternatives, Based on IRR			
● Issue: Can we rank the mutually exclusive projects by the magnitude of IRR?			
n	Project A1		Project A2
0	−$1,000		−$5,000
1	$2,000		$7,000
IRR	100%	>	40%
PW(10%)	$818	<	$1,364

Assuming that you have exactly $5,000 in your investment pool to select either one, would you prefer the first project simply because you expect a higher rate of return?

We can see that project $A2$ is preferred over project $A1$ by the PW measure. On the other hand, the IRR measure gives a numerically higher rating for project $A1$. This inconsistency in ranking occurs because the PW, FW, and AE are **absolute (dollar)** measures of investment worth, whereas the IRR is a **relative (percentage)** measure and cannot be applied in the same way. That is, the IRR measure ignores the **scale** of the investment. Therefore, the answer is no; instead, you would prefer the second project with the lower rate of return, but higher PW. Either the PW or the AE measure would lead to that choice, but comparison of IRRs would rank the smaller project higher. Another approach, referred to as **incremental analysis**, is needed.

7.4.2   Incremental-Investment Analysis

In our previous example, the more costly option requires $4,000 more than the other option, while providing an incremental return of $5,000. To compare these directly, we must look at the total impact on our investment pool, using common terms. Here we look at each project's impact on an investment pool of $5,000. You must consider the following factors:

- If you decide to invest in project $A1$, you will need to withdraw only $1,000 from your investment pool. The remaining $4,000 that is not committed will continue to earn 10% interest. One year later, you will have $2,000 from the outside investment and $4,400 from the investment pool. Thus, with an investment of $5,000, in one year you will have $6,400 (or a 28% return on the $5,000). The equivalent present worth of this wealth change is PW(10%) = −$5,000 + $6,400(P/F, 10%, 1) = $818.

- If you decide to invest in project A2, you will need to withdraw $5,000 from your investment pool, leaving no money in the pool, but you will have $7,000 from your outside investment. Your total wealth changes from $5,000 to $7,000 in a year. The equivalent present worth of this wealth change is PW(10%) = −$5,000 + $7,000(P/F, 10%, 1) = $1,364.

In other words, if you decide to take the more costly option, certainly you would be interested in knowing that this additional investment can be justified at the MARR. The MARR value of 10% implies that you can always earn that rate from other investment sources (e.g., $4,400 at the end of one year for a $4,000 investment). However, in the second option, by investing the additional $4,000, you would make an additional $5,000, which is equivalent to earning at the rate of 25%. Therefore, the incremental investment can be justified. Figure 7.7 summarizes the process of performing incremental-investment analysis.

Now we can generalize the decision rule for comparing mutually exclusive projects. For a pair of mutually exclusive projects (A and B, with B defined as a more costly option), we may rewrite B as

$$B = A + (B - A).$$

In other words, B has two cash flow components: (1) the same cash flow as A and (2) the incremental component (B − A). Therefore, the only situation in which B is preferred to A is when the rate of return on the incremental component (B − A) exceeds the MARR. Therefore, for two mutually exclusive projects, rate-of-return analysis is done by computing the *internal rate of return on incremental investment* (IRR_{B-A}) between the projects. Since we want to consider increments of investment, we compute the cash flow for the difference between the projects by subtracting the cash flow for the lower investment-cost project (A) from that of the higher investment-cost project (B). Then, the decision rule is as follows, where (B − A) is an investment increment (the sign of the first cash flow should be negative).

n	Project A1	Project A2	Incremental Investment (A2 − A1)
0	−$1,000	−$5,000	−$4,000
1	$2,000	$7,000	$5,000
ROR	100%	40%	25%
PW(10%)	$818	$1,364	$546

- Assuming a MARR of 10%, you can always earn that rate fom another investment source— e.g., $4,400 at the end of one year for a $4,000 investment.

- By investing the additional $4,000 in A2, you would make an additional $5,000, which is equivalent to earning at the rate of 25%. Therefore, the incremental investment in A2 is justified.

FIGURE 7.7 Illustration of incremental analysis

If IRR_{B-A} > MARR, select B (higher first-cost alternative).

If IRR_{B-A} = MARR, select either one.

If IRR_{B-A} < MARR, select A (lower first-cost alternative).

If a "do-nothing" alternative is allowed, the smaller cost option must be profitable (its IRR must be greater than the MARR) at first. This means that you compute the rate of return for each alternative in the mutually exclusive group and then eliminate the alternatives that have IRRs less than the MARR before applying the incremental-investment analysis. It may seem odd to you how this simple rule allows us to select the right project. Example 7.7 will illustrate the incremental-investment decision rule for you.

Example 7.7 IRR on Incremental Investment: Two Alternatives

John Covington, a college student, wants to start a small-scale painting business during his off-school hours. To economize the start-up business, he decides to purchase some used painting equipment. He has two mutually exclusive options. Do most of the painting by himself by limiting his business to only residential painting jobs ($B1$) or purchase more painting equipment and hire some helpers to do both residential and commercial painting jobs ($B2$). He expects option $B2$ will have a higher equipment cost, but provide higher revenues as well ($B2$). In either case, he expects to fold the business in three years when he graduates from college.

The cash flows for the two mutually exclusive alternatives are given as follows:

n	B1	B2	B2 − B1
0	−$3,000	−$12,000	−$9,000
1	$1,350	$4,200	$2,850
2	$1,800	$6,225	$4,425
3	$1,500	$6,330	$4,830
IRR	25%	17.43%	

With the knowledge that both alternatives are revenue projects, which project would John select at MARR = 10%? (Note that both projects are profitable at 10%.)

SOLUTION

Given: Incremental cash flow between two alternatives; MARR = 10%.

Find: (a) The IRR on the increment and (b) which alternative is preferable.

(a) To choose the best project, we compute the incremental cash flow for $B2 - B1$. Then we compute the IRR on this increment of investment by solving

$$-\$9,000 + \$2,850(P/F, i, 1) + \$4,425(P/F, i, 2) + \$4,830(P/F, i, 3) = 0.$$

(b) We obtain $i^*_{B2-B1} = 15\%$, as plotted in Figure 7.8. By inspection of the incremental cash flow, we know it is a simple investment, so $\text{IRR}_{B2-B1} = i^*_{B2-B1}$. Since $\text{IRR}_{B2-B1} > \text{MARR}$, we select $B2$, which is consistent with the PW analysis. Note that, at $\text{MARR} > 25\%$, neither project would be acceptable.

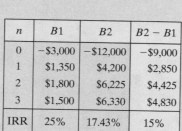

n	B1	B2	B2 − B1
0	−$3,000	−$12,000	−$9,000
1	$1,350	$4,200	$2,850
2	$1,800	$6,225	$4,425
3	$1,500	$6,330	$4,830
IRR	25%	17.43%	15%

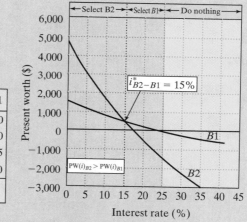

Given MARR = 10%, which project is a better choice?
Since $\text{IRR}_{B2-B1} = 15\% > 10\%$ and $\text{IRR}_{B2} > 10\%$, select $B2$.

FIGURE 7.8 PW profiles for B1 and B2

Comments: Why did we choose to look at the increment $B2 - B1$ instead of $B1 - B2$? We want the first flow of the incremental cash flow series to be negative (investment flow) so that we can calculate an IRR. By subtracting the lower initial-investment project from the higher, we guarantee that the first increment will be an investment flow. If we ignore the investment ranking, we might end up with an increment that involves borrowing cash flow and has no internal rate of return.

The next example indicates that the ranking inconsistency between PW and IRR can also occur when differences in the timing of a project's future cash flows exist, even if their initial investments are the same.

Example 7.8 IRR on Incremental Investment When Initial Flows Are Equal

Consider the following two mutually exclusive investment projects that require the same amount of investment:

n	C1	C2
0	−$9,000	−$9,000
1	$480	$5,800
2	$3,700	$3,250
3	$6,550	$2,000
4	$3,780	$1,561
IRR	18%	20%

Which project would you select, based on rate of return on incremental invest-ment, assuming that MARR = 12%? (Once again, both projects are profitable at 12%.)

SOLUTION

Given: Cash flows for two mutually exclusive alternatives as shown; MARR = 12%.
Find: (a) The IRR on incremental investment and (b) which alternative is preferable.

(a) When initial investments are equal, we progress through the cash flows until we find the first difference and then set up the increment so that this first nonzero flow is negative (i.e., an investment). Thus, we set up the incre-mental investment by taking $(C1 - C2)$:

n	C1 − C2
0	$0
1	−$5,320
2	$450
3	$4,550
4	$2,219

We next set the PW equation equal to zero, as follows:

$$-\$5,320 + \$450(P/F, i, 1) + \$4,550(P/F, i, 2) + \$2,219(P/F, i, 3) = 0.$$

(b) Solving for i yields $i^* = 14.71\%$, which is also an IRR, since the increment is a simple investment. Since $\text{IRR}_{C1-C2} = 14.71\% > \text{MARR}$, we would se-lect C1. If we used PW analysis, we would obtain $\text{PW}(12\%)_{C1} = \$1,443$ and $\text{PW}(12\%)_{C2} = \$1,185$, confirming the preference of C1 over C2.

Comments: When you have more than two mutually exclusive alternatives, they can be compared in pairs by successive examination, as shown in Figure 7.9.

n	D1	D2	D3
0	−$2,000	−$1,000	−$3,000
1	$1,500	$800	$1,500
2	$1,000	$500	$2,000
3	$800	$500	$1,000
IRR	34.37%	40.76%	24.81%

Step 1: Examine the IRR for each project in order to eliminate any project that fails to meet the MARR.

Step 2: Compare $D1$ and $D2$ in pairs:
$$\text{IRR}_{D1-D2} = 27.61\% > 15\%,$$
so select $D1$.

Step 3: Compare $D1$ and $D3$:
$$\text{IRR}_{D3-D1} = 8.8\% < 15\%,$$
so select $D1$.

Here, we conclude that $D1$ is the best alternative.

FIGURE 7.9 IRR on incremental investment: three alternatives

Example 7.9 Incremental Analysis for Cost-Only Projects

Falk Corporation is considering two types of manufacturing systems to produce its shaft couplings over six years: (1) A cellular manufacturing system (CMS) and (2) a flexible manufacturing system (FMS). The average number of pieces to be produced on either system would be 544,000 per year. Operating costs, initial investment, and salvage value for each alternative are estimated as follows:

Items	CMS Option	FMS Option
Investment	$4,500,000	$12,500,000
Total annual operating costs	$7,412,920	$5,504,100
Net salvage value	$500,000	$1,000,000

The firm's MARR is 15%. Which alternative would be a better choice, based on the IRR criterion?

Discussion: Since we can assume that both manufacturing systems would provide the same level of revenues over the analysis period, we can compare these alternatives based on cost only. (These systems are service projects.) Although we cannot compute the IRR for each option without knowing the revenue figures, we can still calculate the IRR on incremental cash flows. Since the FMS option requires a higher initial investment than that of the CMS, the incremental cash flow is the difference (FMS − CMS).

n	CMS Option	FMS Option	Incremental (FMS − CMS)
0	−$4,500,000	−$12,500,000	−$8,000,000
1	−$7,412,920	−$5,504,100	$1,908,820
2	−$7,412,920	−$5,504,100	$1,908,820
3	−$7,412,920	−$5,504,100	$1,908,820
4	−$7,412,920	−$5,504,100	$1,908,820
5	−$7,412,920	−$5,504,100	$1,908,820
6	−$7,412,920 } + $500,000 }	−$5,504,100 } + $1,000,000 }	$2,408,820

SOLUTION

Given: Cash flows shown in Figure 7.10 and $i = 15\%$ per year.
Find: Incremental cash flows, and select the better alternative, based on the IRR criterion.

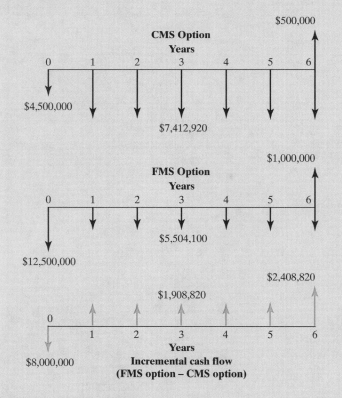

FIGURE 7.10 Cash flow diagrams for comparing cost-only projects

$$PW(i)_{\text{FMS-CMS}} = -\$8,000,000 + \$1,908,820(P/A, i, 5)$$
$$+ \$2,408,820(P/F, i, 6) = 0.$$

Solving for i by Excel yields 12.43%. Since $IRR_{\text{FMS-CMS}} = 12.43\% < 15\%$, we would select CMS. Although the FMS would provide an incremental annual savings of $1,908,820 in operating costs, the savings do not justify the incremental investment of $8,000,000.

Comments: Note that the CMS option was marginally preferred to the FMS option. However, there are dangers in relying solely on the easily quantified savings in input factors—such as labor, energy, and materials—from FMS and in not considering gains from improved manufacturing performance that are more difficult and subjective to quantify. Factors such as improved product quality, increased manufacturing flexibility (rapid response to customer demand), reduced inventory levels, and increased capacity for product innovation are frequently ignored in financial analysis because we have inadequate means for quantifying their benefits. If these intangible benefits were considered, as they ought to be, however, the FMS option could come out better than the CMS option.

7.4.3 Handling Unequal Service Lives

Our purpose is not to encourage you to use the IRR approach to compare projects with unequal lives. Rather, it is to show the correct way to compare them if the IRR approach must be used. In Chapters 5 and 6, we discussed the use of the PW and AE criteria as bases for comparing projects with unequal lives. The IRR measure can also be used to compare projects with unequal lives, as long as we can establish a common analysis period. The decision procedure is then exactly the same as for projects with equal lives. It is likely, however, that we will have a multiple-root problem, which creates a substantial computational burden. For example, suppose we apply the IRR measure to a case in which one project has a five-year life and the other project has an 8-year life, resulting in a least common multiple of 40 years. Moreover, when we determine the incremental cash flows over the analysis period, we are bound to observe many sign changes.[6]

Example 7.10 compares mutually exclusive projects where the incremental cash flow series results in several sign changes.

Example 7.10 Comparing Unequal-Service-Life Problems

Reconsider the unequal-service-life problem given in Example 5.7. Select the desired alternative based on the rate-of-return criterion.

---

[6]This factor leads to the possibility of having many i^*'s. Once again, if you desire to find the true rate of return on this incremental cash flow series, you need to refer to Chapter 7A.

SOLUTION

Given: Cash flows for unequal-service life projects as shown in Figure 5.12, MARR = 15%

Find: Which project should be selected?

Since both models have a shorter life than the required service period (five years), we need to make an explicit assumption of how the service requirement is to be met. With the leasing assumption given in Figure 5.12, the incremental cash flows would look like this:

	A	B	C	D
		A	**B**	
1	Period	Model *A*	Model *B*	Incremental Cash flow (*B–A*)
2				
3	0	−$12,500	−$15,000	−$2,500
4	1	−$5,000	−$4,000	$1,000
5	2	−$5,500	−$4,500	$1,000
6	3	−$4,000	−$5,000	−$1,000
7	4	−$11,500	−$4,000	$7,500
8	5	−$11,500	−$11,500	0
9				
10				
11			Rate of Return	45.67%

Remaining service periods are met by leasing an asset.

Note that there are three sign changes in the incremental cash flows, indicating a possibility of multiple rates of return. However, there is a unique rate of return on this incremental cash flow series, say 45.67%[7]. Since this rate of return on incremental cash flow exceeds the MARR of 15%, Model *B* is a better choice.

Summary

- **Rate of return (ROR)** is the interest rate earned on unrecovered project balances such that an investment's cash receipts make the terminal project balance equal to zero. Rate of return is an intuitively familiar and understandable measure of project profitability that many managers prefer to PW or other equivalence measures.

[7]In Chapter 7A, a project with this type of cash flow series is known as a **pure investment**.

- Mathematically, we can determine the rate of return for a given project's cash flow series by identifying an interest rate that equates the present worth (annual equivalent or future worth) of its cash flows to zero. This break-even interest rate is denoted by the symbol i^*.
- **Internal rate of return (IRR)** is another term for ROR that stresses the fact that we are concerned with the interest earned on the portion of the project that is *internally* invested, not those portions that are released by (borrowed from) the project.
- To apply rate-of-return analysis correctly, we need to classify an investment as either a simple investment or a nonsimple investment. A **simple investment** is defined as an investment in which the initial cash flows are negative and only one sign change in the net cash flow occurs, whereas a **nonsimple investment** is an investment for which more than one sign change in the cash flow series occurs. Multiple i^*'s occur only in nonsimple investments. However, not all nonsimple investments will have multiple i^*'s, as shown in Example 7.10.
- For a simple investment, the solving rate of return (i^*) is the rate of return internal to the project, so the decision rule is as follows:

    If IRR $>$ $MARR$, accept the project.
    If IRR $=$ $MARR$, remain indifferent.
    If IRR $<$ $MARR$, reject the project.

    IRR analysis yields results consistent with PW and other equivalence methods.
- For a nonsimple investment, because of the possibility of having multiple rates of return, it is recommended that the IRR analysis be abandoned and either the PW or AE analysis be used to make an accept-or-reject decision. Procedures are outlined in Chapter 7A for determining the rate of return internal to nonsimple investments. Once you find the IRR (or return on invested capital), you can use the same decision rule for simple investments.
- When properly selecting among alternative projects by IRR analysis, **incremental investment** must be used.

Problems

Note: The symbol i^* represents the interest rate that makes the present worth of the project equal to zero. The symbol IRR represents the *internal rate of return* of the investment. For a simple investment, IRR $= i^*$. For a nonsimple investment, i^* generally is not equal to IRR.

Concept of Rate of Return

7.1 Assume that you are going to buy a new car worth $22,000. You will be able to make a down payment of $7,000. The remaining $15,000 will be financed by the dealer. The dealer computes your monthly payment to be $512 for 48 months of financing. What is the dealer's rate of return on this loan transaction?

7.2 Mr. Smith wishes to sell a bond that has a face value of $1,000. The bond bears an interest rate of 8%, with bond interest payable semiannually. Four years ago, the bond was purchased at $900. At least a 9% annual return on the investment is desired. What must be the minimum selling price now for the bond in order to make the desired return on the investment?

7.3 John Whitney Payson, who purchased a painting by Vincent Van Gogh for $80,000 in 1947, sold it for $53.9 million in 1988. If Mr. Payson had invested his $80,000 in another investment vehicle (such as stock), how much interest would he need to have earned in order to accumulate the same wealth as from the painting investment? Assume for simplicity that the investment period is 40 years and that the interest is compounded annually.

Methods for Finding Rate of Return

7.4 Consider four investments with the following sequences of cash flows:

Net Cash Flow

n	Project A	Project B	Project C	Project D
0	−$18,000	−$20,000	+$34,578	−$56,500
1	$10,000	$32,000	−$18,000	−$2,500
2	$20,000	$32,000	−$18,000	−$6,459
3	$30,000	−$22,000	−$18,000	−$78,345

(a) Identify all the simple investments.
(b) Identify all the nonsimple investments.
(c) Compute the i^* for each investment.
(d) Which project has no rate of return?

7.5 Consider an investment project with the following net cash flow:

Year	Net Cash Flow
0	−$1,500
1	$X
2	$650
3	$X

What would be the value of X if the project's IRR is 10%?

(a) $425
(b) $1,045
(c) $580
(d) $635

7.6 An investor bought 100 shares of stock at a cost of $10 per share. He held the stock for 15 years and then sold it for a total of $4,000. For the first 3 years, he received no dividends. For each of the next 7 years, he received total dividends of $50 per year. For the remaining period, he received total dividends of $100 per year. What rate of return did he make on the investment?

7.7 Consider the following sets of investment projects:

			Project Cash Flow		
n	A	B	C	D	E
0	−$100	−$100		−$200	−$50
1	$60	$70	$20	$120	−$100
2	$900	$70	$10	$40	−$50
3		$40	$5	$40	$0
4		$40	−$180	−$20	$150
5			$60	$40	$150
6			$50	$30	$100
7			$40		$100
8			$30		
9			$20		
10			$10		

(a) Classify each project as either simple or nonsimple.
(b) Compute the i^* for project A, using the quadratic equation.
(c) Obtain the rate(s) of return for each project by plotting the PW as a function of interest rate.

7.8 Consider the following projects:

Net Cash Flow

n	A	B	C	D
0	−$1,000	−$1,000	−$1,700	−$1,000
1	$500	$800	$5,600	$360
2	$100	$600	$4,900	$4,675
3	$100	$500	−$3,500	$2,288
4	$1,000	$700	−$7,000	
5			−$1,400	
6			$2,100	
7			$900	

(a) Classify each project as either simple or nonsimple.

(b) Identify all positive i^*'s for each project.

(c) Plot the present worth as a function of interest rate (i) for each project.

7.9 Consider the following financial data for a project:

Initial investment	$10,000
Project life	8 years
Salvage value	$0
Annual revenue	$5,229
Annual expenses (including income taxes)	$3,000

(a) What is the i^* for this project?

(b) If the annual expenses increase at a 7% rate over the previous year's expenses, but annual revenue is unchanged, what is the new i^*?

(c) Assuming the conditions in part (b), at what annual rate will the annual revenue have to increase in order to maintain the same i^* obtained in part (a)?

7.10 Consider two investments with the following sequences of cash flows:

	Net Cash Flow	
n	**Project A**	**Project B**
0	−$25,000	−$25,000
1	$2,000	$10,000
2	$6,000	$10,000
3	$12,000	$10,000
4	$24,000	$10,000
5	$28,000	$5,000

(a) Compute the i^* for each investment.

(b) Plot the present-worth curve for each project on the same chart, and find the interest rate that makes the two projects equivalent.

Internal-Rate-of-Return Criterion

7.11 Consider an investment project with the following cash flow:

n	Cash Flow
0	−$5,000
1	$0
2	$4,840
3	$1,331

Compute the IRR for this investment. Is this project acceptable at MARR = 10%?

7.12 Consider the following project's cash flow:

n	Net Cash Flow
0	−$2,000
1	$800
2	$900
3	$X

Assume that the project's IRR is 10%.

(a) Find the value of X.

(b) Is this project acceptable at MARR = 8%?

7.13 The InterCell Company wants to participate in the upcoming World Fair. To participate, the firm needs to spend $1 million in year zero to develop a showcase. The showcase will produce a cash flow of $2.5 million at the end of year one. Then, at the end of year two, $1.54 million must be expended to restore the land on which the showcase was presented to its original condition. Therefore, the project's expected net cash flows are as follows (in thousands of dollars):

n	Net Cash Flow
0	−$1,000
1	$2,500
2	−$1,540

(a) Plot the present worth of this investment as a function of i.

(b) Compute the i^*s for this investment.

(c) Would you accept this investment at MARR = 14%?

7.14 Champion Chemical Corporation is planning to expand one of its propylene-manufacturing facilities. At $n = 0$, a piece of property costing $1.5 million must be purchased for the expanded plant site. The building, which needs to be expanded during the first year, costs $3 million. At the end of the first year, the company needs to spend about $4 million on equipment and other start-up costs. Once the plant becomes operational, it will generate revenue in the amount of $3.5 million during the first operating year. This amount will increase at an annual rate of 5% over the previous year's revenue for the next nine years. After 10 years, the sales revenue will stay constant for another three years before the operation is phased out. (The plant will have a project life of 13 years after construction.) The expected salvage value of the land would be about $2 million, the building about $1.4 million, and the equipment about $500,000. The annual operating and maintenance costs are estimated to be about 40% of the sales revenue each year. What is the IRR for this investment? If the company's MARR is 15%, determine whether this expansion is a good investment. (Assume that all figures include the effect of the income tax.)

7.15 Recent technology has made possible a computerized vending machine that can grind coffee beans and brew fresh coffee on demand. The computer also makes possible such complicated functions as changing $5 and $10 bills and tracking the age of an item and then moving the oldest stock to the front of the line, thus cutting down on spoilage. With a price tag of $4,500 for each unit, Easy Snack has estimated the cash flows in millions of dollars over the product's six-year useful life, including the initial investment, as follows:

n	Net Cash Flow
0	−$20
1	$8
2	$17
3	$19
4	$18
5	$10
6	$3

(a) If the firm's MARR is 18%, is this product worth marketing, based on the IRR criterion?

(b) If the required investment remains unchanged, but the future cash flows are expected to be 10% higher than the original estimates, how much increase in IRR do you expect?

(c) If the required investment has increased from $20 million to $22 million, but the expected future cash flows are projected to be 10% smaller than the original estimates, how much decrease in IRR do you expect?

Incremental-Investment Analysis for Comparing Mutually Exclusive Alternatives

7.16 Consider the following two investment situations:

- In 1970, when Wal-Mart Stores, Inc., went public, an investment of 100 shares cost $1,650. That investment was worth $12,283,904 after 32 years (2002), with a rate of return of around 32%.
- In 1980, if you bought 100 shares of Fidelity Mutual Funds, it would have cost $5,245. That investment would have been worth $289,556 after 22 years.

Which of the following statements is correct?

(a) If you bought only 50 shares of the Wal-Mart stocks in 1970 and kept them for 32 years, your rate of return would be 0.5 times 32%.

(b) If you bought 100 shares of Fidelity Mutual Funds in 1980, you would have made a profit at an annual rate of 30% on the funds remaining invested.

(c) If you bought 100 shares of Wal-Mart in 1970, but sold them after 10 years (assume that the Wal-Mart stocks grew at an annual rate of 32% for the first 10 years) and immediately put all the proceeds into Fidelity Mutual Funds, then after 22 years the total worth of your investment would be around $1,462,854.

(d) None of the above.

7.17 Consider two investments with the following sequences of cash flows:

| | Net Cash Flow | |
n	Project A	Project B
0	-$120,000	-$100,000
1	$20,000	$15,000
2	$25,000	$15,000
3	$120,000	$110,000

(a) Compute the IRR for each investment.

(b) At MARR = 15%, determine the acceptability of each project.

(c) If A and B are mutually exclusive projects, which project would you select, based on the rate of return on incremental investment?

7.18 With $10,000 available, you have two investment options. The first option is to buy a certificate of deposit from a bank at an interest rate of 10% annually for five years. The second choice is to purchase a bond for $10,000 and invest the bond's interest payments in the bank at an interest rate of 9%. The bond pays 10% interest annually and will mature to its face value of $10,000 in five years. Which option is better? Assume your MARR is 9% per year.

7.19 A manufacturing firm is considering the following mutually exclusive alternatives:

| | Net Cash Flow | |
n	Project A	Project B
0	-$2,000	-$3,000
1	$1,400	$2,400
2	$1,640	$2,000

Determine which project is a better choice at MARR = 15%, based on the IRR criterion.

7.20 Consider the following two mutually exclusive alternatives:

| | Net Cash Flow | |
n	Project A1	Project A2
0	-$10,000	-$12,000
1	$5,000	$6,100
2	$5,000	$6,100
3	$5,000	$6,100

(a) Determine the IRR on the incremental investment in the amount of $2,000.

(b) If the firm's MARR is 10%, which alternative is the better choice?

7.21 Consider the following two mutually exclusive investment alternatives:

| | Net Cash Flow | |
n	Project A1	Project A2
0	-$16,000	-$20,000
1	$7,500	$5,000
2	$7,500	$15,000
3	$7,500	$8,000
IRR	19.19%	17.65%

(a) Determine the IRR on the incremental investment in the amount of $4,000. (Assume that MARR = 10%.)

(b) If the firm's MARR is 10%, which alternative is the better choice?

7.22 You are considering two types of automobiles. Model A costs $18,000, and Model B costs $15,624. Although the two models are essentially the same, Model A can be sold for $9,000 after four years of use, while Model B can be sold for $6,500 after the same amount of time. Model A commands a better resale value because its styling is popular among young college students. Determine the rate of return on the incremental investment of $2,376. For what range of values of your MARR is Model A preferable?

7.23 A plant engineer is considering two types of solar water-heating systems:

Item	Model A	Model B
Initial cost	$5,000	$7,000
Annual savings	$700	$1,000
Annual maintenance	$100	$50
Expected life	20 years	20 years
Salvage value	$400	$500

The firm's MARR is 10%. Based on the IRR criterion, which system is the better choice?

7.24 Fulton National Hospital is reviewing ways of cutting the stocking costs of medical supplies. Two new stockless systems are being considered to lower the hospital's holding and handling costs. The hospital's industrial engineer has compiled the relevant financial data for each system as follows (dollar values are in millions):

Item	Current Practice	Just-in-Time System	Stockless Supply
Start-up cost	$0	$2.5	$5
Annual stock-holding cost	$3	$1.4	$0.2
Annual operating cost	$2	$1.5	$1.2
System life	8 years	8 years	8 years

The system life of eight years represents the contract period with the medical suppliers. If the hospital's MARR is 10%, which system is more economical?

7.25 Consider the following investment projects:

	Net Cash Flow		
n	Project 1	Project 2	Project 3
0	−$1,000	−$5,000	−$2,000
1	$500	$7,500	$1,500
2	$2,500	$600	$2,000

Assume that MARR = 15%.

(a) Compute the IRR for each project.
(b) If the three projects are mutually exclusive investments, which project should be selected, based on the IRR criterion?

7.26 Consider the following two investment alternatives:

	Net Cash Flow	
n	Project A	Project B
0	−$10,000	−$20,000
1	$5,500	$0
2	$5,500	$0
3	$5,500	$40,000
IRR	30%	?
PW(15%)	?	$6,300

The firm's MARR is known to be 15%.

(a) Compute the IRR of Project B.
(b) Compute the PW of Project A.
(c) Suppose that Projects A and B are mutually exclusive. Using the IRR, which project would you select?

7.27 The GeoStar Company, a leading wireless-communication-device manufacturer, is considering three cost-reduction proposals in its batch-job shop manufacturing operations. The company already calculated rates of return for the three projects, along with some incremental rates of return. A_0 denotes the do-nothing alternative. The required investments are $420,000 for A_1, $550,000 for A_2, and $720,000 for A_3. If the firm's MARR is 15%, which system should be selected?

Incremental Investment	Incremental Rate of Return
$A_1 - A_0$	18%
$A_2 - A_0$	20%
$A_3 - A_0$	25%
$A_2 - A_1$	10%
$A_3 - A_1$	18%
$A_3 - A_2$	23%

7.28 An electronic-circuit-board manufacturer is considering six mutually exclusive cost-reduction projects for its PC-board manufacturing plant. All have lives of 10 years and zero salvage values. The required investment, the estimated after-tax reduction in annual disbursements, and the gross rate of return are given for each alternative in the following table:

Proposal A_i	Required Investment	After-Tax Savings	Rate of Return
A_1	$60,000	$22,000	35.0%
A_2	$100,000	$28,200	25.2%
A_3	$110,000	$32,600	27.0%
A_4	$120,000	$33,600	25.0%
A_5	$140,000	$38,400	24.0%
A_6	$150,000	$42,200	25.1%

The rate of return on incremental investments is given for each project as follows:

Incremental Investment	Incremental Rate of Return
$A_2 - A_1$	9.0%
$A_3 - A_2$	42.8%
$A_4 - A_3$	0.0%
$A_5 - A_4$	20.2%
$A_6 - A_5$	36.3%

Which project would you select, based on the rate of return on incremental investment, if it is stated that the MARR is 15%?

7.29 Baby Doll Shop currently manufactures wooden parts for dollhouses. Its sole worker is paid $8.10 an hour and, using a handsaw, can produce a year's required production (1,600 parts) in just eight weeks of 40 hours per week. That is, the worker averages five parts per hour when working by hand. The shop is considering the purchase of a power band saw with associated fixtures in order to improve the productivity of this operation. Three models of power saw could be purchased: Model A (economy version), Model B (high-powered version), and Model C (deluxe high-end version). The major operating difference between these models is their speed of operation. The investment costs, including the required fixtures and other operating characteristics, are summarized as follows:

Category	By Hand	Model A	Model B	Model C
Production rate (parts/hour)	5	10	15	20
Labor hours required (hours/year)	320	160	107	80
Annual labor cost (@ $8.10/hour)	2,592	1,296	867	648
Annual power cost		$400	$420	$480
Initial investment		$4,000	$6,000	$7,000
Salvage value		$400	$600	$700
Service life (years)		20	20	20

Assume that MARR = 10%. Are there enough savings to make it economical to purchase any of the power band saws? Which model is most economical, based on the rate-of-return principle? (Assume that any effect of income tax has been already considered in the dollar estimates.) (*Source*: This problem is adapted with the permission of Professor Peter Jackson of Cornell University.)

Unequal Service Lives

7.30 Consider the following two mutually exclusive investment projects:

	Net Cash Flow	
n	Project A	Project B
0	−$100	−$200
1	60	120
2	50	150
3	50	
IRR	28.89%	21.65%

Assume that MARR = 15%. Which project would be selected under an infinite planning horizon with project repeatability likely, based on the IRR criterion?

7.31 Consider the following two mutually exclusive investment projects:

	Net Cash Flow	
n	**Project A1**	**Project A2**
0	−$10,000	−$15,000
1	$5,000	$20,000
2	$5,000	
3	$5,000	

(a) To use the IRR criterion, what assumption must be made in order to compare a set of mutually exclusive investments with unequal service lives?

(b) With the assumption defined in part (a), determine the range of MARR that will indicate the selection of Project A1.

Short Case Studies with Excel

7.32 Critics have charged that the commercial nuclear power industry does not consider the cost of "decommissioning" or "mothballing" a nuclear power plant when doing an economic analysis and that the analysis is therefore unduly optimistic. As an example, consider the Tennessee Valley Authority's "Bellefont" twin nuclear generating facility under construction at Scottsboro, in northern Alabama. The first cost is $1.5 billion (the present worth at the start of operations), the estimated life is 40 years, the annual operating and maintenance costs the first year are assumed to be 4.6% of the first cost in the first year and are expected to increase at the fixed rate of 0.05% of the first cost each year, and annual revenues have been estimated to be three times the annual operating and maintenance costs throughout the life of the plant. Based on this information, comment on the following statements:

(a) The criticism of overoptimism in the economic analysis caused by omitting mothballing costs is not justified, since the addition of a cost to mothball the plant equal to 50% of the first cost decreases the 10% rate of return only to approximately 9.9%.

(b) If the estimated life of the plant is more realistically taken as 25 years instead of 40 years, then the criticism is justified. By reducing the life to 25 years, the rate of return of approximately 9% without a mothballing cost drops to approximately 7.7% when a cost to mothball the plant equal to 50% of the first cost is added to the analysis.

7.33 The B&E Cooling Technology Company, a maker of automobile air conditioners, faces an impending deadline to phase out the traditional chilling technique, which uses chlorofluorocarbons (CFCs), a family of refrigerant chemicals believed to attack the earth's protective ozone layer. B&E has been pursuing other means of cooling and refrigeration. As a near-term solution, its engineers recommend a cold technology known as absorption chiller, which uses plain water as a refrigerant and semiconductors that cool down when charged with electricity. B&E is considering two options:

• Option 1: Retrofitting the plant now to adapt the absorption chiller and continuing to be a market leader in cooling technology. Because of untested technology in the large scale, it may cost more to operate the new facility while the new system is being learned.

• Option 2: Deferring the retrofitting until the federal deadline, which is three years away. With expected improvement in cooling technology and technical know-how, the retrofitting cost will be cheaper, but there will be tough market competition, and revenues would be less than that in Option 1.

The financial data for the two options are as follows:

Item	Option 1	Option 2
Investment timing	Now	3 years from now
Initial investment	$6 million	$5 million
System life	8 years	8 years
Salvage value	$1 million	$2 million
Annual revenue	$15 million	$11 million
Annual O&M costs	$6 million	$7 million

(a) What assumptions must be made in order to compare these two options?

(b) If B&E's MARR is 15%, which option is the better choice, based on the IRR criterion?

7.34 An oil company is considering changing the size of a small pump that currently is operational in an oil field. If the current pump is kept, it will extract 50% of the known crude oil reserve in the first year of operation and the remaining 50% in the second year. A pump larger than the current pump will cost $1.6 million, but it will extract 100% of the known reserve in the first year. The total oil revenues over the two years is the same for both pumps: $20 million. The advantage of the large pump is that it allows 50% of the revenues to be realized a year earlier than the small pump. The two options are summarized as follows:

Item	Current Pump	Larger Pump
Investment, year 0	$0	$1.6 million
Revenue, year 1	$10 million	$20 million
Revenue, year 2	$10 million	$0

If the firm's MARR is known to be 20%, what do you recommend, based on the IRR criterion?

7.35 You have been asked by the president of your company to evaluate the proposed acquisition of a new injection-molding machine for the firm's manufacturing plant. Two types of injection-molding machines have been identified with the following estimated cash flows:

	Net Cash Flow	
n	**Machine A**	**Machine B**
0	−$30,000	−$40,000
1	$20,000	$43,000
2	$18,200	$5,000
IRR	18.1%	18.1%

You return to your office, quickly retrieve your old engineering economics text, and then begin to smile: Aha—this is a classic rate-of-return problem! Now, using a calculator, you find out that both machines have about the same rate of return: 18.1%. This rate-of-return figure seems to be high enough for project justification, but you recall that the ultimate justification should be done in reference to the firm's MARR. You call the accounting department to find out the current MARR the firm should use for project justification. "Oh boy, I wish I could tell you, but my boss will be back next week, and he can tell you what to use," says the accounting clerk. A fellow engineer approaches you and says, "I couldn't help overhearing you talking to the clerk. I think I can help you. You see, both machines have the same IRR, and on top of that, machine A requires less investment, but returns more cash flows ($ −$30,000 + $20,000 + $18,200 = $8,200, and −$40,000 + $43,000 + $5,000 = $8,000$); thus, machine A dominates machine B. For this type of decision problem, you don't need to know a MARR!"

(a) Comment on your fellow engineer's statement.

(b) At what range of MARR would you recommend the selection of machine B?

.7A

Resolution of Multiple Rates of Return

To comprehend the nature of multiple i^*s, we need to understand the investment situation represented by any cash flow. The net-investment test will indicate whether the i^* computed represents the true rate of return earned on the money invested in a project while the money is actually in the project. As we shall see, the phenomenon of multiple i^*'s occurs only when the net-investment test fails. When multiple positive rates of return for a cash flow are found, in general, none is suitable as a measure of project profitability, and we must proceed to the next analysis step: introducing an external rate of return.

7A-1 Net-Investment Test

A project is said to be a **net investment** when the project balances computed at the project's $i*$ values, $PB(i*)_n$, are all less than or equal to zero throughout the life of the investment, with $A_0 < 0$. The investment is *net* in the sense that the firm does not overdraw on its return at any point and hence is *not indebted* to the project. This type of project is called a **pure investment**. [On the other hand, **pure borrowing** is defined as the situation where $PB(i*)_n$ values are all positive or zero throughout the life of the loan, with $A_0 > 0$.] *Simple investments will always be pure investments.* Therefore, if a nonsimple project passes the net-investment test (i.e., it is a pure investment), then the accept-or-reject decision rule will be the same as in the simple-investment case given in Section 7.3.2.

If any of the project balances calculated at the project's $i*$ is positive, the project is not a pure investment. A positive project balance indicates that, at some time during the project life, the firm acts as a borrower $[PB(i*)_n > 0]$ rather than an investor $[PB(i*)_n < 0]$ in the project. This type of investment is called a **mixed investment**.

Example 7A.1 Pure versus Mixed Investments

Consider the following four investment projects with known $i*$ values:

n	A	B	C	D
0	−$1,000	−$1,000	−$1,000	−$1,000
1	−$1,000	$1,600	$500	$3,900
2	$2,000	−$300	−$500	−$5,030
3	$1,500	−$200	$2,000	$2,145
$i*$	33.64%	21.95%	29.95%	(10%, 30%, 50%)

Determine which projects are pure investments.

SOLUTION

Given: Four projects with cash flows and $i*$s as shown in the foregoing table.

Find: Which projects are pure investments.

We will first compute the project balances at the projects' respective i^*s. If multiple rates of return exist, we may use the largest value of i^* greater than zero.[1]

Project A:

$$PB(33.64\%)_0 = -\$1,000;$$
$$PB(33.64\%)_1 = -\$1,000(1 + 0.3364) + (-\$1,000) = -\$2,336.40;$$
$$PB(33.64\%)_2 = -\$2,336.40(1 + 0.3364) + \$2,000 = -\$1,122.36;$$
$$PB(33.64\%)_3 = -\$1,122.36(1 + 0.3364) + \$1,500 = 0.$$

$(-, -, -, 0)$: Passes the net-investment test (pure investment).

Project B:

$$PB(21.95\%)_0 = -\$1,000;$$
$$PB(21.95\%)_1 = -\$1,000(1 + 0.2195) + \$1,600 = \$380.50;$$
$$PB(21.95\%)_2 = +\$380.50(1 + 0.2195) - \$300 = \$164.02;$$
$$PB(21.95\%)_3 = +\$164.02(1 + 0.2195) - \$200 = 0.$$

$(-, +, +, 0)$: Fails the net-investment test (mixed investment).

Project C:

$$PB(29.95\%)_0 = -\$1,000;$$
$$PB(29.95\%)_1 = -\$1,000(1 + 0.2995) + \$500 = -\$799.50;$$
$$PB(29.95\%)_2 = -\$799.50(1 + 0.2995) - \$500 = -\$1,538.95;$$
$$PB(29.95\%)_3 = -\$1,538.95(1 + 0.2995) + \$2,000 = 0.$$

$(-, -, -, 0)$: Passes the net-investment test (pure investment).

Project D:

There are three rates of return. We can use any of them for the net-investment test. We use the third rate given, 50%, as follows:

$$PB(50\%)_0 = -\$1,000;$$
$$PB(50\%)_1 = -\$1,000(1 + 0.50) + \$3,900 = \$2400;$$
$$PB(50\%)_2 = +\$2,400(1 + 0.50) - \$5,030 = -\$1,430;$$
$$PB(50\%)_3 = -\$1,430(1 + 0.50) + \$2,145 = 0.$$

$(-, +, -, 0)$: Fails the net-investment test (mixed investment).

---

[1] In fact, it does not matter which rate we use in applying the net-investment test. If one value passes the net-investment test, they will all pass. If one value fails, they will all fail.

Comments: As shown in Figure 7A.1, Projects *A* and *C* are the only pure investments. Project *B* demonstrates that the existence of a unique i^* is a necessary, but not sufficient, condition for a pure investment.

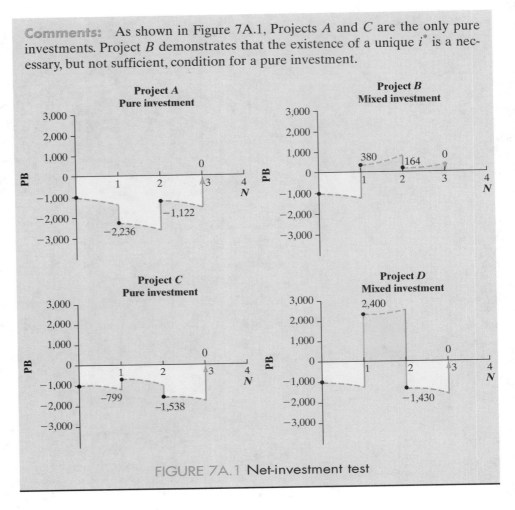

FIGURE 7A.1 Net-investment test

7A-2 The Need for an External Interest Rate

Even for a nonsimple investment, in which there is only one positive rate of return, the project may fail the net-investment test, as demonstrated by Project *B* in Example 7A.1. In this case, the unique i^* still may not be a true indicator of the project's profitability. That is, when we calculate the project balance at an i^* for mixed investments, we notice an important point. Cash borrowed (released) from the project is assumed to earn the same interest rate through external investment as money that remains internally invested. In other words, in solving for a cash flow for an unknown interest rate, it is assumed that money released from a project can be reinvested to yield a rate of return equal to that received from the project. In fact, we have been making this assumption whether a cash flow produces a unique positive i^* or not. Note that money is borrowed from the project only when $PB(i^*) > 0$, and the magnitude of the borrowed amount is the project balance. When $PB(i^*) < 0$, no money is borrowed, even though the cash flow may be positive at that time.

In reality, it is not always possible for cash borrowed (released) from a project to be reinvested to yield a rate of return equal to that received from the project. Instead, it is likely that the rate of return available on a capital investment in the business is much different—usually higher—from the rate of return available on other external investments. Thus, it may be necessary to compute project balances for a project's cash flow at two rates of interest—one on the internal investment and one on the external investments. As we will see later, by separating the interest rates, we can measure the **true rate of return** of any internal portion of an investment project.

Because the net-investment test is the only way to accurately predict project borrowing (i.e., external investment), its significance now becomes clear. In order to calculate accurately a project's true IRR, we should always test a solution by the net-investment test and, if the test fails, take the further analytical step of introducing an external interest rate. Even the presence of a unique positive i^* is a necessary, but not sufficient, condition to predict net investment, so if we find a unique value we should still subject the project to the net-investment test.

7A-3 Calculation of Return on Invested Capital for Mixed Investments

A failed net-investment test indicates a combination of internal and external investment. When this combination exists, we must calculate a rate of return on the portion of capital that remains invested internally. This rate is defined as the **true IRR** for the mixed investment and is commonly known as the **return on invested capital (RIC)**.

How do we determine the true IRR of a mixed investment? Insofar as a project is not a net investment, the money from one or more periods when the project has a net outflow of money (positive project balance) must later be returned to the project. This money can be put into the firm's investment pool until such time when it is needed in the project. The interest rate of this investment pool is the interest rate at which the money can, in fact, be invested outside the project.

Recall that the PW method assumed that the interest rate charged to any funds withdrawn from a firm's investment pool would be equal to the MARR. In this book, we will use the MARR as an established external interest rate (i.e., the rate earned by money invested outside of the project). We can then compute the true IRR, or RIC, as a function of the MARR by finding the value of IRR that will make the terminal project balance equal to zero. (This definition implies that the firm wants to fully recover any investment made in the project and pays off any borrowed funds at the end of the project life.) This way of computing rate of return is an accurate measure of the profitability of the project represented by the cash flow. The following procedure outlines the steps for determining the IRR for a mixed investment:

Step 1: Identify the MARR (or external interest rate).

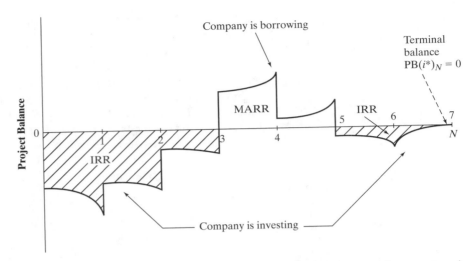

FIGURE 7A.2 Computational logic to find the true IRR for a mixed investment

Step 2: Calculate $PB(i, MARR)_n$, or simply PB_n, according to the following rule:

$$PB(i, MARR)_0 = A_0.$$

$$PB(i, MARR)_1 = \begin{cases} PB_0(1 + i) + A_1, \text{ if } PB_0 < 0. \\ PB_0(1 + MARR) + A_1, \text{ if } PB_0 > 0. \end{cases}$$

$$\vdots$$

$$PB(i, MARR)_n = \begin{cases} PB_{n-1}(1 + i) + A_n, \text{ if } PB_{n-1} < 0. \\ PB_{n-1}(1 + MARR) + A_n, \text{ if } PB_{n-1} > 0. \end{cases}$$

(As defined in the text, A_n stands for the net cash flow at the end of period n. Note also that the terminal project balance must be zero.)

Step 3: Determine the value of i by solving the terminal-project-balance equation:

$$PB(i, MARR)_N = 0.$$

That interest rate is the IRR for the mixed investment.

Using the MARR as an external interest rate, we may accept a project if the IRR exceeds MARR and should reject the project otherwise. Figure 7A.2 summarizes the IRR computation for a mixed investment.

Example 7A.2 IRR for a Nonsimple Project: Mixed Investment

Reconsider the defense contractor's flight-simulator project in Example 7.6. The project was a nonsimple and mixed investment. To simplify the decision-making process, we abandoned the IRR criterion and used the PW to make an accept-or-reject decision. Apply the procedures outlined in this chapter to find the true IRR, or return on invested capital, of this mixed investment:

(a) Compute the IRR (RIC) for this project, assuming MARR = 15%.

(b) Make an accept-or-reject decision, based on the results in part (a).

SOLUTION

Given: Cash flow shown in Example 7.6; MARR = 15%.

Find: (a) IRR and (b) determine whether to accept the project.

(a) As calculated in Example 7.6, the project has multiple rates of return (10% and 20%). This project is obviously not a net investment, as shown in the following table:

Net-Investment Test Using $i^* = 20\%$			
n	**0**	**1**	**2**
Beginning balance	$0	−$1,000	$1,100
Return on investment	$0	−$200	$220
Payment	−$1,000	$2,300	−$1,320
Ending balance	−$1,000	$1,100	$0
(Unit: $1,000)			

Because the net-investment test indicates external as well as internal investment, neither 10% nor 20% represents the true internal rate of return of this project. Since the project is a mixed investment, we need to find the true IRR by applying the steps shown previously.

At $n = 0$, there is a net investment to the firm so that the project-balance expression becomes

$$\text{PB}(i, 15\%)_0 = -\$1,000,000.$$

The net investment of $1,000,000 that remains invested internally grows at i for the next period. With the receipt of $2,300,000 in year 1, the project balance becomes

$$\text{PB}(i, 15\%)_1 = -\$1,000,000(1 + i) + \$2,300,000$$
$$= \$1,300,000 - \$1,000,000i$$
$$= \$1,000,000(1.3 - i).$$

At this point, we do not know whether $\text{PB}(i, 15\%)_1$ is positive or negative. We need to know this information in order to test for net investment and the presence of a unique i^*. Net investment depends on the value of i, which we want to determine. Therefore, we need to consider two situations, (1) $i < 1.3$ and (2) $i > 1.3$:

- **Case 1:** $i < 1.3 \rightarrow \text{PB}(i, 15\%)_1 > 0$.

  Since this condition indicates a positive balance, the cash released from the project would be returned to the firm's investment pool to grow at the MARR until it is required to be put back in the project. By the end of year two, the cash placed in the investment pool would have grown at the rate of 15% [to $1,000,000(1.3 - i)(1 + 0.15)$] and must equal the investment into the project of $1,320,000 required at that time. Then the terminal balance must be

  $$\text{PB}(i, 15\%)_2 = \$1,000,000(1.3 - i)(1 + 0.15) - \$1,320,000$$
  $$= \$175,000 - \$1,150,000i = 0.$$

  Solving for i yields

  $$\text{IRR} = 0.1522, \text{ or } 15.22\%.$$

- **Case 2:** $i > 1.3 \rightarrow \text{PB}(i, 15\%)_1 < 0$.

  The firm is still in an investment mode. Therefore, the balance at the end of year one that remains invested will grow at a rate of i for the next period. Because of the investment of $1,320,000 required in year two and the fact that the net investment must be zero at the end of the project life, the balance at the end of year two should be

  $$\text{PB}(i, 15\%)_2 = \$1,000,000(1.3 - i)(1 + i) - \$1,320,000$$
  $$= -\$20,000 + \$300,000i - \$1,000,000i^2 = 0.$$

  Solving for i gives

  $$\text{IRR} = 0.1, \text{ or } 0.2 < 1.3,$$

  which violates the initial assumption ($i > 1.3$). Therefore, Case 1 is the correct situation.

(b) Case 1 indicates that IRR > MARR, so the project would be acceptable, resulting in the same decision obtained in Example 7.6 by applying the PW criterion.

Comments: In this example, we could have seen by inspection that Case 1 was correct. Since the project required an investment as the final cash flow, the project balance at the end of the previous period (year one) had to be positive in order for the final balance to equal zero. However, inspection does not typically work for more complex cash flows. In general, it is much easier to find the true IRR by using the Cash Flow Analyzer. Figure 7A.3 illustrates how you may obtain the true IRR by using the Cash Flow Analyzer. In doing so, you may need the following steps:

- **Step 1:** Enter cash flow information into the "Cash Flow Input" field.
- **Step 2:** Go to the "Rate of Return" box and specify the MARR for RIC (or external interest rate), say, 15%.

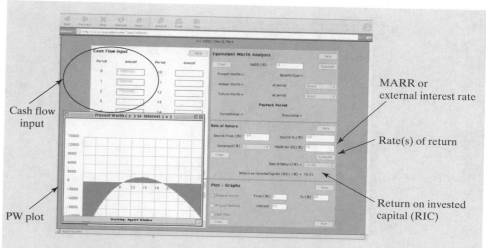

Cash flow
input

PW plot

MARR or
external interest rate

Rate(s) of return

Return on invested
capital (RIC)

FIGURE 7A.3 Computing the true IRR (or RIC) by using the Cash Flow Analyzer that can be accessed by the book's website at http://www.prenhall.com/park

- **Step 3:** Press the "Compute" button.
- **Step 4:** Check the rate(s) of return in the "Rate of Return (%)" box. If multiple rates of return exist, all of them will be displayed in the same box.
- **Step 5:** Find the RIC in the "Return on Invested Capital (RIC) (%)" box. In our example, this value is 15.22%. If this figure is the same as the rate of return, you will have a pure investment. For a mixed investment, the RIC will be different from any of the rate-of-return figures.

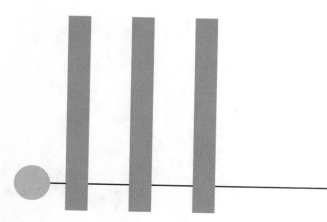

Development of Project Cash Flows

Know What It Costs to Own a Piece of Equipment?

Kimberly Linton is a design engineer employed by SIEMEN Westinghouse, a German firm that designs various instrumentation and controls for power plants. To enhance the firm's competitive position in the marketplace, management has decided to purchase a computer-aided design (CAD) system that features 3D solid modeling and full integration with sophisticated simulation and analysis capabilities. As a part of the design team, Kimberly is excited at the prospect that the design of die-cast molds, the testing of product variations, and the simulation of processing and service conditions can be streamlined by use of this state-of-the-art system. In fact, the more Kimberly thinks about it, the more she wonders why this purchase wasn't made earlier.

Accounting for Depreciation and Income Taxes

Now ask yourself: How does the cost of this system affect the financial position of the firm? In the long run, the system promises to create greater wealth for the organization by improving design productivity, increasing product quality, and cutting down design lead time. In the short run, however, the high cost of this system will negatively impact the organization's bottom line, because it involves high initial costs that are only gradually rewarded by the benefits of the system.

Another consideration should come to mind. This state-of-the-art equipment must inevitably wear out over time, and even if its productive service extends over many years, the cost of maintaining its high level of functioning will increase as the individual pieces of hardware wear out and need to be replaced. Of even greater concern is the question of how long this system will be state of the art. When will the competitive advantage the firm has just acquired become a competitive disadvantage, due to obsolescence?

One of the facts of life that organizations must deal with and account for is that fixed assets lose their value—even as they continue to function and contribute to the engineering projects that use them. This loss of value, called **depreciation**, can involve deterioration and obsolescence. Why do engineers need to understand the concept of asset depreciation? All cash flows described in Chapters 5 through 7 are cash flows after taxes. In order to determine the effects of income taxes on project cash flows, we need to understand how a company calculates the profit (or net income) from undertaking a project, where depreciation expenses play a very critical role. The main function of **depreciation accounting** is to account for the cost of fixed assets in a pattern that matches their decline in value over time. The cost of the CAD system we have just described, for example, will be allocated over several years in the firm's financial statements so that its pattern of costs roughly matches its pattern of service. In this way, as we shall see, depreciation accounting enables the firm to stabilize the statements of financial position that it distributes to stockholders and the outside world.

On a project level, engineers must be able to assess how the practice of depreciating fixed assets influences the investment value of a given project. To make this assessment, they need to estimate the allocation of capital costs over the life of the project, which requires an understanding of the conventions and techniques that accountants use to depreciate assets. In this chapter, we will review the conventions and techniques of asset depreciation.

We begin by discussing the nature of acquiring fixed assets, the significance of depreciation, and income taxes. In engineering economics, the term **cost** is used in many ways. We then focus our attention almost exclusively on the rules and laws that govern asset depreciation and the methods that accountants use to allocate depreciation expenses. Knowledge of these rules will prepare you to assess the depreciation of assets acquired in engineering projects.

8-1 Accounting Depreciation

The acquisition of fixed assets is an important activity for a business organization. This condition is true whether the organization is starting up or whether it is acquiring new assets to remain competitive. Like other disbursements, the costs of these fixed assets must be recorded as expenses on a firm's balance sheet and income statement. However, unlike costs such as maintenance, material, and labor, the costs of fixed assets are not treated simply as expenses to be accounted for in the year that they are acquired. Rather, these assets are **capitalized**; that is, their costs are distributed by subtracting them as expenses from gross income—one part at a time over a number of periods. The systematic allocation of the initial cost of an asset in parts over a time, known as its depreciable life, is what we mean by **accounting depreciation**. Because accounting depreciation is the standard of the business world, we sometimes refer to it more generally as **asset depreciation**.

The process of depreciating an asset requires that we make several preliminary determinations: (1) What is the cost of the asset? (2) What is the asset's value at the end of its useful life? (3) What is the depreciable life of the asset? (4) What method of depreciation do we choose? In this section, we will discuss each of these factors.

8.1.1 Depreciable Property

As a starting point, it is important to recognize what constitutes a **depreciable asset**, that is, a property for which a firm may take depreciation deductions against income. For the purposes of U.S. tax law, any depreciable property has the following characteristics.

1. It must be used in business or held for the production of income.
2. It must have a definite service life, which must be longer than one year.
3. It must be something that wears out, decays, gets used up, becomes obsolete, or loses value from natural causes.

Depreciable property includes buildings, machinery, equipment, vehicles, and some intangible properties.[1] Inventories are not depreciable property, because they are held primarily for sale to customers in the ordinary course of business. If an asset has no definite service life, the asset cannot be depreciated. For example, *you can never depreciate land.*

As a side note, we should add that, while we have been focusing on depreciation within firms, individuals may also depreciate assets as long as the assets meet the conditions listed previously. For example, an individual may depreciate an automobile if the vehicle is used exclusively for business purposes.

8.1.2 Cost Basis

The **cost basis** of an asset represents the total cost that is claimed as an expense over an asset's life, i.e., the sum of the annual depreciation expenses. Cost basis generally includes the actual cost of an asset and all incidental expenses, such as freight, site preparation, and installation. This total cost, rather than the cost of the asset only, must be the basis for depreciation charged as an expense over an asset's life. Besides being used in figuring depreciation deductions, an asset's cost basis is used in calculating the gain or loss to the firm if the asset is ever sold or salvaged.

---

[1]Intangible property is property that has value, but cannot be seen or touched. Some intangible properties are (1) copyrights and patents, (2) customer and subscription lists, (3) designs and patterns, and (4) franchises. Generally, you can either amortize or depreciate intangible property.

Example 8.1 Cost Basis

Rockford Corporation purchased an automatic hole-punching machine priced at $62,500. The vendor's invoice included a sales tax of $3,263. Lanier also paid the inbound transportation charges of $725 on the new machine, as well as a labor cost of $2,150 to install the machine in the factory. In addition, Lanier had to prepare the site before installation, at a cost of $3,500. Determine the cost basis for the new machine for depreciation purposes.

SOLUTION

Given: Invoice price = $62,500, freight = $725, installation cost = $2,150, and site preparation = $3,500.

Find: The cost basis.

The cost of the machine that is applicable for depreciation is computed as follows:

Cost of new hole-punching machine	$62,500
Freight	$725
Installation labor	$2,150
Site preparation	$3,500
Cost of machine (cost basis)	$68,875

Comments: Why do we include all the incidental charges relating to the acquisition of a machine in its cost? Why not treat these incidental charges as expenses of the period in which the machine is acquired? The matching of costs and revenue is a basic accounting principle. Consequently, the total costs of the machine should be viewed as an asset and allocated against the future revenue that the machine will generate. All costs incurred in acquiring the machine are costs of the services to be received from using the machine.

8.1.3 Useful Life and Salvage Value

How long will an asset be useful to the company? What do statutes and accounting rules mandate in determining an asset's depreciable life? These questions must be answered when determining an asset's depreciable life, i.e., the number of years over which the asset is to be depreciated.

Historically, depreciation accounting included choosing a depreciable life that was based on the service life of an asset. Determining the service life of an asset, however, was often very difficult, and the uncertainty of these estimates often led to disputes between taxpayers and the Internal Revenue Service (IRS). To alleviate the problems, the IRS publishes guidelines on lives for categories of assets, known as **Asset Depreciation Ranges**, or **ADRs**. These guidelines specify a range of lives for classes of assets, based on historical data, allowing taxpayers to choose a depreciable life within the specified range for a given asset. An example of ADRs for some assets is given in Table 8.1.

	Asset Depreciation Range (years)		
Assets Used	**Lower Limit**	**Midpoint Life**	**Upper Limit**
Office furniture, fixtures, and equipment	8	10	12
Information systems (computers)	5	6	7
Airplanes	5	6	7
Automobiles and taxis	2.5	3	3.5
Buses	7	9	11
Light trucks	3	4	5
Heavy trucks (concrete ready mixer)	5	6	7
Railroad cars and locomotives	12	15	18
Tractor units	5	6	7
Vessels, barges, tugs, and water transportation systems	14.5	18	21.5
Industrial steam and electrical generation or distribution systems	17.5	22	26.5
Manufacturer of electrical machinery	8	10	12
Manufacturer of electronic components, products, and systems	5	6	7
Manufacturer of motor vehicles	9.5	12	14.5
Telephone distribution plant	28	35	42

TABLE 8.1 Asset Depreciation Ranges (ADRs)

The salvage value of an asset is an asset's estimated value at the end of its life; it is the amount eventually recovered through sale, trade-in, or salvage. The eventual salvage value of an asset must be estimated when the depreciation schedule for the asset is established. If this estimate subsequently proves to be inaccurate, then an adjustment must be made.

8.1.4 Depreciation Methods: Book and Tax Depreciation

One important distinction regarding the general definition of accounting depreciation should be introduced. Most firms calculate depreciation in two different ways, depending on whether the calculation is (1) intended for financial reports (**book depreciation method**), such as for the balance sheet or income statement, or (2) intended for the Internal Revenue Service (IRS) for the purpose of calculating taxes (**tax depreciation method**). (See Figure 8.1.) In the United States, this distinction is totally legitimate under IRS regulations, as it is in many other countries. Calculating depreciation differently for financial reports and for tax purposes allows for the following benefits:

- The book depreciation enables firms to report depreciation to stockholders and other significant outsiders based on the matching concept. Therefore, the actual loss in value of the assets is generally reflected.

- Book Depreciation
    - Used in reporting net income to investors and stockholders
    - Used in pricing decisions
- Tax Depreciation
    - Used in calculating income taxes for the IRS
    - Used in engineering economics

FIGURE 8.1 Types of depreciation and their primary purposes

- The tax depreciation method allows firms to benefit from the tax advantages of depreciating assets more quickly than would be possible using the matching concept. In many cases, tax depreciation allows firms to defer paying income taxes. This deferral does not mean that they pay less tax overall, because the total depreciation expense accounted for over time is the same in either case. However, because tax depreciation methods generally permit a higher depreciation in earlier years than do book depreciation methods, the tax benefit of depreciation is enjoyed earlier, and firms generally pay lower taxes in the initial years of an investment project. Typically, this factor leads to a better cash flow position in early years, with the added cash leading to greater future wealth because of the time value of the funds.

8-2   Book Depreciation Methods

Consider a machine purchased for $10,000 with an estimated life of five years and estimated salvage value of $2,000. The objective of depreciation accounting is to charge this net cost of $8,000 as an expense over the five-year period. How much should be charged as an expense each year? Three different methods can be used to calculate the periodic depreciation allowances for financial reporting: (1) the straight-line method, (2) the declining-balance method, and (3) the unit-of-production method. In engineering economic analysis, we are interested primarily in depreciation in the context of income-tax computation. Nonetheless, a number of reasons make the study of book depreciation methods useful. First, many product pricing decisions are based on book depreciation methods. Second, tax depreciation methods are based largely on the same principles that are used in book depreciation methods. Third, firms continue to use book depreciation methods for financial reporting to stockholders and outside parties. Finally, book depreciation methods are still used for state income-tax purposes in many states and foreign countries.

8.2.1   Straight-Line Method

The **straight-line method (SL)** of depreciation interprets a fixed asset as an asset that provides its services in a uniform fashion. That is, the asset provides an equal amount of service in each year of its useful life. In other words, the depreciation rate is $1/N$, where N is the depreciable life. Example 8.2 illustrates the straight-line depreciation concept.

Example 8.2 Straight-Line Depreciation

Consider the following data on an automobile:

Cost basis of the asset $(I) = \$10,000$;
Useful life $(N) = 5$ years;
Estimated salvage value $(S) = \$2,000$.

Compute the annual depreciation allowances and the resulting book values, using the straight-line depreciation method.

SOLUTION

Given: $I = \$10,000$, $S = \$2,000$, and $N = 5$ years.
Find: D_n and B_n for $n = 1$ to 5.

The straight-line depreciation rate is $\frac{1}{5}$, or 20%. Therefore, the annual depreciation charge is

$$D_n = (0.20)(\$10,000 - \$2,000) = \$1,600.$$

Then the asset would have the following book values during its useful life, where B_n represents the book value before the depreciation charge for year n.

$$B_n = I - (D_1 + D_2 + D_3 + \cdots + D_n).$$

This situation is illustrated in Figure 8.2.

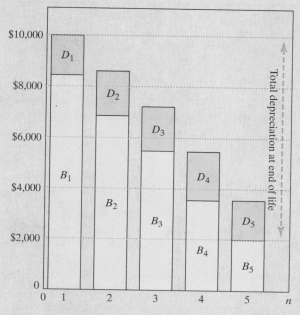

n	D_n	B_n
0		$10,000
1	$1,600	$8,400
2	$1,600	$6,800
3	$1,600	$5,200
4	$1,600	$3,600
5	$1,600	$2,000

$I = \$10,000$
$N = 5$ years
$S = \$2,000$
$D = (I - S)/N$

FIGURE 8.2 Straight-line depreciation method

8.2.2 Declining-Balance Method

The second concept recognizes that the stream of services provided by a fixed asset may decrease over time; in other words, the stream may be greatest in the first year of an asset's service life and least in its last year. This pattern may occur because the mechanical efficiency of an asset tends to decline with age, because maintenance costs tend to increase with age, or because of the increasing likelihood that better equipment will become available and make the original asset obsolete. This reasoning leads to a method that charges a larger fraction of the cost as an expense of the early years than of the later years. This method, the **declining-balance method**, is the most widely used.

Depreciation Rate The declining-balance method of calculating depreciation allocates a fixed fraction of the beginning book balance each year. The fraction α is obtained using the straight-line depreciation rate $(1/N)$ as a basis:

$$\alpha = (1/N)(\text{multiplier}). \tag{8.1}$$

The most commonly used multipliers in the United States are 1.5 (called 150% DB) and 2.0 (called 200% DDB, or double-declining-balance). So, a 200%-DB method specifies that the depreciation rate will be 200% of the straight-line rate. As N increases, α decreases, thereby resulting in a situation in which depreciation is highest in the first year and then decreases over the asset's depreciable life.

Example 8.3 Declining Balance Depreciation

Consider the following accounting information for a computer system:

Cost basis of the asset $(I) = \$10,000$;
Useful life $(N) = 5$ years;
Estimated salvage value $(S) = \$2,000$.

Compute the annual depreciation allowances and the resulting book values, using the double-declining-balance depreciation method (Figure 8.3).

SOLUTION

Given: $I = \$10,000$, $S = \$2,000$, and $N = 5$ years.
Find: D_n and B_n for $n = 1$ to 5.

The book value at the beginning of the first year is $10,000 and the declining-balance rate α is $\frac{1}{5}(2) = 40\%$. Then the depreciation deduction for the first year will be $4,000 ($40\% \times \$10,000 = \$4,000$). To figure out the depreciation deduction in the second year, we must first adjust the book value for the amount of depreciation we deducted in the first year. The first year's depreciation is subtracted from the beginning book value ($\$10,000 - \$4,000 = \$6,000$). This amount is then multiplied by the rate of depreciation ($\$6,000 \times 40\% = \$2,400$). By continuing the process, we obtain D_3. However, in year 4, B_4 would be less than $S = \$2,000$ if the full deduction (864) is taken. *Tax law does not permit us*

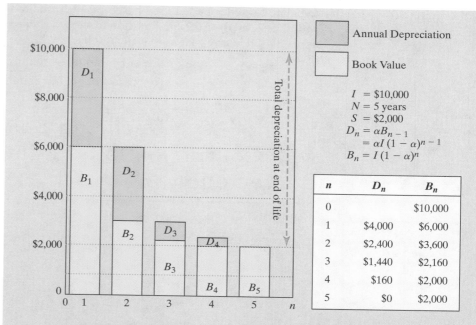

$$I = \$10,000$$
$$N = 5 \text{ years}$$
$$S = \$2,000$$
$$D_n = \alpha B_{n-1}$$
$$= \alpha I (1-\alpha)^{n-1}$$
$$B_n = I (1-\alpha)^n$$

n	D_n	B_n
0		$10,000
1	$4,000	$6,000
2	$2,400	$3,600
3	$1,440	$2,160
4	$160	$2,000
5	$0	$2,000

FIGURE 8.3 Double-declining-balance method

to depreciate assets below their salvage value. Therefore, we adjust D_4 to $160, making $B_4 = \$2,000$. D_5 is zero, and B_5 remains at $2,000. The following table provides a summary of these calculations:

End of Year	D_n	B_n
1	0.4($10,000) = $4,000	$10,000 − $4,000 = $6,000
2	0.4($6,000) = $2,400	$6,000 − $2,400 = $3,600
3	0.4($3,600) = $1,440	$3,600 − $1,440 = $2,160
4	0.4($2,160) = $864 → $160	$2,160 − $160 = $2,000
5	0	$2,000 − $0 = $2,000
	Total = $8,000	

Switching Policy When $B_N > S$, we are faced with a situation in which we have not depreciated the entire cost of the asset and thus have not taken full advantage of depreciation's tax-deferring benefits. If you would prefer to reduce the book value of an asset to its salvage value as quickly as possible, it can be done by switching from DB depreciation to SL depreciation whenever SL depreciation results in larger depreciation charges and therefore a more rapid reduction in the book value of the asset. The switch from DB to SL depreciation can take place in any of the n years, the objective being to identify the optimal year to switch. The switching rule is as follows: if DB depreciation in any year is less than (or equal to) the depreciation amount calculated by SL depreciation based on the remaining years, switch to and remain with the SL method for the duration of the asset's depreciable life. The straight-line depreciation in any year n is calculated by:

$$D_n = \frac{\text{Book value at beginning of year } n - \text{salvage value}}{\text{Remaining useful life at beginning of year } n}. \tag{8.2}$$

Example 8.4 Declining Balance with Conversion to Straight-Line Depreciation ($B_N > S$)

Suppose the asset given in Example 8.3 has a zero salvage value instead of $2,000, i.e.:

$$\text{Cost basis of the asset } (I) = \$10,000;$$
$$\text{Useful life } (N) = 5 \text{ years};$$
$$\text{Salvage value } (S) = \$0;$$
$$\alpha = (1/5)(2) = 40\%.$$

Determine the optimal time to switch from DB to SL depreciation and the resulting depreciation schedule.

SOLUTION

Given: $I = \$10,000$, $S = 0$, $N = 5$ years, and $\alpha = 40\%$.
Find: Optimal conversion time; D_n and B_n for $n = 1$ to 5.

We will first proceed by computing the DDB depreciation for each year as before. The result is shown in Table 8.2(a). Then we compute the SL depreciation

TABLE 8.2 Switching Policy from DDB to SL depreciation

$S = 0.$

(a) Without switching			(b) With switching to SL depreciation		
n	Depreciation	Book Value	n	Depreciation	Book Value
1	$10,000(0.4) = \$4,000$	$6,000	1	$4,000	$6,000
2	$6,000(0.4) \ = \$2,400$	$3,600	2	$6,000/4 = \$1,500 < \$2,400$	$3,600
3	$3,600(0.4) \ = \$1,440$	$2,160	3	$3,600/3 = \$1,200 < \$1,440$	$2,160
4	$2,160(0,4) \ = \quad \$864$	$1,296	4	$2,160/2 = \$1,080 > \quad \$864$	$1,080
5	$1,296(0.4) \ = \quad \$518$	$778	5	$1,080/1 = \$1,080 > \quad \$518$	$0

Note: If we don't switch methods, we do not depreciate the entire cost of the asset and thus do not take full advantage of depreciation's tax-deferring benefits.

for each year, using Eq. (8.2). We compare the SL depreciation with the DDB depreciation for each year and use the decision rule for when to change. The result is shown in Table 8.2(b). The switching should take place in the fourth year.

8.2.3 Units-of-Production Method

Straight-line depreciation can be defended only if the machine is used for exactly the same amount of time each year. What happens when a punch-press machine is run 1,670 hours one year and 780 the next, or when some of its output is shifted to a new machining center? This situation leads us to consider another depreciation concept that views the asset as a bundle of service units rather than as a single unit, as in the SL and DB methods. However, this concept does not assume that the service units will be consumed in a time-phased pattern. The cost of each service unit is the net cost of the asset divided by the total number of such units. The depreciation charge for a period is then related to the number of service units consumed in that period. This definition leads to the **units-of-production method**. By this method, the depreciation in any year is given by

$$D_n = \frac{\text{Service units consumed during year } n}{\text{Total service units}} (I - S). \qquad (8.3)$$

When using the units-of-production method, depreciation charges are made proportional to the ratio of actual output to the total expected output; usually, this ratio is figured in machine hours. Advantages of using this method include the fact that depreciation varies with production volume, and therefore the method gives a more accurate picture of machine usage. A disadvantage of the units-of-production method is that the collection of data on machine use and the accounting methods are somewhat tedious. This method can be useful for depreciating equipment used to exploit natural resources, if the resources will be depleted before the equipment wears out. It is not, however, considered a practical method for general use in depreciating industrial equipment.

Example 8.5 Units-of-Production Depreciation

A truck for hauling coal has an estimated net cost of $55,000 and is expected to give service for 250,000 miles, resulting in a $5,000 salvage value. Compute the allowed depreciation amount for truck usage of 30,000 miles.

SOLUTION

Given: $I = \$55,000$, $S = \$5,000$, total service units $= 250,000$ miles, and usage for this year $= 30,000$ miles.

Find: Depreciation amount in this year.

The depreciation expense in a year in which the truck traveled 30,000 miles would be

$$\frac{30,000 \text{ miles}}{250,000 \text{ miles}}(\$55,000 - \$5,000) = \left(\frac{3}{25}\right)(\$50,000) = \$6,000.$$

8-3   Tax Depreciation Methods

Prior to the Economic Recovery Act of 1981, taxpayers could choose among several methods when depreciating assets for tax purposes. The most widely used methods were the straight-line method and the declining-balance method. The subsequent imposition of the Accelerated Cost Recovery System (ACRS) and the Modified Accelerated Cost Recovery System (MACRS) superseded these methods for use in tax purposes. Currently, the MACRS is the method used to determine the allowed depreciation amount in calculating income taxes. The foregoing history is summarized in Table 8.3.

8.3.1   MACRS Recovery Periods

Historically, for tax purposes as well as for accounting, an asset's depreciable life was determined by its estimated useful life; it was intended that an asset would be fully depreciated at approximately the end of its useful life. The MACRS scheme, however, totally abandoned this practice, and simpler guidelines were set that created several classes of assets, each with a more or less arbitrary lifespan called a **recovery period**. *Note:* These recovery periods do not necessarily correspond to expected useful lives.

As shown in Table 8.4, the MACRS scheme includes eight categories of assets: 3, 5, 7, 10, 15, 20, 27.5, and 39 years. Under the MACRS, *the salvage value of property is always treated as zero.* The MACRS guidelines are summarized as follows:

- Investments in some short-lived assets are depreciated over three years by using 200% DB and then switching to SL depreciation.

TABLE 8.3 History of Tax Depreciation Methods

Tax Depreciation

- **Purpose:** Used to compute income taxes for the IRS.
- Assets placed in service prior to 1981:
    Use book depreciation methods (SL, DB, or SOYD*).
- Assets placed in service from 1981 to 1986:
    Use ACRS Table.
- Assets placed in service after 1986:
    Use MACRS Table.

*SOYD: Sum-of-Years' Digit Method—this method is no longer used as a book depreciation method.

TABLE 8.4 MACRS Property Classifications (ADR = Asset Depreciation Range)

Recovery Period	ADR Midpoint Class	Applicable Property
3 years	ADR ≤ 4	Special tools for manufacture of plastic products, fabricated metal products, and motor vehicles
5 years	$4 < ADR \leq 10$	Automobiles, light trucks, high-tech equipment, equipment used for R&D, computerized telephone switching systems
7 years	$10 < ADR \leq 16$	Manufacturing equipment, office furniture, fixtures
10 years	$16 < ADR \leq 20$	Vessels, barges, tugs, railroad cars
15 years	$20 < ADR \leq 25$	Waste-water plants, telephone-distribution plants, or similar utility property
20 years	$25 \leq ADR$	Municipal sewers, electrical power plant
27.5 years		Residential rental property
39 years		Nonresidential real property including elevators and escalators

- Computers, automobiles, and light trucks are written off over 5 years by using 200% DB and then switching to SL depreciation.

- Most types of manufacturing equipment are depreciated over 7 years, but some long-lived assets are written off over 10 years. Most equipment write-offs are calculated by using the 200%-DB method and then switching to SL depreciation, an approach that allows for faster write-offs in the first few years after an investment is made.

- Sewage-treatment plants and telephone-distribution plants are written off over 15 years by using 150% DB and then switching to SL depreciation.

- Sewer pipes and certain other very long-lived equipment are written off over 20 years by using 150% DB and then switching to SL depreciation.

- Investments in residential rental property are written off in straight-line fashion over 27.5 years. On the other hand, nonresidential real estate (commercial buildings) is written off by the SL method over 39 years.

8.3.2 MACRS Depreciation: Personal Property

Under depreciation methods discussed previously, the rate at which the value of an asset actually declined was estimated, and this rate was used for tax depreciation. Thus, different assets were depreciated along different paths over time. The MACRS method, however, establishes prescribed depreciation rates, called **recovery allowance percentages**, for all assets within each class. These rates, as set forth in 1986 and 1993, are shown in Table 8.5.

TABLE 8.5 MACRS Depreciation Schedules for Personal Property with Half-Year Convention

Year \ Class Depreciation Rate	3 200% DB	5 200% DB	7 200% DB	10 200% DB	15 150% DB	20 150% DB
1	33.33	20.00	14.29	10.00	5.00	3.750
2	44.45	32.00	24.49	18.00	9.50	7.219
3	14.81*	19.20	17.49	14.40	8.55	6.677
4	7.41	11.52*	12.49	11.52	7.70	6.177
5		11.52	8.93*	9.22	6.93	5.713
6		5.76	8.92	7.37	6.23	5.285
7			8.93	6.55*	5.90*	4.888
8			4.46	6.55	5.90	4.522
9				6.56	5.91	4.462*
10				6.55	5.90	4.461
11				3.28	5.91	4.462
12					5.90	4.461
13					5.91	4.462
14					5.90	4.461
15					5.91	4.462
16					2.95	4.461
17						4.462
18						4.461
19						4.462
20						4.461
21						2.231

● *Year to switch from declining balance to straight line. Source: IRS Publication 534. *Depreciation* U.S. Government Printing Offices, Washington, DC. December, 1995.

● For illustration, MACRS percentages of 3-year property, beginning with the first tax year and ending with the fourth year, are computed as follows: Straight-line rate = 1/3, 200% DB rate = $2\left(\frac{1}{3}\right) = 0.6667$, $S = 0$.

Year	Calculation	MACRS percentage
1	$\frac{1}{2}$ year DDB dep = 0.5(0.6667)	= $\boxed{33.33\%}$
2	DDB dep = 0.6667(1 − 0.3333)	= $\boxed{44.45\%}$
	SL dep = $(^1/_{2.5})(1 - 0.3333)$	= 26.67%
3	DDB dep = 0.6667(1 − 0.7778)	= 14.81%
	SL dep = $(^1/_{1.5})(1 - 0.7778)$	= $\boxed{14.81\%}$
4	$\frac{1}{2}$ SL dep = (0.5)(14.81%)	= $\boxed{7.41\%}$

Note that SL dep $\geqq$ DDB dep in year 3 and so we switch to SL.

The yearly recovery, or depreciation expense, is determined by multiplying the asset's depreciation base by the applicable recovery-allowance percentage:

- **Half-Year Convention:** The MACRS recovery percentages shown in Table 8.5 use the **half-year convention**, i.e., it is assumed that all assets are placed in service at midyear and that they will have *zero* salvage value. As a result, only a half-year of depreciation is allowed for the first year that property is placed in service. With half of one year's depreciation being taken in the first year, a full year's depreciation is allowed in each of the remaining years of the asset's recovery period, and the remaining half-year's depreciation is incurred in the year following the end of the recovery period. A half-year of depreciation is also allowed for the year in which the property is disposed of, or is otherwise retired from service, anytime before the end of the recovery period.

- **Switching from the DB Method to the SL Method:** The MACRS asset is depreciated initially by the DB method and then by the SL method. Consequently, the MACRS scheme adopts the switching convention illustrated in Section 8.2.2.

To demonstrate how the MACRS depreciation percentages were calculated by the IRS, using the half-year convention, let's consider Example 8.6.

Example 8.6 MACRS Depreciation: Personal Property

A taxpayer wants to place in service a $10,000 asset that is assigned to the five-year MACRS class. Compute the MACRS percentages and the depreciation amounts for the asset.

SOLUTION

Given: Five-year asset, half-year convention, $\alpha = 40\%$, and $S = 0$.
Find: MACRS depreciation percentages D_n for a $10,000 asset.

We can calculate the depreciation amounts from the percentages shown in Table 8.5. In practice, the percentages are taken directly from Table 8.5, which is supplied by the IRS.

MACRS			
Year N	Percentage (%)	Depreciation Basis	Depreciation Amount (D_n)
1	20 ×	$10,000 =	$2,000
2	32 ×	$10,000 =	$3,200
3	19.20 ×	$10,000 =	$1,920
4	11.52 ×	$10,000 =	$1,152
5	11.52 ×	$10,000 =	$1,152
6	5.76 ×	$10,000 =	$576

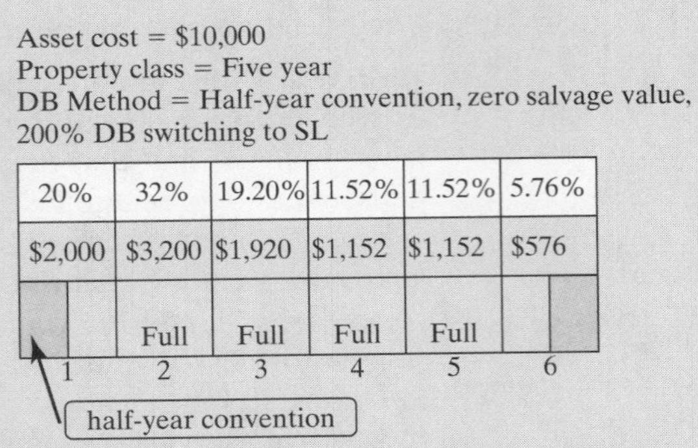

Asset cost = $10,000
Property class = Five year
DB Method = Half-year convention, zero salvage value,
200% DB switching to SL

20%	32%	19.20%	11.52%	11.52%	5.76%
$2,000	$3,200	$1,920	$1,152	$1,152	$576
	Full	Full	Full	Full	
1	2	3	4	5	6

half-year convention

FIGURE 8.4 MACRS depreciation method

The results are also shown in Figure 8.4.

Note that when an asset is disposed of before the end of recovery period, only half of the normal depreciation is allowed. If, for example, the $10,000 asset were to be disposed of in year 2, the MACRS deduction for that year would be $1,600.

Comments: Another way to calculate the MACRS depreciation allowances is to use the depreciation generator in the book's website (http://www.prenhall.com/park; click on "Analysis Tools" to find the on-line depreciation calculator).

8.3.3 MACRS Depreciation: Real Property

Real properties are classified into two categories: (1) residential rental property and (2) commercial building or properties. When depreciating such property, the straight-line method and the **midmonth convention** are used. For example, a property placed in service in March would be allowed 9.5 months depreciation for year one. If it is disposed of before the end of the recovery period, the depreciation percentage must take into account the number of months the property was in service during the year of its disposal. Residential properties are depreciated over 27.5 years, and commercial properties are depreciated over 39 years.

Example 8.7 MACRS Depreciation: Real Property

On May 1, Jack Costanza paid $100,000 for a residential rental property. This purchase price represents $80,000 for the cost of the building and $20,000 for the cost of the land. Three years and five months later, on October 1, he sold the property for $130,000. Compute the MACRS depreciation for each of the four calendar years during which he owned the property.

SOLUTION

Given: Residential real property with cost basis = $80,000; the building was put into service on May 1.

Find: The depreciation in each of the four tax years the property was in service.

In this example, the midmonth convention assumes that the property is placed in service on May 15, which gives 7.5 months of depreciation in the first year. Remembering that only the building (not the land) may be depreciated, we compute the depreciation over a 27.5-year recovery period, using the SL method:

Year	Calculation	D_n	Recovery Percentages
1	$\left(\dfrac{7.5}{12}\right)\dfrac{\$80,000 - 0}{27.5} =$	$1,818	2.273%
2	$\dfrac{\$80,000 - 0}{27.5} =$	$2,909	3.636%
3	$\dfrac{\$80,000 - 0}{27.5} =$	$2,909	3.636%
4	$\left(\dfrac{9.5}{12}\right)\dfrac{\$80,000 - 0}{27.5} =$	$2,303	2.879%

Notice that the midmonth convention also applies to the year of disposal. As for personal property, calculations for real property generally use the precalculated percentages as found at the book's website.

8-4 How to Determine "Accounting Profit"

As we have seen in Chapters 4 through 7, we make our investment decisions based on the net project cash flows. The net project cash flows are the cash flows after taxes. In order to calculate the amount of taxes involved in project evaluation, we need to understand how businesses determine taxable income and thereby net income (profit).

Firms invest in a project because they expect it to increase their wealth. If the project does this—that is, if project revenues exceed project costs—we say it has generated a **profit**, or **income**. If the project reduces a firm's wealth—that is, if project costs exceed project revenues—we say that the project has resulted in a **loss**. One of the most important roles of the accounting function within an organization is to measure the amount of profit or loss a project generates each year, or in any other relevant time period. Any profit generated will be taxed. The accounting measure of a project's after-tax profit during a particular time period is known as **net income**.

8.4.1   Treatment of Depreciation Expenses

Whether you are starting or maintaining a business, you will probably need to acquire assets (such as buildings and equipment). The cost of this property becomes part of your business expenses. The accounting treatment of capital expenditures differs from the treatment of manufacturing and operating expenses, such as cost of goods sold and business operating expenses.

As mentioned earlier in the chapter, **capital expenditures must be capitalized**, i.e., they must be systematically allocated as expenses over their depreciable lives. Therefore, when you acquire a piece of property that has a productive life extending over several years, you cannot deduct the total costs from profits in the year the asset was purchased. Instead, a depreciation allowance is established over the life of the asset, and an appropriate portion of that allowance is included in the company's deductions from profit each year. Because it plays a role in reducing taxable income, depreciation accounting is of special concern to a company. In the next section, we will investigate the relationship between depreciation and net income.

8.4.2   Calculation of Net Income

Accountants measure the net income of a specified operating period by subtracting expenses from revenues for that period. For projects, these terms can be defined as follows:

1. **Project revenue** is the income earned by a business as a result of providing products or services to customers. Revenue comes from sales of merchandise to customers and from fees earned by services performed for clients or others.
2. **Project expenses** are costs incurred in doing business to generate the revenues of the specified operating period. Some common expenses are the cost of goods sold (labor, material, inventory, and supplies), depreciation, the cost of employees' salaries, the operating costs (such as the cost of renting buildings and the cost of insurance coverage), and income taxes.

The aforementioned business expenses are accounted for in a straightforward fashion on a company's income statement and balance sheet; the amount paid by the organization for each item translates dollar for dollar into the expenses listed in financial reports for the period. One additional category of expenses, the purchase of new assets, is treated by depreciating the total cost gradually over time. Because capital goods are given this unique accounting treatment, depreciation is accounted for as a separate expense in financial reports. In the next section, we will discuss how depreciation accounting is reflected in net-income calculations. Once we define the project revenue and expenses, the next step is to determine the corporate taxable income, which is defined as follows:

$$\text{Taxable income} = \text{gross income (i.e., revenues)} - \text{expenses.}$$

Once taxable income is calculated, income taxes are determined as follows:

$$\text{Income taxes} = (\text{tax rate}) \times (\text{taxable income}).$$

(We will discuss how we determine the applicable tax rate in Section 8.5.) We then calculate net income as follows:

$$\text{Net income} = \text{taxable income} - \text{income taxes.}$$

Gross Income
Expenses
    Cost of goods sold
    Depreciation
    Operating expenses
Taxable income
Income taxes
Net income

FIGURE 8.5 Tabular approach to finding the net income

A more common format is to present the net income in the tabular income statement given in Figure 8.5. If the gross income and other expenses remain the same, any decrease in depreciation deduction will increase the amount of taxable income and thus the income taxes, but result in a higher net income. On the other hand, any increase in depreciation deduction would result in a smaller amount of income taxes, but a lower net income.

Example 8.8 Net Income within a Year

A company buys a numerically controlled (NC) machine for $28,000 (year zero) and uses it for five years, after which time it is scrapped. The allowed depreciation deduction during the first year is $4,000, as the equipment falls into the seven-year MACRS-property category. (The first-year depreciation rate is 14.29%.) The cost of the goods produced by this NC machine should include a charge for the depreciation of the machine. Suppose the company estimates the following revenues and expenses, including the depreciation for the first operating year:

Gross income = $50,000;
Cost of goods sold = $20,000;
Depreciation on NC machine = $4,000;
Operating expenses = $6,000.

If the company pays taxes at the rate of 40% on its taxable income, what is its net income during the first year from the project?

SOLUTION

Given: Gross income and expenses as stated; income-tax rate = 40%.
Find: Net income.

At this point, we will defer discussion of how the tax rate (40%) is determined and treat the rate as given. We consider the purchase of the machine to have been made at the end of year zero, which is also the beginning of year one. (Note that our example explicitly assumes that the only depreciation charges for year one are those for the NC machine, a situation that may not be typical.)

Item	Amount
Gross income (revenues)	$50,000
Expenses	
Cost of goods sold	$20,000
Depreciation	$4,000
Operating expenses	$6,000
Taxable income	$20,000
Taxes (40%)	$8,000
Net income	$12,000

Comments: In this example, the inclusion of a depreciation expense reflects the true cost of doing business. This expense is meant to correspond to the amount of the total cost of the machine that has been put to use or "used up" during the first year. This example also highlights some of the reasons that income-tax laws govern the depreciation of assets. If the company were allowed to claim the entire $28,000 as a year-one expense, a discrepancy would exist between the one-time cash outlay for the machine's cost and the gradual benefits of its productive use. This discrepancy would lead to dramatic variations in the firm's net income, and net income would become a less accurate measure of the organization's performance. On the other hand, failing to account for this cost would lead to increased reported profit during the accounting period. In this situation, the profit would be a "false profit" in that it would not accurately account for the usage of the machine. Depreciating the cost over time allows the company a logical distribution of costs that matches the utilization of the machine's value.

8.4.3 Operating Cash Flow versus Net Income

Certain expenses are not really cash outflows. Depreciation and amortization are the best examples of this type of expense. Even though depreciation (or amortization expense) is deducted from revenue for tax or book purposes on a yearly basis, no cash is paid to anyone, except when the asset was purchased.

We just saw in Example 8.8 that the annual depreciation allowance has an important impact on both taxable income and net income. However, although depreciation has a direct impact on net income, it is *not* a cash outlay; as such, it is important to distinguish between annual income in the presence of depreciation and annual cash flow.

The situation described in Example 8.8 serves as a good vehicle to demonstrate the difference between depreciation costs as expenses and the cash flow generated by the purchase of a fixed asset. In this example, cash in the amount of $28,000 was expended in year zero, but the $4,000 depreciation charged against the income in year one is not a cash outlay. Figure 8.6 summarizes the difference.

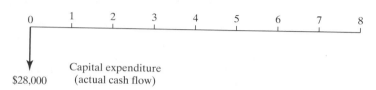

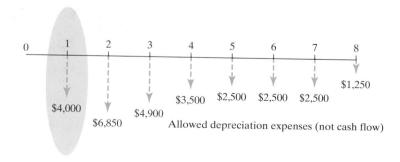

FIGURE 8.6 Capital expenditure versus depreciation expenses

Net income (**accounting profit**) is important for accounting purposes, but **cash flows** are more important for project-evaluation purposes. However, as we will now demonstrate, net income can provide us with a starting point to estimate the cash flow of a project.

The procedure for calculating net income is identical to that used for obtaining net cash flow (after tax) from operations, with the exception of depreciation, which is excluded from the net cash flow computation. (Depreciation is needed only for computing income taxes.) Assuming that revenues are received and expenses are paid in cash, we can obtain the net (operating) cash flow by adding the **noncash expense** (i.e., depreciation) to net income, which cancels the operation of subtracting it from revenues:

Operating cash flow = net income + noncash expense (i.e., depreciation).

Example 8.9 Cash Flow from Operation versus Net Income

For the situation described in Example 8.8, assume that (1) all sales are cash sales and (2) all expenses except depreciation were paid during year one. How much cash would be generated from operations?

SOLUTION

Given: Net-income components as in Example 8.8.
Find: Cash flow from operation.

We can generate a cash flow statement by simply examining each item in the income statement and determining which items actually represent receipts or disbursements. Some of the assumptions listed in the statement of the problem make this process simpler. We summarize our findings as follows:

Item	Income	Cash Flow
Gross income (revenues)	$50,000	$50,000
Expenses		
Cost of goods sold	$20,000	−$20,000
Depreciation	$4,000	
Operating expenses	$6,000	−$6,000
Taxable income	$20,000	
Taxes (40%)	$8,000	−$8,000
Net income	$12,000	−
Cash flow from operation		$16,000

The second column shows the income statement, while the third column shows the statement on a cash flow basis. The sales of $50,000 are all cash sales. Costs other than depreciation were $26,000; these costs were paid in cash, leaving $24,000. Depreciation is not a cash flow; that is, the firm did not pay out $4,000 in depreciation expenses. Taxes, however, are paid in cash, so the $8,000 for taxes must be deducted from the $24,000, leaving a net cash flow from operation of $16,000. Figure 8.7 illustrates in graphical format how the net cash flow is related to the net income.

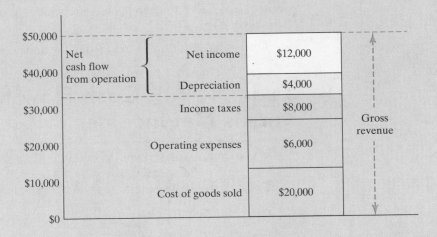

FIGURE 8.7 Relationship between net income and operating cash flow

8-5   Corporate Taxes

Now that we have learned what elements constitute taxable income, we turn our attention to the process of computing income taxes. The corporate tax rate is applied to the taxable income of a corporation, which is defined as its gross income minus allowable deductions. As we briefly discussed in Section 8.1, the allowable deductions include the cost of goods sold, salaries and wages, rent, interest, advertising, depreciation, amortization, depletion, and various tax payments other than federal income tax.

8.5.1   Income Taxes on Operating Income

The corporate-tax-rate structure for 2003 is relatively simple. As shown in Table 8.6, there are four basic rate brackets (15%, 25%, 34%, and 35%) plus two surtax rates (5% and 3%) based on taxable incomes. U.S. tax rates are progressive; that is, businesses with lower taxable incomes are taxed at lower rates than those with higher taxable incomes.

Marginal Tax Rate   **Marginal tax rate** is defined as the rate applied to the last dollar of income earned. Income of up to $50,000 is taxed at a 15% rate (meaning that, if your taxable income is less than $50,000, your marginal tax rate is 15%); income between $50,000 and $75,000 is taxed at a 25% rate; and income over $75,000 is taxed at a 34% rate. An additional 5% surtax (resulting in 39%) is imposed on a corporation's taxable income in excess of $100,000, the maximum additional tax being limited to $11,750 (235,000 × 5%). This surtax provision phases out the benefit of graduated rates for corporations with taxable incomes between $100,000 and $335,000. Another 3% surtax rate is imposed on a corporate taxable income in the range $15,000,001 to $18,333,333. Corporations with incomes in excess of $18,333,333, in effect, pay a flat tax of 35%. As shown in Table 8.6, the corporate tax is also progressive up to $18,333,333 in taxable income, but is essentially constant thereafter.

TABLE 8.6 U.S. Corporate Tax Schedule (2003)

Taxable income	Marginal Surtax Tax rate	Tax computation
0–$50,000	15%	$0 + 0.15(Δ)
$50,001–$75,000	25%	$7,500 + 0.25(Δ)
$75,000–$100,000	34%	$13,750 + 0.34(Δ)
$100,001–$335,000	34% + 5%	$22,250 + 0.39(Δ)
$335,001–$10,000,000	34%	$113,900 + 0.34(Δ)
$10,000,001–$15,000,000	35%	$3,400,000 + 0.35(Δ)
$15,000,001–$18,333,333	35% + 3%	$5,150,000 + 0.38(Δ)
$18,333,334 and up	35%	$6,416,666 + 0.35(Δ)

(Δ) denotes the taxable income in excess of the lower bound of each tax bracket.

Effective (Average) Tax Rate Effective tax rates can be calculated from the data in Table 8.6. For example, if your corporation had a taxable income of $16,000,000 in 2001, then the income tax owed by the corporation for that year would be calculated as follows:

Marginal and Effective (Average) Tax Rate for a Taxable Income of $16,000,000			
Taxable Income	**Marginal Tax Rate**	**Amount of Taxes**	**Cumulative Taxes**
First $50,000	15%	$7,500	$7,500
Next $25,000	25%	$6,250	$13,750
Next $25,000	34%	$8,500	$22,250
Next $235,000	39%	$91,650	$113,900
Next $9,665,000	34%	$3,286,100	$3,400,000
Next $5,000,000	35%	$1,750,000	$5,150,000
Remaining $1,000,000	38%	$380,000	$5,530,000

$$\text{Average tax rate} = \frac{\$5,530,000}{\$16,000,000} = 34.56\%$$

Or, using the tax formulas in Table 8.6, we obtain

$$\$5,150,000 + 0.38(\$16,000,000 - \$15,000,000) = \$5,530,000$$

for the total amount of taxes paid. The effective tax rate would then be

$$\$5,530,000/\$16,000,000 = 0.3456, \text{ or } 34.56\%,$$

as opposed to the marginal rate of 38%. In other words, on average, the company paid 34.56 cents for each taxable dollar it generated during the accounting period.

Example 8.10 Corporate Taxes

A mail-order computer company sells personal computers and peripherals. The company leased showroom space and a warehouse for $20,000 a year and installed $290,000 worth of inventory-checking and packaging equipment. The allowed depreciation expense for this capital expenditure ($290,000 total) amounted to $58,000. The store was completed and operations began on January 1. The company had a gross income of $1,250,000 for the calendar year. Supplies and all operating expenses other than the lease expense were itemized as follows:

Merchandise sold in the year	$600,000
Employee salaries and benefits	$150,000
Other supplies and expenses	$90,000
	$840,000

Compute the taxable income for this company. How much will the company pay in federal income taxes for the year? What is its average corporate tax rate?

SOLUTION

Given: Income, foregoing cost information, and depreciation amount.
Find: Taxable income, amount paid in federal income taxes, and average corporate tax rate.

First we compute the taxable income as follows:

Gross revenues	$1,250,000
Expenses	$840,000
Lease expense	$20,000
Depreciation	$58,000
Taxable income	$332,000

Note that capital expenditures are not deductible expenses. Since the company is in the 39% marginal tax bracket, the income tax can be calculated by using the corresponding formula given in Table 8.6, $22,250 + 0.39(X - 100,000)$:

$$\text{Income tax} = \$22,250 + 0.39(\$332,000 - \$100,000) = \$112,730.$$

The firm's current marginal tax rate is 39%, but its average corporate tax rate is

$$\$112,730/\$332,000 = 33.95\%.$$

8.5.2 Gain Taxes on Asset Disposals

When a depreciable asset used in business is sold for an amount that differs from its book value, the gains or losses have an important effect on income taxes. The gains or losses are found as

$$\text{Gains(losses)} = \text{salvage value} - \text{book value},$$

where the salvage value represents the proceeds from the sale (selling price) less any selling expense or removal cost.

The gains, commonly known as **depreciation recapture**, are taxed as ordinary income under current tax law. In the unlikely event that an asset is sold for an amount greater than its cost basis, the gains (salvage value − book value) are divided into two parts for tax purposes:

$$\text{Gains} = \text{salvage value} - \text{book value}$$
$$= \underbrace{(\text{salvage value} - \text{cost basis})}_{\text{Capital gains}} + \underbrace{(\text{cost basis} - \text{book value})}_{\text{Ordinary gains}}.$$

Recall that cost basis is the purchase cost of an asset plus any incidental costs, such as freight and installation. As illustrated in Figure 8.8,

$$\text{Capital gains} = \text{salvage value} - \text{cost basis}$$

and

$$\text{Ordinary gains} = \text{cost basis} - \text{book value}.$$

This distinction is necessary only when capital gains are taxed at the capital-gains tax rate and ordinary gains (or depreciation recapture) are taxed at the ordinary income-tax rate. Current tax law does not provide a special low rate of taxation for capital gains. Currently, capital gains are treated as ordinary income, but the maximum tax rate is set at the U.S. statutory rate of 35%. Nevertheless, the statutory structure for capital gains has been retained in the tax code. This provision could allow Congress to restore preferential treatment for capital gains at some future time.

To calculate a gain or loss, we first need to determine the book value of the depreciable asset at the time of disposal. Taxable gains are defined as the difference between the salvage value and the book value. If the salvage value is greater than the cost basis, these taxable gains can be further divided into capital gains and ordinary gains, as shown in Figure 8.8. These ordinary gains are also known as depreciation recaptures. The government views these gains as ordinary income due to too much depreciation taken by the taxpayer, and the ordinary income-tax rate is applied to these gains.

For a MACRS property, one important consideration at the time of disposal is whether the property is disposed of *during* or *before* its specified recovery period. Moreover, with the half-year convention, which is now mandated by all MACRS depreciation methods, the year of disposal is charged one-half of that year's annual

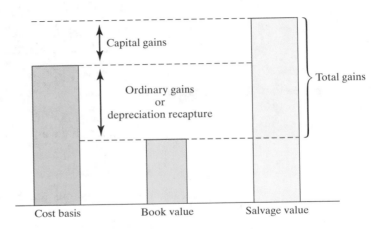

FIGURE 8.8 Determining ordinary gains and capital gains

If a MACRS asset is disposed of during the *recovery period,*

- Personal property: <u>the half-year convention</u> is applied to the depreciation amount for the year of disposal.

Asset is disposed of during year 4

- Real property: <u>the mid-month convention</u> is applied to the month of disposal.

FIGURE 8.9 Disposal of depreciable asset

depreciation amount, should the disposal occur during the recovery period. Figure 8.9 highlights the rules that govern disposal of depreciable assets.

Example 8.11 Gains and Losses on Depreciable Assets

A company purchased a drill press costing $230,000 in year zero. The drill press, classified as seven-year recovery property, has been depreciated by the MACRS method. If it is sold at the end of three years for (1) $150,000 or (2) $100,000, compute the gains (losses) for each situation. Assume that both capital gains and ordinary income are taxed at 34%.

SOLUTION

Given: Seven-year MACRS asset, cost basis = $230,000, sold three years after purchase

Find: Gains or losses, tax effects and net proceeds from the sale if sold for $150,000 or $100,000

In this example, we first compute the current book value of the machine. From the MACRS depreciation schedule in Table 8.5, the allowed annual depreciation percentages for the first three years of a seven-year MACRS property are 14.29%, 24.49%, and 17.49%, respectively. Since the asset is disposed of before the end of its recovery period, the depreciation amount in year three will be reduced by half. The total depreciation and final book value are calculated as follows:

$$\text{Total allowed depreciation} = \$230,000(0.1429 + 0.2449 + 0.1749/2)$$
$$= \$109,308;$$

$$\text{Book value} = \$230,000 - \$109,308$$
$$= \$120,693.$$

- **Case 1:** $S = \$150,000.$

  Since book value < salvage value < cost basis, there are no capital gains to consider.

Drill press: $230,000
Project life: Three years
MACRS: Seven year property class
Salvage value: $150,000 at the end of year three

Full	Full	Half				
14.29	24.49	17.49	12.49	8.92	8.92	8.92

Total depreciation = $230,000(0.1439 + 0.2449 + 0.1749/2) = $109,308
Book value = $230,000 − $109,308 = $120,693

Gains = salvage value − book value = $150,000 − $120,693
= $29,308
Gains tax (34%) = 0.34 ($29,308) = $9,965
Net proceeds from sale = $150,000 − $9,965 = $140,035

FIGURE 8.10 Gain or loss on a depreciable asset

All gains are ordinary gains:

$$\text{Ordinary gains} = \text{salvage value} - \text{book value}$$
$$= \$150,000 - \$120,693$$
$$= \$29,308.$$

So, with an ordinary-gains tax of 34% for this bracket, we find that the amount of tax paid on the gains is

$$0.34(\$29,308) = \$9,965.$$

Thus,

$$\text{Net proceeds from sale} = \text{salvage value} - \text{gains tax}$$
$$= \$150,000 - \$9,965$$
$$= \$140,035.$$

The computation process is summarized in Figure 8.10.

- **Case 2:** $S = \$100,000$.

Since the book value is $120,693, the amount of loss will be $20,693. The anticipated tax savings will be $20,693(0.34) = $7,036. Therefore, the net proceeds from sale will be $107,036.

Summary

- Explicit consideration of taxes is a necessary aspect of any complete economic study of an investment project.
- Since we are interested primarily in the measurable financial aspects of depreciation, we consider the effects of depreciation on two important measures of an organization's financial position, **net income** and **cash flow from operations**. Once we understand that depreciation has a significant influence on the income and cash position of a firm, we will be able to appreciate fully

the importance of using depreciation as a means to maximize the value both of engineering projects and of the organization as a whole.

- Machine tools and other manufacturing equipment, and even factory buildings themselves, are subject to wear over time. However, it is not always obvious how to account for the cost of their replacement. How we determine the estimated service life of a machine, and the method used to calculate the cost of operating it, can have significant effects on an asset's management.

- The entire cost of replacing a machine cannot be properly charged to any one year's production; rather, the cost should be spread (or capitalized) over the years in which the machine is in service. The cost charged to operations of an asset during a particular year is called **depreciation**. Several different meanings and applications of depreciation have been presented in this chapter. From an engineering economics point of view, our primary concern is with **accounting depreciation**—the systematic allocation of an asset's value over its depreciable life.

- Accounting depreciation can be broken into two categories:

  1. **Book depreciation**—the method of depreciation used for financial reports and pricing products;
  2. **Tax depreciation**—governed by tax legislation, the method of depreciation used for calculating taxable income and income taxes.

- The four components of information required in order to calculate depreciation are

  1. the cost basis of the asset,
  2. the salvage value of the asset,
  3. the depreciable life of the asset, and
  4. the method of the asset's depreciation.

- Because it employs accelerated methods of depreciation and shorter-than-actual depreciable lives, the **MACRS (Modified Accelerated Cost Recovery System)** gives taxpayers a break by allowing them to take earlier and faster advantage of the tax-deferring benefits of depreciation.

- Many firms select straight-line depreciation for book depreciation because of its relative ease of calculation.

- Given the frequently changing nature of depreciation and tax law, we must use whatever percentages, depreciable lives, and salvage values were mandated *at the time an asset was acquired*.

- For corporations, the U.S. tax system has the following characteristics:

  1. Tax rates are progressive: The more you earn, the more you pay.
  2. Tax rates increase in stair-step fashion: There are four brackets and two additional surtax brackets, giving a total of six brackets.
  3. Allowable exemptions and deductions may reduce the overall tax assessment.

- Three distinct terms to describe taxes were used in this chapter: **marginal tax rate**, which is the rate applied to the last dollar of income earned; **effective** (average) **tax rate**, which is the ratio of income tax paid to net income; and **incremental tax rate**, which is the average rate applied to the incremental income generated by a new investment project.

● **Capital gains** are currently taxed as ordinary income, and the maximum rate is capped at 35%. **Capital losses** are deducted from capital gains; net remaining losses may be carried backward and forward for consideration in years other than the current tax year.

Problems

Note: Unless otherwise specified, use current tax rates for corporate taxes. Check the website described in the preface for the most current tax rates for corporations.

Depreciation Concept

8.1 A machine now in use was purchased four years ago at a cost of $5,000. It has a book value of $1,300. It can be sold now for $2,300, or it could be used for three more years, at the end of which time it would have no salvage value. Assuming it is kept for three more years, what is the amount of economic loss during this ownership?

Cost Basis

8.2 General Service Contractor Company paid $120,000 for a house and lot. The value of the land was appraised at $65,000, and the value of the house at $55,000. The house was then torn down at an additional cost of $5,000 so that a warehouse could be built on the lot at a cost of $50,000. What is the total value of the property with the warehouse? For depreciation purposes, what is the cost basis for the warehouse?

8.3 To automate one of their production processes, Milwaukee Corporation bought three flexible manufacturing cells at a price of $500,000 each. When they were delivered, Milwaukee paid freight charges of $25,000 and handling fees of $12,000. Site preparation for these cells cost $35,000. Six foremen, each earning $15 an hour, worked five 40-hour weeks to set up and test the manufacturing cells. Special wiring and other materials applicable to the new manufacturing cells cost $1,500. Determine the cost basis (the amount to be capitalized) for these cells.

8.4 A new drill press was purchased for $95,000 by trading in a similar machine that had a book value of $25,000. Assuming that the trade-in allowance was $20,000 and that $75,000 cash was paid for the new asset, what is the cost basis of the new asset for depreciation purposes?

8.5 A lift truck priced at $35,000 is acquired by trading in a similar lift truck and paying cash for the remaining balance. Assuming that the trade-in allowance is $10,000 and the book value of the asset traded in is $6,000, what is the cost basis of the new asset for the computation of depreciation for tax purposes?

Book Depreciation Methods

8.6 Consider the following data on an asset:

Cost of the asset, I	$100,000
Useful life, N	5 years
Salvage value, S	$10,000

Compute the annual depreciation allowances and the resulting book values, using the following methods:

(a) the straight-line depreciation method
(b) the double-declining-balance method

8.7 A firm is trying to decide whether to keep an item of construction equipment another year. The firm is using the DDB method for book purposes, and this is the fourth year of ownership of the equipment. The item cost $150,000 when it was new. What is the depreciation in year three? Assume that the depreciable life of the equipment is eight years, with zero salvage value.

8.8 Consider the following data on an asset:

Cost of the asset, I	$30,000
Useful life, N	7 years
Salvage value, S	$0

Compute the annual depreciation allowances and the resulting book values by using the DDB method and then switching to the SL method.

8.9 The double-declining-balance method is to be used for an asset with a cost of $80,000, estimated salvage value of $22,000, and estimated useful life of six years.

(a) What is the depreciation for the first three tax years, assuming that the asset was placed in service at the beginning of the year?

(b) If switching to the straight-line method is allowed, when is the optimal time to switch?

8.10 Compute the double-declining-balance depreciation schedule for an asset with the following data:

Cost of the asset, I	$60,000
Useful life, N	8 years
Salvage value, S	$5,000

8.11 Compute the 150% DB depreciation schedule for an asset with the following data:

Cost of the asset, I	$12,000
Useful life, N	5 years
Salvage value, S	$2,000

(a) What is the value of α?

(b) What is the amount of depreciation for the first full year of use?

(c) What is the book value of the asset at the end of the fourth year?

8.12 Upjohn Company purchased new packaging equipment with an estimated useful life of five years. The cost of the equipment was $20,000, and the salvage value was estimated to be $3,000 at the end of year five. Compute the annual depreciation expenses through the five-year life of the equipment under each of the following methods of book depreciation:

(a) the straight-line method

(b) the double-declining-balance method (limit the depreciation expense in the fifth year to an amount that will cause the book value of the equipment at year end to equal the $3,000 estimated salvage value)

8.13 A secondhand bulldozer acquired at the beginning of the fiscal year at a cost of $58,000 has an estimated salvage value of $8,000 and an estimated useful life of 12 years. Determine the following:

(a) the amount of annual depreciation computed by the straight-line method

(b) the amount of depreciation for the third year computed by the double-declining-balance method

Units-of-Production Method

8.14 If a truck for hauling coal has an estimated net cost of $85,000 and is expected to give service for 250,000 miles, resulting in a salvage value of $5,000. Compute the allowed depreciation amount for the truck usage amounting to 55,000 miles.

8.15 A diesel-powered generator with a cost of $60,000 is expected to have a useful operating life of 50,000 hours. The expected salvage value of this generator is $8,000. In its first operating year, the generator was operated for 5,000 hours. Determine the depreciation for the year.

8.16 Ingot Land Company owns four trucks dedicated primarily to its landfill business. The company's accounting record indicates the following:

| Description | Truck | | | |
	A	B	C	D
Purchase cost	$50,000	$25,000	$18,500	$35,600
Salvage value	$5,000	$2,500	$1,500	$3,500
Useful life (miles)	200,000	120,000	100,000	200,000
Accumulated depreciation as year begins	$0	$1,500	$8,925	$24,075
Miles driven during year	25,000	12,000	15,000	20,000

Determine the amount of depreciation for each truck during the year.

Tax Depreciation

8.17 Zerex Paving Company purchased a hauling truck on January 1, 2003, at a cost of $32,000. The truck has a useful life of eight years, with an estimated salvage value of $5,000. The straight-line method is used for book purposes. For tax purposes, the truck would be depreciated using MACRS over its five-year class life. Determine the annual depreciation amount to be taken over the useful life of the hauling truck for both book and tax purposes.

8.18 The Harris Foundry Company purchased new casting equipment in 2003 at a cost of $180,000. Harris also paid $35,000 to have the equipment delivered and installed. The casting machine has an estimated useful life of 12 years, but it will be depreciated using MACRS over its seven-year class life.

(a) What is the cost basis of the casting equipment?

(b) What will be the depreciation allowance in each year of the seven-year class life for the casting equipment?

8.19 A piece of equipment uses the seven-year MACRS recovery period. Compute the book value for tax purposes at the end of three years. The cost basis is $100,000.

8.20 A piece of machinery purchased at a cost of $68,000 has an estimated salvage value of $9,000 and an estimated useful life of five years. It was placed in service on May 1 of the current fiscal year, which ends on December 31. The asset falls into a seven-year MACRS-property category. Determine the annual depreciation amounts over the machinery's useful life.

8.21 Suppose that a taxpayer places in service a $10,000 asset that is assigned to a new six-year MACRS property class with the half-year convention. Develop the MACRS deductions, assuming a 200% declining-balance rate switching to straight-line depreciation.

8.22 On April 1, Leo Smith paid $170,000 for a residential rental property. This purchase price represents $130,000 for the building and $40,000 for the land. Five years later, on November 1, he sold the property for $200,000. Compute the MACRS depreciation for each of the five calendar years during which he had the property.

8.23 In 2003, you purchased a spindle machine (seven-year MACRS property class) for $26,000, which you placed in service in January. Compute the depreciation allowances for the machine.

8.24 In 2003, three assets were purchased and placed in service by a firm:

Asset Type	Date Placed in Service	Cost Base	MACRS Property Class
Car	Feb. 17	$15,000	five year
Furniture	Mar. 25	$5,000	seven year
Copy machine	Apr. 3	$10,000	five year

Compute the depreciation allowances for each asset.

8.25 On October 1, you purchased a residential home in which to locate your professional office for $150,000. The appraisal is divided into $30,000 for the land and $120,000 for the building.

(a) In your first year of ownership, how much can you deduct for depreciation for tax purposes?

(b) Suppose that you sold the property at $187,000 at the end of the fourth year of ownership. What is the book value of the property?

8.26 Consider the data in the following two tables:

First cost	$80,000
Book depreciation life	7 years
MACRS property class	7 year
Salvage value	$24,000

Depreciation Schedule

n	A	B	C	D
1	$8,000	$22,857	$11,429	$22,857
2	$8,000	$16,327	$19,592	$16,327
3	$8,000	$11,661	$13,994	$11,661
4	$8,000	$5,154	$9,996	$8,330
5	$8,000	$0	$7,140	$6,942
6	$8,000	$0	$7,140	$6,942
7	$8,000	$0	$7,140	$6,942
8	$0	$0	$3,570	$0

Identify the depreciation method used for each depreciation schedule as one of the following:

- double-declining-balance (DDB) depreciation;
- straight-line depreciation;
- DDB with conversion to straight-line depreciation, assuming a zero salvage value;
- MACRS seven-year depreciation with the half-year convention.

8.27 A manufacturing company has purchased three assets:

	Asset Type		
Item	Lathe	Truck	Building
Initial cost	$45,000	$25,000	$800,000
Book life	12 yr	200,000 miles	50 years
MACRS class	7 year	5 year	39 year
Salvage value	$3,000	$2,000	$100,000
Book depreciation	DDB	Unit production (UP)	SL

The truck was depreciated using the units-of-production method. Usage of the truck was 22,000 miles and 25,000 miles during the first two years, respectively.

(a) Calculate the book depreciation for each asset for the first two years.

(b) Calculate the tax depreciation for each asset for the first two years.

(c) If the lathe is to be depreciated over the early portion of its life by using the DDB method and then by switching to the SL method for the remainder of its life, when should the switch occur?

8.28 Flint Metal Shop purchased a stamping machine for $147,000 on March 1, 2000. It is expected to have a useful life of 10 years, salvage value of $27,000, production of 250,000 units, and number of working hours of 30,000. During 2003, Flint used the stamping machine for 2,450 hours to produce 23,450 units. From the information given, compute the book depreciation expense for 2003 under each of the following methods:

(a) straight-line

(b) units of production

(c) working hours

(d) double-declining-balance (without conversion to straight-line depreciation)

(e) double-declining-balance (with conversion to straight-line depreciation)

8.29 At the beginning of the fiscal year, Borland Company acquired new equipment at a cost of

$65,000. The equipment has an estimated life of five years and an estimated salvage value of $5,000.

(a) Determine the annual depreciation (for financial reporting) for each of the five years of estimated useful life of the equipment, the accumulated depreciation at the end of each year, and the book value of the equipment at the end of each year by (1) the straight-line method and (2) the double-declining-balance method.

(b) Determine the annual depreciation for tax purposes, assuming that the equipment falls into the seven-year MACRS property class.

(c) Assume that the equipment was depreciated under MACRS for a seven-year property class. In the first month of the fourth year, the equipment was traded in for similar equipment priced at $82,000. The trade-in allowance on the old equipment was $10,000, and cash was paid for the balance. What is the cost basis of the new equipment for computing the amount of depreciation for income-tax purposes?

8.30 A company purchased a new forging machine to manufacture disks for airplane turbine engines. The new press cost $3,500,000, and it falls into the seven-year MACRS property class. The company has to pay property taxes for ownership of this forging machine at a rate of 1.2% on the beginning book value of each year to the local township.

(a) Determine the book value of the asset at the beginning of each tax year.

(b) Determine the amount of property taxes over the machine's depreciable life.

Accounting Profits (Net Income)

8.31 Tiger Construction Company had a gross income of $20,000,000 in tax year one, $3,000,000 in salaries, $4,000,000 in wages, $800,000 in depreciation expenses, a loan principal payment of $200,000, and a loan interest payment of $210,000. Determine the net income of the company in tax year one.

8.32 A consumer electronics company was formed to sell a portable handset system. The company purchased a warehouse and converted it into a manufacturing plant for $2,000,000 (including the purchase price of the warehouse). It completed installation of assembly equipment worth $1,500,000 on December 31. The plant began operation on January 1. The company had a gross income of $2,500,000 for the calendar year. Manufacturing costs and all operating expenses, excluding the capital expenditures, were $1,280,000. The depreciation expenses for capital expenditures amounted to $128,000.

(a) Compute the taxable income of this company.

(b) How much will the company pay in federal income taxes for the year?

8.33 Valdez Corporation will commence operations on January 1, 2004. The company projects the following financial performance during its first year of operation:

- Sales revenues are estimated at $1,500,000.
- Labor, material, and overhead costs are projected at $600,000.
- The company will purchase a warehouse worth $500,000 in February. To finance this warehouse, on January 1 the company will issue $500,000 of long-term bonds, which carry an interest rate of 10%. The first interest payment will occur on December 31.
- For depreciation purposes, the purchase cost of the warehouse is divided into $100,000 for the land and $400,000 for the building. The building is classified into the 39-year MACRS real-property class and will be depreciated accordingly.
- On January 5, the company will purchase $200,000 of equipment that has a five-year MACRS class life.

(a) Determine the total depreciation expenses allowed in 2004.

(b) Determine Valdez's tax liability in 2004.

Net Income versus Cash Flow

8.34 Quick Printing Company had sales revenue of $1,250,000 from operations during tax year one. Here are some operating data on the company for that year:

Labor expenses	$550,000
Materials costs	$185,000
Depreciation expenses	$32,500
Interest income on time deposit	$6,250
Bond interest income	$4,500
Interest expenses	$12,200
Rental expenses	$45,000
Dividend payment to Quick's shareholders	$40,000
Proceeds from sale of old equipment that had a book value of $20,000	$23,000

(a) What is Quick's taxable gains?
(b) What is Quick's taxable income?
(c) What is Quick's marginal and effective (average) tax rate?
(d) What is Quick's net cash flow after tax?

8.35 In a particular year Elway Aerospace Company had gross revenues of $1,200,000 from operations. The following financial transactions were posted during the year:

Manufacturing expenses (including depreciation of $45,000)	$450,000
Operating expenses (excluding interest expenses)	$120,000
A new short-term loan from a bank	$50,000
Interest expenses on borrowed funds (old and new)	$40,000
Dividends paid to common stockholders	$80,000
Old equipment sold	$60,000

The old equipment had a book value of $75,000 at the time of sale.

(a) What is Elway's income tax liability?
(b) What is Elway's operating income?
(c) What is the net cash flow?

Gains or Losses

8.36 Consider a five-year MACRS asset that was purchased at $60,000. (Note that a five-year MACRS property class is depreciated over six years, due to the half-year convention. The applicable salvage values would be $20,000 in year three, $10,000 in year five, and $5,000 in year six.) Compute the gain or loss amounts if the asset were disposed of in the following years:

(a) year three
(b) year five
(c) year six

8.37 An electrical appliance company purchased an industrial robot costing $300,000 in year zero. The industrial robot, to be used for welding operations, is classified as a seven-year MACRS recovery property. If the robot is to be sold after five years, compute the amounts of gains (losses) for the following three salvage values (assume that both capital gains and ordinary incomes are taxed at 34%):

(a) $10,000
(b) $125,460
(c) $200,000

8.38 LaserMaster, Inc., a laser-printing service company, had sales revenues of $1,250,000 during tax year 2003. The following table provides other financial information relating to the tax year:

Labor expenses	$550,000
Material costs	$185,000
Depreciation	$32,500
Interest income	$6,250
Interest expenses	$12,200
Rental expenses	$45,000
Proceeds from the sale of old printers	$23,000

The sold printers had a combined book value of $20,000 at the time of sale.

(a) Determine the taxable income for the tax year.

(b) Determine the taxable gains for the tax year.

(c) Determine the amount of income taxes and gains taxes (or loss credits) for the tax year.

8.39 A machine now in use that was purchased three years ago at a cost of $4,000 has a book value of $2,000. It can be sold now for $2,500, or it could be used for three more years, at the end of which time it would have no salvage value. The annual O&M costs amount to $10,000 for the machine. If the machine is sold, a new machine can be purchased at an invoice price of $14,000 to replace the present equipment. Freight will amount to $800, and the installation cost will be $200. The new machine has an expected service life of five years and will have no salvage value at the end of that time. With the new machine, the expected direct cash savings amount to $8,000 the first year and $7,000 in O&M for each of the next two years. Corporate income taxes are at an annual rate of 40%, and the net capital gain is taxed at the ordinary income-tax rate. The present machine has been depreciated according to a straight-line method, and the proposed machine would be depreciated on a seven-year MACRS schedule. Consider each of the following questions independently:

(a) If the old asset is to be sold now, what would be the amount of its equivalent book value?

(b) For depreciation purposes, what would be the first cost of the new machine (depreciation base)?

(c) If the old machine is to be sold now, what would be the amount of taxable gains and gains taxes?

(d) If the old machine is sold for $5,000 now instead of $2,500, what would be the amount of gains tax?

(e) If the old machine had been depreciated by using 175% DB and then switching to SL depreciation, what would be the current book value?

(f) If the old machine were not replaced by the new one and has been depreciated by the 175% DB method, when would be the time to switch from DB to SL depreciation?

Short Case Studies with Excel

8.40 Chuck Robbins owns and operates a small unincorporated electrical service business, Robbins Electrical Service (RES). Chuck is married with two children, so he claims four exemptions on his tax return. As business grows steadily, tax considerations are important to him. Therefore, Chuck is considering incorporation of the business. Under either form of the business, the family will initially own 100% of the firm. Chuck plans to finance the firm's expected growth by drawing a salary just sufficient for his family's living expenses and by retaining all other income in the business. He estimates the income and expenses over the next three years to be as follows:

	Year 1	Year 2	Year 3
Gross income	$80,000	$95,000	$110,000
Salary	$40,000	$45,000	$50,000
Business expenses	$15,000	$20,000	$30,000
Personal exemptions	$12,000	$12,000	$12,000
Itemized deductions	$6,000	$8,000	$10,000

Which form of business (corporation or sole ownership) will allow Chuck to pay the lowest taxes (and retain the most income) during the three years? Personal income-tax brackets and amounts of personal exemption are updated yearly, so you need to consult the IRS tax manual for the tax rates as well as for the exemptions that are applicable to the tax years.

8.41 Electronic Measurement and Control Company (EMCC) has developed a laser speed detector that emits infrared light invisible to humans and radar detectors alike. For full-scale commercial marketing, EMCC needs to invest $5 million in new manufacturing facilities. The system is priced at $3,000 per unit. The company expects to sell 5,000 units annually over the next five years. The new manufacturing facilities will be depreciated according to a seven-year MACRS property class. The expected salvage value of the manufacturing facilities at the end of five years is $1.6 million. The manufacturing cost for the detector is $1,200 per unit, excluding depreciation expenses. The operating and maintenance costs are expected to run to $1.2 million per year. EMCC has a combined federal and state income-tax rate of 35%, and undertaking this project will not change this current marginal tax rate.

(a) Determine the incremental taxable income, income taxes, and net income due to undertaking this new product for the next five years.

(b) Determine the gains or losses associated with the disposal of the manufacturing facilities at the end of five years.

8.42 Diamonid is a start-up diamond-coating company planning to manufacture a microwave plasma reactor that synthesizes diamonds. Diamonid anticipates that the industry demand for diamonds will skyrocket over the next decade for use in industrial drills, high-performance microchips, and artificial human joints, among other things. Diamonid has decided to raise $50 million through the issuance of common stocks for investment in a plant ($10 million) and equipment ($40 million). Each reactor can be sold at a price of $100,000 per unit. Diamonid can expect to sell 300 units per year during the next eight years. The unit manufacturing cost is estimated at $30,000, excluding depreciation. The operating and maintenance cost for the plant is estimated at $12 million per year. Diamonid expects to phase out the operation at the end of eight years and revamp the plant and equipment to adopt a new diamond-manufacturing technology. At that time, Diamonid estimates that the salvage values for the plant and equipment will be about 60% and 10% of the original investments, respectively. The plant and equipment will be

depreciated according to the guidelines for the 39-year real-property class (placed in service in January) and the seven-year MACRS property class, respectively. Diamonid pays 5% of state and local income taxes on its taxable income.

(a) If the 2003 corporate tax system continues over the project life, determine the combined state and federal income-tax rate for each year.

(b) Determine the gains or losses at the time of plant revamping.

(c) Determine the net income each year over the plant's life.

8.43 Julie Magnolia has $50,000 cash to invest for three years. Two types of bonds are available for consideration. She can buy a tax-exempt Arizona state bond that pays interest of 9.5% per year, or she can buy a corporate bond. Julie's marginal tax rate is 25% for both ordinary income and capital gains. Assume that any investment decision considered will not change her marginal tax bracket.

(a) If Julie were looking for a corporate bond that was just as safe as the state bond, what interest rate on the corporate bond is required so that Julie would be indifferent between the two bonds? There are no capital gains or losses at the time of trading the bond.

(b) In (a), suppose at the time of trading (year three) that the corporate bond is expected to be sold at a price 5% higher than its face value. What interest rate on the corporate bond is required so that Julie would be indifferent between the two bonds?

(c) Alternatively, Julie can invest the amount in a tract of land that could be sold at $75,000 (after paying the real-estate commission) at the end of year three. Is this investment better than the state bond?

Improving Checkout Service at Home Depot[1]

Home Depot, Inc., moving to quicken customer checkouts, plans to install self-serve aisles in about half its stores by the end of next year. Under the system, customers can scan their own merchandise and pay the total themselves with a debit card, credit card, or cash.

The move is part of a larger revamping of Home Depot's cashier operations. And it comes as the home-improvement giant overhauls dated computer operations throughout its 1,487-store system.

In general, consumers are becoming more comfortable with self-serve checkouts, now available in grocery stores and certain retailers, Home Depot said. The Atlanta-based company said it will be the first major home-improvement retailer to roll them out, aiming to place self-serve checkouts in about 300 high-volume stores by the end of this year and another 500 stores by the end of next year. At high-volume stores, particularly on Saturday mornings, lines can frustrate some shoppers.

Home Depot has tested self-checkout in about 17 stores and says its do-it-yourself customers have stepped right up. "What we're seeing is they like to be in control over the transaction," said Rob Lynch, a vice president of operations for the company. "It's surprising what customers take up there." Among the bulky items checked through: trash cans, big tubs, plastic gutter hosing, and two-by-fours.

When checking large or bulky items, customers can peel off the bar-code labels and run them across stationary scanners, or a Home Depot employee can help them. In general, one employee—armed with a cordless scanner—will assist and monitor four self-check aisles.

Customers also can use the self-serve aisles to pay for bulky items that are stacked up outside the stores, like pine straw. They do this by using a touch screen to identify the item and then ordering the item and paying for it. Home Depot said that, with proper supervision of the self-serve aisles, it can keep shoplifting to a minimum. Also, the self-checkout units can automatically weigh merchandise as customers pass it through, which helps store security monitor what the customers are doing.

[1] Source: Dan Morse, "Home Depot Aims to Quicken Checkout with Self-Service," The Wall Street Journal, December 3, 2001, Page B3.

Project Cash Flow Analysis

The self-checkout devices are part of a larger project that Home Depot calls FAST, or Front-End Accuracy and Service Transformation, which will cost in excess of "tens of millions" of dollars. As part of this project, live cashiers will have more intuitive touch-screen machines that are designed to reduce the amount of merchandise that slips through without being rung up.

According to Home Depot, it costs tens of millions dollars to install the self-checkout systems. The question is how to estimate the projected savings from the systems. Certainly, the system will reduce the number of cashiers to hire, reduce the amount of shoplifting that takes place, and improve customer satisfaction. Projecting cash flows is the most important—and the most difficult—step in the analysis of a capital project. Typically, a capital project initially requires investment outlays and only later produces annual net cash inflows. A great many variables are involved in forecasting cash flows, and many individuals, ranging from engineers to cost accountants and marketing people, participate in the process. This chapter provides the general principles on which the determination of a project's cash flows is based.

9-1   Understanding Project Cost Elements

First, we need to understand the types of costs that must be considered in estimating project cash flows. Because there are many types of costs, each is classified differently according to the immediate needs of management. For example, engineers may want cost data in order to prepare external reports, to prepare planning budgets, or to make decisions. Also, each different use of cost data demands a different classification and definition of cost. For example, the preparation of external financial reports requires the use of historical cost data, whereas decision making may require current cost or estimated future cost data.

9.1.1   Classifying Costs for Manufacturing Environments

Our initial focus in this chapter is on manufacturing companies, because their basic activities (such as acquiring raw materials, producing finished goods, and marketing) are commonly found in most other businesses. Therefore, an understanding of the costs of a manufacturing company can be very helpful for understanding costs in other types of business organizations.

Manufacturing Costs   The several types of manufacturing costs incurred by a typical manufacturer are illustrated in Figure 9.1. In converting raw materials into finished goods, a manufacturer incurs various costs of operating a factory. Most

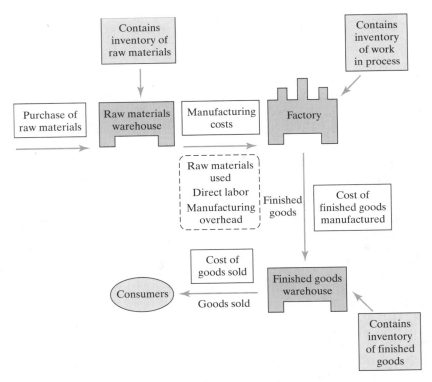

FIGURE 9.1 **Various types of manufacturing costs**

manufacturing companies divide manufacturing costs into three broad categories—direct materials, direct labor, and manufacturing overhead:

- **Direct Materials:** Direct raw materials refer to any materials that are used in the final product and that can be easily traced into it. Some examples are wood in furniture, steel in bridge construction, paper in printed products, and fabric in clothing. It is also important to note that the finished product of one company can become the raw materials of another company. For example, the computer chips produced by Intel are a raw material used by Dell in its personal computers.

- **Direct Labor:** Just as the term "direct materials" refers to materials costs for the final product, direct labor refers to those labor costs that go into the fabrication of a product. The labor costs of assembly-line workers, for example, would be direct labor costs, as would the labor costs of welders in metal-fabricating industries, carpenters and bricklayers in home-building businesses, and machine operators in various manufacturing operations.

- **Manufacturing Overhead:** Manufacturing overhead, the third type of manufacturing cost, includes all costs of manufacturing except direct materials and direct labor. In particular, it includes such items as indirect materials,[2] indirect labor,[3] maintenance and repairs on production equipment, heat and light, property taxes, depreciation, insurance on manufacturing facilities, and overtime premium. The most important thing to note about manufacturing overhead is the fact that, unlike direct materials and direct labor, it is not easily traceable to specific units of output. In addition, many manufacturing overhead costs do not change as output changes, as long as the production volume stays within the capacity.

Nonmanufacturing Costs   There are two additional types of cost incurred in supporting any manufacturing operation. They are (1) operating costs, such as warehouse leasing and vehicle rentals, and (2) marketing (or selling) and administrative expenses. Marketing or selling costs include all costs necessary to secure customer orders and get the finished product or service into the hands of the customer. Cost breakdowns of these types provide data for control over selling and administrative functions in the same way that manufacturing-cost breakdowns provide data for control over manufacturing functions. For example, a company incurs costs for the following nonmanufacturing items:

- **Overhead:** Heat and light, property taxes, depreciation, and similar items associated with its selling and administrative functions.

- **Marketing:** Advertising, shipping, sales travel, sales commissions, and sales salaries.

---

[2]Sometimes, it may not be worth the effort to trace the costs of relatively insignificant materials to the finished products. Such minor items would include the solder used to make electrical connections in a computer circuit board or the glue used to bind this textbook. Materials such as solder and glue are called indirect materials and are included as part of manufacturing overhead.

[3]Sometimes, we may not be able to trace some of the labor costs to the creation of product. We treat this type of labor cost as a part of manufacturing overhead, along with indirect materials. Indirect labor includes the wages of janitors, supervisors, material handlers, and night security guards. Although the efforts of these workers are essential to production, it would be either impractical or impossible to trace their costs to specific units of product. Therefore, we treat such labor costs as indirect labor.

 • **Administrative functions:** Executive compensation, general accounting, public relations, and secretarial support. Administrative costs include all executive, organizational, and clerical costs associated with the general management of an organization.

9.1.2 Classifying Costs for Financial Statements

For purposes of preparing financial statements, we often classify costs as either period costs or product costs. To understand the difference between period costs and product costs, we must introduce the matching concept essential to accounting studies. In financial accounting, the **matching principle** states that *the costs incurred to generate particular revenue should be recognized as expenses in the same period that the revenue is recognized.* This matching principle is the key to distinguishing between period costs and product costs. Some costs are matched against periods and become expenses immediately. Other costs, however, are matched against products and do not become expenses until the products are sold, which may be in the following accounting period.

Period Costs   **Period costs** are those costs that are charged to expenses in the period in which the expenses are incurred. The underlying assumption is that the associated benefits are received in the same period as the expenses are incurred. Some specific examples of period costs are all general and administrative expenses, selling expenses, insurance, and income-tax expenses. Therefore, advertising costs, executive salaries, sales commissions, public-relations costs, and the other nonmanufacturing costs discussed earlier would all be period costs. Such costs are not related to the production and flow of manufactured goods, but are deducted from revenue in the income statement. In other words, period costs will appear on the income statement as expenses in the time period in which they occur.

Product Costs   Some costs are better matched against products than they are against periods. Costs of this type—called **product costs**—consist of the costs involved in the purchase or manufacturing of goods. In the case of manufactured goods, these costs consist of direct materials, direct labor, and manufacturing overhead. Product costs are not viewed as expenses; rather, they are the cost of creating inventory. Thus, product costs are considered an asset until the related goods are sold. At this point of sale, the costs are released from inventory as expenses (typically called *cost of goods sold*) and matched against sales revenue. Since product costs are assigned to inventories, they are also known as *inventory costs*. In theory, product costs include all manufacturing costs—that is, all costs relating to the manufacturing process. Product costs appear on financial statements when the inventory, or final good, is sold, not when the product is manufactured.

Cost Flows in a Manufacturing Company   To understand product costs more fully, we now look briefly at the flow of costs in a manufacturing company. By doing so, we will be able to see how product costs move through the various accounts and affect the balance sheet and the income statement in the course of the manufacture and sale of goods. The flows of period costs and product costs through the financial statements are illustrated in Figure 9.2. All product costs filter

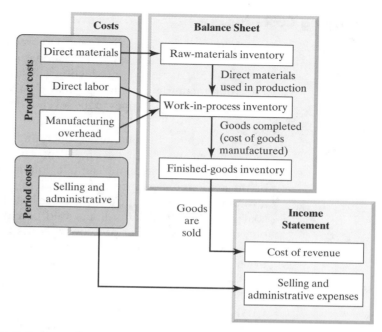

FIGURE 9.2 Cost flows and classifications in a manufacturing company

through the balance-sheet statement in the name of "inventory cost." If a product is sold, its inventory costs in the balance-sheet statement are transferred to the income statement in the name of "cost of goods sold." There are three types of inventory cost reflected in the balance sheet:

- **Raw-materials inventory:** This balance-sheet entry represents the unused portion of the raw materials on hand at the end of the fiscal year.

- **Work-in-process inventory:** This balance-sheet entry consists of the partially completed goods on hand in the factory at year end. When raw materials are used in production, their costs are transferred to the work-in-process inventory account as direct materials. Note that direct-labor costs and manufacturing overhead costs are also added directly to the work-in-process entry. The work-in-process concept can be viewed as the assembly line in a manufacturing plant, where workers are stationed and where products slowly take shape as they move from one end of the assembly line to the other.

- **Finished-goods inventory:** This balance-sheet entry shows the cost of finished goods on hand and awaiting sale to customers at year end. As goods are completed, accountants transfer the corresponding cost in the work-in-process account into the finished-goods account. Here, the goods await sale to a customer. As goods are sold, their cost is transferred from finished goods into cost of goods sold (or cost of revenue). At this point, we finally treat the various material, labor, and overhead costs that were involved in the manufacture of the units being sold as *expenses* in the income statement.

9.1.3 Classifying Costs for Predicting Cost Behavior

In project cash flow analysis, we need to predict how a certain cost will behave in response to a change in activity. For example, a manager may want to estimate the impact that a 5% increase in production will have on the company's total wages before a decision to alter production is made. **Cost behavior** describes how a cost item will react or respond to changes in the level of business activity.

Volume Index   In general, the operating costs of any company are likely to respond in some way to changes in its operating volume. In studying cost behavior, we need to determine some measurable volume or activity that has a strong influence on the amount of cost incurred. The unit of measure used to define volume is called a **volume index**. A volume index may be based on production inputs (such as tons of coal processed, direct labor-hours used, or machine-hours worked) or on production outputs (such as number of kilowatt-hours generated). For example, for a vehicle, the number of miles driven per year may be used as a volume index. Once we identify a volume index, we try to find out how costs change in response to changes in this volume index.

Cost Behaviors   Accounting systems typically record the cost of resources acquired and track their subsequent usage. Fixed costs and variable costs are the two most common cost behavior patterns. The costs in an additional category known as "mixed (semivariable) costs" contain two parts: The first part of cost is fixed, and the other part is variable as the volume of output varies.

Fixed Costs   The costs of providing a company's basic operating capacity are known as its **fixed costs** or **capacity costs**. For a cost item to be classified as fixed, it must have a relatively wide span of output for which costs are expected to remain constant. (See Figure 9.3.) This span is called the **relevant range**. In other words, fixed costs do not change within a given time period, although volume may change. For our previous automobile example, the annual insurance premium, property tax,

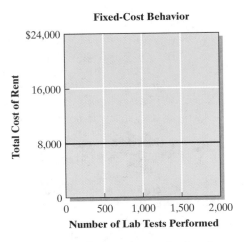

FIGURE 9.3 Fixed-cost behavior

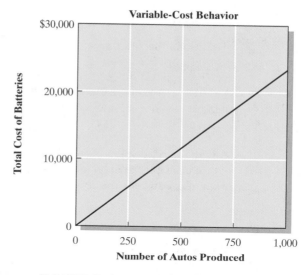

FIGURE 9.4 Variable-cost behavior

and license fee are fixed costs, since they are independent of the number of miles driven per year. Some typical examples of fixed costs are building rents; depreciation of buildings, machinery, and equipment; and salaries of administrative and production personnel.

Variable Costs   In contrast to fixed operating costs, variable operating costs have a close relationship to the level of volume. (See Figure 9.4.) If, for example, volume increases 10%, a total variable cost will also increase by approximately 10%. Gasoline is a good example of a variable automobile cost, as fuel consumption is directly related to miles driven. Similarly, the tire replacement cost will also increase as a vehicle is driven more. In a typical manufacturing environment, direct labor and material costs are major variable costs. The difference between the unit sales price and the unit variable cost is known as **marginal contribution**. This means each unit sold contributes toward absorbing the company's fixed costs.

Mixed Costs   Some costs do not fall precisely into either the fixed or the variable category, but contain elements of both. We refer to these costs as mixed costs (or **semivariable costs**). In our automobile example, **depreciation** (loss of value) is a mixed cost. Some depreciation occurs simply from the passage of time, regardless of how many miles a car is driven, and this amount represents the fixed portion of depreciation. On the other hand, the more miles an automobile is driven a year, the faster it loses its market value, and this amount represents the variable portion of depreciation. A typical example of a mixed cost in manufacturing is the cost of electric power. Some components of power consumption, such as lighting, are independent of operating volume, while other components are likely to vary directly with volume (e.g., number of machine-hours operated).

9-2 Why Do We Need to Use Cash Flows in Economic Analysis?

Traditional accounting stresses net income as a means of measuring a firm's profitability, but it is desirable to discuss why cash flows are relevant data to be used in project evaluation. As seen in Section 9.1.2, net income is an accounting measure based, in part, on the **matching concept**. Costs become expenses as they are matched against revenue. The actual timing of cash inflows and outflows is ignored.

Over the life of a firm, net incomes and net cash inflows will usually be the same. However, the timing of incomes and cash inflows can differ substantially. Given the time value of money, it is better to receive cash now rather than later, because cash can be invested to earn more cash. (You cannot invest net income.) For example, consider the following income and cash flow schedules of two firms over two years:

		Company A	Company B
Year 1	Net income	$1,000,000	$1,000,000
	Cash flow	$1,000,000	$0
Year 2	Net income	$1,000,000	$1,000,000
	Cash flow	$1,000,000	$2,000,000

Both companies have the same amount of net income and cash sum over two years, but Company A returns $1 million cash yearly, while Company B returns $2 million at the end of the second year. Company A could invest the $1 million it receives at the end of the first year at 10%, for example. In this case, while Company B receives only $2 million in total at the end of the second year, Company A receives $2.1 million in total.

Cash Flow vs. Net Income

Net income: Net income is an accounting means of measuring a firm's profitability, using the matching concept. Costs become expenses as they are matched against revenue. The actual timing of cash inflows and outflows is ignored.

Cash flow: Given the time value of money, it is better to receive cash now rather than later, because cash can be invested to earn more money. This factor is the reason that cash flows are relevant data to use in project evaluation.

9-3   Income-Tax Rate to Be Used in Economic Analysis

As we have seen in Chapter 8, average (effective) income-tax rates for corporations vary with the level of taxable income from 0 to 35%. Suppose that a company now paying a tax rate of 25% on its current operating income is considering a profitable investment. What tax rate should be used in calculating the taxes on the investment's projected income? As we will explain, the choice of the rate depends on the incremental effect the investment has on taxable income. In other words, *the tax rate to use is the rate that applies to the additional taxable income projected in the economic analysis*.

To illustrate, consider ABC Corporation, whose taxable income from operations is expected to be $70,000 for the current tax year. ABC's management wishes to evaluate the incremental tax impact of undertaking a particular project during the same tax year. The revenues, expenses, and taxable incomes before and after the project are estimated as follows:

	Before	After	Incremental
Gross revenue	$200,000	$240,000	$40,000
Salaries	$100,000	$110,000	$10,000
Wages	$30,000	$40,000	$10,000
Taxable income	$70,000	$90,000	$20,000

Because the income-tax rate is progressive, the tax effect of the project cannot be isolated from the company's overall tax obligations. The base operations of ABC without the project are projected to yield a taxable income of $70,000. With the new project, the taxable income increases to $90,000. Using the tax computation formula in Table 8.6, we find that the corporate income taxes with and without the project, respectively, are as follows:

$$\text{Income tax with the project} = \$13,750 + 0.34(\$90,000 - \$75,000)$$
$$= \$18,850;$$
$$\text{Income tax without the project} = \$7,500 + 0.25(\$70,000 - \$50,000)$$
$$= \$12,500.$$

The additional income tax is then $18,850 − $12,500 = $6,350. The $6,350 tax on the additional $20,000 of taxable income, a rate of 31.75%, is an incremental rate. This is the rate we should use in evaluating the project in isolation from the rest of ABC's operations. As shown in Figure 9.5, the 31.75% is not an arbitrary figure, but a weighted average of two distinct marginal rates. Because the new project pushes ABC into a higher tax bracket, the first $5,000 it generates is taxed at 25%; the remaining $15,000 it generates is taxed in the higher bracket, at 34%. Thus, we could have calculated the incremental tax rate as follows:

$$0.25(\$5,000/\$20,000) + 0.34(\$15,000/\$20,000) = 31.75\%.$$

$$0.25(\$5,000/\$20,000) + 0.34(\$15,000/\$20,000) = 31.75\%$$

	Before	After	Incremental
Taxable income	$70,000	$90,000	$20,000
Income taxes	$12,500	$18,850	$6,350
Average tax rate	17.86%	20.94%	
Incremental tax rate			31.75%

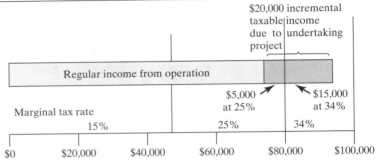

FIGURE 9.5 Illustration of incremental tax rate

The average tax rates before and after the new project being considered would be as shown in the following table:

	Before	**After**	**Incremental**
Taxable income	$70,000	$90,000	$20,000
Income taxes	$12,500	$18,850	$6,350
Average tax rate	17.86%	20.94%	
Incremental tax rate			31.75%

    However, in conducting an economic analysis of an individual project, neither one of the company-wide average rates is appropriate; we want the incremental rate applicable to just the new project for use in generating its cash flows.

    A corporation with continuing base operations that place it consistently in the highest tax bracket will have both marginal and average federal tax rates of 35%. For such firms, the tax rate on an additional investment project is, naturally, 35%. If state income taxes are considered, the combined state and federal marginal tax rate may be close to 40%. But for corporations in lower tax brackets, and those that fluctuate between losses and profits, marginal and average tax rates are likely to vary. For such corporations, estimating a prospective incremental tax rate for a new investment project may be difficult. The only solution may be to perform scenario analysis, which is to examine, for each potential situation, how much the income tax fluctuates from undertaking the project. (In other words, calculate the total taxes and the incremental taxes for each scenario.)

9-4   Incremental Cash Flows from Undertaking a Project

When a company purchases a fixed asset such as equipment, it makes an investment. The company commits funds today with the expectation of earning a return on those funds in the future. For a fixed asset, the future return is in the form of cash flows generated by the profitable use of the asset. In evaluating a capital investment, we are concerned only with those cash flows that result directly from the investment. These cash flows, called **differential** or **incremental cash flows**, represent the change in the firm's total cash flow that occurs as a direct result of the investment. In this section, we will look into some of the cash flow elements common to most investments. Once the cash flow elements are determined (both inflows and outflows), we may group them into three areas based on their use or sources: (1) cash flow elements associated with operations, (2) cash flow elements associated with investment activities (such as capital expenditures), and (3) cash flow elements associated with project financing (such as borrowing). The main purpose of grouping cash flows in this way is to provide information about the operating, investing, and financing activities of a project.

9.4.1   Operating Activities

In general, cash flows from operations include current sales revenues, cost of goods sold, operating expenses, and income taxes. Cash flows from operations should generally reflect the cash effects of transactions entering into the determination of net income. The interest portion of a loan repayment is a deductible expense allowed when determining net income and is included in the operating activities. Since we usually look only at yearly flows, it is logical to express all cash flows on a yearly basis.

As we discussed in Section 8.4.3, we can determine the net cash flow from operations by using either (1) the net income or (2) the cash flow by computing income taxes in a separate step. When we use net income as the starting point for cash flow determination, we should add any noncash expenses (mainly depreciation and amortization expenses) to net income in order to estimate the net cash flow from the operation.

Recall that depreciation (or amortization) is not a cash flow, but is deducted, along with operating expenses and lease costs, from gross income to find taxable income and therefore taxes. Accountants calculate net income by subtracting taxes from taxable income. But depreciation—which, again, is not a cash flow—was subtracted to find taxable income, so it must be added back into net income if we wish to use the net-income figure as an intermediate step along the path to determining after-tax cash flow. Mathematically, it is easy to show that the two approaches are identical:

Cash flow from operation = net income + (depreciation and amortization).

9.4.2   Investing Activities

In general, three types of investment flows are associated with buying a piece of equipment: the original investment, the salvage value at the end of the equipment's useful life, and the **working-capital investment** (or recovery). Here, *the investment in working capital typically refers to the investment made in nondepreciable assets*, such as carrying raw-material inventories. We will assume that our outflows for both capital investment and working-capital investment take place in year zero. It is possible, however, that both investments will not occur instantaneously, but rather over a few months as the project gets into gear; we could then use year one as an investment year. (Capital expenditures may occur over several years before a large investment project becomes fully operational. In this case, we should enter all expenditures as they occur.) For a small project, either method of timing these flows is satisfactory, because the numerical differences are likely to be insignificant.

9.4.3 Financing Activities

Cash flows classified as financing activities include (1) the amount of borrowing and (2) the repayment of principal. Recall that interest payments are tax-deductible expenses, so they are classified as operating, not financing, activities.

The net cash flow for a given year is simply the sum of the net cash flows from operating, investing, and financing activities. Figure 9.6 can be used as a road map when you set up a cash flow statement, because it classifies each type of cash flow element as an operating, investing, or financing activity.

9–5   Developing Project Cash Flow Statements

In this section, we will illustrate through a series of numerical examples how we actually prepare a project's cash flow statement; a generic version is shown in Figure 9.6, where we first determine the net income from operations and then adjust the net income by adding any noncash expenses, mainly depreciation (or amortization). We will also consider a case in which a project generates a negative taxable income for an operating year.

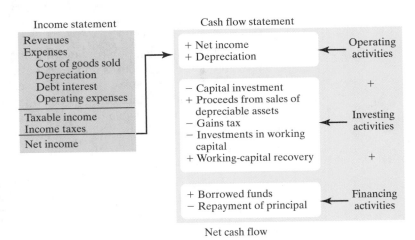

FIGURE 9.6 A typical format used in presenting a net cash flow statement

9.5.1   When Projects Require Only Operating and Investing Activities

We will start with the simple case of generating after-tax cash flows for an investment project with only operating and investment activities. In the next section, we will add complexities to this problem by including financing borrowing activities.

Example 9.1 Cash Flow Statement with Only Operating and Investing Activities

A computerized machining center has been proposed for a small tool manufacturing company. If the new system, which costs $125,000, is installed, it will generate annual revenues of $100,000 and will require $20,000 in annual labor, $12,000 in annual material expenses, and another $8,000 in annual overhead (power and utility) expenses. It also requires an investment in working capital in the amount of $23,331, which will be recovered at the end of year five.

The automation facility would be classified as a seven-year MACRS property. The company expects to phase out the facility at the end of five years, at which time it will be sold for $50,000. Find the year-by-year after-tax net cash flow for the project at a 40% marginal tax rate, and determine the after-tax net present worth of the project at the company's cost of capital of 15%.

Discussion: We can approach the problem in two steps by using the format shown in Figure 9.6 to generate an income statement and then a cash flow statement. In year zero (that is, at present), we have an investment cost of $125,000 for the equipment.[4] This cost will be depreciated in years one through five. The revenues and costs are uniform annual flows in years one through five. Once we find depreciation allowances for each year, we can easily compute the results for years one through four, which have fixed revenue and expense entries along with the variable depreciation charges. In year five, we will need to incorporate the salvage value and any gains tax from the asset's disposal.

We will employ the business convention that no signs (positive or negative) be used in preparing the income statement, except in the situation where we have a negative taxable income or tax savings. In this situation, we will use () to denote a negative entry. However, in preparing the cash flow statement, we will observe explicitly the sign convention: A positive sign indicates a cash inflow; a negative sign or () indicates a cash outflow.

SOLUTION

Given: Aforementioned cash flow information.
Find: After-tax cash flow.

Table 9.1 presents the cash flow statement.

---

[4]We will assume that the asset is purchased and placed in service at the beginning of year one (or the end of year zero) and that the first year's depreciation will be claimed at the end of year one.

TABLE 9.1 Cash Flow Statement for the Automated Machining Center Project

	A	B	C	D	E	F	G
1	**Income Statement**						
2		0	1	2	3	4	5
3							
4	Revenues		$100,000	$100,000	$100,000	$100,000	$100,000
5	Expenses:						
6	Labor		20,000	20,000	20,000	20,000	20,000
7	Material		12,000	12,000	12,000	12,000	12,000
8	Overhead		8,000	8,000	8,000	8,000	8,000
9	Depreciation		17,863	30,613	21,863	15,613	5,581
10							
11	Taxable Income		$42,137	$29,387	$38,137	$44,387	$54,419
12	Income Taxes (40%)		16,855	11,755	15,255	17,755	21,768
13							
14	Net Income		$25,282	$17,632	$22,882	$26,632	$32,651
15							
16	**Cash Flow Statement**						
17							
18	Operating Activities:						
19	Net Income		25,282	17,632	22,882	26,632	32,651
20	Depreciation		17,863	30,613	21,863	15,613	5,581
21	Investment Activities:						
22	Investment	(125,000)					
23	Salvage						50,000
24	Gains Tax						(6,613)
25	Working Capital	(23,331)					23,331
26							
27	Net Cash Flow	$(148,331)	$43,145	$48,245	$44,745	$42,245	$104,950

The following notes explain the essential items in Table 9.1:

- **Depreciation calculation**

    1. If the equipment is held for all eight years, we can depreciate it with respective percentages of 14.29%, 24.49%, 17.49%, 12.49%, 8.93%, 8.92%, 8.93%, and 4.46% for a seven-year property. (See Table 8.4.)

    2. If the asset is sold at the end of the fifth tax year (during the recovery period), the applicable depreciation amounts would be $17,863, $30,613, $21,863, $15,613, and $5,581. Since the asset is disposed of in the fifth tax year, the last year's depreciation, which would ordinarily be $11,163, is halved, because of the half-year convention.

    We now have a value for our unknown D_n, which will enable us to complete the statement for years one through five. The results of these simple calculations appear in Table 9.1.

- **Salvage value and gain taxes**

    In year five, we must deal with two aspects of the asset's disposal: salvage value and gains (both ordinary and capital). We list the estimated salvage value as a positive cash flow. Taxable gains are calculated as follows:

    1. The total depreciation in years one through five is $17,863 + $30,613 + $21,863 + $15,613 + $5,581 = $91,533.

    2. The book value at the end of year five is the cost basis minus the total depreciation, or $125,000 − $91,533 = $33,467.

    3. The gains on the sale are the salvage value minus the book value, or $50,000 − $33,467 = $16,533. (The salvage value is less than the cost basis, so all the gain is ordinary.)

    4. The tax on the ordinary gains is $16,533 × 40% = $6,613. This is the amount placed in the table under "Gains tax."

- **An alternative way of preparing the cash flow statement**

    A tabular format widely used in traditional engineering economics texts is shown in Table 9.2. Unlike with the income-statement approach shown in Table 9.1, you may skip the entire process of computing the net income, thereby computing income taxes directly. However, most firms prepare the cash flow statement based on the format similar to Table 9.1, as they want to know both net income as well as cash flow figures. Therefore, we will use the cash flow statement based on the format in Table 9.1 whenever possible throughout this text.

- **Investment analysis**

    Once we obtain the project's after-tax net cash flows, we can determine their equivalent present worth at the firm's discount rate. Since this series does not

TABLE 9.2 Traditional Net Cash Flow Statement

A	B	C	D	E	F	G	H	I	J
Year End	Investment & Salvage Value	Revenue	Labor	Materials	Overhead	Depreciation	Taxable Income	Income Taxes	Net Cash Flow
0	−$125,000								−$148.331
	−$23,331								
1		$100,000	$20,000	$12,000	$8,000	$17,863	$42,137	$16,855	$43,145
2		$100,000	$20,000	$12,000	$8,000	$30,613	$29,387	$11,755	$48,245
3		$100,000	$20,000	$12,000	$8,000	$21,863	$38,137	$15,255	$44,745
4		$100,000	$20,000	$12,000	$8,000	$15,613	$44,387	$17,755	$42,245
5	+$23,331	$100,000	$20,000	$12,000	$8,000	$5,581	$54,419	$21,678	$61,563
	$50,000						$16,525	$6,613	$43,387

Note that

$H = C - D - E - F - G$
$I = 0.4 \times H$
$J = B + C - D - E - F - I$

Information required to calculate the income taxes

contain any patterns to simplify our calculations, we must find the net present worth of each payment. Using $i = 15\%$, we have

$$PW(15\%) = -\$148,331 + \$43,145(P/F, 15\%, 1) + \$48,245(P/F, 15\%, 2)$$
$$+ \$44,745(P/F, 15\%, 3) + \$42,245(P/F, 15\%, 4)$$
$$+ \$104,950(P/F, 15\%, 5)$$
$$= \$31,420.$$

This result means that investing $148,331 in the automated facility would bring in enough revenue to recover the initial investment and the cost of funds, with a surplus of $31,420.

9.5.2 When Projects Are Financed with Borrowed Funds

Many companies use a mixture of debt and equity to finance their physical plant and equipment. The ratio of total debt to total investment, generally called the **debt ratio**, represents the percentage of the total initial investment provided by borrowed funds. For example, a debt ratio of 0.3 indicates that 30% of the initial investment is borrowed, and the rest is provided from the company's earnings (also known as **equity**). Since interest is a tax-deductible expense, companies in high tax brackets may incur lower after-tax interest costs by financing through debt. (The method of loan repayment can also have a significant impact on taxes.)

Example 9.2 Cash Flow Statement: with Financing (Borrowing) Activities

Rework Example 9.1, assuming that $62,500 of the $125,000 paid for the investment is obtained through debt financing (debt ratio = 0.5). The loan is to be repaid in equal annual installments at 10% interest over five years. The remaining $62,500 will be provided by equity (e.g., from retained earnings).

SOLUTION

Given: Same scenario as in Example 9.1, but $62,500 is borrowed, repaid in equal installments over five years at 10%.

Find: Net after-tax cash flows in each year.

We first need to compute the size of the annual loan repayment installments:

$$\$62,500(A/P, 10\%, 5) = \$16,487.$$

Next, we determine the repayment schedule of the loan by itemizing both the interest and the principal represented in each annual repayment:

Amount financed: $62,500, or 50% of total capital expenditure
Financing rate: 10% per year
Annual installment: $16,487, or $A = \$62,500\ (A/P, 10\%, 5)$

End of Year	Beginning Balance	Interest Payment	Principal Payment	Ending Balance
1	$62,500	$6,250	$10,237	$52,263
2	$52,263	$5,226	$11,261	$41,002
3	$41,002	$4,100	$12,387	$28,615
4	$28,615	$2,861	$13,626	$14,989
5	$14,989	$1,499	$14,988	$0

$16,487

The resulting after-tax cash flow is detailed in Table 9.3. The present-worth equivalent of the after-tax cash flow series is

$$PW(15\%) = -\$85,351 + \$29,158(P/F, 15\%, 1) + \cdots$$
$$+ \$89,063(P/F, 15\%, 5)$$
$$= \$44,439.$$

TABLE 9.3 Cash Flow Statement for Automated Machining Center Project with Debt Financing

	A	B	C	D	E	F	G	H
		0	1	2	3	4	5	
1								
2	Income Statement							
3								
4	Revenues		$100,000	$100,000	$100,000	$100,000	$100,000	
5	Expenses:							
6	Labor		20,000	20,000	20,000	20,000	20,000	
7	Material		12,000	12,000	12,000	12,000	12,000	
8	Overhead		8,000	8,000	8,000	8,000	8,000	
9	Depreciation		17,863	30,613	21,863	15,613	5,581	
10	Debt Interest		6,250	5,226	4,100	2,861	1,499	
11	Taxable Income		$35,887	$24,161	$34,037	$41,526	$52,920	
12	Income Taxes (40%)		14,355	9,664	13,615	16,610	21,168	
13	Net Income		$21,532	$14,497	$20,422	$24,916	$31,752	
14								
15	Cash Flow Statement							
16								
17	Operating Activities:							
18	Net Income		21,532	14,497	20,422	24,916	31,752	Items related
19	Depreciation		17,863	30,613	21,863	15,613	5,581	to financing
20	Investment Activities:							activities
21	Investment	(125,000)						
22	Salvage						50,000	
23	Gains Tax						(6,613)	
24	Working Capital	(23,331)					23,331	
25	Financing Activities:							
26	Borrowed Funds	62,500						
27	Principal Repayment		(10,237)	(11,261)	(12,387)	(13,626)	(14,988)	
28	Net Cash Flow	$(85,831)	$29,158	$33,849	$29,898	$26,903	$89,063	

When this amount is compared with the amount found in the case that involved no borrowing, $31,420, we see that debt financing actually increases the present worth by $13,019. This surprising result is caused largely by the firm being able to borrow the funds at a cheaper rate 10% than its MARR (opportunity cost rate) of 15%. However, we should be careful in interpreting the result. It is true, to some extent, that firms can usually borrow money at lower rates than their MARR. However, if the firm can borrow money at a significantly lower rate, this factor also affects the firm's MARR, because the borrowing rate is one of the elements used in determining the MARR. Therefore, a significant difference between present values with borrowing and without borrowing is not expected in practice.

9–6   Effects of Inflation on Project Cash Flows

We will now introduce inflation into some investment projects. We are especially interested in two elements of project cash flows: depreciation expenses and interest expenses. These two elements are essentially immune to the effects of inflation, as they are always given in actual dollars. We will also consider the complication of how to proceed when multiple price indexes have been used to generate various project cash flows. Capital projects requiring increased levels of **working capital** suffer from inflation because additional cash must be invested to maintain new price levels. For example, if the cost of inventory increases, additional outflows of cash are required in order to maintain appropriate inventory levels over time. A similar phenomenon occurs with funds committed to account receivables. These additional working-capital requirements can significantly reduce a project's profitability or rate of return.

9.6.1   Depreciation Allowance under Inflation

Because depreciation expenses are calculated on some base-year purchase amount, they do not increase over time to keep pace with inflation. Thus, they lose some of their value to defer taxes, because inflation drives up the general price level and hence taxable income. Similarly, the selling prices of depreciable assets can increase with the general inflation rate, and, because any gains on salvage values are taxable, they can result in increased taxes. Example 9.3 illustrates how a project's profitability changes under an inflationary economy.

Example 9.3 Effects of Inflation on Projects with Depreciable Assets

Reconsider the automated machining center investment project described in Example 9.1. A summary of the financial facts in the absence of inflation is as follows:

Project:	Automated machining center
Required investment:	$125,000
Investment in working capital:	$23,331
Project life:	five years
Salvage value:	$50,000
Depreciation method:	seven-year MACRS
Annual revenues:	$100,000 per year
Annual expenses:	
Labor	$20,000 per year
Material	$12,000 per year
Overhead	$8,000 per year
Marginal tax rate:	40%
Inflation-free interest rate (i'):	15%

The after-tax cash flow for the automated machining center project was given in Table 9.1, and the net present worth of the project in the absence of inflation was calculated to be $31,420.

What will happen to this investment project if the general inflation rate during the next five years is expected to increase by 5% annually? Sales, operating costs, and working-capital requirements are assumed to increase accordingly. Depreciation will remain unchanged, but taxes, profits, and thus cash flow will be higher. The firm's inflation-free interest rate (i') is known to be 15%.

(a) Determine the PW of the project.
(b) Determine the real rate of return for the project.

Discussion: All cash flow elements, except depreciation expenses, are assumed to be in constant dollars. Since income taxes are levied on actual taxable income, we will use the actual-dollar analysis, which requires that all cash flow elements be expressed in actual dollars. We make the following observations:

- For the purposes of this illustration, all inflationary calculations are made as of year end.
- Cash flow elements such as sales, labor, material, overhead, and selling price of the asset will be inflated at the general inflation rate. For example, whereas annual sales had been estimated at $100,000, under conditions of inflation they become 5% greater in year one (or $105,000), 10.25% greater in year two (110,250), and so forth.
- Future cash flows in actual dollars for other elements can be obtained in a similar way.
- No change occurs in the investment in year zero or in the depreciation expenses, since these items are unaffected by expected future inflation.

- Investment in working capital will change under inflation. Working-capital levels can be maintained only by an additional infusion of cash. For example, the $23,331 investment in working capital made in year zero will be recovered at the end of the first year, assuming a one-year recovery cycle. However, because of 5% inflation, the required working capital for the second year increases to $24,498. Thus, in addition to reinvesting the $23,331 revenues, $1,167 must be made. The $24,498 will be recovered at the end of the second year. However, the project will need a 5% increase, or $25,723, for the third year, and so forth.

- The selling price of the asset is expected to increase at the general inflation rate. Therefore, the salvage value in actual dollars will be

$$\$50,000(1 + 0.05)^5 = \$63,814.$$

This increase in salvage value will also increase the taxable gains as the book value remains unchanged. The calculations for both the book value and gains tax are shown in Table 9.4.

SOLUTION

Given: Financial data as shown in Table 9.1, but with a general inflation rate of 5%.
Find: (a) PW of the after-tax project cash flows and (b) real rate of return of the project.

(a) Table 9.4 shows the after-tax cash flows in actual dollars. Since we are dealing with cash flows in actual dollars, we need to find the market interest rate. The market interest rate to use is $i = 0.15 + 0.05 + (0.05)(0.15) = 20.75\%$.

Since PW(20.75%) = $23,441 > 0$, the investment is still economically attractive.

(b) If you calculate the rate of return of the project based on actual dollars, it will be 22.20%. Since the market interest rate is 20.75%, the project is still justifiable. If you want to know the real (inflation-free) rate of return, you can use the following relationship:

$$i' = \frac{i - \bar{f}}{1 + f} = \frac{0.2220 - 0.05}{1 + 0.05} = 16.38\%.$$

Since the inflation-free MARR is 15%, the project also can be justified based on real dollars.

Comments: Note that the PW in the absence of inflation was $31,420 in Example 9.1. The $7,979 decline (known as inflation loss) in the PW under inflation, illustrated in this example, is due entirely to income-tax considerations and *working-capital drains*. The depreciation expense is a charge against taxable income, which reduces the amount of taxes paid and, as a result, increases the cash flow attributable to an investment by the amount of taxes saved. But the depreciation expense under existing tax laws is based on historic cost. As time

TABLE 9.4 Cash Flow Statement for the Automated Machining Center Project under Inflation

	A	B	C	D	E	F	G	H
1	**Income Statement**							
2		Inflation Rate	0	1	2	3	4	5
3								
4	Revenues	5%		$105,000	$110,250	$115,763	$121,551	$127,628
5	Expenses:							
6	Labor	5%		21,000	22,050	23,153	24,310	25,526
7	Material	5%		12,600	13,230	13,892	14,586	15,315
8	Overhead	5%		8,400	8,820	9,261	9,724	10,210
9	Depreciation			17,863	30,613	21,863	15,613	5,581
10								
11	Taxable Income			$45,137	$35,537	$47,595	$57,317	$70,996
12	Income Taxes (40%)			18,055	14,215	19,038	22,927	28,398
13								
14	Net Income			$27,082	$21,322	$28,557	$34,390	$42,598
15								
16	**Cash Flow Statement**							
17								
18	Operating Activities:							
19	Net Income			27,082	21,322	28,557	34,390	42,598
20	Depreciation			17,863	30,613	21,863	15,613	5,581
21	Investment Activities:							
22	Investment		(125,000)					
23	Salvage	5%						63,814
24	Gains Tax							(12,139)
25	Working Capital	5%	(23,331)	(1,167)	(1,225)	(1,287)	(1,351)	28,361
26								
27	Net Cash Flow		$(148,331)	$43,778	$50,710	$49,133	$48,652	$128,215

goes by, the depreciation expense is charged to taxable income in dollars of declining purchasing power; as a result, the "real" cost of the asset is not totally reflected in the depreciation expense. Depreciation costs are thereby understated, and the taxable income is thus overstated, resulting in higher taxes.

9.6.2 Handling Multiple Inflation Rates

As we noted previously, the inflation rate f_j represents a rate applicable to a specific segment j of the economy. For example, if we are estimating the future cost of a piece of machinery, we should use the inflation rate appropriate for that item. Furthermore, we may need to use several rates in order to accommodate the different costs and revenues in our analysis. The following example introduces the complexity of multiple inflation rates.

Example 9.4 Applying Specific Inflation Rates

In this example, we will rework Example 9.3, using different annual indices (differential inflation rates) in the prices of cash flow components. Suppose that we expect the general rate of inflation, $\bar{f}$, to average 6% during the next five years. We also expect that the selling price (salvage value) of the equipment will increase 3% per year, that wages (labor) and overhead will increase 5% per year, and that the cost of material will increase 4% per year. The working-capital requirement will increase at 5% per year. We expect sales revenue to climb at the general inflation rate. Table 9.5 shows the relevant calculations, based on the income-statement format. For simplicity, all cash flows and inflation effects are assumed to occur at year end. Determine the net present worth of this investment, using the adjusted-discount-rate method.

SOLUTION

Given: Financial data as shown in Table 9.5, but with multiple inflation rates.
Find: PW of the after-tax project cash flows.

To use actual dollars to evaluate the present worth, we must adjust the original discount rate of 15%, which is an inflation-free interest rate, i'. The appropriate interest rate to use is the market interest rate:

$$i = i + \bar{f} + i'\bar{f} = 0.15 + 0.06 + (0.15)(0.06) = 21.90\%.$$

The equivalent present worth is obtained as follows:

$$
\begin{aligned}
\text{PW}(21.90\%) = \ & -\$148{,}331 + \$44{,}450(P/F, 21.90\%, 1) \\
& + \$52{,}127(P/F, 21.90\%, 2) + \cdots \\
& + \$128{,}851(P/F, 21.90\%, 5) \\
= \ & \$22{,}903.
\end{aligned}
$$

TABLE 9.5 Cash Flow Statement for Automated Machining Center Project under Inflation (Multiple Price Indices)

	A	B	C	D	E	F	G	H
1	**Income Statement**							
2		Inflation Rate	0	1	2	3	4	5
3								
4	Revenues	6%		$106,000	$112,360	$119,102	$126,248	$133,823
5	Expenses:							
6	Labor	5%		21,000	22,050	23,153	24,310	25,526
7	Material	4%		12,480	12,979	13,498	14,038	14,600
8	Overhead	5%		8,400	8,820	9,261	9,724	10,210
9	Depreciation			17,863	30,613	21,863	15,613	5,581
10								
11	Taxable Income			$46,257	$37,898	$51,327	$62,562	$77,906
12	Income Taxes (40%)			18,503	15,159	20,531	25,025	31,162
13								
14	Net Income			$27,754	$22,739	$30,796	$37,537	$46,744
15								
16	**Cash Flow Statement**							
17								
18	Operating Activities:							
19	Net Income			27,754	22,739	30,796	37,537	46,744
20	Depreciation			17,863	30,613	21,863	15,613	5,581
21	Investment Activities:							
22	Investment		(125,000)					
23	Salvage	3%						57,964
24	Gains Tax							(9,799)
25	Working Capital	5%	(23,331)	(1,167)	(1,225)	(1,287)	(1,351)	28,361
26								
27	Net Cash Flow		$(148,331)	$44,450	$52,127	$51,372	$51,799	$128,851
28	(in actual dollars)							

9–7   Discount Rate to Be Used in After-Tax Economic Analysis: Cost of Capital

In most of the capital-budgeting examples in the earlier chapters, we assumed that the projects under consideration were financed entirely with equity funds. In those cases, the **cost of capital** may have represented only the firm's required return on equity. However, most firms finance a substantial portion of their capital budget with long-term debt (bonds), and many also use preferred stock as a source of capital. In these cases, a firm's cost of capital must reflect the average cost of the various sources of long-term funds that the firm uses, not only the cost of equity. In this section, we will discuss the ways in which the cost of each individual type of financing (retained earnings, common stock, preferred stock, and debt) can be estimated, given a firm's target capital structure.[5]

9.7.1   Cost of Equity

Whereas debt and preferred stocks are contractual obligations that have easily determined costs, it is not easy to measure the cost of equity. In principle, the cost of equity capital involves an **opportunity cost**. In fact, the firm's after-tax cash flows belong to the stockholders. Management may either pay out these earnings in the form of dividends or retain the earnings and reinvest them in the business. If management decides to retain earnings, an opportunity cost is involved; stockholders could have received the earnings as dividends and invested this money in other financial assets. Therefore, the firm should earn on its retained earnings at least as much as the stockholders themselves could earn in alternative, but comparable, investments.

What rate of return can stockholders expect to receive on retained earnings? This question is difficult to answer, but the value sought is often regarded as the rate of return stockholders require on a firm's common stock. If a firm cannot invest retained earnings so as to earn at least the rate of return on equity, it should pay these funds to these stockholders and let them invest directly in other assets that do provide this return. In general, the expected return on any risky asset is composed of three factors:[6]

$$\begin{pmatrix} \text{Expected return} \\ \text{on risky asset} \end{pmatrix} = \begin{pmatrix} \text{Risk-free} \\ \text{interest rate} \end{pmatrix} + \begin{pmatrix} \text{Inflation} \\ \text{premium} \end{pmatrix} + \begin{pmatrix} \text{Risk} \\ \text{premium} \end{pmatrix}.$$

This equation says that the owner of a risky asset should expect to earn a return from three sources:

- The first is compensation from the opportunity cost incurred in holding the asset. This is the risk-free interest rate.
- The second is compensation for the declining purchasing power of the investment over time, known as the inflation premium.
- The third is compensation for bearing risk, known as the risk premium.

[5]Estimating or calculating the cost of capital in any precise fashion is a very difficult task.
[6]This section is based on the material from Robert C. Higgins, *Analysis for Financial Management*, 5th edition, New York, Irwin/McGraw-Hill, 1998.

Fortunately, we do not need to treat the first two terms as separate factors, because together they equal the expected return on a default-free bond such as a government bond. In other words, owners of government bonds expect a return from the first two sources, but not the third. So, we can simplify the previous equation as

$$\begin{pmatrix} \text{Expected return} \\ \text{on risky asset} \end{pmatrix} = \begin{pmatrix} \text{Interest rate on a} \\ \text{government bond} \end{pmatrix} + \begin{pmatrix} \text{Risk} \\ \text{premium} \end{pmatrix}.$$

When investors are contemplating buying a firm's stock, they have two primary things in mind: (1) cash dividends and (2) gains (share appreciation) at the time of sale. From a conceptual standpoint, investors determine market values of stocks by discounting expected future dividends at a rate that takes into account any future growth. Since investors seek growing companies, a desired growth factor for future dividends is usually included in the calculation.

The cost of equity is the risk-free interest rate (for example, 20-year U.S. Treasury Bills that return around 6%) plus a premium for taking a risk as to whether a return will be received. The risk premium is the average return on the market, typically the return for Standard & Poor's 500 large U.S. stocks, or S&P 500, (say, 12.5%) less the risk-free cost of debt. This premium is multiplied by *beta*, an approximate measure of stock price volatility. **Beta** (β) quantifies risk by measuring one firm's stock price relative to all the market's stock prices as a whole. A number greater than one means that, on average, the stock is more volatile than the market; a number less than one means that, on average, the stock is less volatile than the market. The values for beta are commonly found for most publicly traded stocks in various sources such as Value Line.[7] The cost of equity (i_e) is quantified by

$$i_e = r_f + \beta[r_M - r_f], \tag{9.1}$$

where

r_f = risk-free interest rate (commonly referenced to U.S. Treasury bond yield, inflation adjusted), and

r_M = market rate of return (commonly referenced to average return on S&P 500 stock index funds, inflation adjusted).

Note that this amount is almost always higher than the cost of debt. This is because the U.S. Tax Code allows the deduction of interest expense, but does not allow the deduction of the cost of equity, which could be considered more subjective and complex. Example 9.5 illustrates how we may determine the cost of equity.

---

[7]Value Line reports are presently available for over 5,000 public companies, and that number is growing. The Value Line reports contain the following information: (1) total assets, (2) total liabilities, (3) total equity, (4) long-term debt as a percent of capital, (5) equity as a percent of capital, (6) financial strength (which is used to determine interest rates), (7) beta, and (8) return on invested capital.

Example 9.5 Determining the Cost of Equity

Alpha Corporation needs to raise $10 million for plant modernization. Alpha's target capital structure calls for a debt ratio of 0.4, indicating that $6 million has to be financed from equity. The pertinent information is as follows:

- Alpha is planning to raise $6 million from the financial markets.
- Alpha's beta is known to be 1.8, which is greater than one, indicating that the firm is perceived to be more risky than the market average.
- The risk-free interest rate is 6%, and the average market return is 13%. (These interest rates are adjusted to reflect inflation in the economy.)

Determine the cost of equity to finance the plant modernization.

SOLUTION

Given: $r_M = 13\%, r_f = 6\%$, and $\beta = 1.8$.

Find: i_e.

We calculate i_e as follows:

$$i_e = 0.06 + 1.8(0.13 - 0.06) = 18.60\%.$$

Comments: What does this 18.60% represent? It means that, if Alpha finances the project entirely from its equity funds, the project must earn at least an 18.60% return on investment to be worthwhile.

9.7.2 Cost of Debt

Now let us consider the calculation of the specific cost that is to be assigned to the debt component of the weighted-average cost of capital. The calculation is relatively straightforward and simple. Two types of debt financing are term loans and bonds. Because the interest payments on both are tax deductible, they reduce the effective cost of debt. To determine the after-tax cost of debt (i_d), we can evaluate the expression

$$i_d = \left(\frac{c_s}{c_d}\right)k_s(1 - t_m) + \left(\frac{c_b}{c_d}\right)k_b(1 - t_m), \tag{9.2}$$

where

c_s = the amount of the term loan,
k_s = the before-tax interest rate on the term loan,
t_m = the firm's marginal tax rate,
k_b = the before-tax interest rate on the bond,
c_b = the amount of bond financing, and $c_s + c_b = c_d$.

Example 9.6 illustrates the process of computing the cost of debt for the Alpha Corporation scenario introduced in Example 9.5.

Example 9.6 Determining the Cost of Debt

For the case in Example 9.5, suppose that Alpha has decided to finance the remaining $4 million by securing a term loan and issuing 20-year $1,000 par bonds under the following conditions:

Source	Amount	Interest Fraction	Interest Rate
Term loan	$1.33 million	0.333	12% per year
Bonds	$2.67 million	0.667	10.74% per year

Alpha's marginal tax rate is 38%, and it is expected to remain constant in the future. Determine the after-tax cost of debt.

SOLUTION

Given: $k_s = 12\%$, $k_b = 10.74\%$, $c_s/c_d = 0.333$, $c_b/c_d = 0.667$, $t_m = 38\%$.

Find: i_d.

The after-tax cost of debt is the interest rate on debt, multiplied by $(1 - t_m)$. In effect, the government pays part of the cost of debt, because interest is tax deductible. Now we are ready to compute the after-tax cost of debt as follows:

$$i_d = (0.333)(0.12)(1 - 0.38) + (0.667)(0.1074)(1 - 0.38) = 6.92\%.$$

9.7.3 Calculating the Cost of Capital

With the specific cost of each financing component determined, now we are ready to calculate the tax-adjusted weighted-average cost of capital, based on total capital. Then we will define the marginal cost of capital that should be used in project evaluation.

Weighted-Average Cost of Capital   Assuming that a firm raises capital based on the target capital structure and that the target capital structure remains unchanged in the future, we can determine a **tax-adjusted weighted-average cost of capital** (or, simply stated, the **cost of capital**). As illustrated in Figure 9.7, this cost of capital represents a composite index reflecting the cost of raising funds from different sources. The cost of capital is defined as

$$k = \frac{i_d c_d}{V} + \frac{i_e c_e}{V}, \tag{9.3}$$

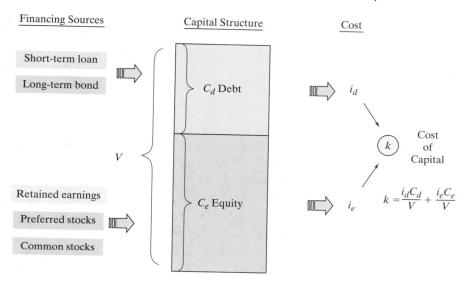

FIGURE 9.7 Illustration of weighted cost of capital

where

c_d = total debt capital (such as bonds) in dollars,

c_e = total equity capital in dollars,

$V = c_d + c_e$,

i_e = average equity interest rate per period, considering all equity sources,

i_d = after-tax average borrowing interest rate per period, considering all debt sources, and

k = tax-adjusted weighted-average cost of capital.

Note that the cost of equity is already expressed in terms of after-tax cost, because any return to holders of either common stock or preferred stock is made after the payment of income taxes.

Marginal Cost of Capital   Now we know how to calculate the cost of capital. Could a typical firm raise unlimited new capital at the same cost? The answer is no. As a practical matter, as a firm tries to attract more new dollars, the cost of raising each additional dollar will at some point rise. As this increase occurs, the weighted-average cost of raising each additional dollar also rises. Thus, the **marginal cost of capital** is defined as the cost of obtaining another dollar of new capital, and this value rises as more and more capital is raised during a given period. In evaluating an investment project, we use the concept of marginal cost of capital. The formula to find the marginal cost of capital is exactly the same as Eq. (9.3). However, the costs of debt and equity in Eq. (9.3) are the interest rates on new debt and equity, not outstanding (or combined) debt or equity, respectively; in other words, we are interested in the marginal cost of capital. Our primary concern with the cost of capital is to use it in evaluating a new investment project. The rate at which the firm has borrowed in the past is less important for this purpose. Example 9.7 works through the computations for finding the cost of capital (k).

Example 9.7 Calculating the Marginal Cost of Capital

Reconsider Examples 9.5 and 9.6. The marginal income-tax rate (t_m) for Alpha is expected to remain at 38% in the future. Assuming that Alpha's capital structure (debt ratio) also remains unchanged in the future, determine the cost of capital (k) of raising $10 million in addition to Alpha's existing capital.

SOLUTION

Given: With c_d = $4 million, c_e = $6 million, V = $10 million, i_d = 6.92%, i_e = 18.60%, and Eq. (9.3).
Find: Marginal cost of capital (k).

We calculate the marginal cost of capital as follows:

$$k = \frac{(0.0692)(4)}{10} + \frac{(0.1860)(6)}{10} = 13.93\%.$$

This 13.93% would be the marginal cost of capital that a company with this capital structure would expect to pay to raise $10 million, and this is the discount rate to be used in evaluating Alpha's new investment project. In other words, the project must return greater than 13.93% in order to be justified.

9.7.4 Choice of a MARR in After-Tax Cash Flow Analysis

Thus far, we have said little about what interest rate, or minimum attractive rate of return (MARR), is suitable for use in a particular investment situation. Choosing the MARR is a difficult problem; no single rate is always appropriate. In this section, we will discuss briefly how to select a MARR for project evaluation. Then we will examine the relationship between capital budgeting and the cost of capital.

Choice of MARR When Project Financing Is Known   When cash flow computations reflect interest, taxes, and debt repayment, what is left is called **net equity flow**. If the goal of a firm is to maximize the wealth of its stockholders, why not focus only on the after-tax cash flow to equity, instead of on the flow to all suppliers of capital? Focusing on only the equity flows will permit us to use the cost of equity as the appropriate discount rate. In a typical engineering economic analysis environment, we often assume the financing source is known, so that the project cash flows represent *net equity flows*; thus, the MARR used represents the **cost of equity (i_e)**.

For example, the after-tax cash flows calculated in Table 9.3 represent a **net equity flow**. In this case, the discount rate of 15% represents the cost of equity (i_e). If the debt flows were not considered explicitly in Table 9.3, the appropriate discount rate to use is the cost of capital (k). In other words, because the interest expense as well as debt repayment flows are ignored, the resulting net cash flows should be discounted using k, not i_e.

Choice of MARR When Project Financing Is Unknown

You might well ask why, if we use the cost of equity (i_e) exclusively, of what use is the cost of capital (k)? The answer to this question is that, by using the value of k, we may evaluate investments without explicitly treating the debt flows (both interest and principal). In practice, it is common to evaluate various projects without considering the sources of financing explicitly. The underlying assumption is that you will raise the capital according to the target capital structure if you decide to fund the project. In this case, we make a tax adjustment to the discount rate by employing the effective after-tax cost of debt. This approach recognizes that the net-interest cost is effectively transferred from the tax collector to the creditor, in the sense that there is a dollar-for-dollar reduction in taxes up to this amount of interest payments. Therefore, debt financing is treated implicitly. This method would be appropriate when debt financing is not identified with individual investments, but rather enables the company to engage in a set of investments.

Summary

- Most manufacturing companies divide **manufacturing costs** into three broad categories: *direct materials, direct labor*, and *manufacturing overhead*. **Nonmanufacturing costs** are classified into two categories: *marketing* or *selling costs*, and *administrative costs*.

- For the purpose of valuing inventories and determining expenses for the balance sheet and income statement, costs are classified as either **product costs** or **period costs**.

- For the purpose of predicting **cost behavior**—how costs will react to changes in activity—managers commonly classify costs into two categories: variable and fixed costs.

- Cash flow (not net income) must be considered to evaluate the economic merit of any investment project.

- The tax rate to use in economic analysis is the incremental tax rate from undertaking a project, not the overall tax rate or average tax rate.

- Identifying and estimating relevant project cash flows is perhaps the most challenging aspect of engineering economic analysis. All cash flows can be organized into one of the following three categories:

  1. operating activities;
  2. investing activities;
  3. financing activities.

- The **income-statement approach** is typically used in organizing project cash flows. This approach groups cash flows according to whether they are associated with operating, investing, or financing functions.

- Project cash flows may be stated in one of two forms:

  actual dollars (A_n): dollars that reflect the inflation or deflation rate;
  constant dollars (A'_n): year-zero dollars.

- Interest rates for project evaluation may be stated in one of two forms:

  market interest rate (i): a rate that combines the effects of interest and inflation—used with **actual-dollar analysis**;
  inflation-free interest rate (i'): a rate from which the effects of inflation have been removed—used with constant-dollar analysis.

- The selection of an appropriate MARR, in the absence of capital limits, depends generally upon the cost of capital—the rate the firm must pay to various sources for the use of capital.

  - The cost of equity is used when debt-financing methods and repayment schedules are known explicitly.
  - The cost of capital (k) is used when exact financing methods are unknown, but a firm keeps its capital structure on target. In this situation, a project's after-tax cash flows contain no debt cash flows such as an interest payment.

Problems

Classifying Costs

9.1 Identify which of the following transactions and events are product costs and which are period costs:

- Storage and material handling costs for raw materials
- Gains or losses on disposal of factory equipment
- Lubricants for machinery and equipment used in production
- Depreciation of a factory building
- Depreciation of manufacturing equipment
- Depreciation of the company president's automobile
- Leasehold costs for land on which factory buildings stand
- Inspection costs of finished goods
- Direct labor costs
- Raw materials costs
- Advertising expenses

Cost Behavior

9.2 Identify which of the following costs are fixed and which are variable:

- Wages paid to temporary workers
- Property taxes on a factory building
- Property taxes on an administrative building
- Sales commission
- Electricity for machinery and equipment in a plant
- Heat and air conditioning for a plant
- Salaries paid to design engineers
- Regular maintenance on machinery and equipment
- Basic raw materials used in production
- Factory fire insurance

9.3 The accompanying figures depict a number of cost behavior patterns that might be found in a company's cost structure. The vertical axis on each graph represents total cost, and the horizontal axis on each graph represents level of activity (volume). For each of the given situations, identify the graph that illustrates the cost pattern involved. Any graph may be used more than once. (Adapted originally from the CPA exam; also found in R. H. Garrison and E. W. Noreen, *Managerial Accounting*, 8th edition, Irwin, 1997, copyright © Richard D. Irwin, p. 271.)

(a) Electricity bill—a flat-rate fixed charge, plus a variable cost after a certain number of kilowatt-hours are used.

(b) City water bill, which is computed as follows:

First 1,000,000 gallons	$1,000 flat or less
Next 10,000 gallons	$0.003 per gallon used
Next 10,000 gallons	$0.006 per gallon used
Next 10,000 gallons	$0.009 per gallon used
Etc.	etc.

(c) Depreciation of equipment, where the amount is computed by the straight-line method. When the depreciation rate was established, it was anticipated that the obsolescence factor would be greater than the wear-and-tear factor.

(d) Rent on a factory building donated by the city, where the agreement calls for a fixed-fee payment unless 200,000 labor-hours or more are worked, in which case no rent need be paid.

(e) Cost of raw materials, where the cost decreases by 5 cents per unit for each of the first 100 units purchased, after which it remains constant at $2.50 per unit.

(f) Salaries of maintenance workers, where one maintenance worker is needed for every 1,000 machine hours or less (that is, 0 to 1,000 hours require one maintenance worker, 1,001 to 2,000 hours require two maintenance workers, etc.).

(g) Cost of raw materials used.

(h) Rent on a factory building donated by the county, where the agreement calls for rent of $100,000 less $1 for each direct labor-hour worked in excess of 200,000 hours, but a minimum rental payment of $20,000 must be paid.

(i) Use of a machine under a lease, where a minimum charge of $1,000 is paid for up to 400 hours of machine time. After 400 hours of machine time, an additional charge of $2 per hour is paid, up to a maximum charge of $2,000 per period.

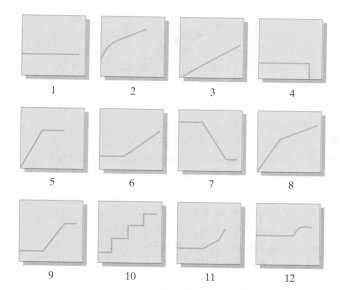

Cost Concepts Relevant to Decision Making

9.4 The Morton Company produces and sells two products, *A* and *B*. Financial data related to producing these two products are summarized as follows:

	Product **A**	Product **B**
Selling price	$10.00	$12.00
Variable costs	$5.00	$10.00
Fixed costs	$2,000	$600

(a) If these products are sold in the ratio of 4*A* for 3*B*, what is the break-even point?

(b) If the product mix has changed to 5A for 5B, what would happen to the break-even point?

(c) In order to maximize the profit, which product mix should be pushed?

(d) If both products must go through the same manufacturing machine and there are only 30,000 machine hours available per period, which product should be pushed? Assume that product A requires 0.5 hour per unit and B requires 0.25 hour per unit.

9.5 Pearson Company manufactures a variety of electronic PCBs (printed circuit boards) that go into cellular phones. The company has just received an offer from an outside supplier to provide the electrical soldering for the company's Motorola product line (Z-7 PCB, slim-line). The quoted price is $4.80 per unit. Pearson is interested in this offer, since its own PCB soldering operations are at peak capacity. The details of the supplier's offer and Pearson's current soldering operation are as follows:

● **Outsourcing Option:** The company estimates that if the supplier's offer were accepted, the direct-labor and variable- overhead costs of the Z-7 slim-line would be reduced by 15% and the direct-materials cost would be reduced by 20%.

● **In-House Production Option:** Under the present operation, Pearson Company manufactures all of its own PCBs from start to finish. The Z-7 slim-lines are sold through Motorola at $20 per unit. Fixed-overhead charges to the Z-7 slim-line total $20,000 each year. A breakdown of the costs of producing one unit is as follows:

Direct materials	$7.50
Direct labor	$5.00
Manufacturing overhead	$4.00
Total cost	$16.50

The manufacturing overhead of $4.00 per unit includes both variable and fixed manufacturing overhead, based on production of 100,000 units each year.

(a) Should Pearson Company accept the outside supplier's offer?

(b) What is the maximum unit price that Pearson Company would be willing to pay the outside supplier?

Marginal Tax Rate in Project Evaluation

9.6 Buffalo Ecology Corporation expects to generate a taxable income of $250,000 from its regular business in 2001. The company is also considering a new venture: cleaning up oil spills made by fishing boats in lakes. This new venture is expected to generate an additional taxable income of $150,000.

(a) Determine the firm's marginal tax rates before and after the venture.

(b) Determine the firm's average tax rates before and after the venture.

9.7 Boston Machine Shop expects to have an annual taxable income of $270,000 from its regular business over the next six years. The company is also considering the proposed acquisition of a new milling machine during year zero. The machine's installed price is $200,000. The machine falls into the MACRS five-year class, and it will have an estimated salvage value of $30,000 at the end of six years. The machine is expected to generate an additional before-tax revenue of $80,000 per year.

(a) Determine the company's annual marginal tax rates over the next six years with the machine.

(b) Determine the company's annual average tax rates over the next six years with the machine.

9.8 Major Electrical Company expects to have an annual taxable income of $450,000 from its residential accounts over the next two years. The company is also bidding on a two-year wiring service job for a large apartment complex. This commercial service requires the purchase of a new truck equipped with wire-pulling tools at a cost of $50,000. The equipment falls into the MACRS five-year class and will be retained for future use (instead of being sold) after two years, indicating no gain or loss on this property. The project will bring in an additional annual revenue of $200,000, but it is expected to incur additional annual operating costs of $100,000. Compute the marginal tax rates applicable to the project's operating profits for the next two years.

9.9 A small manufacturing company has an estimated annual taxable income of $95,000. Due to an increase in business, the company is considering purchasing a new machine that will generate an additional (before-tax) annual revenue of $50,000 over the next five years. The new machine requires an investment of $100,000, which will be depreciated under the five-year MACRS method.

(a) What is the increment in income tax caused by the purchase of the new machine in tax year one?

(b) What is the incremental tax rate associated with the purchase of the new equipment in year one?

9.10 Simon Machine Tools Company is considering the purchase of a new set of machine tools to process special orders over the next three years. The following financial information is available:

- Without the project: The company expects to have a taxable income of $300,000 each year from its regular business over the next three years.
- With the project: This three-year project requires the purchase of a new set of machine tools at a cost of $50,000. The equipment falls into the MACRS three-year class. The tools will be sold at the end of the project life for $10,000. The project will be bringing in an additional annual revenue of $80,000, but it is expected to incur additional annual operating costs of $20,000.

(a) What are the additional taxable incomes (from undertaking the project) during years one through three, respectively?

(b) What are the additional income taxes (from undertaking the new orders) during years one through three, respectively?

Generating Net Cash Flows: Incremental Cash Flows

9.11 An asset in the five-year MACRS property class cost $100,000 and has a zero estimated salvage value after six years of use. The asset will generate annual revenues of $300,000 and will require $100,000 in annual labor costs and $50,000 in annual material expenses. There are no other revenues and expenses. Assume a tax rate of 40%.

(a) Compute the after-tax cash flows over the project life.

(b) Compute the PW at MARR = 12%. Is this investment acceptable?

9.12 An auto-part manufacturing company is considering the purchase of an industrial robot to do spot welding, which is currently done by skilled labor. The initial cost of the robot is $235,000, and the annual labor savings are projected to be $122,000. If purchased, the robot will be depreciated under MACRS as a five-year recovery property. This robot will be used for seven years, after which the firm expects to sell the robot for $50,000. The company's marginal tax rate is 38% over the project period. Determine the net after-tax cash flows for each period over the project life.

9.13 A highway contractor is considering buying a new trench excavator that costs $200,000 and can dig a three-foot-wide trench at the rate of 16 feet per hour. With adequate maintenance, the production rate will remain constant for the first 1,200 hours of operation and then decrease by two feet per hour for each additional 400 hours thereafter. The expected average annual use is 400 hours, and maintenance and operating costs will be $15 per hour. The contractor will depreciate the equipment by using a five-year MACRS. At the end of five years, the excavator will be sold for $40,000. Assuming that the contractor's marginal tax rate is 34% per year, determine the annual after-tax cash flow.

9.14 Tampa Construction Company builds residential solar homes. Because of an anticipated increase in business volume, the company is considering the acquisition of a loader at a cost of $54,000. This acquisition cost includes delivery charges and applicable taxes. The firm has estimated that if the loader is acquired, the following additional revenues and operating expenses (excluding depreciation) should be expected:

End of Year	Additional Operating Revenue	Additional Operating Expenses, Excluding Depreciation	Allowed Tax Depreciation
1	$66,000	$29,000	$10,800
2	$70,000	$28,400	$17,280
3	$74,000	$32,000	$10,368
4	$80,000	$38,800	$6,221
5	$64,000	$31,000	$6,221
6	$50,000	$25,000	$3,110

The projected revenue is assumed to be in cash in the year indicated, and all the additional operating expenses are expected to be paid in the year in which they are incurred. The estimated salvage value for the loader at the end of sixth year is $8,000. The firm's incremental (marginal) tax rate is 35%. What is the after-tax cash flow if the loader is acquired?

9.15 A Los Angeles company is planning to market an answering device for people working alone who want the prestige that comes with having a secretary, but who cannot afford one. The device, called Tele-Receptionist, is similar to a voicemail system. It uses digital recording technology to create the illusion that a person is operating the switchboard at a busy office. The company purchased a 40,000 ft^2 building

and converted it into an assembly plant for $600,000 ($100,000 for the land and $500,000 for the building). Installation of the assembly equipment, worth $500,000, will be completed on December 31. The plant will begin operation on January 1. The company expects to have a gross annual income of $2,500,000 over the next 5 years. Annual manufacturing costs and all operating expenses (excluding depreciation) are projected to be $1,280,000. For depreciation purposes, the assembly-plant building will be classified as 39-year real property and the assembly equipment as a seven-year MACRS property. The property value of the land and the building at the end of year five will appreciate as much as 15% over the initial purchase cost. The residual value of the assembly equipment is estimated to be about $50,000 at the end of year five. The firm's marginal tax rate is expected to be about 40% over the project period. Determine the project's after-tax cash flows over the period of five years.

9.16 A facilities engineer is considering a $50,000 investment in an energy management system (EMS). The system is expected to save $10,000 annually in utility bills for N years. After N years, the EMS will have a zero salvage value. In an after-tax analysis, how many years would N need to be in order for the engineer's company to earn a 10% return on the investment? Assume MACRS depreciation with a three-year class life and a 35% tax rate.

9.17 A corporation is considering purchasing a machine that will save $130,000 per year before taxes. The cost of operating the machine, including maintenance, is $20,000 per year. The machine will be needed for five years, after which it will have a zero salvage value. MACRS depreciation will be used, assuming a three-year class life. The marginal income-tax rate is 40%. If the firm wants 12% IRR after taxes, how much can it afford to pay for this machine?

9.18 The Manufacturing Division of Ohio Vending Machine Company is considering its Toledo plant's request for a half-inch-capacity automatic screw-cutting machine to be included in the division's 2001 capital budget:

- Name of project: Mazda Automatic Screw Machine
- Project cost: $68,701
- Purpose of project: To reduce the cost of some of the parts that are now being subcontracted by this plant, to cut down on inventory by shortening lead time, and to better control the quality of the parts. The proposed equipment includes the following cost basis:

Machine cost	$48,635
Accessory cost	$8,766
Tooling	$4,321
Freight	$2,313
Installation	$2,110
Sales tax	$2,556
Total cost	$68,701

- Anticipated savings: As shown in the following table.
- Tax depreciation method: seven-year MACRS.
- Marginal tax rate: 40%.
- MARR: 15%.

(a) Determine the net after-tax cash flows over the project life of six years. Assume a salvage value of $3,500.

(b) Is this project acceptable, based on the PW criterion?

(c) Determine the IRR for this investment.

9.19 American Aluminum Company is considering making a major investment of $150 million ($5 million for land, $45 million for buildings, and $100 million for manufacturing equipment and facilities) to develop a stronger, lighter material, called aluminum lithium, that will make aircraft sturdier and more fuel efficient. Aluminum lithium, which has been sold commercially for only a few

Item	Hours Present M/C labor	Proposed M/C labor	Present Method	Proposed Method
Setup		350 hrs		$7,700
Run	2,410 hrs	800 hrs		$17,600
Overhead				
Indirect labor				$3,500
Fringe benefits				$8,855
Maintenance				$1,350
Tooling				$6,320
Repair				$890
Supplies				$4,840
Power				$795
Taxes and insurance				$763
Other relevant costs				
Floor space				$3,210
Subcontracting			$122,468	
Material				$27,655
Other				$210
Total			$122,468	$83,688
Operating advantage				$38,780

years as an alternative to composite materials, will likely be the material of choice for the next generation of commercial and military aircraft, because it is so much lighter than conventional aluminum alloys, which use a combination of copper, nickel, and magnesium to harden aluminum. Another advantage of aluminum lithium is that it is cheaper than composites. The firm predicts that aluminum lithium will account for about 5% of the structural weight of the average commercial aircraft within five years and 10% within 10 years. The proposed plant, which has an estimated service life of 12 years, would have a capacity of about 10 million pounds of aluminum lithium, although domestic consumption of the material is expected to be only three million pounds during the first four years, five million for the next three years, and eight million for the remaining plant life. Aluminum lithium costs $12 a pound to produce, and the firm would expect to sell it at $17 a pound. The building will be depreciated according to the 39-year MACRS real-property class, with the building placed in service on July 1 of the first year. All manufacturing equipment and facilities will be classified as seven-year MACRS property. At the end of project life, the land will be worth $8 million, the building $30 million, and the equipment $10 million. The firm's marginal tax rate is 40% and its capital gains tax rate is 35%.

(a) Determine the net after-tax cash flows.
(b) Determine the IRR for this investment.
(c) Determine whether the project is acceptable if the firm's MARR is 15%.

9.20 An automaker is considering installing a three-dimensional (3-D) computerized car-styling system at a cost of $200,000 (including hardware and software). With the 3-D computer modeling system, designers will have the ability to view their design from many angles and to fully account for the space required for the engine and passengers. The digital information used to create the computer model can be revised in consultation with engineers, and the data can be used to run milling machines that make physical models quickly and precisely. The automaker expects to decrease turnaround time by 22% for new automobile models (from configuration to final design). The expected savings is $250,000 per year. The training and operating maintenance cost for the new system is expected to be $50,000 per year. The system has a five-year useful life and can be depreciated as a five-year MACRS class. The system will have an estimated salvage value of $5,000. The automaker's marginal tax rate is 40%. Determine the annual cash flows for this investment. What is the return on investment for this project? The firm's MARR is 12%.

9.21 Ampex Corporation produces a wide variety of tape cassettes for commercial and government markets. Due to increased competition in VHS cassette production, Ampex is concerned about pricing its product competitively. Currently, Ampex has 18 loaders that load cassette tapes into half-inch VHS cassette shells. Each loader is manned by one operator per shift. Ampex currently produces 25,000 half-inch tapes per week and operates 15 shifts per week, 50 weeks per year.

As a means of reducing the unit cost, Ampex can purchase cassette shells for $0.15 less (each) than it can currently produce them. A supplier has guaranteed a price of $0.77 per cassette for the next three years. However, Ampex's current loaders will not be able to load the proposed shells properly. To accommodate the vendor's shells, Ampex must purchase eight KING-2500 VHS loaders at a cost of $40,000 each. For these new machines to operate properly, Ampex must also purchase $20,827 worth of conveyor equipment, the cost of which will be included in the overall depreciation base of $340,827. The new machines are much faster and will handle more than the current demand of 25,000 cassettes per week. The new loaders will require two people per machine per shift, three shifts per day, five days a week. The new machines will fall into a seven-year MACRS equipment class and will have an approximate life of eight years. At the end of the project life, Ampex expects the market value for each loader to be $3,000.

The average pay of the needed new employees is $8.27 per hour, and adding 23% for benefits will make it $10.17 per hour. The new loaders are simple to operate; therefore, the training impact of the alternative is minimal. The operating cost, including maintenance, is expected to stay the same for the new loaders. This cost will not be considered in the analysis. The cash inflows from the project will be a material savings per cassette of $0.15 and the labor savings of two employees per shift. This gives an annual savings in materials and labor costs of $187,500 and $122,065, respectively. If the new loaders are purchased, the old machines will be shipped to other plants for standby use. Therefore, no gains will be realized. Ampex's combined marginal tax rate is running at 40%.

(a) Determine the after-tax cash flows over the project life.

(b) Determine the IRR for this investment.

(c) Is this investment profitable at MARR = 15%?

Effects of Borrowing on Project Cash Flows

9.22 Consider a project with an initial investment of $300,000, which must be financed at an interest rate of 12% per year. Assuming that the required repayment period is six years, determine the repayment schedule by identifying the principal as well as the interest payments for each of the following repayment methods:

(a) Equal repayment of the principal

(b) Equal repayment of the interest

(c) Equal annual installments

9.23 The Huron Development Company is considering buying an overhead pulley system. The new system has a purchase price of $100,000, an estimated useful life and MACRS class life

of five years, and an estimated salvage value of $30,000. It is expected to allow the company to economize on electric-power usage, labor, and repair costs, as well as to reduce the number of defective products made. A total annual savings of $45,000 will be realized if the new pulley system is installed. The company is in the 30% marginal tax bracket. The initial investment will be financed with 40% equity and 60% debt. The before-tax debt interest rate, which combines both short-term and long-term financing, is 15%, with the loan to be repaid in equal annual installments over the project life.

(a) Determine the after-tax cash flows.

(b) Evaluate this investment project by using a MARR of 20%.

(c) Evaluate this investment project based on the IRR criterion.

9.24 Ann Arbor Die Casting Company is considering the installation of a new process machine for its manufacturing facility. The machine costs $250,000 installed, will generate additional revenues of $80,000 per year, and will save $50,000 per year in labor and material costs. The machine will be financed by a $150,000 bank loan repayable in three equal annual principal installments, plus 9% interest on the outstanding balance. The machine will be depreciated using seven-year MACRS. The useful life of this process machine is 10 years, at which time it will be sold for $20,000. The combined marginal tax rate is 40%.

(a) Find the year-by-year after-tax cash flow for the project.

(b) Compute the IRR for this investment.

(c) At MARR = 18%, is this project economically justifiable?

9.25 Consider the following financial information about a retooling project at a computer manufacturing company:

- The project costs $2 million and has a five-year service life.

- The project can be classified as a seven-year property under the MACRS rule.
- At the end of fifth year, any assets held for the project will be sold. The expected salvage value will be about 10% of the initial project cost.
- The firm will finance 40% of the project money from an outside financial institution at an interest rate of 10%. The firm is required to repay the loan with five equal annual payments.
- The firm's incremental (marginal) tax rate on this investment is 35%.
- The firm's MARR is 18%.

Use the foregoing financial information to complete the following tasks:

(a) Determine the after-tax cash flows.

(b) Compute the annual equivalent cost for this project.

9.26 A manufacturing company is considering the acquisition of a new injection-molding machine at a cost of $100,000. Because of a rapid change in product mix, the need for this particular machine is expected to last only eight years, after which time the machine is expected to have a salvage value of $10,000. The annual operating cost is estimated to be $5,000. The addition of this machine to the current production facility is expected to generate an annual revenue of $40,000. The firm has only $60,000 available from its equity funds, so it must borrow the additional $40,000 required at an interest rate of 10% per year, with repayment of principal and interest in eight equal annual amounts. The applicable marginal income tax rate for the firm is 40%. Assume that the asset qualifies as a seven-year MACRS property class.

(a) Determine the after-tax cash flows.

(b) Determine the PW of this project at MARR = 14%.

9.27 Suppose an asset has a first cost of $6,000, a life of five years, a salvage value of $2,000 at the end of five years, and a net annual before-tax revenue of $1,500. The firm's marginal tax rate is 35%. The asset will be depreciated by three-year MACRS.

(a) Determine the cash flow after taxes, assuming that the first cost will be entirely financed by the equity funds.

(b) Rework part (a), assuming that the entire investment would be financed by a bank loan at an interest rate of 9%.

(c) Given a choice between paying the first cost up front and using the financing method of part (b), show calculations to justify your choice of which is the better one at an interest rate of 9%.

9.28 A construction company is considering the proposed acquisition of a new earthmover. The purchase price is $100,000, and an additional $25,000 is required in order to modify the equipment for special use by the company. The equipment falls into the MACRS seven-year classification (tax life), and it will be sold after five years (project life) for $50,000. Purchase of the earthmover will have no effect on revenues, but it is expected to save the firm $60,000 per year in before-tax operating costs, mainly labor. The firm's marginal tax rate is 40%. Assume that the initial investment is to be financed by a bank loan at an interest rate of 10%, payable annually. Determine the after-tax cash flows and the worth of investment for this project if the firm's MARR is known to be 12%.

9.29 Air South, a leading regional airline that is now carrying 54% of all the passengers that pass through the Southeast, is considering the possibility of adding a new long-range aircraft to its fleet. The aircraft being considered for purchase is the Boeing DC-9-532 "Funjet," which is quoted at $60 million per unit. Boeing requires a 10% down payment at the time of delivery, and the balance is to be paid over a 10-year period at an interest rate of 12%, compounded annually. The actual payment schedule calls for only interest payments over the 10-year period, with the original principal amount to be paid off at the end of the 10th year. Air South expects to generate $35 million per year by adding this aircraft to its current fleet, but also estimates an operating and maintenance cost of $20 million per year. The aircraft is expected to have a 15-year service life, with a salvage value of 15% of the original purchase price. If Air South purchases the aircraft, it will be depreciated by the seven-year MACRS property classifications. The firm's combined federal and state marginal tax rate is 38%, and its MARR is 18%.

(a) Determine the cash flow associated with the debt financing.

(b) Is this project acceptable?

Comparing Mutually Exclusive Alternatives

9.30 The headquarters building owned by a rapidly growing company is not large enough for current needs. A search for enlarged quarters revealed two new alternatives that would provide sufficient room, enough parking, and the desired appearance and location:

- Option 1: Lease for $144,000 per year.
- Option 2: Purchase for $800,000, including a $150,000 cost for land.
- Option 3: Remodel the current headquarters building.

It is believed that land values will not decrease over the ownership period, but the value of all structures will decline to 10% of the purchase price in 30 years. Annual property tax payments are expected to be 5% of the purchase price. The present headquarters building is already paid for and is now valued at $300,000. The land it is on is appraised at $60,000. The structure can be remodeled at a cost of $300,000 to make it comparable with the other alternatives. However, the remodeled building will occupy part of the existing parking lot. An adjacent, privately owned parking lot can be leased for 30 years under an agreement that the first year's rental of $9,000 will increase by $500 each year. The annual property taxes on the remodeled property will again be 5% of the present valuation plus the cost to remodel. The study period for the comparison is 30 years, and the desired rate of return on investments is 12%. Assume that the firm's marginal tax rate is 40% and the new building and remodeled structure will be depreciated under MACRS using a real-property recovery period of 39 years. If the annual upkeep costs are the same for all three alternatives, which one is preferable?

9.31 An international manufacturer of prepared food items needs 50,000,000 kWh of electrical energy a year, with a maximum demand of 10,000 kW. The local utility presently charges $0.085 per kWh, a rate considered high throughout the industry. Because the firm's power consumption is so large, its engineers are considering installing a 10,000-kW plant have steam-turbine plant. Three types of plant have been proposed ($ units in thousands):

	Plant A	Plant B	Plant C
Average station heat rate (BTU/kWh)	16,500	14,500	13,000
Total investment (boiler, turbine, electrical, and structures)	$8,530	$9,498	$10,546
Annual operating cost:			
Fuel	$1,128	$930	$828
Operating labor	$616	$616	$616
Maintenance	$150	$126	$114
Supplies	$60	$60	$60
Insurance and property taxes	$10	$12	$14

The service life of each plant is expected to be 20 years. The plant investment will be subject to a 20-year MACRS property classification. The expected salvage value of the plant at the end of its useful life is about 10% of the original investment. The firm's MARR is known to be 12%. The firm's marginal income-tax rate is 39%.

(a) Determine the unit power cost ($/kWh) for each plant.

(b) Which plant would provide the most economical power?

9.32 The Jacob Company needs to acquire a new lift truck for transporting final products to their warehouse. One alternative is to purchase the lift truck for $40,000, which will be financed by a bank at an interest rate of 12%. The loan must be repaid in four equal installments, payable at the end of each year. Under the borrow-to-purchase arrangement, Jacob would have to maintain the truck at an annual cost of $1,200, payable at year end. Alternatively, Jacob could lease the truck on a four-year contract for a lease payment of $11,000 per year. Each annual lease payment must be made at the beginning of each year. The truck would be maintained by the lessor. The truck falls into the five-year MACRS classification, and it has a salvage value of $10,000, which is the expected market value after four years, at which time Jacob plans to replace the truck irrespective of whether he leases or buys. Jacob has a marginal tax rate of 40% and a MARR of 15%.

(a) What is Jacob's cost of leasing in present worth?

(b) What is Jacob's cost of owning in present worth?

(c) Should the truck be leased or purchased?

9.33 Janet Wigandt, an electrical engineer for Instrument Control, Inc. (ICI), has been asked to perform a lease–buy analysis on a new pin-inserting machine for ICI's PC-board manufacturing operation. The details of the two options are as follows:

- Buy Option: The equipment costs $120,000. To purchase it, ICI could obtain a term loan for the full amount at 10% interest with four equal annual installments (end-of-year payment). The machine falls into a five-year MACRS property classification. Annual revenues of $200,000 and annual operating costs of $40,000 are anticipated. The machine requires annual maintenance at a cost of $10,000. Because technology is changing rapidly in pin-inserting machinery, the salvage value of the machine is expected to be only $20,000.

- Lease Option: Business Leasing, Inc. (BLI) is willing to write a four-year operating lease on the equipment for payments of $44,000 at the beginning of each year. Under this operating-lease arrangement, BLI will maintain the asset, so the annual maintenance cost of $10,000 will be saved. ICI's marginal tax rate is 40%, and its MARR is 15% during the analysis period.

(a) What is ICI's present-worth (incremental) cost of owning the equipment?

(b) What is ICI's present-worth (incremental) cost of leasing the equipment?

(c) Should ICI buy or lease the equipment?

9.34 The Boggs Machine Tool Company has decided to acquire a pressing machine. One alternative is to lease the machine on a three-year contract for a lease payment of $15,000 per year, with payments to be made at the beginning of each year. The lease would include maintenance. The second alternative is to purchase the machine outright for $100,000, financing the investment with a bank loan for the net purchase price and amortizing the loan over a three-year period at an interest rate of 12% per year (annual payment = $41,635).

Under the borrow-to-purchase arrangement, the company would have to maintain the machine at an annual cost of $5,000, payable at year end. The machine falls into the five-year MACRS classification, and it has a salvage value of $50,000, which is the expected market value at the end of year three. After three years the company plans to replace the machine irrespective of whether it leases or buys. Boggs has a tax rate of 40% and a MARR of 15%.

(a) What is Boggs's PW cost of leasing?

(b) What is Boggs's PW cost of owning?

(c) From the financing analysis in parts (a) and (b), what are the advantages and disadvantages of leasing and owning?

9.35 Enterprise Capital Leasing Company is in the business of leasing tractors for construction companies. The firm wants to set a three-year lease payment schedule for a tractor purchased at $53,000 from the equipment manufacturer. The asset is classified as a five-year MACRS property. The tractor is expected to have a salvage value of $22,000 at the end of three years of rental. Enterprise will require the lessee to make a security deposit in the amount of $1,500 that is refundable at the end of the lease term. Enterprise's marginal tax rate is 35%. If Enterprise wants an after-tax return of 10%, what lease payment schedule should be set?

9.36 The Pittsburgh division of Vermont Machinery, Inc., manufactures drill bits. One of the production processes for a drill bit requires tipping, whereby carbide tips are inserted into the bit to make it stronger and more durable. This tipping process usually requires four or five operators, depending on the weekly work load. The same operators are also assigned to the stamping operation, where the size of the drill bit and the company's logo are imprinted into the bit. Vermont is considering acquiring three automatic tipping machines to replace the manual tipping and stamping operations. If the tipping process is automated, the division's engineers will have to redesign the shapes of the carbide tips to be used in the machine. The new design requires less carbide, resulting in material savings. The following financial data have been compiled:

- Project life: six years.
- Expected annual savings: reduced labor,

$56,000; reduced material, $75,000; other benefits (reduced carpal tunnel syndrome and related problems), $28,000; reduced overhead, $15,000.

- Expected annual O&M costs: $22,000.
- Tipping machines and site preparation: equipment costs (for three machines), including delivery, $180,000; site preparation, $20,000.
- Salvage value: $30,000 (total for the three machines) at the end of six years.
- Depreciation method: seven-year MACRS.
- Investment in working capital: $25,000 at the beginning of the project year, which will be fully recovered at the end of the project year.
- Other accounting data: marginal tax rate of 39%; MARR of 18%.

To raise $200,000, Vermont is considering the following financing options:

- Option 1: Finance the tipping machines by using retained earnings.
- Option 2: Secure a 12% term loan over six years (six equal annual installments).
- Option 3: Lease the tipping machines. Vermont can obtain a six-year financial lease on the equipment (maintenance costs are taken care of by the lessor) for payments of $55,000 at the beginning of each year.

(a) Determine the net after-tax cash flows for each financing option.

(b) What is Vermont's PW cost of owning the equipment by borrowing?

(c) What is Vermont's PW cost of leasing the equipment?

(d) Recommend the best course of action for Vermont.

Effects of Inflation on Project Cash Flows

9.37 Consider the following expected after-tax cash flow for a project and the expected annual general inflation rate during the project period:

End of Year	Expected	
	Cash Flow (in Actual $)	General Inflation Rate
0	−$45,000	
1	$26,000	6.5%
2	$26,000	7.7%
3	$26,000	8.1%

(a) Determine the average annual general inflation rate over the project period.

(b) Convert the cash flows in actual dollars into equivalent constant dollars with the base year zero.

(c) If the annual inflation-free interest rate is 5%, what is the present worth of the cash flow? Is this project acceptable?

9.38 Gentry Machines, Inc., has just received a special job order from one of its clients. The following financial data on the order have been collected:

- This two-year project requires the purchase of a special-purpose piece of equipment for $55,000. The equipment falls into the MACRS five-year class.
- The machine will be sold at the end of two years for $27,000 (today's dollars).
- The project will bring in an additional annual revenue of $114,000 (actual dollars), but it is expected to incur an additional annual operating cost of $53,800 (today's dollars).
- To purchase the equipment, the firm expects to borrow $50,000 at 10% over a two-year period (equal annual payments of $28,810 [actual dollars]). The remaining $5,000 will be taken from the firm's retained earnings.
- The firm expects a general inflation of 5% per year during the project period. The firm's marginal tax rate is 40%, and its market interest rate is 18%.

(a) Compute the after-tax cash flows in actual dollars.

(b) What is the equivalent present worth of this amount at time zero?

9.39 Hugh Health Product Corporation is considering the purchase of a computer to control plant packaging for a spectrum of health products. The following data have been collected:

- First cost = $120,000 to be borrowed at 9% interest, where only interest is paid each year, and the principal is due in a lump sum at end of year two.
- Economic service life (project life) = six years.
- Estimated selling price in year-zero dollars = $15,000.
- Depreciation = five-year MACRS property.
- Marginal income-tax rate = 40% (remains constant).
- Annual revenue = $145,000 (today's dollars).
- Annual expense (not including depreciation and interest) = $82,000 (today's dollars).
- Market interest rate = 18%.

(a) With an average general inflation rate of 5% expected during the project period, which will affect all revenues, expenses, and the salvage value, determine the cash flows in actual dollars.

(b) Compute the net present worth of the project under inflation.

(c) Compute the net-present-worth loss (gain) due to inflation.

(d) Compute the present-worth loss (or gain) due to borrowing.

9.40 Norcross Textile Company is considering automating its piece-goods screen-printing system at a cost of $20,000. The firm expects to phase out this automated printing system at the end of five years, because of changes in style. At that time, the firm could scrap the system for $2,000 in today's dollars. The expected net savings provided by the automation are in today's dollars (constant dollars) as follows:

End of Year	Cash Flow (in Constant $)
1	$15,000
2	$17,000
3–5	$14,000

The system qualifies as a five-year MACRS property and will be depreciated accordingly. The expected average general inflation rate over the next five years is approximately 5% per year. The firm will finance the entire project by borrowing at 10%. The scheduled repayment of the loan will be as follows:

End of Year	Principal Payment	Interest Payment
1	$6,042	$2,000
2	$6,647	$1,396
3	$7,311	$731

The firm's market interest rate for project evaluation during this inflation-ridden time is 20%. Assume that the net savings and the selling price will be responsive to this average inflation rate. The firm's marginal tax rate is known to be 40%.

(a) Determine the after-tax cash flows of this project in actual dollars.

(b) Determine the PW reduction (or gains) in profitability caused by inflation.

9.41 The J. F. Manning Metal Co. is considering the purchase of a new milling machine during year zero. The machine's base price is $135,000, and it will cost another $15,000 to modify it for special use by the firm, resulting in a $150,000 cost base for depreciation. The machine falls into the MACRS seven-year property class. The machine will be sold after three years for $80,000 (actual dollars). Use of the machine will require an increase in net working capital (inventory) of $10,000 at the beginning of the project year. The machine will have no effect on revenues, but it is expected to save the firm $80,000

(today's dollars) per year in before-tax operating costs, mainly labor. The firm's marginal tax rate is 40%, and this rate is expected to remain unchanged over the project's duration. However, the company expects that the labor cost will increase at an annual rate of 5% and that the working-capital requirement will grow at an annual rate of 8%, caused by inflation. The selling price of the milling machine is not affected by inflation. The general inflation rate is estimated to be 6% per year over the project period. The firm's market interest rate is 20%.

(a) Determine the project cash flows in actual dollars.

(b) Determine the project cash flows in constant (time zero) dollars.

(c) Is this project acceptable?

Rate-of-Return Analysis under Inflation

9.42 Fuller Ford Company is considering purchasing a vertical drill machine. The machine will cost $50,000 and will have an eight-year service life. The selling price of the machine at the end of eight years is expected to be $5,000 in today's dollars. The machine will generate annual revenues of $20,000 (today's dollars), but it expects to have an annual expense (excluding depreciation) of $8,000 (today's dollars). The asset is classified as a seven-year MACRS property. The project requires a working-capital investment of $10,000 at year zero. The marginal income-tax rate for the firm is averaging 35%. The firm's market interest rate is 18%.

(a) Determine the internal rate of return of this investment.

(b) Assume that the firm expects a general inflation rate of 5%, but that it also expects an 8% annual increase in revenue and working capital and a 6% annual increase in expenses, caused by inflation. Compute the real (inflation-free) internal rate of return. Is this project acceptable?

9.43 You have $10,000 cash, which you want to invest. Normally, you would deposit the money in a savings account that pays an annual interest rate of 6%. However, you are now considering the possibility of investing in a bond. Your marginal tax rate is 30% for both ordinary income and capital gains. You expect the general inflation rate to be 3% during the investment period. You can buy a high-grade municipal bond costing $10,000 that pays interest of 9% ($900) per year. This interest is not taxable. A comparable high-grade corporate bond is also available that is just as safe as the municipal bond, but that pays an interest rate of 12% ($1,200) per year. This interest is taxable as ordinary income. Both bonds mature at the end of year five.

(a) Determine the real (inflation-free) rate of return for each bond.
(b) Without knowing your MARR, can you make a choice between these two bonds?

9.44 Air Florida is considering two types of engines for use in its planes, each of which has the same life, same maintenance costs, and same repair record:

- Engine *A* costs $100,000 and uses 50,000 gallons per 1,000 hours of operation at the average service load encountered in passenger service.
- Engine *B* costs $200,000 and uses 32,000 gallons per 1,000 hours of operation at the same service load.

Both engines are estimated to operate for 10,000 service hours before any major overhaul of the engines is required. If fuel currently costs $1.25 per gallon and its price is expected to increase at a rate of 8% because of inflation, which engine should the firm install for an expected 2,000 hours of operation per year? The firm's marginal income-tax rate is 40%, and the engine will be depreciated based on the units-of-production method. Assume that the firm's market interest rate is 20%. It is estimated that both engines will retain a market value of 40% of their initial cost (actual dollars) if they are sold on market after 10,000 hours of operation.

(a) Using the present-worth criterion, which project would you select?
(b) Using the annual-equivalence criterion, which project would you select?
(c) Using the future-worth criterion, which project would you select?

9.45 Johnson Chemical Company has just received a special subcontracting job from one of its clients. This two-year project requires the purchase of a special-purpose painting sprayer for $60,000. This equipment falls into the MACRS five-year class. After the subcontracting work is completed, the painting sprayer will be sold at the end of two years for $40,000 (actual dollars). The painting system will require an increase in net working capital (spare-parts inventory, such as spray nozzles) of $5,000. This investment in working capital will be fully recovered at the end of project termination. The project will bring in an additional annual revenue of $120,000 (today's dollars), but it is expected to incur an additional annual operating cost of $60,000 (today's dollars). It is projected that, because of inflation, there will be sales-price increases at an annual rate of 5% (which implies that annual revenues will also increase at an annual rate of 5%). An annual increase of 4% for expenses and working-capital requirements is expected. The company has a marginal tax rate of 30%, and it uses a market interest rate of 15% for project evaluation during the inflationary period. The firm expects a general inflation of 8% during the project period.

(a) Compute the after-tax cash flows in actual dollars.
(b) What is the rate of return on this investment (real earnings)?
(c) Is this special order profitable?

9.46 Land Development Corporation is considering the purchase of a bulldozer. The bulldozer will cost $100,000 and have an estimated salvage value of $30,000 at the end of six years. The asset will generate annual before-tax revenues of $80,000 over the next six years. The asset is classified as a five-year MACRS property. The marginal tax rate is 40%, and the firm's market interest rate is known to be 18%. All dollar figures represent constant dollars at time zero and are responsive to the general inflation rate $\bar{f}$.

(a) With $\bar{f} = 6\%$, compute the after-tax cash flows in actual dollars.

(b) Determine the real rate of return of this project on an after-tax basis.

(c) Suppose that the initial cost of the project will be financed through a local bank at an interest rate of 12%, with an annual payment of $24,323 over six years. With this additional condition, rework part (a).

(d) Based on your answer to part (a), determine the PW loss due to inflation.

(e) Based on your answer to part (c), determine how much the project has to generate in additional before-tax annual revenues in actual dollars (equal amount) in order to make up the inflation loss.

9.47 Wilson Machine Tools, Inc., a manufacturer of fabricated metal products, is considering the purchase of a high-tech computer-controlled milling machine at a cost of $95,000. The cost of installing the machine, preparing the site, wiring, and rearranging other equipment is expected to be $15,000. This installation cost will be added to the cost of the machine in order to determine the total cost basis for depreciation. Special jigs and tool dies for the particular product will also be required, at a cost of $10,000. The milling machine is expected to last 10 years, but the jigs and dies for only five years. Therefore, another set of jigs and dies has to be purchased at the end of five years. The milling machine will have a $10,000 salvage value at the end of its life, and the special jigs and dies are worth only $300 as scrap metal at any time in their lives. The machine is classified as a seven-year MACRS property, and the special jigs and dies are classified as a three-year MACRS property. With the new milling machine, Wilson expects an additional annual revenue of $80,000 from increased production. The additional annual production costs are estimated as follows: materials, $9,000; labor, $15,000; energy, $4,500; miscellaneous O&M costs, $3,000. Wilson's marginal income-tax rate is expected to remain at 35% over the project life of 10 years. All dollar figures are in today's dollars. The firm's market interest rate is 18%, and the expected general inflation rate during the project period is estimated at 6%.

(a) Determine the project cash flows in the absence of inflation.

(b) Determine the internal rate of return for the project, based on your answer to part (a).

(c) Suppose that Wilson expects the following price increases during the project period: materials at 4% per year, labor at 5% per year, and energy and other O&M costs at 3% per year. To compensate for these increases in prices, Wilson is planning to increase annual revenue at the rate of 7% per year by charging its customers a higher price. No changes in salvage value are expected for the machine, as well as for the jigs and dies. Determine the project cash flows in actual dollars.

(d) Based on your answer to part (c), determine the real (inflation-free) rate of return of the project.

(e) Determine the economic loss (or gain) in present worth caused by inflation.

Cost of Capital

9.48 Calculate the after-tax cost of debt under each of the following conditions:

(a) Interest rate, 12%; tax rate, 25%.

(b) Interest rate, 14%; tax rate, 34%.

(c) Interest rate, 15%; tax rate, 40%.

9.49 The estimated beta (β) of a firm is 1.7. The market return (r_m) is 14%, and the risk-free rate (r_f) is 7%. Estimate the cost of equity (i_e).

9.50 The Callaway Company's cost of equity (i_e) is 22%. Its before-tax cost of debt is 13%, and its marginal tax rate is 40%. The firm's capital structure calls for a debt-to-equity ratio of 45%. Calculate Callaway's cost of capital (k).

9.51 An automobile company is contemplating issuing stock to finance investment in producing a new sports-utility vehicle. The annual return to the market portfolio is expected to be 15%, and the current risk-free interest rate is 5%. The company's analysts further believe that the expected return to the project will be 20% annually. What is the maximum beta value that would induce the auto-maker to issue the stock?

Short Case Studies with Excel

9.52 The National Parts, Inc., an auto-parts manufacturer, is considering purchasing a rapid prototyping system to reduce prototyping time for form, fit, and function applications in automobile-parts manufacturing. An outside consultant has been called in to estimate the initial hardware requirement and installation costs. He suggests the following costs:

- Prototyping equipment: $185,000.
- Posturing apparatus: $10,000.
- Software: $15,000.
- Maintenance: $36,000 per year to the equipment manufacturer for maintenance services.
- Resin: Annual liquid-polymer consumption is 400 gallons at $350 per gallon.
- Site preparation: $2,000—Some facility changes are required when installing the rapid prototyping system. (Certain liquid resins contain a toxic substance, so the work area must be well vented.)

The expected life of the system is six years, with an estimated salvage value of $30,000. The proposed system is classified as a five-year MACRS property. A group of computer consultants must be hired to develop customized software to run on these systems. These software development

costs will be $20,000 and can be expensed during the first tax year. The new system will reduce prototype development time by 75% and material waste (resin) by 25%. This reduction in development time and material waste will save the firm $314,000 and $35,000 annually, respectively. The firm's expected marginal tax rate over the next six years will be 40%. The firm's interest rate is 20%.

(a) Assuming that the entire initial investment will be financed from the firm's retained earnings (equity financing), determine the after-tax cash flows over the investment life. Compute the PW of this investment.

(b) Assuming that the entire initial investment will be financed through a local bank at an interest rate of 13%, compounded annually, determine the net after-tax cash flows for the project. Compute the PW of this investment.

(c) Suppose that a financial lease is available for the prototype system at $62,560 per year, payable at the beginning of each year. Compute the PW of this investment with lease financing.

(d) Select the best financing option, based on the rate of return on incremental investment.

9.53 National Office Automation, Inc. (NOAI) is a leading developer of imaging systems, controllers, and related accessories. The company's product line consists of systems for desktop publishing, automatic identification, advanced imaging, and office-information markets. The firm's manufacturing plant in Ann Arbor, Michigan, consists of eight different functions: cable assembly, board assembly, mechanical assembly, controller integration, printer integration, production repair, customer repair, and shipping. NOAI is considering the process of transporting pallets loaded with eight packaged desktop printers from the printer integration department to the shipping department. Several alternatives have been examined to minimize operating and maintenance costs. The two most feasible alternatives are described as follows:

- Option 1: Use gas-powered lift trucks to transport pallets of packaged printers from printer integration to shipping. The trucks will also be used to return printers that must be reworked. The trucks can be leased at a cost of $5,465 per year. With a maintenance contract costing $6,317 per year, the dealer will maintain the trucks. A fuel cost of $1,660 per year is also expected. Each truck requires a driver for each of the three shifts, at a total cost of $58,653 per year for labor. It is also estimated that transportation by truck would cause damages to material and equipment totaling $10,000 per year.

- Option 2: Install an automatic guided vehicle system (AGVS) to transport pallets of packaged printers from printer integration to shipping and to return products that require reworking. The AGVS, using an electric powered cart and embedded wire-guidance system, would do the same job that the trucks would do, but without drivers. The total investment costs, including installation, are $159,000. NOAI could obtain a term loan for the full investment amount ($159,000) at a 10% interest rate. The loan would be amortized over five years, with payments made at the end of each year. The AGVS falls into the seven-year MACRS classification, and it has an estimated service life of 10 years, with no salvage value. If the AGVS is installed, a maintenance contract would be obtained at a cost of $20,000, payable at the beginning of each year.

The firm's marginal tax rate is 35%, and its MARR is 15%.

(a) Determine the net cash flows for each alternative over 10 years.

(b) Compute the incremental cash flows (Option 2 − Option 1), and determine the rate of return on this incremental investment.

(c) Determine the best course of action, based on the rate-of-return criterion.

9.54 Recent biotechnological research has made possible the development of a sensing device that implants living cells onto a silicon chip. The chip is capable of detecting physical and chemical changes in cell processes. Proposed uses include researching the mechanisms of disease on a cellular level, developing new therapeutic drugs, and replacing the use of animals in cosmetic and drug testing. Biotech Device Corporation (BDC) has just perfected a process for mass producing the chip. The following information has been compiled for the board of directors:

- BDC's marketing department plans to target sales of the device to larger chemical and drug manufacturers. BDC estimates that annual sales would be 2,000 units, if the device were priced at $95,000 per unit (dollars of the first operating year).

- To support this level of sales volume, BDC would need a new manufacturing plant. Once the "go" decision is made, this plant could be built and made ready for production within one year. BDC would need a 30-acre tract of land that would cost $1.5 million. If the decision were to be made, the land could be purchased on December 31, 2003. The building would cost $5 million and would be depreciated according to the MACRS 39-year class. The first payment of $1 million would be due to the contractor on December 31, 2004, and the remaining $4 million would be due on December 31, 2005.

- The required manufacturing equipment would be installed late in 2005 and would be paid for on December 31, 2005. BDC would have to purchase the equipment at an estimated cost of $8 million, including transportation, plus a further $500,000 for installation. The equipment would fall into the MACRS seven-year class.

- The project would require an initial investment of $1 million in working capital. This initial working-capital investment would be made on December 31, 2005; on December 31 of each following year, net working capital would be increased by an amount equal to 15% of any sales increase expected during the coming year. The investments in working capital would be fully recovered at the end of the project year.

- The project's estimated economic life is six years (excluding the two-year construction period). At that time, the land is expected to have a market value of $2 million, the building a value of $3 million, and the equipment a value of $1.5 million. The estimated variable manufacturing costs would total 60% of the dollar sales. Fixed costs, excluding depreciation, would be $5 million for the first year of operations. Since the plant would begin operations on January 1, 2006, the first operating cash flows would occur on December 31, 2006.

- Sales prices and fixed overhead costs, other than depreciation, are projected to increase with general inflation, which is expected to average 5% per year over the six-year life of the project.

- To date, BDC has spent $5.5 million on research and development (R&D) associated with the cell-implanting research. The company has already expensed $4 million R&D costs. The remaining $1.5 million will be amortized over six years (i.e., the annual amortization expense would be $250,000). If BDC decides not to proceed with the project, the $1.5 million R&D cost could be written off on December 31, 2003.

- BDC's marginal tax rate is 40%, and its market interest rate is 20%. Any capital gains will also be taxed at 40%.

(a) Determine the after-tax cash flows of the project in actual dollars.

(b) Determine the inflation-free (real) IRR of the investment.

(c) Would you recommend that the firm accept the project?

9.55 American Chemical Corporation (ACC) is a multinational manufacturer of industrial chemical products. ACC has made great progress in energy-cost reduction and has implemented several cogeneration projects in the United States and Puerto Rico, including the completion of a 35-megawatt (MW) unit in Chicago and a 29-MW unit in Baton Rouge. The division of ACC being considered for one of its more recent cogeneration projects is a chemical plant located in Texas. The plant has a power usage of 80 million kilowatt-hours (kWh) annually. However, on average, it uses 85% of its 10-MW capacity, which would bring the average power usage to 68 million kWh annually. Texas Electric presently charges $0.09 per kWh of electric consumption for the ACC plant, a rate that is considered high throughout the industry. Because ACC's power consumption is so large, the purchase of a cogeneration unit is considered to be desirable. Installation of the cogeneration unit would allow ACC to generate its own power and to avoid the annual $6,120,000 expense to Texas Electric. The total initial investment cost would be $10,500,000: $10,000,000 for the purchase of the power unit itself—a gas-fired 10-MW Allison 571—and engineering, design, and site preparation. The remaining $500,000 includes the purchase of interconnection equipment, such as poles and distribution lines, that will be used to interface the cogenerator with the existing utility facilities.

As ACC's management has decided to raise the $10.5 million by selling bonds, the company's engineers have estimated the operating costs of the cogeneration project. The annual cash flow is composed of many factors: maintenance costs, overhaul costs, expenses for standby power, and other miscellaneous expenses. Maintenance costs are projected to be approximately $500,000 per year. The unit must be over-hauled every three years, at a cost of $1.5 million per overhaul. Standby power is the service provided by the utility in the event of a cogeneration-unit trip or scheduled maintenance outage. Unscheduled outages are expected to occur four times annually, with each outage averaging two hours in duration, at an annual expense of $6,400. In addition, overhauling the unit takes approximately 100 hours and occurs every three years, requiring another triennial standby-power cost of $100,000. Miscellaneous expenses, such as additional personnel and insurance, are expected to total $1 million. Fuel (spot gas) will be consumed at a rate of 8,000 BTU per kWh, including the heat-recovery cycle. At $2.00 per million BTU, the annual fuel cost will reach $1,280,000. Due to obsolescence, the expected life of the cogeneration project will be 12 years, after which Allison will pay ACC $1 million for salvage of all equipment.

Revenues will be incurred from the sale of excess electricity to the utility company at a negotiated rate. Since the chemical plant will consume on average 85% of the unit's 10-MW output, 15% of the output will be sold at $0.04 per kWh, bringing in an annual revenue of $480,000. ACC's marginal tax rate (combined federal and state) is 36%, and its minimum required rate of return for any cogeneration project is 27%. The anticipated costs and revenues are summarized as follows:

Initial Investment	
Cogeneration unit and engineering, design, and site preparation (15-year MACRS class)	$10,000,000
Interconnection equipment (5-year MACRS class)	$500,000
Salvage value after 12 years of use	$1,000,000
Annual Expenses	
Maintenance	$500,000
Miscellaneous (additional personnel and insurance)	$1,000,000
Standby power	$6,400
Fuel	$1,280,000
Other Operating Expenses	
Overhaul every three years	$1,500,000
Standby power during overhaul	$100,000
Revenues	
Sale of excess power to Texas Electric	$480,000

(a) If the cogeneration unit and other connecting equipment could be financed by issuing corporate bonds at an interest rate of 9%, compounded annually, determine the net cash flow from the cogeneration project.

(b) If the cogeneration unit can be leased, what would be the maximum annual lease amount that ACC is willing to pay?

.10

The Chad Cameroon Petroleum Development and Pipeline Project: Risky Business[1]

The $3.7 billion Chad Cameroon Pipeline Project represents one of the largest private sector investments in Africa. The project developer is an international consortium consisting of Exxon (40%), Chevron (25%), and Malaysia's Petronas (35%). The project is being funded by both the International Finance Corporation (IFC) and the World Bank. The project involves the drilling of 300 oil wells in the Doba basin of southern Chad and the construction of a 1070 km export pipeline to the Atlantic coast of Cameroon. An offshore export terminal facility will be built and connected to the port of Kribi in South Cameroon by a 12 km underwater pipeline. The project is expected to produce 225,000 barrels of oil per day during peak production. Project revenue as a whole will be $12 billion. Expected government revenues are $1.7 billion for Chad and $505 million for Cameroon over the 28-year operating period. According to the World Bank, the project will "spur further oil exploration and development in Chad and Cameroon."

[1] *Source*: International Finance Corporation (IFC), Summary of Project Information (SPI), Chad-Cameroon Petroleum Development and Pipeline Project, March 3, 2000.

Handling Project Uncertainty

In their own assessment, World Bank and IFC rate the overall risk of the project as "significant" and acknowledge "significant adverse impacts." However, the claimed rationale behind World Bank and IFC support for the project is the alleviation of poverty in both Chad and Cameroon through increased state revenues from oil. Many critics of the project, however, question whether the project will in fact achieve these goals, given its structure and the situation on the ground, and the mechanistic assumption that revenue generation is necessarily equated with poverty reduction. In addition to the risks to the companies, the project presents significant risks to local people in both Chad and Cameroon in terms of environmental degradation, adverse health impacts, and human-rights abuses.

What appears to be the greatest risk of undertaking this project for the oil companies? What is the risk to the IFC in its financing of the project? In previous chapters, cash flows from projects were assumed to be known with complete certainty; our analysis was concerned with measuring the economic worth of projects and selecting the best investment projects. Although these types of analyses can provide a reasonable decision basis in many investment situations, we should certainly recognize that most projects involve uncertainty.

In this type of situation, management rarely has precise expectations about the future cash flows to be derived from a particular project. In fact, the best that a firm can reasonably expect to do is to estimate the range of possible future costs and benefits and the relative chances of achieving a reasonable return on the investment. We use the term **risk** to describe an investment project where cash flows are not known in advance with certainty, but for which an array of outcomes and their probabilities (odds) can be considered. We will also use the term **project risk** to refer to variability in a project's PW. Higher project risk reflects a higher anticipated variability in a project's PW. In essence, we can see *risk as the potential for loss*. This chapter begins by exploring the origins of project risk.

10-1 Origins of Project Risk

When deciding whether or not to make a major capital investment, such as introducing a new product, a number of issues must be considered and estimated. The factors to be estimated include the total market for the product; the market share that the firm can attain; the growth in the market; the cost of producing the product; the selling price of the product; the life of the product; the cost and life of the equipment needed; and the effective tax rates. Many of these factors are subject to substantial uncertainty. A common approach is to make single-number "best estimates" for each of the uncertain factors and then to calculate measures of profitability, such as PW or rate of return for the project. This approach has two drawbacks:

1. No guarantee can ever ensure that the best estimates will ever match actual values.

2. No provision is made to measure the risk associated with the investment or the project risk. In particular, managers have no way of determining either the probability that a project will lose money or the probability that it will generate large profits.

Because cash flows can be so difficult to estimate accurately, project managers frequently consider a range of possible values for cash flow elements. If a range of values for individual cash flows is possible, it follows that a range of values for the PW of a given project is also possible. Clearly, the analyst will want to try to gauge the probability and reliability of individual cash flows occurring and, consequently, the level of certainty about overall project worth.

Quantitative statements about risk are given as numerical probabilities or as values for likelihood (odds) of occurrence. Probabilities are given as decimal fractions in the interval 0.0 to 1.0. An event or outcome that is certain to occur has a probability of 1.0. As the probability of an event approaches 0, the event becomes increasingly less likely to occur. The assignment of probabilities to the various outcomes of an investment project is generally called **risk analysis**.

10–2   Methods of Describing Project Risk

We may begin analyzing project risk by first determining the uncertainty inherent in a project's cash flows. We can do this analysis in a number of ways, which range from making informal judgments to performing complex economic and statistical analyses. In this section, we will introduce three methods of describing project risk: (1) sensitivity analysis, (2) break-even analysis, and (3) scenario analysis. Each method will be explained with reference to a single example (Boston Metal Company). We also introduce the method for conducting sensitivity analysis for mutually exclusive alternatives.

10.2.1   Sensitivity Analysis

One way to glean a sense of the possible outcomes of an investment is to perform a sensitivity analysis. **Sensitivity analysis** determines the effect on the PW of variations in the input variables (such as revenues, operating cost, and salvage value) used to estimate after-tax cash flows. In calculating cash flows, some items have a greater influence on the final result than others. In some problems, the most significant item is easily identified. For example, the estimate of sales volume can have a major impact in a project's PW, especially in new product introductions. We may want to identify the items that have an important influence on the final results so that they can be subjected to special scrutiny.

Sensitivity analysis is sometimes called "what-if analysis," because it answers questions such as what if incremental sales are only 1,000 units, rather than 2,000 units? Then what will the PW be? Sensitivity analysis begins with a base-case situation, which is developed using the most likely value for each input. We then change the specific variable of interest by several specified percentages above and below the most likely value, while holding other variables constant. Next, we calculate a new PW for each of these values. A convenient and useful way to present the results of a sensitivity analysis is to plot **sensitivity graphs**. The slopes of the lines show how sensitive the PW is to changes in each of the inputs: The steeper the slope, the more sensitive the PW is to a change in a particular variable. Sensitivity graphs identify the crucial variables that affect the final outcome the most. We will use Example 10.1 to illustrate the concept of sensitivity analysis.

Example 10.1   Sensitivity Analysis

Boston Metal Company (BMC), a small manufacturer of fabricated metal parts, must decide whether to compete to become the supplier of transmission housings for Gulf Electric. Gulf Electric produces transmission housings in its own in-house manufacturing facility, but it has almost reached its maximum production capacity. Therefore, Gulf is looking for an outside supplier. To compete, BMC must design a new fixture for the production process and purchase a new forge. The available details for this purchase are as follows:

- The new forge would cost $125,000. This total includes retooling costs for the transmission housings.
- If BMC gets the order, it may be able to sell as many as 2,000 units per year to Gulf Electric for $50 each and variable production costs, such as direct-labor and direct-material costs, will be $15 per unit. The increase in fixed costs, other than depreciation, will amount to $10,000 per year.
- The firm expects that the proposed transmission-housings project will have about a five-year project life. The firm also estimates that the amount ordered by Gulf Electric for the first year will also be ordered in each of the subsequent four years. (Due to the nature of contracted production, the annual demand and unit price would remain the same over the project after the contract is signed.)
- The initial investment can be depreciated on a MACRS basis over a seven-year period, and the marginal income-tax rate is expected to remain at 40%. At the end of five years, the forge is expected to retain a market value of about 32% of the original investment.
- Based on this information, the engineering and marketing staffs of BMC have prepared the cash flow forecasts shown in Table 10.1. Since the PW is positive ($40,168) at the 15% opportunity cost of capital (MARR), the project appears to be worth undertaking.

What Makes BMC Managers Worry: BMC's managers are uneasy about this project, because too many uncertain elements have not been considered in the analysis:

- If it decides to compete for the project, BMC must invest in the forging machine in order to provide Gulf Electric with samples as a part of the bidding process. If Gulf Electric does not like BMC's samples, BMC stands to lose its entire investment in the forging machine.
- If Gulf likes BMC's samples, but feels that they are overpriced, BMC would be under pressure to bring the price in line with those of competing firms. Even the possibility that BMC would get a smaller order must be considered, as Gulf may use its overtime capacity to produce extra units instead of purchasing the entire amount it desires. BMC is also not certain about the variable- and fixed-cost projections.

Recognizing these uncertainties, the managers want to assess a variety of possible scenarios before making a final decision. Put yourself in BMC's management position and describe how you might address the uncertainty associated with the project. In doing so, perform a sensitivity analysis for each variable and develop a sensitivity graph.

Discussion: Table 10.1 shows BMC's expected, but ultimately uncertain, cash flows. In particular, BMC is not confident in its revenue forecasts. The managers

TABLE 10.1	After-Tax Cash Flow for BMC's Transmission-Housings Project, Based on Most Likely Estimates						
	A	B	C	D	E	F	G
1	**Income Statement**						
2		0	1	2	3	4	5
3	Revenues:						
4	Unit Price		$50	$50	$50	$50	$50
5	Demand (units)		2000	2000	2000	2000	2000
6	Sales Revenue		$100,000	$100,000	$100,000	$100,000	$100,000
7	Expenses:						
8	Unit Variable Cost		$15	$15	$15	$15	$15
9	Variable Cost		30,000	30,000	30,000	30,000	30,000
10	Fixed Cost		10,000	10,000	10,000	10,000	10,000
11	Depreciation		17,863	30,613	21,863	15,613	5,581
12							
13	Taxable Income		$42,137	$29,387	$38,137	$44,387	$54,419
14	Income Taxes (40%)		16,855	11,755	15,255	17,755	21,768
15							
16	Net Income		$25,282	$17,632	$22,882	$26,632	$32,651
17							
18	**Cash Flow Statement**						
19	Operating Activities:						
20	Net Income		25,282	17,632	22,882	26,632	32,651
21	Depreciation		17,863	30,613	21,863	15,613	5,581
22	Investment Activities:						
23	Investment	(125,000)					
24	Salvage						40,000
25	Gains Tax						(2,613)
26							
27	Net Cash Flow	$(125,000)	$43,145	$48,245	$44,745	$42,245	$76,619
28							

think that if competing firms enter the market BMC will lose a substantial portion of the projected revenues by not being able to increase its bidding price. Before undertaking the project described, the company needs to identify the key variables that will determine whether the project will succeed or fail. The marketing department has estimated annual revenue as follows:

$$\text{Annual revenue} = (\text{production demand}) \times (\text{unit price})$$
$$= (2,000)(\$50) = \$100,000.$$

The engineering department has estimated variable costs such as that labor and material per unit at $15. Since the projected sales volume is 2,000 units per year, the total variable cost is $30,000.

After defining the unit sales, unit price, unit variable cost, fixed cost, and salvage value, we conduct a sensitivity analysis with respect to these key input variables. This is done by varying each of the estimates by a given percentage and determining what effect the variation in that item will have on the final results. If the effect is large, the result is sensitive to that item. Our objective is to locate the most sensitive item(s).

SOLUTION

Sensitivity analysis: We begin the sensitivity analysis with a consideration of the base-case situation, which reflects the best estimate (expected value) for each input variable. In developing Table 10.2, we changed a given variable by 20% in 5% increments above and below the base-case value and calculated new PWs, while other variables were held constant. Now we ask a series of what-if questions: What if sales are 20% below the expected level? What if operating costs rise? What if the unit price drops from $50 to $45? Table 10.2 summarizes the results of varying the values of the key input variables.

Sensitivity graph: Figure 10.1 shows the transmission-housings project's sensitivity graphs for five of the key input variables. The base-case PW is plotted on

TABLE 10.2 Sensitivity Analysis for Five Key Input Variables								
Deviation	**−20%**	**−15%**	**−10%**	**−5%**	**0%**	**5%**	**10%**	**15%**	**20%**
Unit price	$57	$9,999	$20,055	$30,111	$40,169	$50,225	$60,281	$70,337	$80,393
Demand	$12,010	$19,049	$26,088	$33,130	$40,169	$47,208	$54,247	$61,286	$68,325
Variable cost	$52,236	$49,219	$46,202	$43,186	$40,169	$37,152	$34,135	$31,118	$28,101
Fixed cost	$44,191	$43,185	$42,179	$41,175	$40,169	$39,163	$38,157	$37,151	$36,145
Salvage value	$37,782	$38,378	$38,974	$39,573	$40,169	$40,765	$41,361	$41,957	$42,553

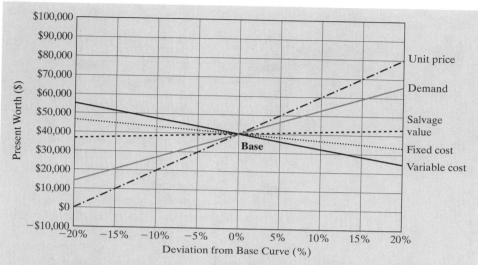

FIGURE 10.1 Sensitivity graph for BMC's transmission-housings project

the ordinate of the graph at the value 1.0 on the abscissa (or 0% deviation). Next, the value of product demand is reduced to 0.95% of its base-case value, and the PW is recomputed, with all other variables held at their base-case value. We repeat the process by either decreasing or increasing the relative deviation from the base case. The lines for the variable unit price, variable unit cost, fixed cost, and salvage value are obtained in the same manner.

In Figure 10.1, we see that the project's PW is (1) very sensitive to changes in product demand and unit price, (2) fairly sensitive to changes in the variable costs, and (3) relatively insensitive to changes in the fixed cost and the salvage value. Graphic displays such as the one in Figure 10.1 provide a useful means to communicate the relative sensitivities of the different variables on the corresponding PW value. However, sensitivity graphs do not explain any interactions among the variables or the likelihood of realizing any specific deviation from the base case. Certainly, it is conceivable that a project might not be very sensitive to changes in either of two items, but very sensitive to combined changes in them.

10.2.2 Sensitivity Analysis for Mutually Exclusive Alternatives

In Figure 10.1, each variable is uniformly adjusted by ±20%, and all variables are plotted on the same chart. This uniform adjustment can be too simplistic an assumption; in many situations, each variable can have a different range of uncertainty. Also, plotting all variables on the same chart could be confusing if there are too many variables to consider. When we perform sensitivity analysis for mutually exclusive alternatives, it may be more effective to plot the PWs (or any other measures, such as AEs) of all alternatives over the range of each input; in other words, create one plot for each input, with units of the input on the horizontal axis. Example 10.2 illustrates this approach.

Example 10.2 Sensitivity Analysis for Mutually Exclusive Alternatives

A local U.S. Postal Service office is considering purchasing a 4,000 lb forklift truck, which will be used primarily for processing incoming as well as outgoing postal packages. Forklift trucks traditionally have been fueled by either gasoline, liquid propane gas (LPG), or diesel fuel. Battery-powered electric forklifts, however, are increasingly popular in many industrial sectors, because of their economic and environmental benefits. Therefore, the postal service is interested in comparing the four different types of forklifts. Purchase costs as well as annual operating and maintenance costs for each type of forklift are provided by a local utility company and the Lead Industries Association. Annual fuel and maintenance costs are measured the discount of number of shifts per year, where one shift is equivalent to eight hours of operation.

The postal service is unsure of the number of shifts per year, but it expects it should be somewhere between 200 and 260 shifts. Since the U.S. Postal Service does not pay income taxes, no depreciation or tax information is required. The U.S. government uses 10% as the discount rate for any project evaluation of this nature. Develop a sensitivity graph that shows how the best choice out of alternatives changes as a function of number of shifts per year.

SOLUTION

Two annual cost components are pertinent to this problem: (1) ownership cost (capital cost) and (2) operating cost (fuel and maintenance cost). Since the operating cost is already given on an annual basis, we only need to determine the equivalent annual ownership cost for each alternative.

A comparison of the variables of the four forklift types is given in the following table:

	Electrical Power	LPG	Gasoline	Diesel Fuel
Life expectancy	7 years	7 years	7 years	7 years
Initial cost	$30,000	$21,000	$20,000	$25,000
Salvage value	$3,000	$2,000	$2,000	$2,200
Maximum shifts per year	260	260	260	260
Fuel consumption/shift	32 kWh	12 gal	11 gal	7 gal
Fuel cost/unit	$0.05/kWh	$1.00/gal	$1.20/gal	$1.10/gal
Fuel cost/shift	$1.60	$12.00	$13.20	$7.70
Annual maintenance cost:				
Fixed cost	$500	$1,000	$1,200	$1,500
Variable cost/shift	$5	$6	$7	$9

(a) Ownership cost (capital cost): Using the capital-recovery-with-return formula developed in Eq. (6.3), we compute the following:

Electrical power: $CR(10\%) = (\$30,000 - \$3,000)(A/P, 10\%, 7)$
$+ (0.10)\$3,000$
$= \$5,845;$

LPG: $CR(10\%) = (\$21,000 - \$2,000)(A/P, 10\%, 7)$
$+ (0.10)\$2,000$
$= \$4,103;$

Gasoline: $CR(10\%) = (\$20,000 - \$2,000)(A/P, 10\%, 7)$
$+ (0.10)\$2,000$
$= \$3,897;$

Diesel fuel: $CR(10\%) = (\$25,000 - \$2,200)(A/P, 10\%, 7)$
$+ (0.10)\$2,200$
$= \$4,903.$

(b) Annual operating cost: We can express the annual operating cost as a function of number of shifts per year (M) by combining the variable- and fixed-cost portions of the fuel and maintenance expenditures:

Electrical power: $\$500 + (1.6 + 5)M = \$500 + 6.6M;$

LPG: $\$1,000 + (12 + 6)M = \$1,000 + 18M;$

Gasoline: $\$800 + (13.2 + 7)M = \$800 + 20.2M;$

Diesel fuel: $\$1,500 + (7.7 + 9)M = \$1,500 + 16.7M.$

(c) Total equivalent annual cost: This value is the sum of the ownership cost and operating cost:

Electrical power: $AE(10\%) = \$6,345 + 6.6M;$

LPG: $AE(10\%) = \$5,103 + 18M;$

Gasoline: $AE(10\%) = \$4,697 + 20.2M;$

Diesel fuel: $AE(10\%) = \$6,403 + 16.7M.$

In Figure 10.2, these four annual-equivalence costs are plotted as a function of the number of shifts, M. It appears that the economics of the electric forklift truck can be justified as long as the number of annual shifts exceeds approximately 121. If the number of shifts is less than 121, the gasoline truck becomes the most economically viable option. The diesel option is not a viable alternative for any range of M.

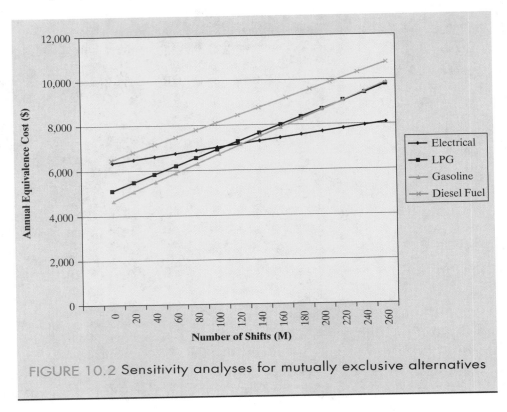

FIGURE 10.2 Sensitivity analyses for mutually exclusive alternatives

10.2.3 Break-Even Analysis

When we perform a sensitivity analysis for a project, we are asking how serious the effect of lower revenues or higher costs will be on the project's profitability. Managers sometimes prefer to ask how much sales can decrease below forecasts before the project begins to lose money. This type of analysis is known as **break-even analysis**. To illustrate the procedure of break-even analysis based on PW, we use the Goal Seek function from Excel. Note that this break-even value calculation is similar to the calculation used for the internal rate of return when we want to find the interest rate that makes the PW equal zero, as well as many other "cutoff values" where a choice changes.

Example 10.3 Break-Even Analysis with Excel

From the sensitivity analysis in Example 10.1, BMC's managers are convinced that the PW is most sensitive to changes in annual sales volume. Determine the required annual sales volume (units) to break even.

SOLUTION

The after-tax cash flows shown in Table 10.3 are basically the same as in Table 10.1. The table is simply an Excel spreadsheet in which the cash flow entries are a

TABLE 10.3 Break-Even Analysis Using Excel's Goal Seek Function

	A	B	C	D	E	F	G
1	Example 10.3 Break-Even Analysis						
2							
3	Input Data (Base);			Output Analysis:			
4							
5	Unit Price ($)	$50		Output (PW)		$0	
6	Demand	1429.39					
7	Var. cost ($/unit)	$15					
8	Fixed cost ($)	$10,000		By changing cell		Set cell	
9	Salvage ($)	$40,000					
10	Tax rate (%)	40%					
11	MARR (%)	15%					
12							
13							
14		0	1	2	3	4	5
15	Income Statement						
16	Revenues:						
17	Unit Price		$50	$50	$50	$50	$50
18	Demand (units)		1429.39	1429.39	1429.39	1429.39	1429.39
19	Sales Revenue		$71,470	$71,470	$71,470	$71,470	$71,470
20	Expenses:						
21	Unit Variable Cost		$15	$15	$15	$15	$15
22	Variable Cost		21,441	21,441	21,441	21,441	21,441
23	Fixed Cost		10,000	10,000	10,000	10,000	10,000
24	Depreciation		17,863	30,613	21,863	15,613	5,581
25							
26	Taxable Income		$22,166	$9,416	$18,166	$24,416	$34,448
27	Income Taxes (40%)		8,866	3,766	7,266	9,766	13,779
28							
29	Net Income		$13,299	$5,649	$10,899	$14,649	$20,669
30							
31	Cash Flow Statement						
32	Operating Activities:						
33	Net Income		13,299	5,649	10,899	14,649	20,669
34	Depreciation		17,863	30,613	21,863	15,613	5,581
35	Investment Activities:						
36	Investment	(125,000)					
37	Salvage						40,000
38	Gains Tax						(2,613)
39							
40	Net Cash Flow	($125,000)	$31,162	$36,262	$32,762	$30,262	$63,636
41							

Sample cell formulas: Cell F6 (PW): = NPV(B11, C40:G40) + B40
Cell G38 (Gains Tax): = −0.4*(G37−(B36−SUM(C24:G24)))

function of the input variables. Here, what we are looking for is the amount of annual sales (demand) that just makes the PW zero.

By using the Goal Seek function, we want to set the PW (cell $F$5) to zero by changing the demand value (cell $B$6). Pressing the OK button will produce the results shown in Table 10.3, indicating that the project will break even when the number of demand units reaches exactly 1,429.39, or 1,430 units.

Goal Seek	?	X
Set cell:	$F$5	
To value:	0	
By changing cell:	$B$6	
OK	Cancel	

10.2.4 Scenario Analysis

Although both sensitivity and break-even analyses are useful, they have limitations. Often, it is difficult to specify precisely the relationship between a particular variable and the PW. The relationship is further complicated by interdependencies among the variables. Holding operating costs constant while varying unit sales may ease the analysis, but in reality, operating costs do not behave in this manner. Yet, it may complicate the analysis too much to permit movement in more than one variable at a time.

A scenario analysis is a technique that considers the sensitivity of the PW to changes in both key variables and the range of likely variable values. For example, the decision maker may consider two extreme cases: a worst-case scenario (low unit sales, low unit price, high variable cost per unit, high fixed cost, and so on) and a best-case scenario. The PWs under the worst and the best conditions are then calculated and compared with the expected, or base-case, PW. Example 10.4 illustrates a plausible scenario analysis for BMC's transmission-housings project.

Example 10.4 Scenario Analysis

Consider again BMC's transmission-housings project from Example 10.1. Assume that the company's managers are fairly confident of their estimates of all of the project's cash flow variables, except the estimates for unit sales. Further, assume that they regard a decline in unit sales to below 1,600 or a rise above 2,400 as extremely unlikely. Thus, decremental annual sales of 400 units define the lower bound, or the worst-case scenario, whereas incremental annual sales of 400 units define the upper bound, or the best-case scenario. (Remember that the most likely value was 2,000 in annual unit sales.) Discuss the worst- and best-case scenarios, assuming that the unit sales for all five years would be equal.

Discussion: To carry out the scenario analysis, we ask the marketing and engineering staffs to give optimistic (best case) and pessimistic (worst case) estimates for the key variables. Then we use the worst-case variable values to obtain the worst-case PW and the best-case variable values to obtain the best-case PW.

SOLUTION

The results of our analysis are summarized as follows:

Variable Considered	Worst-Case Scenario	Most Likely Scenario	Best-Case Scenario
Unit demand	1,600	2,000	2,400
Unit price	$48	$50	$53
Variable cost	$17	$15	$12
Fixed cost	$11,000	$10,000	$8,000
Salvage value	$30,000	$40,000	$50,000
PW(15%)	−$5,856	$40,169	$104,295

We see that the base case produces a positive PW, the worst case produces a negative PW, and the best case produces a large positive PW.

10–3   Including Risk in Investment Evaluation

Once you have an idea of the degree of risk inherent in an investment, the next step is to incorporate this information into your evaluation of the proposed project. There are two fundamental approaches. The first approach is to consider the risk elements directly through probabilistic assessments. The second approach is to adjust the discount rate to reflect any perceived risk in the project's cash flows. We will consider both approaches briefly. However, the risk-adjusted discount-rate approach is more commonly practiced in the real world, as the method is mathematically much simpler than the probabilistic approach.

10.3.1   Probabilistic Approach

In principle, investment risk is concerned with the range of possible outcomes from an investment; the greater this range, the greater is the risk. Figure 10.3 illustrates this intuitive notion. It shows the possible rates of return that might be earned on

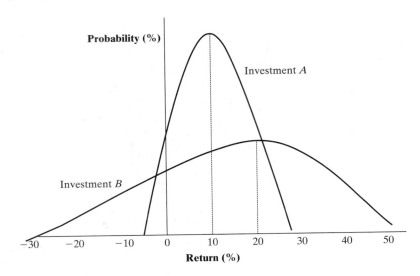

FIGURE 10.3 Illustration of investment risk: Investment A has a lower expected return but a lower risk than investment B

two investments in the form of bell-shaped curves. According to the figure, the expected return on investment A is about 10%, and the corresponding figure for investment B is about 20%.

As we would define **expected return** as the probability-weighted average of possible returns, this expected figure represents the central tendency of the random outcome (in our case, the return). To take a simple example, if three returns are possible—6, 9, and 18%—and if the chance of each occurring is 0.40, 0.30, and 0.30, respectively, the investment's expected return is calculated as follows:

Expected return $(\mu) = (0.40 \times 6\%) + (0.30 \times 9\%) + (0.30 \times 18\%) = 10.5\%$.

Basically, **risk** refers to the bunching of possible returns about an investment's expected return. If there is considerable bunching, as with investment A, the investment is low risk. With investment B, there is considerably less clustering of returns about the expected return, so it has higher risk. The way to measure this clustering tendency is to calculate a probability-weighted average of the deviations of possible returns from the expected return. One such average is the **standard deviation** (σ) of returns.

To illustrate the calculation of the standard deviation of returns, we calculate the differences between the possible returns and the expected return in the foregoing example as $(6\% - 10.5\%)$, $(9\% - 10.5\%)$, and $(18\% - 10.5\%)$. Because some of these deviations are positive and others are negative, they would tend to cancel one another out if we added them directly. So we square them to ensure the same sign, calculate the probability-weighted average of the squared deviations [a value known as the variance (σ^2)], and then find the square root:

Standard deviation $(\sigma) = (25.65)^{1/2} = 5.065\%$.

Event	Deviations	Weighted Deviations
1	$(6\% - 10.5\%)^2$	$0.40 \times (6\% - 10.5\%)^2$
2	$(9\% - 10.5\%)^2$	$0.30 \times (9\% - 10.5\%)^2$
3	$(18\% - 10.5\%)^2$	$0.30 \times (18\% - 10.5\%)^2$
		$(\sigma^2) = 25.65$

What we can tell here is that risk corresponds to the *dispersion,* or *uncertainty,* in possible outcomes. We also know that statistical techniques exist to measure this dispersion. In our example, the smaller standard deviation means a considerable bunching, or less risk. When comparing investments with the same expected returns, conservative, or risk-averse investors would prefer the investment with the smaller standard deviation of returns.

Having defined risk and risk aversion in at least a general sense, we might be interested in estimating the amount of risk present in a particular investment opportunity. For an investment project, if we can determine the expected cash flow as well as the variability of the cash flow in each period over the project life, we may be able to aggregate the risk over the project life in terms of net present value PW(r) as

$$E[\text{PW}(r)] = \sum_{n=0}^{N} \frac{E(A_n)}{(1 + r)^n} \tag{10.1}$$

and

$$V[\text{PW}(r)] = \sum_{n=0}^{N} \frac{V(A_n)}{(1 + r)^{2n}}, \tag{10.2}$$

where

r	= risk-free discount rate,
A_n	= cash flow in period n,
$E(A_n)$	= expected cash flow in period n,
$V(A_n)$	= variance of the cash flow in period n,
$E[\text{PW}(r)]$	= expected net present value of the project, and
$V[\text{PW}(r)]$	= variance of the net present value of the project.

In defining Eq. (10.2), we are also assuming the independence of cash flows, meaning that knowledge of what will happen to one particular period's cash flow will not allow us to predict what will happen to cash flows in other periods. Borrowing again from statistics, we are assuming statistically mutually independent project cash flows. In case we cannot assume such a statistical independence among cash flows, we need to consider the degree of dependence among the cash flows.[2]

---

[2] As we seek further refinement in our decision under risk, we may consider dependent random variables or the degree of dependence through correlation coefficients, which is beyond the scope of this text. See Park, C. S., and Sharp-Bette, G. P., *Advanced Engineering Economics,* New York: John Wiley, 1990 (Chapters 10 and 11).

Example 10.5 Computing the Mean and Variance of an Investment Opportunity

Assume that a project is expected to produce the following cash flows in each year, each cash flow is independent of one another, and the risk-free rate is 6%: Find the expected PW as well as the variance of the PW.

Period	Expected Cash Flow	Estimated Standard Deviation
0	−$2,000	$100
1	$1,000	$200
2	$2,000	$500

SOLUTION

Given: Periodic estimated project cash flows (means and variances); risk-free interest rate.

Find: $E[\text{PW}(r)]$ and $V[\text{PW}(r)]$.

Using Eqs. (10.1) and (10.2), we find that the expected PW and the variance of the PW are :

$$E[\text{PW}(6\%)] = -\$2,000 + \frac{\$1,000}{1.06} + \frac{\$2,000}{1.06^2} = \$723$$

and

$$V[\text{PW}(6\%)] = \$100^2 + \frac{\$200^2}{1.06^2} + \frac{\$500^2}{1.06^4} = \$243,623,$$

respectively. Thus, the standard deviation is $494.

Comments: How is this information used in decision making? It is assumed that most probability distributions are completely described by six standard deviations—three standard deviations above and three standard deviations below the mean. Therefore, the true PW of this project would almost certainly fall between −$759 and $2,205, as shown in Figure 10.4. If the PW below 3σ from the mean is still positive, we may say that the project is quite safe. If that figure is negative, it is then up to the decision maker to determine whether it is worth investing in the project, given the expected mean and standard deviation of the project.

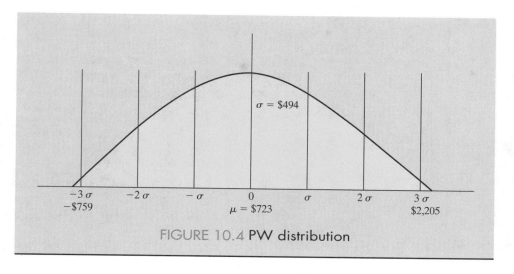

FIGURE 10.4 PW distribution

10.3.2 Risk-Adjusted Discount Rate Approach

An alternative approach in considering risk in project evaluation is to adjust the discount rate to reflect the degree of perceived investment risk. The most common way to do this is to add an increment to the discount rate—that is, discount the expected value of the risky cash flows at a rate that includes a premium for risk. The size of the risk premium naturally increases with the perceived risk of the investment.

What makes the risk-adjusted discount rate approach appealing in practical application is its simplicity. Most CFOs (chief financial officers) have at least a rough idea of how an investment's required rate of return should vary with perceived risk. For example, they know from the historical data that, over many years, common stocks have yielded an average annual return about 7% higher than the return on government bonds (risk-free return). If the present return on government bonds is 8%, it is plausible to expect an investment that is about as risky as common stocks to yield a return of about 15%. Similarly, CFOs know that an investment promising a return of 40% is attractive unless its risk is extraordinary high. Granted, such reasoning is imprecise; nonetheless, it does lend some objectivity to risk assessment. To illustrate the use of such risk-adjusted discount rates, consider Example 10.6.

Example 10.6 Investment Evaluation with Risk-Adjusted Discount Rate Approach

You are considering a $1 million investment promising risky cash flows with an expected value of $250,000 annually for 10 years. What is the investment's PW when the risk-free interest rate is 8% and management has decided to use a 6% risk premium to compensate for the uncertainty of the cash flows?

SOLUTION

Given: Initial investment = $1 million, expected annual cash flow = $250,000, $N = 10$ years, $r = 8\%$, and risk premium = 6%.

Find: PW and whether it is worth investing.

First find the risk-adjusted discount rate:

$$8\% + 6\% = 14\%.$$

Then calculate the net present value, using this risk-adjusted discount rate:

$$\text{PW}(14\%) = -\$1 \text{ million} + \$250,000(P/A, 14\%, 10) = \$304,028.$$

Because the PW is positive, the investment is attractive even after adjusting for risk.

Comments: Note how the risk-adjusted discount rate reduces the investment's appeal. If the investment were riskless, its PW at an 8% discount rate would be $677,520, but because a higher risk-adjusted rate is deemed appropriate, the PW falls by almost $373,492. Certainly, as the perceived project risk increases beyond the 8% risk premium, management would require an inducement higher than this amount before it is willing to make the investment.

10–4   Investment Strategies under Uncertainty

Once you understand the implications of project risk, you need to come up with an investment strategy, that tells you what to do as far as putting together an appropriate investment portfolio. The next question is how you go about actually implementing the decisions you have made. Because investing is an inexact science, *it is better to be approximately right than precisely wrong*. This is the approach taken in this chapter. The technique that is commonly practiced in financial investment is the concept of project diversification. The same concept also applies to the creation of an investment project portfolio.

10.4.1 Trade-Off between Risk and Reward

When it comes to investing, trying to weigh risk and reward can be a challenging task. Investors do not know the actual returns that project assets will deliver, or the difficulties that will occur along the way. Risk and reward are the two key words that will form the foundation for much of this section. This is what investing is all about—the trade-off between the opportunity to earn higher returns and the consequences of trying to do so. The greater the risk, the more you stand to gain or lose. There is no such thing as a truely risk-free investment. So the real task is not to try to find "risk-free" investments; strictly speaking, there aren't. The challenge is to decide what level of risk you are willing to assume and then, having decided on your risk tolerance, to understand the implications of that choice. Your range of investment choices—and their relative risk factors—may be classified into three types of investment groups: cash, debt, and equities.

10.4.2 Broader Diversification Reduces Risk

Even if you find risk exciting sometimes, you will probably sleep better if you have your money spread among different assets; don't put all your eggs in one basket. Your best protection against risk is diversification—spreading your investments around instead of investing in only one thing. For example, you can balance cash investments like CDs and money-market funds with stocks, bonds, and mutual funds. Even within equity investments, you can buy stocks of small, growing companies while also investing in large and well-established companies. What would you gain from this diversification practice? Well, you would hope to reduce the effect of market volatility on your holdings.

Usually, when returns are low in one area, you would like to see returns go up in another area. We may explain the concept of diversification graphically as shown in Figure 10.5. This figure shows three different investment scenarios explained as follows:

- **Case 1—Invest in two assets with similar return characteristics:** Suppose you have the two types of investments shown in Figure 10.5(a). These two investments are similar in their pattern of return; that is, they fluctuate to the same degree such that as one goes up or down, so does the other.[3] In other words, you will experience a great deal of volatility while you are in the market. If we keep both investments, our expected rate of return is simply the weighted average of their returns.

- **Case 2—Invest in two assets with dissimilar return characteristics:** Suppose you find the set of investments shown in Figure 10.5(b). The investments both have the same potential return, but the returns come at opposite times. That is, as one goes down, the other goes up.[4] Again, the return would be the weighted average of the returns of the two investments, but we may control the risk (fluctuation) considerably. All the negative returns of one investment would be offset by the positive returns of the other.

- **Case 3—Invest in multiple assets with dissimilar return characteristics:** Of course, in the real world, we are not likely to find either of the foregoing scenarios. The more likely situation is that larger portfolios will have a number of assets in them with differing, but not necessarily opposite, patterns of return. Then the results could be as shown in Figure 10.5(c). The overall yield of the portfolio is the weighted average of the individual assets, but the fluctuation—the risk—is dampened. It is therefore possible to achieve a higher rate of return without considerably increasing the risk by building a multiple-asset portfolio. This is exactly what we achieve in asset investing through diversification.

10.4.3 Broader Diversification Increases Expected Return

As we observe in Figure 10.5, diversification reduces risk. However, there is far more to the power of diversification than simply spreading your assets over a

---

[3]Mathematicians refer to such a relationship as a *perfect positive* cross correlation.

[4]This relationship is technically called a *perfect negative* cross correlation.

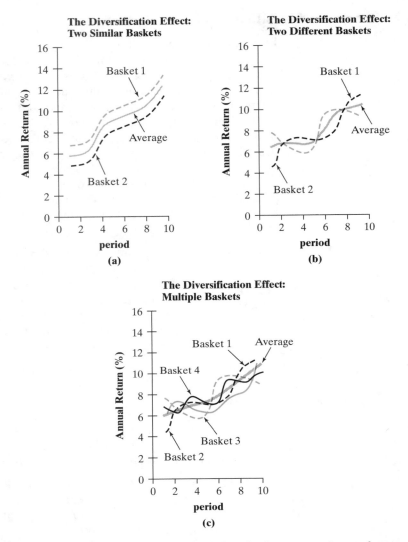

FIGURE 10.5 Reducing investment risks by asset diversification

number of investments. Well-diversified portfolios contain various mixes of stocks, bonds, mutual funds, and cash equivalents like treasury bills. With such portfolios, over lengthy periods, you do not have to sacrifice much in the way of returns in order to get that reduced volatility. Finding the right mix depends on your assets, your age, and your risk tolerance. Diversification also means regularly evaluating your assets and realigning the investment mix. For example, if your stocks increase in value, they will make up a larger percentage of your portfolio. To maintain a certain level of risk tolerance, you may want to decrease your stock holdings and increase your cash or bond holdings. Example 10.7 illustrates the difference an asset allocation makes for a long-term investor.

Example 10.7 Broader Diversification Increases Return

Suppose you have $10,000 in cash and are considering the following two options for investing money:

- Option 1: You put the entire $10,000 into a secure-investment mutual fund consisting of a long-term U.S. Treasury bond with a yield of 7%.
- Option 2: You split the $10,000 into equal amounts of $2,000 and diversify among five investment opportunities with varying degrees of risk from, say, extremely risky to very conservative and with potential returns ranging from −100% to 15%.

Amount	Investment	Expected Return
$2,000	Buying lottery tickets	−100% (?)
$2,000	Under the mattress	0%
$2,000	Term deposit (CD)	5%
$2,000	Corporate bond	10%
$2,000	Mutual fund (stocks)	15%

Given these two options, which would you choose?

Discussion: The $10,000 bond investment, which earns 7%, poses virtually no risk. With the diversified approach, you are going to lose the first $2,000 with virtual certainty, make nothing on the next $2,000, make only 5% on the third $2,000, make 10% on the fourth $2,000, and make 15% on the last $2,000. At first glance, you might think Option 1 is a more rational strategy, because of the very limited risk potential. Indeed, that might well be the best alternative for many short-term investors. However, say that we add one more element to our scenario: our time horizon is 25 years. Would that make a difference?

SOLUTION

Given: Rate of return for each investment option.
Find: Value of the investments at the end of 25 years.

As you will see, the longer the time horizon, the better choice certain investments, such as stocks (represented by the mutual fund in this example), become. First, we can find the value of the government bond in 25 years as follows:

$$F = \$10,000(F/P, 7\%, 25) = \$54,274.$$

Similarly, we can find the value for each investment class in Option 2:

Option	Amount	Investment	Expected Return	Value in 25 Years
1	$10,000	Bond	7%	$54,274
	$2,000	Lottery tickets	−100%	$0
	$2,000	Mattress	0%	$2,000
2	$2,000	Term deposit (CD)	5%	$6,773
	$2,000	Corporate bond	10%	$21,669
	$2,000	Mutual fund (stocks)	15%	$65,838
				$96,280

At the outset, Option 2 appears to be a losing proposition, but you would end up with about 77% more money despite the fact that the first two choices you made were, at best, unproductive. Of course, you can come up a counterexample where Option 1 would be a better choice. However, the message is clear: Diversification among properly chosen assets can increase the return without taking unnecessary risks as long as you keep the assets invested in the market over a long period of time.

Comments: In an extreme case, if you invested the entire $10,000 in the stock, the expected return over 25 years would be $10,000(1 + 0.15)^{25} = $329,190$, which is a lot more than the return expected from the diversification. Therefore, it is commonly suggested that if you are a long-term investor, you can increase your exposure to more risky assets, such as stocks.

Summary

- Often, cash flow amounts and other aspects of investment project analysis are uncertain. Whenever such uncertainty exists, we are faced with the difficulty of **project risk**—the possibility that an investment project will not meet our minimum requirements for acceptability and success.
- Three of the most basic tools for assessing project risk are as follows:

  1. **Sensitivity analysis**—a means of identifying the project variables that, when varied, have the greatest effect on project acceptability.
  2. **Break-even analysis**—a means of identifying the value of a particular project variable that causes the project to exactly break even.
  3. **Scenario analysis**—a means of comparing a base-case, or expected, project measurement (such as PW) with the measurement(s) for one or more additional scenarios, such as best and worst case, to identify the extreme and most likely project outcomes.

- In considering the risk elements in project evaluation, there are two common approaches. The first approach is to describe the riskiness of the project cash flows in terms of probability distributions. Then use the risk-free interest rate to determine the net-present-value distribution. Once you obtain the PW distribution (mean and variance), you need to determine whether the expected value of the PW distribution is large enough to undertake the risk perceived in the project, which is revealed by the variance of the PW distribution. The second approach is to adjust the discount rate to reflect your perceived risk in terms of the risk premium and then use this adjusted rate to discount the expected cash flows. In practice, the risk-adjusted discount-rate approach is much more popular, as the risk assessment process is much simpler than the probabilistic approach.

- Once you set your risk tolerance, you are establishing an upper-bound limit on the portfolio's long-term expected rate of return.

- There is no such thing as a risk-free investment. The challenge is to decide what level of risk you are willing to assume and then, having decided on your risk tolerance, to understand the implications of that choice.

- There is far more to the power of **diversification** than simply spreading your assets over a number of investments to reduce risk. By combining assets with different patterns of return, it is possible to achieve a higher rate of return without significantly increasing risk.

Problems

Sensitivity Analysis

10.1 Ford Construction Company is considering the acquisition of a new earthmover. The mover's base price is $70,000, and it will cost another $15,000 to modify it for special use by the company. This earthmover falls into the MACRS five-year class. It will be sold after four years for $30,000. The earthmover purchase will have no effect on revenues, but it is expected to save the firm $32,000 per year in before-tax operating costs, mainly labor. The firm's marginal tax rate (federal plus state) is 40%, and its MARR is 15%.

(a) Is this project acceptable, based on the most likely estimates given in the problem?

(b) If the firm's MARR is increased to 20%, what would be the required savings in labor so that the project remains profitable?

10.2 Minnesota Metal Forming Company has just invested $500,000 of fixed capital in a manufacturing process that is estimated to generate an after-tax annual cash flow of $200,000 in each of the next five years. At the end of year five, no further market for the product and no salvage value for the manufacturing process are expected. If a manufacturing problem delays plant start-up for one year (leaving only four years of process life), what additional after-tax cash flow will be needed to maintain the same internal rate of return if no delay had occurred?

10.3 A real estate developer seeks to determine the most economical height for a new office building. The building will be sold after five years. The relevant net annual revenues and salvage values are as follows:

	Height			
	2 Floors	**3 Floors**	**4 Floors**	**5 Floors**
First cost (net after tax)	$500,000	$750,000	$1,250,000	$2,000,000
Lease revenue	$199,100	$169,200	$149,200	$378,150
Net resale value (after tax)	$600,000	$900,000	$2,000,000	$3,000,000

(a) The developer is uncertain about the interest rate (i) to use, but is certain that it is in the range of 5% to 20%. For each building height, find the range of values of i for which that building height is the most economical.

(b) Suppose that the developer's interest rate is known to be 15%. What would be the cost, in terms of PW, of an error in overestimation of resale value such that the true value is 10% lower than the original estimate?

10.4 A special purpose milling machine was purchased four years ago for $20,000. It was estimated at that time that this machine would have a life of 10 years and a salvage value of $1,000, with a cost of removal of $1,500. These estimates are still good. This machine has annual operating costs of $2,000, and its current book value is $13,000. If the machine is retained over the life of the asset, the remaining annual depreciation schedule would be $2,000 for years five through 10. A new, more efficient machine will reduce operating costs to $1,000, but will require an investment of $12,000. The life of the new machine is estimated to be six years, with a salvage value of $2,000. The new machine would fall into the five-year MACRS property class. An offer of $6,000 for the old machine has been made, and the purchaser would pay for the removal of the machine. The firm's marginal tax rate is 40%, and its required minimum rate of return is 10%.

(a) What incremental cash flows will occur for the next six years as a result of replacing the old machine? Should the old machine be replaced now?

(b) Suppose that the operating costs for the old milling machine would increase at an annual rate of 5% over the remaining service life. With this change in future operating costs for the old machine, would your answer to (a) change?

(c) What is the minimum trade-in value for the old machine so that both alternatives are economically equivalent?

10.5 A local telephone company is considering the installation of a new phone line for a new row of apartment complexes. Two types of cables are being considered: conventional copper wire and fiber optics. Transmission by copper-wire cables, although cumbersome, involves much less complicated and expensive support hardware than fiber optics. The local company may use five different types of copper-wire cables: cable with 100 pairs, 200 pairs, 300 pairs, 600 pairs, and 900 pairs of wire per cable. In calculating the cost per length of cable, the following equation is used:

Cost per length = [cost per foot

+ cost per pair (number of pairs)](length).

We know that 22-gauge copper wire costs $1.692 per foot and that the cost per pair = $0.013. The annual cost of the cable as a percent of the first cost is 18.4%. The life of the system is 30 years.

In fiber optics, a cable is referred to as a ribbon. One ribbon contains 12 fibers. The fibers are grouped in fours; therefore, one ribbon contains three groups of four fibers. Each group of four fibers can produce 672 lines (equivalent to 672 pairs of wires), and since each ribbon contains three groups, the total capacity of the ribbon is 2,016 lines. To use fiber optics to transmit signals, many modulators, wave guides, and terminators are needed to convert the signals from electric currents to modulated light waves. Fiber optic ribbon costs $15,000 per mile. At each end of the ribbon, three terminators are needed, one for each group of four fibers, at a cost of $30,000 per terminator. Twenty-one modulating systems are needed at each end of the ribbon, at a cost of $12,092 for a unit in the central office and $21,217 for a unit in the field. Every 22,000 feet, a repeater is required in order to keep the modulated light waves in the ribbon at an intelligible intensity for detection. The unit cost of this repeater is $15,000. The annual cost, including income taxes, for the 21 modulating systems is 12.5% of the first cost of the units. The annual cost of the ribbon itself is 17.8% of its first cost. The life of the whole system is 30 years. (All figures represent after-tax costs.)

(a) Suppose that the apartments are located five miles from the phone company's central switching system and that about 2,000 telephones will be required. This scenario would necessitate either 2,000 pairs of copper wire or one fiber optic ribbon and related hardware. If the telephone company's interest rate is 15%, which option is more economical?

(b) Suppose that the apartments are located 10 miles from the phone company's central switching system. Which option is more economically attractive? What about if the apartments are 25 miles from the central switching system?

10.6 A small manufacturing firm is considering the purchase of a new boring machine to modernize one of its production lines. Two types of boring machines are available on the market. The lives of machine A and machine B are eight years and ten years, respectively. The machines have the following receipts and disbursements:

Item	Machine A	Machine B
First cost	$6,000	$8,500
Service life	8 years	10 years
Salvage value	$500	$1,000
Annual O&M costs	$700	$520
Depreciation (MACRS)	7 year	7 year

Assume an after-tax MARR of 10% and a marginal tax rate of 30%.

(a) Which machine would be most economical to purchase under an infinite planning horizon? Explain any assumption that you need to make about future alternatives.

(b) Determine the break-even annual O&M costs for machine A so that the present worths of machines A and B are the same.

(c) Suppose that the required service life of the machine is only five years. The estimated salvage values at the end of the required service period are estimated to be $3,000 for machine A and $3,500 for machine B. Which machine is more economical?

10.7 The management of Langdale Mill is considering replacing a number of old looms in the mill's weave room. The looms to be replaced are two 86-inch President looms, sixteen 54-inch President looms, and twenty-two 72-inch Draper X-P2 looms. The company may either replace the old looms with new ones of the same kind or buy 21 new shutterless Pignone looms. The first alternative requires the purchase of 40 new President and Draper looms and the scrapping of the old looms. The second alternative involves scrapping the 40 old looms, relocating 12 Picanol looms, and constructing a concrete floor, plus purchasing the 21 Pignone looms and various related equipment. The known financial details for both alternatives are as follows:

Description	Alternative 1	Alternative 2
Machinery and related equipment	$2,119,170	$1,071,240
Removal cost of old looms and site preparation	$26,866	$49,002
Salvage value of old looms	$62,000	$62,000
Annual sales increase with new looms	$7,915,748	$7,455,084
Annual labor cost	$261,040	$422,080
Annual O&M cost	$1,092,000	$1,560,000
Depreciation (MACRS)	7 year	7 year
Project life	8 years	8 years
Salvage value	$169,000	$54,000

The firm's MARR is 18%. This figure is set by corporate executives, who feel that various investment opportunities available for the mills will guarantee a rate of return on investment of at least 18%. The mill's marginal tax rate is 40%.

(a) Perform a sensitivity analysis on the project's data, varying the revenue, labor cost, annual maintenance cost, and the MARR. Assume that each of these variables can deviate from its base-case expected value by ±10%, ±20%, and ±30%.

(b) From the results of part (a), prepare and interpret sensitivity diagrams.

10.8 Mike Lazenby, an industrial engineer at Energy Conservation Service, has found that the anticipated profitability of a newly developed water-heater temperature control device can be measured by net present worth as

PW = 4.028V(2X − $11) − 77,860,

where V is the number of units produced and sold and X is the sales price per unit. Mike also found that the V parameter value could occur anywhere over the range of 1,000 to 6,000 units

and the X parameter value anywhere between $20 and $45 per unit. Develop a sensitivity graph as a function of number of units produced and sales price per unit.

10.9 Burlington Motor Carriers, a trucking company, is considering the installation of a two-way mobile satellite messaging service on its 2,000 trucks. Based on tests done last year on 120 trucks, the company found that satellite messaging could cut 60% of its $5 million bill for long-distance communications with truck drivers. More importantly, because of this system, the drivers reduced the number of "deadhead" miles—those driven without paying loads—by 0.5%. Applying that improvement to all 230 million miles covered by the Burlington fleet each year would produce an extra $1.25 million in savings.

Equipping all 2,000 trucks with the satellite hookup will require an investment of $8 million and the construction of a message-relaying system costing $2 million. The equipment and on-board devices will have a service life of eight years and negligible salvage value; they will be depreciated under the five-year MACRS class. Burlington's marginal tax rate is about 38%, and its required minimum attractive rate of return is 18%.

(a) Determine the annual net cash flows from the project.

(b) Perform a sensitivity analysis on the project's data, varying savings in the telephone bill and savings in deadhead miles. Assume that each of these variables can deviate from its base-case expected value by ±10%, ±20%, and ±30%.

(c) Prepare sensitivity diagrams and interpret the results.

Break-Even Analysis

10.10 Susan Campbell is thinking about going into the motel business near Disney World in Orlando, FL. The cost to build a motel is $2,200,000. The lot costs $600,000. Furniture and furnishings cost $400,000 and should be recovered in seven years (seven-year MACRS property), while the cost of the motel building should be recovered in 39 years (39-year MACRS real property placed in service on January 1). The land will appreciate at an annual rate of 5% over the project period, but the building will have zero salvage value after 25 years. When the motel is full (100% capacity), it takes in (receipts) $4,000 per day for 365 days per year. The motel has fixed operating expenses, exclusive of depreciation, of $230,000 per year. The variable operating expenses are $170,000 at 100% capacity and vary directly with percent capacity down to $0 at 0% capacity. If the interest rate is 10%, compounded annually, at what percentage capacity must this motel operate at in order to break even? (Assume that Susan's tax rate is 31% and the project life is 25 years.)

10.11 A plant engineer wishes to know which of two types of lightbulbs should be used to light a warehouse. The bulbs currently used cost $45.90 per bulb and last 14,600 hours before burning out. The new bulb ($60 per bulb) provides the same amount of light and consumes the same amount of energy, but lasts twice as long. The labor cost to change a bulb is $16.00. The lights are on 19 hours a day, 365 days a year. If the firm's MARR is 15%, what is the maximum price (per bulb) the engineer should be willing to pay to switch to the new bulb? (Assume that the firm's marginal tax rate is 40%.)

10.12 Robert Cooper is considering the purchase of a piece of business rental property containing stores and offices at a cost of $250,000. Cooper estimates that annual receipts from rentals will be $35,000 and that annual disbursements, other than income taxes, will be about $12,000. The property is expected to appreciate at the annual rate of 5%. Cooper expects to retain the property for 20 years once it is acquired. Then it will be depreciated based on the 39-year real property class (MACRS), assuming that the property would be placed in service on January 1. Cooper's marginal tax rate is 30%, and his MARR is 15%. What would be the minimum annual total of rental receipts that would make the investment break even?

10.13 The city of Opelika was having a problem locating land for a new sanitary landfill site when the Alabama Energy Extension Service (AEES) offered the solution of burning the solid waste to generate steam. At the same time, Uniroyal Tire Company seemed to be having a similar problem disposing of solid waste in the form of rubber tires. It was determined that there would be about 200 tons per day of waste to be burned; this figure includes both municipal and industrial waste (Opelika's waste and Uniroyal's tires, respectively). At AEES's suggestion, the city is considering building a waste-fired steam plant, which would cost $6,688,800. To finance the construction cost, the city would issue resource-recovery revenue bonds in the amount of $7,000,000 at an interest rate of 11.5%. Bond interest is payable annually. The differential amount between the actual construction costs and the amount of bond financing ($7,000,000 − $6,688,800 = $311,200) will be used to settle the bond discount and expenses associated with the bond financing. The expected life of the steam plant is 20 years. The expected salvage value is estimated to be about $300,000. The expected labor costs would be $335,000 per year. The annual operating and maintenance costs (including fuel, electricity, maintenance, and water) are expected to be $175,000. This plant would generate 16,560 pounds of waste after incineration, which will have to be disposed of as landfill. At the present rate of $19.45 per pound, this disposal will cost the city a total of about $322,000 per year.

The revenues for the steam plant will come from two sources: (1) steam sales and (2) disposal tipping fees. The city expects 20% downtime per year for the waste-fired steam plant. With an input of 200 tons per day and 3.01 pounds of steam per pound of refuse, a maximum of 1,327,453 pounds of steam can be produced per day. However, with 20% downtime, the actual output would be 1,061,962 pounds of steam per day. The initial steam charge will be approximately $4.00 per thousand pounds. This rate would bring in $1,550,520 in steam revenue the first year. The tipping fee is used in conjunction with the sale of steam to offset the total plant cost. It is the goal of the Opelika steam plant to phase out the tipping fee as soon as possible. The tipping fee will be $20.85 per ton in the first year of plant operation and will be phased out at the end of the eighth year. The scheduled tipping-fee assessment is as follows:

Year	Tipping Fee
1	$976,114
2	$895,723
3	$800,275
4	$687,153
5	$553,301
6	$395,161
7	$208,585

(a) At an interest rate of 10%, would the steam plant generate sufficient revenue to recover the initial investment?

(b) At an interest rate of 10%, what would be the minimum charge (per thousand pounds) for steam sales to make the project break even?

(c) Perform a sensitivity analysis as a function of the plant's downtime.

10.14 A firm is considering two different methods of solving a production problem. Both methods are expected to be obsolete in six years. Method A would cost $80,000 initially and have annual operating costs of $22,000. Method B would cost $52,000 initially and have annual operating costs of $17,000. The salvage value realized with Method A would be $20,000 and with Method B would be $15,000. Method A would generate $16,000 of revenue a year more than Method B. Investments in both methods depreciate as a five-year MACRS property class. The firm's marginal income tax rate is 40%. The firm's MARR is 20%. What would be the required additional annual revenue for Method A such that the firm would be indifferent in its choice of method?

10.15 Rocky Mountain Publishing Company is considering introducing a new morning newspaper in Denver. Its direct competitor charges $0.25 per copy at retail, with $0.05 going to the retailer. For the level of news coverage the company desires, it determines the fixed cost of editors, reporters, rent, press-room expenses, and wire-service charges to be $300,000 per month. The variable cost of ink and paper is $0.10 per copy, but advertising revenues of $0.05 per paper will be generated. To print the morning paper, the publisher has to purchase a new printing press, which will cost $600,000. The press machine will be depreciated according to the method for the seven-year MACRS class. The press machine will be used for 10 years, at which time its salvage value will be about $100,000. Assume 25 weekdays of printing in a month, a 40% tax rate, and a 13% MARR. How many copies per day must be sold in order to break even at a selling price of $0.25 per paper at retail?

Probabilistic Analysis

10.16 A corporation is trying to decide whether to buy the patent for a product designed by another company. The decision to buy will mean an investment of $8 million, and the demand for the product is not known. If demand is light, the company expects a return of $1.3 million each year for three years. If the demand is moderate, the return will be $2.5 million each year for four years, and a high demand will mean a return of $4 million each year for four years. It is estimated that the probability of a high demand is 0.4 and the probability of a light demand is 0.2. The firm's interest rate (risk free) is 12%. Calculate the expected present worth of the investment. On this basis, should the company make the investment? (All figures represent after-tax values.)

10.17 Consider the following investment cash flows over a two-year life:

n	$E(A_n)$	$V(A_n)$
0	−$1,000	0
1	$500	200^2
2	$1,500	300^2

If all cash flows are mutually independent, compute $E(\text{PW})$ and $V(\text{PW})$ at $i = 10\%$.

10.18 Assume that we can estimate a project's cash flows as follows:

n	Expected Flow $E(A_n)$	Estimate of Standard Deviation σ_n
0	−$300	$20
1	$120	$10
2	$150	$15
3	$150	$20
4	$110	$25
5	$100	$30

In this case, each annual flow can be represented by a random variable with known mean and variance. Further assume complete independence among the cash flows.

(a) Find the expected PW and the variance of this project at $i = 10\%$.

(b) If your risk-adjusted discount rate is 18%, is this project justifiable?

10.19 The risk-free interest rate is 6%, and the market risk premium is 7%. The beta (β) of the firm under analysis is 1.4. With expected net cash flows are estimated at $50,000 per year for five years. The required investment outlay on this project is $120,000.

(a) What is the required risk-adjusted return on the project?

(b) Should the project be accepted?

10.20 Kellogg Company is considering an investment project with the following parameters, where all cost and revenue figures are estimated in constant dollars:

• The project requires the purchase of a $9,000 asset, which will be used for only two years (project life). The project also requires an investment of $2,000 in working capital, and this amount will be fully recovered at the end of project year.

• The salvage value of this asset at the end of two years is expected to be $4,000.

• The annual revenue and the general inflation rate are discrete random variables, but can be described by the following probability distributions:

Annual Revenue (X)	Probability	General Inflation Rate (Y)	Probability
$10,000	0.30	3%	0.25
$20,000	0.40	5%	0.50
$30,000	0.30	7%	0.25

Both random variables are statistically independent.

• The investment will be classified as a three-MACRS property (tax life).

• It is assumed that the revenues, salvage value, and working capital are responsive to the general inflation rate.

• The revenue and inflation rate dictated during the first year will prevail over the remaining project period.

• The marginal income tax rate for the firm is 40%. The firm's inflation-free interest rate (i') is 10%.

(a) Determine the PW as a function of X.

(b) Compute the expected PW of this investment.

(c) Compute the variance of the PW of this investment.

Comparing Risky Projects

10.21 Juan Carols is considering two investment projects whose PWs are described as follows:

• Project 1: $\text{PW}(10\%) = 2X(X - Y)$, where X and Y are statistically independent discrete random variables with the following distributions:

X		Y	
Event	Probability	Event	Probability
$20	0.60	$10	0.40
$40	0.40	$20	0.60

● Project 2:

PW(10%)	Probability
$0	0.24
$400	0.20
$1,600	0.36
$2,400	0.20

The cash flows between the two projects are assumed to be statistically independent.

(a) Develop the PW distribution for Project 1.

(b) Compute the mean and variance of the PW for Project 1.

(c) Compute the mean and variance of the PW for Project 2.

(d) Suppose that Projects 1 and 2 are mutually exclusive. Which project would you select?

10.22 A financial investor has an investment portfolio. A bond in his investment portfolio will mature next month and provide him with $25,000 to reinvest. The choices for reinvestment have been narrowed down to the following two options:

● Option 1: Reinvest in a foreign bond that will mature in one year. This transaction will entail a brokerage fee of $150. For simplicity, assume that the bond will provide interest over the one-year period of $2,450, $2,000, or $1,675 and that the probabilities of these occurrences are assessed to be 0.25, 0.45, and 0.30, respectively.

● Option 2: Reinvest in a $25,000 certificate with a savings and loan association. Assume that this certificate has an effective annual rate of 7.5%.

Which form of reinvestment should the investor choose in order to maximize his expected financial gain?

10.23 A manufacturing firm is considering two mutually exclusive projects. Both projects have an economic service life of one year, with no salvage value. The first cost of Project 1 is $1,000, and the first cost of Project 2 is $800. The net year-end revenue for each project is given as follows:

	Project 1	
	Probability	Revenue
Net revenue given in PW	0.2	$2,000
	0.6	$3,000
	0.2	$3,500

	Project 2	
	Probability	Revenue
Net revenue given in PW	0.3	$1,000
	0.4	$2,500
	0.3	$4,500

Assume that both projects are statistically independent of each other.

(a) If you make decisions by maximizing the expected PW, which project would you select?

(b) If you also consider the variance of the projects, which project would you select?

10.24 A business executive is trying to decide whether to undertake one of two contracts or neither one. He has simplified the situation somewhat and feels that it is sufficient to estimate that the contracts are specified as follows:

Contract A		Contract B	
PW	**Probability**	**PW**	**Probability**
$100,000	0.2	$40,000	0.3
$50,000	0.4	$10,000	0.4
$0	0.4	−$10,000	0.3

(a) Should the executive undertake either one of the contracts? If so, which one? What would he choose if he were to make decisions by maximizing his expected PW?

(b) What would be the probability that Contract A would result in a larger profit than that of Contract B?

10.25 Two machines are being considered for a cost-reduction project:

- Machine A has a first cost of $60,000 and a salvage value (after tax) of $22,000 at the end of six years of service life. Probabilities of annual after-tax operating costs of this machine are estimated as follows:

Annual O&M Costs	Probability
$5,000	0.20
$8,000	0.30
$10,000	0.30
$12,000	0.20

- Machine B has a first cost of $35,000, and its estimated salvage value (after tax) at the end of four years of service is negligible. Probabilities of its annual after-tax operating costs are estimated as follows:

Annual O&M Costs	Probability
$8,000	0.10
$10,000	0.30
$12,000	0.40
$14,000	0.20

The MARR on this project is 10%. The required service period of these machines is estimated to be 12 years, and no technological advance in either machine is expected.

(a) Assuming statistical independence of the two projects' cash flows, calculate the mean and variance for the equivalent annual cost of operating each machine.

(b) From the results of part (a), calculate the probability that the annual cost of operating Machine A will exceed the annual cost of operating Machine B.

10.26 Two mutually exclusive investment projects are under consideration. It is assumed that the cash flows are statistically independent random variables with means and variances estimated as follows:

End of Year	Project A		
	Mean	**Variance**	
0	−$5,000	$1,000^2	
1	$4,000	$1,000^2	
2	$4,000	$1,500^2	

End of year	Project B		
	Mean	**Variance**	
0	−$10,000	$2,000^2	
1	$6,000	$1,500^2	
2	$8,000	$2,000^2	

(a) For each project, determine the mean and standard deviation for the PW, using an interest rate of 15%.

(b) Based on the results of part (a), which project would you recommend?

Investment Strategies

10.27 Which of the following statements is incorrect?

(a) Holding on to cash is the most risk-free investment option.

(b) To maximize your return on total assets (ignoring financial risk), you must put all your money into the same type of investment category.

(c) Diversification among well-chosen investments can yield a higher rate of return overall, with less risk.

(d) Broader diversification among well-chosen assets could increase return without increasing additional risk.

Short Case Studies with Excel

10.28 Mount Manufacturing Company produces industrial and public safety shirts. As is done in most apparel manufacturing, the cloth must be cut into shirt parts by marking sheets of paper in the way that the particular cloth is to be cut. At present, these sheet markings are done manually, and the annual labor cost of the process is running around $103,718. Mount has the option of purchasing one of two automated marking systems. The two systems are the Lectra System 305 and the Tex Corporation Marking System. The comparative characteristics of the two systems are as follows:

	Most Likely Estimates	
	Lectra System	**Tex System**
Annual labor cost	$51,609	$51,609
Annual material savings	$230,000	$274,000
Investment cost	$136,150	$195,500
Estimated life	6 years	6 years
Salvage value	$20,000	$15,000
Depreciation method (MACRS)	5 year	5 year

The firm's marginal tax rate is 40%, and the interest rate used for project evaluation is 12% after taxes.

(a) Based on the most likely estimates, which alternative is the best?

(b) Suppose that the company estimates the material savings during the first year for each system on the basis of the following probability distributions:

Lectra System	
Material Savings	**Probability**
$150,000	0.25
$230,000	0.40
$270,000	0.35

Tex System	
Material Savings	**Probability**
$200,000	0.30
$274,000	0.50
$312,000	0.20

Further assume that the annual material savings for both Lectra and Tex are statistically independent. Compute the mean and variance for the equivalent annual value of operating each system.

10.29 The following is a comparison of the cost structure of a conventional manufacturing technology (CMT) with that of a flexible manufacturing system (FMS) at one U.S. firm:

	Most Likely Estimates	
	CMT	**FMS**
Number of part types	3,000	3,000
Number of pieces produced/year	544,000	544,000
Variable labor cost/part	$2.15	$1.30
Variable material cost/part	$1.53	$1.10
Total variable cost/part	$3.68	$2.40
Annual overhead costs	$3.15M	$1.95M
Annual tooling costs	$470,000	$300,000
Annual inventory costs	$141,000	$31,500
Total annual fixed operating costs	$3.76M	$2.28M
Investment	$3.5M	$10M
Salvage value	$0.5M	$1M
Service life	10 years	10 years
Depreciation method (MACRS)	7 year	7 year

(a) The firm's marginal tax rate and MARR are 40% and 15%, respectively. Determine the incremental cash flow (FMS - CMT), based on the most likely estimates.

(b) Management feels confident about all input estimates for the CMT. However, the firm does not have any previous experience in operating an FMS. Therefore, many of the input estimates for that system, except the investment and salvage value, are subject to variations. Perform a sensitivity analysis on the project's data, varying the elements of the operating costs. Assume that each of these variables can deviate from its base-case expected value by ±10%, ±20%, and ±30%.

(c) Prepare sensitivity diagrams and interpret the results.

(d) Suppose that probabilities of the variable material cost and the annual inventory cost for the FMS are estimated as follows:

Material Cost

Cost per Part	Probability
$1.00	0.25
$1.10	0.30
$1.20	0.20
$1.30	0.20
$1.40	0.05

Inventory Cost

Annual Inventory Cost	Probability
$25,000	0.10
$31,000	0.30
$50,000	0.20
$80,000	0.20
$100,000	0.20

What are the best and the worst cases of incremental PW?

(e) In part (d), assuming that the random variables of the cost per part and the annual inventory cost are statistically independent, find the mean and variance of the PW for the incremental cash flows.

(f) In parts (d) and (e), what is the probability that the FMS would be a more expensive investment option?

IV.

Special Topics in Engineering Economics

Is Now the Time to Replace the Old Lift Truck? If Not, When?

In March 2002, Steve Hausmann, a production engineer at IKEA–USA, a European-style furniture company, was considering replacing a 1,000 lb capacity industrial forklift truck. The truck was being used to move assembled furniture from the finishing department to the warehouse. Recently, the truck had not been dependable and was frequently out of service while awaiting repairs. Maintenance expenses were rising steadily. When the truck was not available, the company had to rent one. In addition, the forklift truck was diesel operated, and workers in the plant were complaining about the air pollution. If retained, it would have required an immediate engine overhaul to keep it in operable condition. The overhaul would have neither extended the originally estimated service life nor increased the value of the truck.

Two types of forklift trucks were recommended as replacements. One was electrically operated, and the other was gasoline operated. The electric truck would eliminate the air-pollution problem entirely, but would require a battery change twice a day, which would significantly increase the operating cost. If the gasoline-operated truck were to be used, it would require more frequent maintenance.

Mr. Hausmann was undecided about which truck the company should buy at this point. He felt that he should do some homework before approaching upper management for authorization of the replacement. Two questions came to his mind immediately:

1. Should the forklift truck be repaired now or replaced by one of the more advanced, fuel-efficient trucks?
2. If the replacement option were chosen, when should the replacement occur—now or sometime in the future?

Replacement Decisions

The answer to the first question is almost certainly that the truck should be replaced. Even if the truck is repaired now, at some point it will fail to meet the company's needs, because of physical or functional depreciation. If the replacement option was chosen, the company would probably want to take advantage of technological advances that would provide more fuel efficiency and safer lifting capability in newer trucks.

The answer to the second question is less clear. Presumably, the current truck can be maintained in serviceable condition for some time; what will be the economically optimal time to make the replacement? Furthermore, what is the technologically optimal time to switch? By waiting for a more technologically advanced truck to evolve and gain commercial acceptance, the company may be able to obtain even better results tomorrow than by just switching to whatever truck is available on the market today.

In Chapters 5 through 7, we presented methods that helped us choose the best investment alternative. The problems we examined in those chapters concerned primarily profit-adding projects. However, economic analysis is also frequently performed on projects with existing facilities or on profit-maintaining projects. Profit-maintaining projects are projects whose primary purpose is not to increase sales, but rather simply to maintain ongoing operations. In practice, profit-maintaining projects less frequently involve the comparison of new machines; instead, the problem often facing management is whether to buy new and more efficient equipment or to continue to use existing equipment. This class of decision problems is known as the **replacement problem**. In this chapter, we examine the basic concepts and techniques related to replacement analysis.

11-1   Replacement-Analysis Fundamentals

In this section and the following two sections, we examine three aspects of the replacement problem: (1) approaches for comparing defender and challenger, (2) determination of economic service life, and (3) replacement analysis when the required service period is long. The impact of income-tax regulations will be ignored in these sections. In Section 11.4, we revisit these replacement problems, considering the effect of income taxes on them.

11.1.1   Basic Concepts and Terminology

Replacement projects are decision problems involving the replacement of existing obsolete or worn-out assets. The continuation of operations is dependent on these assets. Failure to make an appropriate decision results in a slowdown or shutdown of the operations. The question is, when should the existing equipment be replaced with more efficient equipment? This situation has given rise to the use of the terms **defender** and **challenger**, terms commonly used in the boxing world. In every boxing class, the current defending champion is constantly faced with a new challenger. In replacement analysis, the defender is the existing machine (or system), and the challenger is the best available replacement equipment.

An existing piece of equipment will be removed at some future time, either when the task it performs is no longer necessary or when the task can be performed more efficiently by newer and better equipment. The question is not whether the existing piece of equipment will be removed, but when it will be removed. A variation of this question is, why should we replace existing equipment at this time rather than postponing replacement of the equipment by repairing or overhauling it? Another aspect of the defender–challenger comparison concerns deciding exactly which equipment is the best challenger. If the defender is to be replaced by the challenger, we would generally want to install the very best of the possible alternatives.

Current Market Value   The most common problem encountered in considering the replacement of existing equipment is the determination of what financial information is actually relevant to the analysis. Often, a tendency to include irrelevant information in the analysis is apparent. To illustrate this type of decision problem, let us consider Example 11.1.

Example 11.1 Relevant Information for Replacement Analysis

Macintosh Printing, Inc., purchased a $20,000 printing machine two years ago. The company expected this machine to have a five-year life and a salvage value of $5,000. Unfortunately, the company spent $5,000 last year on repairs, and current operating costs are now running at the rate of $8,000 per year. Furthermore, the anticipated salvage value has now been reduced to $2,500 at the end of the printer's remaining useful life. In addition, the company has found that the current machine has a book value of $14,693 and a market value of $10,000 today. The equipment vendor will allow the company the full amount of the market value as a trade-in on a new machine. What value(s) for the defender is (are) relevant in our analysis?

SOLUTION

In this example, four different quantities relating to the defender are presented:

1. Original cost: The printing machine was purchased for $20,000.
2. Market value: The company estimates the old machine's market value at $10,000.
3. Book value: The original cost minus the accumulated depreciation if the machine sold now is $14,693.
4. Trade-in allowance: This amount is the same as the market value. (In other cases, however, this value could be different from the market value.)

In this example, and in all replacement analyses, the relevant cost is the **current market value** of the equipment. The original cost, repair cost, and trade-in value are irrelevant.[1] A common misconception is that the trade-in value is always the same as the current market value of the equipment and thus could be used to assign a suitable current value to the equipment. This is not always the case. For example, a car dealer typically offers a trade-in value on a customer's old car to reduce the price of a new car. Would the dealer offer the same value on the old car if he or she were not also selling the new one to the customer? Not usually. In many instances, the trade-in allowance is inflated in order to make the deal look good, and the price of the new car is also inflated to compensate for the dealer's trade-in cost. In this type of situation, the trade-in value does not represent the true value of the item, so we should not use it in economic analysis.[2]

Sunk Costs A **sunk cost** is any past cost unaffected by any future investment decision. In Example 11.1, the company spent $20,000 to buy the machine two years ago. Last year, $5,000 more was spent on this machine. So, the total accumulated

---

[1] The original cost and current book value are relevant when you calculate any gains or losses associated with the disposal of the equipment.

[2] If we do make the trade, however, the actual net cash flow at this time, properly used, is certainly relevant.

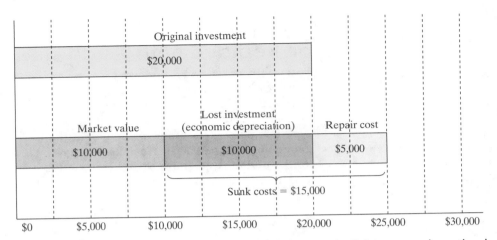

FIGURE 11.1 Sunk costs associated with disposal of the asset described in Example 11.1

expenditure on this machine is $25,000. However, if the machine is sold today, the company can get only $10,000 back (Figure 11.1). It is tempting to think that the company would lose $15,000 in addition to the cost of the new machine if the machine were to be sold and replaced with a new one, but this interpretation is incorrect. In a proper engineering economic analysis, only future costs should be considered; past or sunk costs should be ignored. Thus, the value of the defender that should be used in a replacement analysis should be its current market value, not what it cost when it was originally purchased and not the cost of repairs that have already been made on the machine.

Sunk cost refers to money that has already been spent; no present action can recover it. These costs represent past actions. They are the results of decisions made in the past. In making economic decisions at the present time, one should consider only the possible outcomes of the different available options and pick the one with the best possible future results. Using sunk costs in arguing for one option over the other would only lead to more bad decisions.

Operating Costs   The driving force for replacing existing equipment is that equipment becomes more expensive to operate with time. The total cost of operating a piece of equipment may include repair and maintenance costs, wages for operators, energy consumption costs, and costs of materials. Increases in any one or a combination of these cost items over a period of time may lead us to find a replacement for the existing asset. Since the challenger is usually newer than the defender and often incorporates design improvements and newer technology, it will likely be cheaper to operate than the defender.

Whether or not these savings in operating costs offset the initial investment of buying the challenger thus becomes the focus of our analysis. It is important to focus on those operating costs that differ for the defender and challenger. In many cases, projected labor, energy, and material costs may be identical for each asset—whereas repair and maintenance costs will differ, as older assets require more maintenance. Regardless of which cost items we choose to include in the operating costs, it is essential that the same items are included for both the defender and the challenger. For example, if energy costs are included in the operating costs of the defender, they should also be included in the operating costs of the challenger.

11.1.2 Approaches for Comparing Defender and Challenger

Although replacement projects are a subcategory of the mutually exclusive project we studied in Chapter 5, they do possess unique characteristics that allow us to use specialized concepts and analysis techniques in their evaluation. We consider two basic approaches to analyzing replacement problems: the cash flow approach and the opportunity-cost approach. We start with a replacement problem where both the defender and the challenger have the same useful life, which begins now.

Cash Flow Approach   The cash flow approach can be used as long as the analysis period is the same for all replacement alternatives. In other words, we consider explicitly the actual cash flow consequences for each replacement alternative and compare them, based on either PW or AE values.

Example 11.2 Replacement Analysis Using the Cash Flow Approach

Consider again the scenario in Example 11.1. Suppose that the company has been offered the opportunity to purchase another printing machine for $15,000. Over its three-year useful life, the machine will reduce labor and raw-materials usage sufficiently to cut operating costs from $8,000 to $6,000. This reduction in costs will allow after-tax profits to rise by $2,000 per year. It is estimated that the new machine can be sold for $5,500 at the end of year three. If the new machine were purchased, the old machine would be sold to another company rather than be traded in for the new machine. Suppose that the firm will need either machine (old or new) for only three years and that it does not expect a new, superior machine to become available on the market during the required service period. Assuming that the firm's interest rate is 12%, decide whether replacement is justified now.

SOLUTION

• **Option 1: Keep the defender** If the old machine is kept, there is no additional cash expenditure today. The machine is in perfect operational condition. The annual operating cost for the next three years will be $8,000 per

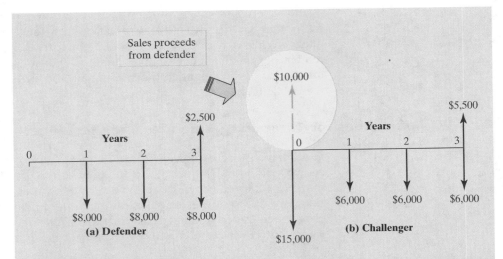

FIGURE 11.2 Comparison of defender and challenger based on the cash flow approach

year, and the machine's salvage value three years from today will be $2,500. The cash flow diagram for the defender is shown in Figure 11.2(a).

- **Option 2: Replace the defender with the challenger** If this option is taken, the defender (designated D) can be sold for $10,000. The cost of the challenger (designated C) is $15,000. Thus, the initial combined cash flow for this option is a cash flow of $-\$15,000 + \$10,000 = -\$5,000$. The annual operating cost of the challenger is $6,000. The salvage value of the challenger after three years will be $5,500. The cash flow diagram for this option is shown in Figure 11.2(b).

We now use these cash flow values to find the present worth of each option and then use this value to find the annual equivalent value for the option. For the defender, we have

$$\text{PW}(12\%)_D = \$2,500(P/F, 12\%, 3) - \$8,000(P/A, 12\%, 3)$$
$$= -\$17,434.90,$$

so

$$\text{AE}(12\%)_D = \text{PW}(12\%)_D(A/P, 12\%, 3)$$
$$= -\$7,259.10.$$

For the challenger, we have

$$\text{PW}(12\%)_C = \$5,500(P/F, 12\%, 3) - \$5,000 - \$6,000(P/A, 12\%, 3)$$
$$= -\$15,495.90,$$

so

$$\text{AE}(12\%)_C = \text{PW}(12\%)_C(A/P, 12\%, 3)$$
$$= -\$6,451.90.$$

Because of the annual difference of $807.20 in favor of the challenger, the replacement should be made now.

Comments: If we had found that the defender should not be replaced now, we would need to address the question of whether the defender should be kept for one or two years (or longer) before being replaced with the challenger. This is a valid question that requires more data on market values over time. We address this situation later, in Section 11.3.

Opportunity-Cost Approach Another way to analyze such a problem is to charge the $10,000 as an **opportunity cost** of keeping the asset.[3] That is, instead of deducting the salvage value from the purchase cost of the challenger, we consider the salvage value as a cash outflow for the defender (or investment required in keeping the defender).

Example 11.3 Replacement Analysis Using the Opportunity-Cost Approach

Rework Example 11.2, using the opportunity-cost approach.

SOLUTION

Recall that the cash flow approach in Example 11.2 credited proceeds in the amount of $10,000 from the sale of the defender toward the $15,000 purchase price of the challenger, and no initial outlay would have been required had the decision been to keep the defender. If the decision to keep the defender had been made under the opportunity-cost approach, however, the $10,000 current salvage value of the defender would have been treated as an incurred cost. Figure 11.3 illustrates the cash flows related to these decision options.

Since the lifetimes are the same, we can use either PW or AE analysis to find the better option. For the defender, we have

$$PW(12\%)_D = -\$10,000 - \$8,000(P/A, 12\%, 3) + \$2,500(P/F, 12\%, 3)$$
$$= -\$27,434.90.$$

---

[3]The opportunity-cost concept should not be confused with the outsider-viewpoint approach, which is commonly described in traditional engineering economics texts. The outsider-viewpoint method approaches a typical replacement problem from the standpoint of a person (an outsider) who has a need for the service that the defender or challenger can provide, but owns neither. This view has some conceptual flaws, however. For example, the outsider purchases the defender at the seller's market price (or seller's tax rate) and assumes the seller's depreciation schedule. In practice, however, if you place a used asset in service, you will be able to depreciate it even though the asset may have been fully depreciated by the previous owner.

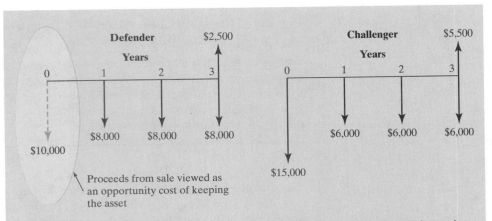

FIGURE 11.3 Comparison of defender and challenger based on the opportunity-cost approach

and

$$AE(12\%)_D = PW(12\%)_D(A/P, 12\%, 3)$$
$$= -\$11,422.64.$$

For the challenger, we have

$$PW(12\%)_C = -\$15,000 - \$6,000(P/A, 12\%, 3) + \$5,500(P/F, 12\%, 3)$$
$$= -\$25,495.90$$

and

$$AE(12\%)_C = PW(12\%)_C(A/P, 12\%, 3)$$
$$= -\$10,615.33.$$

The decision outcome is the same as in Example 11.2, that is, the replacement should be made. Since both the challenger and defender cash flows were adjusted by the same amount (−$10,000) at time zero, this outcome should not come as a surprise.

Comments: Recall that we assumed the same service life for both the defender and the challenger in Examples 11.2 and 11.3. In general, however, old equipment has a relatively short remaining life compared with new equipment, so this assumption is overly simplistic. In the next section, we discuss how to find the economic service life of equipment.

11-2 Economic Service Life

You have probably seen a 50-year-old automobile still in service. Provided it receives the proper repair and maintenance, almost anything can be kept operating for an extended period of time. If it is possible to keep a car operating for an almost indefinite period, why aren't there more old cars on the streets? There are several reasons. For example, some people get tired of driving the same old car. Others want to keep a car as long as it will last, but realize that repair and maintenance costs will become excessive.

In general, we need to consider explicitly how long an asset should be held, once it is placed in service. For instance, a truck rental firm that frequently purchases fleets of identical trucks may wish to make a policy decision on how long to keep a vehicle before replacing it. If an appropriate lifespan is computed, the firm could stagger the schedule of truck purchases and replacements to smooth out annual capital expenditures for overall truck purchases.

As we discussed in Section 6.2, the costs of owning and operating an asset can be divided into two categories: **capital costs** and **operating costs**. Capital costs have two components: initial investment and the salvage value at the time of disposal. The initial investment for the challenger is simply its purchase price. For the defender, we should treat the opportunity cost as its initial investment. We will use N to represent the length of time in years that the asset will be kept, I to represent the initial investment, and S_N to represent the salvage value at the end of the ownership period of N years.

The annual equivalent of capital cost, which is called the capital-recovery cost (refer to Section 6.2), over the period of N years can be calculated with the following equation:

$$CR(i) = I(A/P, i, N) - S_N(A/F, i, N). \tag{11.1}$$

Generally speaking, as an asset becomes older, its salvage value becomes smaller. As long as the salvage value is less than the initial cost, the capital-recovery cost is a decreasing function of N. In other words, the longer we keep an asset, the lower the capital-recovery cost becomes. If the salvage value is equal to the initial cost, then no matter how long the asset is kept, the capital-recovery cost is also constant.

As described earlier, operating and maintenance (O&M) costs tend to increase as a function of the age of the asset. Because of the increasing trend of the O&M costs, the total operating costs of an asset usually increase as the asset ages. We use OC_n to represent the total operating costs in year n of the ownership period and $OC(i)$ to represent the annual equivalent of the operating costs over a lifespan of N years. $OC(i)$ can be expressed as:

$$OC(i) = \left(\sum_{n=1}^{N} OC_n(P/F, i, n) \right)(A/P, i, N). \tag{11.2}$$

The total annual equivalent costs of owning and operating an asset are a summation of the capital-recovery costs and the annual equivalent of the operating costs of the asset:

$$AE = CR(i) + OC(i). \tag{11.3}$$

The economic service life of an asset is the period of useful life that minimizes the annual equivalent costs, in constant dollars, of owning and operating that asset. Based on the foregoing discussions, we need to find the value of N that minimizes AE as expressed in Eq. (11.3). If $CR(i)$ is a decreasing function of N and $OC(i)$ is an increasing function of N, as is often the case, then AE will be a convex function of N with a unique minimum point. (See Figure 11.4.)

- Capital Cost:
$$CR(i) = I(A/P, i, N) - S_N(A/F, i, N)$$

- Operating Cost:
$$OC(i) = \boxed{\sum_{n=1}^{N} OC_n(P/F, i, n)} (A/P, i, N)$$

- Total Cost:
$$AE = CR(i) + OC(i)$$

- Objective: Find n^*
that minimizes AE

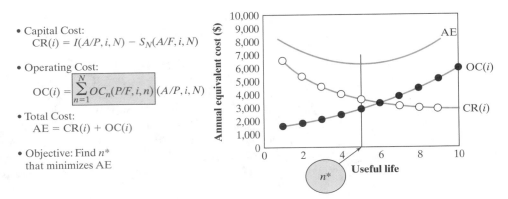

FIGURE 11.4 A schematic illustrating the trends of capital-recovery cost (ownership cost), annual operating cost, and total annual equivalent cost, as a function of asset age

In this book, we assume that AE has a unique minimum point. If the salvage value is constant and equal to the initial cost, and the annual operating cost increases with time, AE is an increasing function of N and attains its minimum at $N = 1$. In this case, we should try to replace the asset as soon as possible. If the annual operating cost is constant and the salvage value is less than the initial cost and decreases with time, AE is a decreasing function of N. In this case, we would try to delay replacement of the asset as much as possible. If the salvage value is constant and equal to the initial cost, and the annual operating costs are constant, AE will also be constant. In this case, the time at which the asset is replaced does not make any economic difference.

If a new asset is purchased and operated for the length of its economic life, the annual equivalent cost is minimized. If we further assume that a new asset of identical price and features can be purchased repeatedly over an indefinite period, we would always replace this kind of asset at the end of its economic life. By replacing perpetually according to an asset's economic life, we obtain the minimum AE cost stream in *constant dollars* over an indefinite period. However, if the identical-replacement assumption cannot be made, we will have to use the methods to be covered in Section 11.3 to perform replacement analysis. The next example explains the computational procedure for determining economic service life.[4]

Example 11.4 Economic Service Life for a Forklift Truck

As a challenger to the forklift truck described in the chapter opening, consider a new electric forklift truck that would cost $18,000, have operating costs of $1,000 in the first year, and have a salvage value of $10,000 at the end

---

[4]Unless otherwise mentioned or in the absence of income taxes, we are assuming that all cash flows are estimated in constant dollars. Therefore, the interest rate used in finding the economic service life represents the inflation-free (or real) interest rate.

of the first year. For the remaining years, operating costs increase each year by 15% over the previous year's operating costs. Similarly, the salvage value declines each year by 25% from the previous year's salvage value. The truck has a maximum life of seven years. Overhauls costing $3,000 and $4,500 will be required during the fifth and seventh year of service, respectively. The firm's required rate of return is 15%. Find the economic service life of this new machine.

Discussion: For an asset whose revenues are either unknown or irrelevant, we compute its economic life based on the costs for the asset and its year-by-year salvage values. To determine an asset's economic service life, we need to compare the options of keeping the asset for one year, two years, three years, and so forth. The option that results in the lowest annual equivalent (AE) cost gives the economic service life of the asset. In this case, our examination proceeds as follows:

- $N = 1$ (one-year replacement cycle): In this case, the machine is bought, used for one year, and sold at the end of year one. The cash flow diagram for this option is shown in Figure 11.5. The annual equivalent cost for this option is calculated as follows:

$$AE(15\%) = \$18,000(A/P, 15\%, 1) + \$1,000 - \$10,000$$
$$= \$11,700.$$

Note that $(A/P, 15\%, 1) = (F/P, 15\%, 1)$ and that the annual equivalent cost is the equivalent cost at the end of year one, since $N = 1$. Because we are calculating the annual equivalent costs, we have treated cost items with a positive sign, while the salvage value has a negative sign.

- $N = 2$ (two-year replacement cycle): In this case, the truck will be used for two years and disposed of at the end of year two. The operating cost in year two is 15% higher than that in year one, and the salvage value at the end of year two is 25% lower than that at the end of year one. The cash flow diagram for this option is also shown in Figure 11.5. The annual equivalent cost over the two-year period is calculated as follows:

$$AE(15\%) = [\$18,000 + \$1,000(P/A, 15\%, 15\%, 2)](A/P, 15\%, 2)$$
$$- \$7,500(A/F, 15\%, 2)$$
$$= \$8,653.$$

- $N = 3$ (three-year replacement cycle): In this case, the truck will be used for three years and sold at the end of year three. The salvage value at the end of year three is 25% lower than that at the end of year two, that is, $\$7,500(1 - 25\%) = \$5,625$. The operating cost per year increases at a rate of 15%. The cash flow diagram for this option is also shown in Figure 11.5. For this case, the annual equivalent cost is calculated as follows:

$$AE(15\%) = [\$18,000 + \$1,000(P/A, 15\%, 15\%, 3)](A/P, 15\%, 3)$$
$$- \$5,625(A/F, 15\%, 3)$$
$$= \$7,406.$$

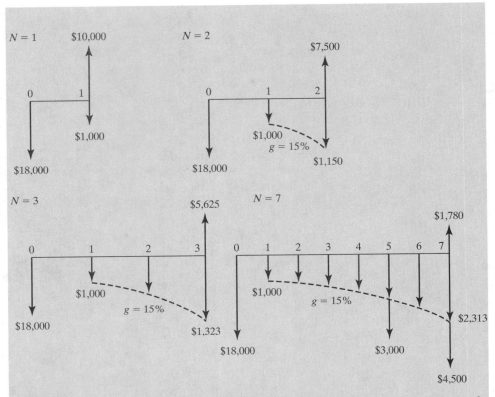

FIGURE 11.5 Cash flow diagrams for the options of keeping the asset for one year, two years, three years, and seven years, where g represents a geometric gradient

● Similarly, we can find the annual equivalent costs for the options of keeping the asset for four years, five years, six years, and seven years. One has to note that there is an additional cost of overhaul in year five. The cash flow diagram for $N = 7$ is shown in Figure 11.5. The computed annual equivalent costs for these four options are as follows:

$$N = 4: \text{AE}(15\%) = \$6,678;$$
$$N = 5: \text{AE}(15\%) = \$6,642;$$
$$N = 6: \text{AE}(15\%) = \$6,258;$$
$$N = 7: \text{AE}(15\%) = \$6,394.$$

From the foregoing calculated AE values for $N = 1, \ldots, 7$, we find that AE(15%) is the smallest when $N = 6$. If the truck were to be sold after six years, it would have an annual cost of $6,258 per year. If it were to be used for a period other than six years, the annual equivalent costs would be higher than $6,258. Thus, a life span of six years for this truck results in the lowest annual cost. We conclude that the economic service life of the truck is six years. By replacing the assets perpetually according to an economic life of six years, we

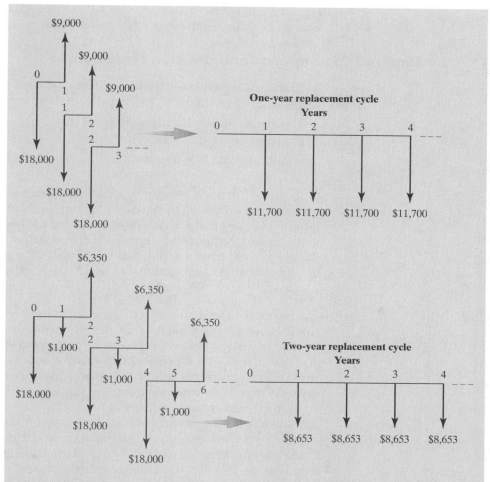

FIGURE 11.6 Conversion of an infinite number of replacement cycles to infinite AE cost streams

obtain the minimum annual equivalent cost stream. Figure 11.6 illustrates the concept of perpetual replacement for one- and two-year replacement cycles. Of course, we should envision a long period of required service for this kind of asset.

11–3   Replacement Analysis When the Required Service Period Is Long

Now that we understand how the economic service life of an asset is determined, the next question is how to use these pieces of information to decide whether now is the time to replace the defender. If now is not the right time, when is the optimal time to replace the defender? Before presenting an analytical approach to answer this question, we consider several important assumptions.

11.3.1   Required Assumptions and Decision Frameworks

In deciding whether now is the time to replace the defender, we need to consider the following three factors:

- planning horizon (study period),
- technology, and
- relevant cash flow information.

Planning Horizon (Study Period)   By planning horizon, we simply mean the service period required by the defender and a sequence of future challengers. The infinite planning horizon is used when we are unable to predict when the activity under consideration will be terminated. In other situations, it may be clear that the project will have a definite and predictable duration. In these cases, replacement policy should be formulated more realistically based on a finite planning horizon.

Technology   Predictions of technological patterns over the planning horizon refer to the development of types of challengers that may replace those under study. A number of possibilities exist in predicting purchase cost, salvage value, and operating cost as dictated by the efficiency of the machine over the life of an asset. If we assume that all future machines will be the same as those now in service, we are implicitly saying that no technological progress in the area will occur. In other cases, we may explicitly recognize the possibility of future machines that will be significantly more efficient, reliable, or productive than those currently on the market. (Personal computers are a good example.) This situation leads to recognition of technological change or obsolescence. Clearly, if the best available machine gets better and better over time, we should certainly investigate the possibility of delaying replacement for a couple of years; this scenario contrasts with situations where technological change is unlikely.

Relevant Cash Flow Information   Many varieties of predictions can be used to estimate the patterns of revenue, cost, and salvage value over the life of an asset. Sometimes revenue is constant, but costs increase and salvage value decreases, over the life of a machine. In other situations, a decline in revenue over equipment life can be expected. The specific situation will determine whether replacement analysis is directed toward cost minimization (with constant revenue) or profit maximization (with varying revenue). We formulate a replacement policy for an asset for which the salvage value does not increase with age.

Decision Criterion   Although the economic life of the defender is defined as the number of years of service that minimizes the annual equivalent cost (or maximizes the annual equivalent revenue), the end of the economic life is not necessarily the *optimal time* to replace the defender. The correct replacement time depends on certain data for the challenger, as well as on certain data for the defender.

As a decision criterion, the AE method provides a more direct solution when the planning horizon is infinite. When the planning horizon is finite, the PW method is more convenient to use. We will develop the replacement-decision procedure for both situations: (1) the infinite planning horizon and (2) the finite planning horizon. We begin by analyzing an infinite planning horizon without technological change. Even though a simplified situation such as this is not likely to occur in real life, the analysis of this replacement situation introduces methods useful for analyzing infinite-horizon replacement problems with technological change.

11.3.2   Replacement Strategies under the Infinite Planning Horizon

Consider the situation where a firm has a machine in use in a process. The process is expected to continue for an indefinite period. Presently, a new machine is on the market that is, in some ways, more effective for the application than the defender. The question is, when, if at all, should the defender be replaced with the challenger?

Under the infinite planning horizon, the service is required for a very long time. Either we continue to use the defender to provide the service or we replace the defender with the best available challenger for the same service requirement. In such cases, the following procedure may be followed in replacement analysis:

1. Compute the economic lives of both defender and challenger. Let's use N_{D^*} and N_{C^*} to indicate the economic lives of the defender and the challenger, respectively. The annual equivalent costs for the defender and the challenger for their respective economic lives are indicated by AE_{D^*} and AE_{C^*}, respectively.

2. Compare AE_{D^*} and AE_{C^*}. If AE_{D^*} is bigger than AE_{C^*}, we know that it is more costly to keep the defender than to replace it with the challenger. Thus, the challenger should replace the defender now. If AE_{D^*} is smaller than AE_{C^*}, it costs less to keep the defender than to replace it with the challenger. Thus, the defender should not be replaced now. The defender should continue to be used at least for the duration of its economic life, as long as there are no technological changes over the economic life of the defender.

3. If the defender should not be replaced now, when should it be replaced? First, we need to continue to use it until its economic life is over. Then we should calculate the cost of using the defender for one more year beyond its economic life. If this cost is greater than AE_{C^*}, the defender should be replaced at the end of its economic life. Otherwise, we should calculate the cost of running the defender for the second year after its economic life. If this cost is bigger than AE_{C^*}, the defender should be replaced one year after its economic life. This process should be continued until you find the optimal replacement time. This approach is called **marginal analysis**; that is, you calculate the incremental cost of operating the defender for just one more year. In other words, we want to see whether the cost of extending the use of the defender for an additional year exceeds the savings resulting from delaying the purchase of the challenger. Here, we have assumed that the best available challenger does not change.

It should be noted that this procedure might be applied dynamically. For example, it may be performed annually for replacement analysis. Whenever there are updated data on the costs of the defender or new challengers available on the market, these new data should be used in the procedure. Example 11.5 illustrates the foregoing procedure.

Example 11.5 Replacement Analysis under the Infinite Planning Horizon

Advanced Electrical Insulator Company is considering replacing a broken inspection machine, which has been used to test the mechanical strength of electrical insulators, with a newer and more efficient one. If repaired, the old machine can be used for another five years, although the firm does not expect to realize any salvage value from scrapping it in five years. Alternatively, the firm can sell the machine to another firm in the industry now for $5,000. If the machine is kept, it will require an immediate $1,200 overhaul to restore it to operable condition. The overhaul will neither extend the service life originally estimated nor increase the value of the inspection machine. The operating costs are estimated at $2,000 during the first year and are expected to increase by $1,500 per year thereafter. Future market values are expected to decline by $1,000 per year.

The new machine costs $10,000 and will have operating costs of $2,000 in the first year, increasing by $800 per year thereafter. The expected salvage value is $6,000 after one year and will decline 15% each year. The company requires a rate of return of 15%. Find the economic life for each option, and determine when the defender should be replaced.

SOLUTION

1. **Economic service life:**

   - **Defender:** If the company retains the inspection machine, it is in effect deciding to overhaul the machine and invest the machine's current market value in that alternative. The opportunity cost of the machine is $5,000. Because an overhaul costing $1,200 is also needed in order to make the machine operational, the total initial investment of the machine is considered to be $5,000 + $1,200 = $6,200. Other data for the defender are summarized as follows:

n	Overhaul	Forecasted Operating Cost	Market Value If Disposed of
0	$1,200		$5,000
1	$0	$2,000	$4,000
2	$0	$3,500	$3,000
3	$0	$5,000	$2,000
4	$0	$6,500	$1,000
5	$0	$8,000	$0

We can calculate the annual equivalent costs in *constant dollars* if the defender is to be kept for one year, two years, three years, and so forth, respectively. For example, the cash flow diagram for $N = 4$ years is shown in Figure 11.7. For $N = 4$ years, we calculate that

$$
\begin{aligned}
\text{AE}(15\%) &= \$6{,}200(A/P, 15\%, 4) + \$2{,}000 \\
&\quad + \$1{,}500(A/G, 15\%, 4) - \$1{,}000(A/F, 15\%, 4) \\
&= \$5{,}961.
\end{aligned}
$$

where G represents the gradient amount.

The other AE cost figures can be calculated with the following equation:

$$
\begin{aligned}
\text{AE}(15\%)_N &= \$6{,}200(A/P, 15\%, N) + \$2{,}000 + \$1{,}500(A/G, 15\%, N) \\
&\quad - \$1{,}000(5 - N)(A/F, 15\%, N) \text{ for } N = 1, 2, 3, 4, \text{ and } 5.
\end{aligned}
$$

For $N = 1$ to 5, the results are as follows:

$$
\begin{aligned}
N &= 1: \text{AE}(15\%) = \$5{,}130; \\
N &= 2: \text{AE}(15\%) = \$5{,}116 \rightarrow \text{minimum AE cost}; \\
N &= 3: \text{AE}(15\%) = \$5{,}500; \\
N &= 4: \text{AE}(15\%) = \$5{,}961; \\
N &= 5: \text{AE}(15\%) = \$6{,}434.
\end{aligned}
$$

When $N = 2$ years, we get the lowest AE value. Thus, the defender's economic life is two years. Using the notation we have defined in the procedure, we have

$$
N_{D^*} = 2 \text{ years}
$$

and

$$
\text{AE}_{D^*} = \$5{,}116.
$$

Actually, after computing AE for $N = 1, 2,$ and 3, we can stop right there. There is no need to compute AE for $N = 4$ and $N = 5$, because

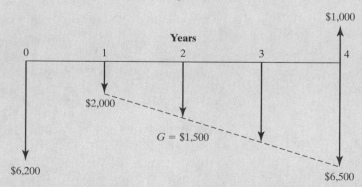

FIGURE 11.7 Cash flow diagram for defender when $N = 4$ years

AE is increasing when $N > 2$, and we have assumed that AE has a unique minimum point.

- **Challenger:** The economic life of the challenger can be determined using the same procedure shown in this example for the defender and in Example 11.4. A summary of the general equation for AE calculation for the challenger is as follows:

$$AE(15\%)_N = \$10,000(A/P, 15\%, N) + \$2,000 + \$800(A/G, 15\%, N)$$
$$-\$6,000(1 - 15\%)^{N-1}(A/F, 15\%, N).$$

You don't have to summarize such an equation when you need to determine the economic life of an asset, as long as you follow the procedure illustrated in Example 11.4.

The obtained results are as follows:

$$N = 1 \text{ year: AE}(15\%) = \$7,500;$$
$$N = 2 \text{ years: AE}(15\%) = \$6,151;$$
$$N = 3 \text{ years: AE}(15\%) = \$5,857;$$
$$N = 4 \text{ years: AE}(15\%) = \$5,826 \rightarrow \text{minimum AE value};$$
$$N = 5 \text{ years: AE}(15\%) = \$5,897.$$

The economic life of the challenger is thus four years, that is,

$$N_{C*} = 4 \text{ years}$$

and

$$AE_{C*} = \$5,826.$$

2. **Should the defender be replaced now?**

Since $AE_{D*} = \$5,116 < AE_{C*} = \$5,826$, the answer is not to replace the defender now. If there are no technological advances in the next few years, the defender should be used for at least $N_{D*} = 2$ more years. However, it is not necessarily best to replace the defender right at the end of its economic life.

3. **When should the defender be replaced?**

If we need to find the answer to this question today, we have to calculate the cost of keeping and using the defender for the third year from today. That is, what is the cost of not selling the defender at the end of year two, using it for the third year, and replacing it at the end of year three? The following cash flows are related to this question:

(a) opportunity cost at the end of year two, which is equal to the market value then: $3,000;

(b) operating cost for the third year: $5,000;

(c) salvage value of the defender at the end of year three: $2,000.

The diagram in Figure 11.8 represents these cash flows.

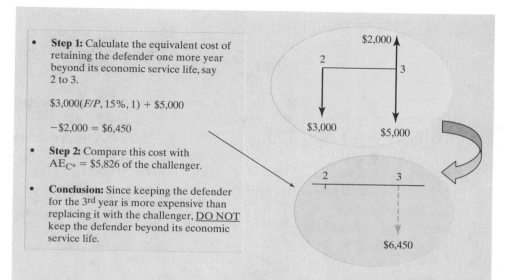

- **Step 1:** Calculate the equivalent cost of retaining the defender one more year beyond its economic service life, say 2 to 3.

  $3,000(F/P, 15\%, 1) + \$5,000$

  $-\$2,000 = \$6,450$

- **Step 2:** Compare this cost with $AE_{C^*} = \$5,826$ of the challenger.

- **Conclusion:** Since keeping the defender for the 3rd year is more expensive than replacing it with the challenger, <u>DO NOT</u> keep the defender beyond its economic service life.

FIGURE 11.8 Illustration of marginal analysis to determine the optimal time for replacing the defender with the challenger

The cost of using the defender for one year beyond its economic life is

$$\$3,000 \times 1.15 + \$5,000 - \$2,000 = \$6,450.$$

Now compare this cost with $AE_{C^*} = \$5,826$. It is greater than AE_{C^*}. So, it is more expensive to keep the defender for the third year than to replace it with the challenger. (See Figure 11.8.) Thus, the conclusion is to replace the defender at the end of year two. If this one-year cost is still smaller than AE_{C^*}, we need to calculate the cost of using the defender for the fourth year and then compare it with AE_{C^*}.

In replacement analysis, it is common for a defender and its challenger to have different economic service lives. The annual-equivalence approach is frequently used in replacement analysis, but it is important to know that we use the AE method in replacement analysis not because we have to deal with the unequal-service-life problem, but rather because the AE approach provides some computational advantage for a special class of replacement problem.

11–4   Replacement Analysis with Tax Considerations

Up to this point, we have covered various concepts and techniques that are useful in replacement analysis. In this section, we illustrate how to use those concepts and techniques to conduct replacement analysis on an after-tax basis.

To apply the concepts and methods covered in Sections 11.1 through 11.3 to after-tax comparisons of defender and challenger, we have to incorporate the gains (or losses) tax effects whenever an asset is disposed of. Whether the defender is kept or the challenger is purchased, we also need to incorporate the tax effects of depreciation allowances in our analysis.

Replacement studies require knowledge of the depreciation schedule and of taxable gains or losses at disposal. Note that the depreciation schedule is determined at the time of asset acquisition, whereas the current tax law determines the gains tax effects at the time of disposal. In this section, we will use Example 11.6 to illustrate how to do replacement analyses on an after-tax basis.

Example 11.6 Replacement Analysis under an Infinite Planning Horizon

Recall Example 11.5, where Advanced Electrical Insulator Company was considering replacing a broken inspection machine. Let's assume the following additional data:

- The old machine has been fully depreciated, so it has zero book value. The machine could be used for another five years, but the firm does not expect to realize any salvage value from scrapping it in five years.
- The new machine falls into the five-year MACRS property class and will be depreciated accordingly. The operating cost will be $2,000 in the first year, but increasing by $1,000 thereafter.

The marginal income-tax rate is 40%, and the after-tax MARR is 15%. Find the useful life for each option, and decide whether the defender should be replaced now or later.

SOLUTION

1. **Economic service life:**

   - **Defender:** The defender is fully depreciated, so all salvage values can be treated as ordinary gains and taxed at 40%. The after-tax salvage values are thus as follows:

n	Current Market Value	After-Tax Salvage Value
0	$5,000	$5,000(1 − 0.40) = $3,000
1	$4,000	$4,000(1 − 0.40) = $2,400
2	$3,000	$3,000(1 − 0.40) = $1,800
3	$2,000	$2,000(1 − 0.40) = $1,200
4	$1,000	$1,000(1 − 0.40) = $600
5	$0	$0
6	$0	$0
⋮	⋮	⋮

If the company retains the inspection machine, it is in effect deciding to overhaul the machine and invest the machine's current market value (after taxes) in that alternative. Although there is no actual cash flow, the firm is withholding from the investment the market value of the inspection machine (opportunity cost). The after-tax O&M costs are as follows:

n	Overhaul	Forecasted O&M Cost	After-tax O&M Cost
0	$1,200		$1,200(1 − 0.40) = $720
1	$0	$2,000	$2,000(1 − 0.40) = $1,200
2	$0	$3,500	$3,500(1 − 0.40) = $2,100
3	$0	$5,000	$5,000(1 − 0.40) = $3,000
4	$0	$6,500	$6,500(1 − 0.40) = $3,900
5	$0	$8,000	$8,000(1 − 0.40) = $4,800

Using the current year's market value as the investment required to retain the defender, we obtain the data in Table 11.1, which indicates that the remaining useful life of the defender is two years, *in the absence of future challengers*. The overhaul (repair) cost of $1,200 in year zero can be treated as a deductible operating expense for tax purposes, as long as it does not add value to the property. (Any repair or improvement expenses that increase the value of the property must be capitalized by depreciating them over the estimated service life.)

• **Challenger:** Because the challenger will be depreciated over its tax life, we must determine the book value of the asset at the end of each period in order to compute the after-tax salvage value. This process is shown in Table 11.2(A). With the after-tax salvage values computed in Table 11.2(A), we are now ready to find the economic service life of the challenger by generating AE value entries. These calculations are summarized in Table 11.2(B). The economic life of the challenger is four years, with an AE(15%) value of $4,065.

2. **Optimal time to replace the defender:**

Since the AE value for the defender's remaining useful life (two years) is $3,070, which is less than $4,065, the decision will be to keep the defender for now. Note, however, that the defender's remaining useful life of two years does not necessarily imply that the defender should be kept for two years before switching to the challenger. The reason for this is that the defender's remaining useful life of two years was calculated without considering what type of challenger would be available in the future. When a challenger's financial data are available, we need to enumerate all timing possibilities for

TABLE 11.1 Cost of Retaining the Defender for N More Years

N	n	O&M	Depreciation	A/T O&M	Depreciation Credit	Net Operating Cost	Investment and Net Salvage	Net A/T Cash Flow	Equivalent Annual Cost if the Defender Is Kept for N More Years		
									Capital Cost	Operating Cost	Total Cost
	0	$1,200		($720)		($720)	($3,000)	($3,720)			
1	1	$2,000		($1,200)		($1,200)	$2,400	$1,200	($1,050)	($2,028)	($3,078)
	0	$1,200		($720)		($720)	($3,000)	($3,720)			
	1	$2,000		($1,200)		($1,200)		($1,200)			
2	2	$3,500		($2,100)		($2,100)	$1,800	($300)	($1,008)	($2,061)	($3,070)
	0	$1,200		($720)		($720)	($3,000)	($3,720)			
	1	$2,000		($1,200)		($1,200)		($1,200)			
	2	$3,500		($2,100)		($2,100)		($2,100)			
3	3	$5,000		($3,000)		($3,000)	$1,200	($1,800)	($968)	($2,332)	($3,300)
	0	$1,200		($720)		($720)	($3,000)	($3,720)			
	1	$2,000		($1,200)		($1,200)		($1,200)			
	2	$3,500		($2,100)		($2,100)		($2,100)			
	3	$5,000		($3,000)		($3,000)		($3,000)			
4	4	$6,500		($3,900)		($3,900)	$600	($3,300)	($931)	($2,646)	($3,576)
	0	$1,200		($720)		($720)	($3,000)	($3,720)			
	1	$2,000		($1,200)		($1,200)		($1,200)			
	2	$3,500		($2,100)		($2,100)		($2,100)			
	3	$5,000		($3,000)		($3,000)		($3,000)			
	4	$6,500		($3,900)		($3,900)		($3,900)			
5	5	$8,000		($4,800)		($4,800)	$0	($4,800)	($895)	($2,965)	($3,860)

TABLE 11.2A Forecasted C&M Costs and Net Proceeds from Sale as a Function of Holding Period

Holding Period	O&M	Permitted Annual Depreciation Amounts over the Holding Period							Total Depreciation	Book Value	Expected Market Value	Taxable Gains	Gains Tax	Net A/T Salvage Value
		1	2	3	4	5	6	7						
1	$2,000	$2,000							$2,000	$8,000	$6,000	($2,000)	($800)	$6,800
2	$3,000	$2,000	$1,600						$3,600	$6,400	$5,100	($1,300)	($520)	$5,620
3	$4,000	$2,000	$3,200	$960					$6,160	$3,840	$4,335	$495	$198	$4,137
4	$5,000	$2,000	$3,200	$1,920	$576				$7,696	$2,304	$3,685	$1,381	$552	$3,133
5	$6,000	$2,000	$3,200	$1,920	$1,152	$576			$8,848	$1,152	$3,132	$1,980	$792	$2,340
6	$7,000	$2,000	$3,200	$1,920	$1,152	$1,152	$576		$10,000	$0	$2,662	$2,662	$1,065	$1,597
7	$8,000	$2,000	$3,200	$1,920	$1,152	$1,152	$576	$0	$10,000	$0	$2,263	$2,263	$905	$1,358

Note: Asset price of $10,000, depreciated under MACRS for five-year property with the half-year convention.

417

TABLE 11.2B Economics of Owning and Operating the Challenger for N More Years

(1)	(2)	(3)	(4)	(5)	(6)	(7)	(8)	(9)	(10)	(11)	(12)
Holding Period		Before-Tax Operating Expenses			After-Tax Cash Flow If the Asset Is Kept for N More Years				Equivalent-Annual Cost If the Challenger Is Kept for N More Years		
N	n	O&M	Depreciation	A/T O&M	Depreciation Credit	Net Operating Cost	Investment and Net Salvage	Net A/T Cash Flow	Capital Cost	Operating Cost	Total Cost
1	0	—		—	—	—	($10,000)	($10,000)			
	1	$2,000	$2,000	($1,200)	$800	($400)	$6,800	$6,400	($4,700)	($400)	($5,100)
2	0	—		—	—	—	($10,000)	($10,000)			
	1	$2,000	$2,000	($1,200)	$800	($400)		($400)			
	2	$3,000	$1,600	($1,800)	$640	($1,160)	$5,620	$4,460	($3,536)	($753)	($4,290)
3	0	—		—	—	—	($10,000)	($10,000)			
	1	$2,000	$2,000	($1,200)	$800	($400)		($400)			
	2	$3,000	$3,200	($1,800)	$1,280	($520)		($520)			
	3	$4,000	$960	($2,400)	$384	($2,016)	$4,137	$2,121	($3,188)	($905)	($4,094)
4	0	—		—	—	—	($10,000)	($10,000)			
	1	$2,000	$2,000	($1,200)	$800	($400)		($400)			
	2	$3,000	$3,200	($1,800)	$1,280	($520)		($520)			
	3	$4,000	$1,920	($2,400)	$768	($1,632)		($1,632)			
	4	$5,000	$576	($3,000)	$230	($2,770)	$3,133	$363	($2,875)	($1,190)	($4,065)
	0	—		—	—	—	($10,000)	($10,000)			
	1	$2,000	$2,000	($1,200)	$800	($400)		($400)			
	2	$3,000	$3,200	($1,800)	$1,280	($520)		($520)			
	3	$4,000	$1,920	($2,400)	$768	($1,632)		($1,632)			
	4	$5,000	$1,152	($3,000)	$461	($2,539)		($2,539)			

418

TABLE 11.2B (Continued)

5	5	$6,000	$576	($3,600)	$230	($3,370)	$2,340	($1,030)	($2,636)	($1,474)	($4,110)
6	0	—					($10,000)	($10,000)			
	1	$2,000	$2,000	($1,200)	$800	($400)		($400)			
	2	$3,000	$3,200	($1,800)	$1,280	($520)		($520)			
	3	$4,000	$1,920	($2,400)	$768	($1,632)		($1,632)			
	4	$5,000	$1,152	($3,000)	$461	($2,539)		($2,539)			
	5	$6,000	$1,152	($3,600)	$461	($3,139)		($3,139)			
	6	$7,000	$576	($4,200)	$230	($3,970)	$1,597	($2,373)	($2,460)	($1,729)	($4,189)
7	0	—					($10,000)	($10,000)			
	1	$2,000	$2,000	($1,200)	$800	($400)		($400)			
	2	$3,000	$3,200	($1,800)	$1,280	($520)		($520)			
	3	$4,000	$1,920	($2,400)	$768	($1,632)		($1,632)			
	4	$5,000	$1,152	($3,000)	$461	($2,539)		($2,539)			
	5	$6,000	$1,152	($3,600)	$461	($3,139)		($3,139)			
	6	$7,000	$576	($4,200)	$230	($3,970)		($3,970)			
	7	$8,000	—	($4,800)	—	($4,800)	$1,358	($3,442)	($2,281)	($2,006)	($4,287)

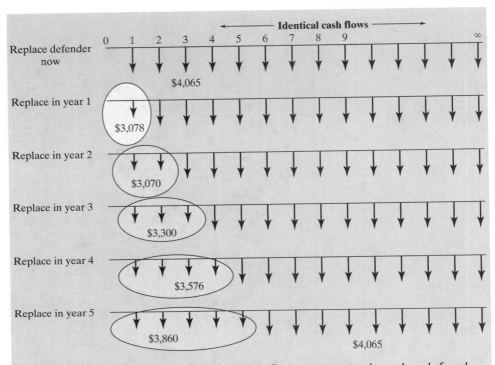

FIGURE 11.9 Equivalent annual cash flow streams when the defender is kept for *n* years followed by infinitely repeated purchases of the challenger every four years

replacement. Since the defender can be used for another five years, six replacement strategies exist:

- Replace now with the challenger.
- Replace in year one with the challenger.
- Replace in year two with the challenger.
- Replace in year three with the challenger.
- Replace in year four with the challenger.
- Replace in year five with the challenger.

If the costs and efficiency of the current challenger remain unchanged in the future years, the possible replacement cash patterns associated with each alternative are shown in Figure 11.9. From the figure, we observe that, on an annual basis, the cash flows after the remaining physical life of the defender are the same.

Before we evaluate the economics of various replacement-decision options, recall the AE values for the defender and the challenger under the assumed service lives (a boxed figure denotes the minimum AE value at $n_0 = 2$ and $n^* = 4$, respectively):

	Annual Equivalent Cost ($)	
n	**Defender**	**Challenger**
1	−$3,078	−$5,100
2	−$3,070	−$4,290
3	−$3,300	−$4,094
4	−$3,576	−$4,065
5	−$3,860	−$4,110
6		−$4,189
7		−$4,287

Instead of using the marginal analysis in Example 11.5, we will use PW analysis, which requires evaluation of infinite cash flow streams. (You will have the same result under marginal analysis.) Immediate replacement of the defender by the challenger is equivalent to computing the PW for an infinite cash flow stream of −$4,065. If we use the capitalized-equivalent-worth approach from Chapter 4 ($CE(i) = A/i$), we obtain the following:

- $n = 0$:

$$PW(15\%)_{n_0=0} = (1/0.15)(-\$4,065)$$
$$= -\$27,100.$$

Suppose we retain the old machine *n* more years and then replace it with the new one.

Now we will compute $PW(i)n_0 = n$.

- $n = 1$:

$$PW(15\%)_{n_0=1} = -\$3,078(P/A, 15\%, 1) - \$27,100(P/F, 15\%, 1)$$
$$= -\$26,242.$$

- $n = 2$:

$$PW(15\%)_{n_0=2} = -\$3,070(P/A, 15\%, 2) - \$27,100(P/F, 15\%, 2)$$
$$= -\$25,482.$$

- $n = 3$:

$$PW(15\%)_{n_0=3} = -\$3,300(P/A, 15\%, 3) - \$27,100(P/F, 15\%, 3)$$
$$= -\$25,353.$$

- $n = 4$:

$$PW(15\%)_{n_0=4} = -\$3,576(P/A, 15\%, 4) - \$27,100(P/F, 15\%, 4)$$
$$= -\$25,704.$$

- $n = 5$:

$$\text{PW}(15\%)_{n_0=5} = -\$3,860(P/A, 15\%, 5) - \$27,100(P/F, 15\%, 5)$$
$$= -\$26,413.$$

This result leads us to conclude that the defender should be kept for three more years. The present worth of $25,353 represents the net cost associated with retaining the defender for three years, replacing it with the challenger, and then replacing the challenger every four years for an indefinite period.

Summary

- In replacement analysis, the **defender** is an existing asset; the **challenger** is the best available replacement candidate.
- The **current market value** is the value to use in preparing a defender's economic analysis. **Sunk costs**—past costs that cannot be changed by any future investment decision—should not be considered in a defender's economic analysis.
- Two basic approaches to analyzing replacement problems are the **cash flow approach** and the **opportunity-cost approach**. The cash flow approach explicitly considers the actual cash flow consequences for each replacement alternative as it occurs. Typically, the net proceeds from the sale of the defender are subtracted from the purchase price of the challenger. The opportunity-cost approach views the net proceeds from the sale of the defender as an opportunity cost of keeping the defender. That is, instead of deducting the salvage value from the purchase cost of the challenger, the salvage value is considered an investment required in order to keep the asset.
- **Economic service life** is the remaining useful life of a defender (or a challenger) that results in the minimum equivalent annual cost or maximum annual equivalent revenue. We should use the respective economic service lives of the defender and the challenger when conducting a replacement analysis.
- Ultimately, in replacement analysis, the question is not *whether* to replace the defender, but *when* to do so. The AE method provides a marginal basis on which to make a year-by-year decision about the best time to replace the defender. As a general decision criterion, the PW method provides a more direct solution to a variety of replacement problems with either an infinite or a finite planning horizon or a technological change in a future challenger.
- The role of **technological change** in asset improvement should be weighed in making long-term replacement plans: If a particular item is undergoing rapid, substantial technological improvements, it may be prudent to shorten or delay replacement (to the extent where the loss in production does not exceed any savings from improvements in future challengers) until a desired future model is available.

Replacement Basics without Considering Income-Tax Effects

11.1 Inland Trucking Company is considering the replacement of a 1,000 lb capacity forklift truck. The truck was purchased three years ago at a cost of $15,000. The diesel-operated forklift truck was originally expected to have a useful life of eight years and a zero estimated salvage value at the end of that period. However, the truck has not been dependable and is frequently out of service while awaiting repairs. The maintenance expenses of the truck have been rising steadily and currently amount to about $3,000 per year. The truck could be sold now for $6,000. If retained, the truck will require an immediate $1,500 overhaul to keep it in operable condition. This overhaul will neither extend the originally estimated service life nor increase the value of the truck. The updated annual operating costs, engine overhaul cost, and market values over the next five years are estimated as follows:

N	O&M	Depreciation	Engine Overhaul	Market Value
−3				
−2		$3,000		
−1		$4,800		
0		$2,880	$1,500	$6,000
1	$3,000	$1,728		$4,000
2	$3,500	$1,728		$3,000
3	$3,800	$864		$1,500
4	$4,500	$0		$1,000
5	$4,800	$0	$5,000	$0

A drastic increase in O&M costs during the fifth year is expected as a result of another overhaul, which will be required in order to keep the truck in operating condition. The firm's MARR is 15%.

(a) If the truck is to be sold now, what will be its sunk cost?

(b) What is the opportunity cost of not replacing the truck now?

(c) What is the equivalent annual cost of owning and operating the truck for two more years?

(d) What is the equivalent annual cost of owning and operating the truck for five more years?

11.2 Komatsu Cutting Technologies is considering replacing one of its CNC machines with a machine that is newer and more efficient. The firm purchased the CNC machine 10 years ago at a cost of $135,000. It had an expected economic life of 12 years at the time of purchase and an expected salvage value of $12,000 at the end of the 12 years. The original salvage estimate is still good, and the machine has a remaining useful life of two years. The firm can sell this old machine now to another firm in the industry for $30,000. The new machine can be purchased for $165,000, including installation costs. It has an estimated useful (economic) life of eight years. The new machine is expected to reduce cash operating expenses by $30,000 per year over its eight-year life. At the end of its useful life, the machine is estimated to be worth only $5,000. The company has a MARR of 12%.

(a) What is the opportunity cost of retaining the old asset?

(b) Compute the cash flows associated with retaining the old machine for the next two years (years one to two).

(c) Compute the cash flows associated with purchasing the new machine for use over the next eight years (years one to eight). (Use the opportunity-cost concept.)

(d) If the firm needs the service of this type of machine for an indefinite period and no technology improvement is expected in future machines, what would be your decision?

11.3 Air Links, a commuter airline company, is considering the replacement of one of its baggage loading–unloading machines with a newer and more efficient one. The relevant details for both machines are as follows:

- The current book value of the old machine is $50,000, and it has a remaining useful life of five years. The salvage value expected from scrapping the old machine at the end of five years is zero, but the company can sell the machine now to another firm in the industry for $10,000.

- The new baggage-handling machine has a purchase price of $120,000 and an estimated useful life of seven years. It has an estimated salvage value of $30,000 and is expected to realize economic savings on electric-power usage, labor, and repair costs and to reduce the amount of damaged luggage. In total, annual savings of $50,000 will be realized if the new machine is installed.

The firm uses a MARR of 15%. Using the opportunity-cost approach, address the following questions:

(a) What is the initial cash outlay required for the new machine?

(b) What are the cash flows for the defender in years zero to five?

(c) Should the airline purchase the new machine?

11.4 Duluth Medico purchased a digital image-processing machine three years ago at a cost of $50,000. The machine had an expected life of eight years at the time of purchase and an expected salvage value of $5,000 at the end of the eight years. However, the old machine has been slow at handling the increased business volume, so management is considering replacing it. A new machine can be purchased for

$75,000, including installation costs. Over its five-year life, the new machine will reduce cash operating expenses by $30,000 per year. Sales are not expected to change. At the end of its useful life, the new machine is estimated to be worthless. The old machine can be sold today for $10,000. The firm's interest rate for project justification is known to be 15%. The firm does not expect a better machine (other than the current challenger) to be available for the next five years. Assume that the economic service life for the new machine and the remaining useful life for the old machine are both five years.

(a) Determine the cash flows associated with each option (keeping the defender versus purchasing the challenger).

(b) Should the company replace the defender now?

11.5 The Northwest Manufacturing Company is currently manufacturing one of its products on a hydraulic stamping press machine. The unit cost of the product is $12, and in the past year 3,000 units were produced and sold for $19 each. It is expected that the future demand of the product and the unit price will remain steady at 3,000 units per year and $19 per unit, respectively. The machine has a remaining useful life of three years and could be sold on the open market now for $5,500. Three years from now, the machine is expected to have a salvage value of $1,200. A new machine would cost $36,500, and the unit manufacturing cost on the new machine is projected to be $11. The new machine has an expected economic life of five years and an expected salvage of $6,300. The appropriate MARR is 12%. The firm does not expect a significant improvement in the machine's technology to occur, and it needs the service of either machine for an indefinite period of time.

(a) Compute the cash flows over the remaining useful life if the firm decides to retain the old machine.

(b) Compute the cash flows over the economic service life if the firm decides to purchase the machine.

(c) Should the new machine be acquired now?

Economic Service Life

11.6 A firm is considering replacing a machine that has been used for making a certain kind of packaging material. The new machine will cost $31,000 installed and will have an estimated economic life of 10 years, with a salvage value of $2,500. Operating costs are expected to be $1,000 per year throughout its service life. The machine currently in use had an original cost of $25,000 four years ago, and its service life (physical life) at the time of purchase was estimated to be seven years, with a salvage value of $5,000. This machine has a current market value of $7,700. If the firm retains the old machine, its updated market values and operating costs for the next four years will be as follows:

Year End	Market Value	Book Value	Operating Costs
0	$7,700	$7,809	
1	$4,300	$5,578	$3,200
2	$3,300	$3,347	$3,700
3	$1,100	$1,116	$4,800
4	$0	$0	$5,850

The firm's minimum attractive rate of return is 12%.

(a) Working with the updated estimates of market values and operating costs over the next four years, determine the remaining useful life of the old machine.

(b) Determine whether it is economical to make the replacement now.

(c) If the firm's decision is to replace the old machine in part (b), then when should the replacement occur?

11.7 The University Resume Service has just invested $8,000 in a new desktop publishing system. From past experience, the owner of the company estimates its after-tax cash returns as

$$A_n = \$8,000 - \$4,000(1 + 0.15)^{n-1}$$

and

$$S_n = \$6,000(1 - 0.3)^n,$$

where A_n represents the net after-tax cash flows from operation during period n and S_n represents the after-tax salvage value at the end of period n.

(a) If the company's MARR is 12%, compute the economic service life of the desktop publishing system.

(b) Explain how the economic service life varies with the interest rate.

11.8 A special-purpose machine is to be purchased at a cost of $15,000. The following table shows the expected annual operating and maintenance cost and the salvage value for each year of service:

Year of Service	O&M Cost	Market Value
1	$2,500	$12,800
2	$3,200	$8,100
3	$5,300	$5,200
4	$6,500	$3,500
5	$7,800	$0

(a) If the interest rate is 10%, what is the economic service life for this machine?

(b) Repeat (a), using $i = 15\%$.

Replacement Decisions when Required Service Life is Long

11.9 A special-purpose turnkey stamping machine was purchased four years ago for $20,000. It was estimated at that time that this machine would have a life of 10 years and a salvage value of $3,000, with a removal cost of $1,500. These estimates are still good. This machine has annual operating costs of $2,000. A new machine, which is more efficient, will reduce the annual operating costs to $1,000, but will require an investment of $20,000, plus $1,000 for installation. The life of the new machine is estimated to be 12 years, with a salvage value of $2,000 and a removal cost of $1,500. An offer of $6,000 has been made for the old machine, and the purchaser is willing to pay for the removal of the machine. Find the economic advantage of replacement or of continuing with the present machine. State any assumptions that you make. (Assume that the MARR = 8%.)

11.10 A five-year-old defender has a current market value of $4,000 and expected O&M costs of $3,000 this year, increasing by $1,500 per year. Future market values are expected to decline by $1,000 per year. The machine can be used for another three years. The challenger costs $6,000 and has O&M costs of $2,000 per year, increasing by $1,000 per year. The machine will be needed for only three years, and the salvage value at the end of the three years is expected to be $2,000. The MARR is 15%.

(a) Determine the annual cash flows for retaining the old machine for three years.

(b) Determine whether now is the best time to replace the old machine. First show the annual cash flows for the challenger.

11.11 Greenleaf Company is considering the purchase of a new set of air-electric quill units to replace an obsolete machine. The current machine has a market value of zero; however, it is in good working order, and it will last physically for at least an additional five years. The new quill units will perform the operation with so much more efficiency that the firm's engineers estimate that labor, material, and other direct costs will be reduced by $3,000 a year if they are installed. The new set of quill units costs $10,000 delivered and installed,

and its economic life is estimated to be five years, with zero salvage value. The firm's MARR is 10%.

(a) What is the investment required in order to keep the old machine?

(b) Compute the cash flow to use in the analysis of each option.

(c) If the firm uses the internal-rate-of-return criterion, would the analysis indicate that the firm should buy the new machine?

11.12 Dean Manufacturing Company is considering the replacement of an old, relatively inefficient vertical drill machine that was purchased seven years ago at a cost of $10,000. The machine had an original expected life of 12 years and a zero estimated salvage value at the end of that period. The current market value of the machine is $1,000. The divisional manager reports that a new machine can be bought and installed for $12,000. Over its five-year life, this machine will expand sales from $10,000 to $11,500 a year and, furthermore, will reduce labor and raw-materials usage sufficiently to cut annual operating costs from $7,000 to $5,000. The new machine has an estimated salvage value of $2,000 at the end of its five-year life. The firm's MARR is 15%.

(a) Should the new machine be purchased now?

(b) What current market value of the old machine would make the two options equal?

11.13 Advanced Robotics Company is faced with the prospect of replacing its old call-switching system, which has been used in the company's headquarters for 10 years. This particular system was installed at a cost of $100,000 and was assumed to have a 15-year life, with no appreciable salvage value. The current annual operating costs are $20,000 for this old system, and these costs are presumed to be the same for the rest of its life. A sales representative from North Central Bell is trying to sell this company a computerized switching system. The new system would require an investment of $200,000 for installation. The economic life of this computerized system is estimated to be 10 years, with a salvage value of $18,000, and the system will reduce annual operating costs to $5,000. No detailed agreement has been made with the sales representative about the disposal of the old system. Determine the range of resale values associated with the old system that would justify installation of the new system at a MARR of 14%.

11.14 A company is currently producing chemical compounds by a process installed 10 years ago at a cost of $100,000. It was assumed that the process would have a 20-year life, with a zero salvage value. The current market value of the equipment, however, is $60,000, and the initial estimate of its economic life is still good. The annual operating costs associated with this process are $18,000. A sales representative from U.S. Instrument Company is trying to sell a new chemical-compound-making process to the company. This new process will cost $200,000; have a service life of 10 years, with a salvage value of $20,000; and reduce annual operating costs to $4,000. Assuming the company desires a return of 12% on all investments, should it invest in the new process?

11.15 Eight years ago, a firm purchased a lathe for $45,000. The operating expenses for the lathe are $8,700 per year. An equipment vendor offers the firm a new machine for $53,500, with operating costs of $5,700 per year. An allowance of $8,500 would be made for the old machine on purchase of the new one. The old machine is expected to be scrapped at the end of five years. The new machine's economic service life is five years, with a salvage value of $12,000. The new machine's O&M cost is estimated to be $4,200 for the first year, increasing at an annual rate of $500 thereafter. The firm's MARR is 12%. Which option would you recommend?

11.16 The New York Taxi Cab Company has just purchased a new fleet of 2003 models. Each brand-new cab cost $20,000. From past experience, the

company estimates after-tax cash returns for each cab as

$$A_n = \$65,800 - 30,250(1 + 0.15)^{n-1}$$

and

$$S_n = \$20,000(1 - 0.15)^n,$$

where, again, A_n represents the net after-tax cash flows from operation during period n and S_n represents the after-tax salvage value at the end of period n. The management views the replacement process as a constant and infinite chain.

(a) If the firm's MARR is 10% and it expects no major technological and functional change in future models, what is the optimal time period (constant replacement cycle) to replace its cabs? (Ignore inflation.)

(b) What is the internal rate of return for a cab if it is retired at the end of its economic service life? What is the internal rate of return for a sequence of identical cabs if each cab in the sequence is replaced at the optimal time?

11.17 Four years ago, a firm purchased an industrial batch oven for $23,000. The oven has been depreciated over a 10-year life and has a $1,000 salvage value. If sold now, the machine will bring in $2,000. If sold at the end of the year, it will bring in $1,500. Annual operating costs for subsequent years are $3,800. A new machine will cost $50,000 and have a 12-year life, with a $3,000 salvage value. The operating cost for the new machine will be $3,000 as of the end of each year, where the $6,000-per-year savings are due to better quality control. If the firm's MARR is 10%, should the new machine be purchased now?

11.18 Georgia Ceramic Company has an automatic glaze sprayer that has been used for the past 10 years. The sprayer can be used for another 10 years and will have a zero salvage value at that time. The annual operating and maintenance costs for the sprayer amount to $15,000 per year. Due to an increase in business, a new sprayer must be purchased, either in addition to or as a replacement for the old sprayer:

- Option 1: If the old sprayer is retained, a new, smaller capacity sprayer will be purchased at a cost of $48,000; this new sprayer will have a $5,000 salvage value in 10 years and annual operating and maintenance costs of $12,000. The old sprayer has a current market value of $6,000.
- Option 2: If the old sprayer is sold, a new sprayer of larger capacity will be purchased for $84,000. This sprayer will have a $9,000 salvage value in 10 years and annual operating and maintenance costs of $24,000.

Which option should be selected at MARR = 12%?

Replacement Analysis with Tax Considerations

11.19 Rework Problem 11.1, assuming the following additional information: The asset is classified as a five-year MACRS property and has a book value of $5,760 if disposed of now. The firm's marginal tax rate is 40%, and its after-tax MARR is 15%.

11.20 Rework Problem 11.3, assuming the following additional information:

- The current book value of the old machine is $50,000. By the conventional straight-line methods, the old machine is being depreciated toward a zero salvage value, or by $10,000 per year.
- The new machine will be depreciated under a seven-year MACRS class.
- The company's marginal tax rate is 40%, and the firm uses an after-tax MARR of 15%.

11.21 Rework Problem 11.4, assuming the following additional information:

- The old machine has been depreciated using the MACRS under the five-year property class.
- The new machine will be depreciated under a five-year MACRS class.
- The marginal tax rate is 35%, and the firm's after-tax MARR is 15%.

Economic Service Life with Tax Considerations

11.22 Rework Problem 11.6, assuming the following additional information:

- The current book value of the old machine is $7,809. The anticipated book values for the next four years are as follows: year 1: $5,578; year 2: $3,347; year 3: $1,116; and year 4: $0.
- The new machine will be depreciated under a seven-year MACRS class.
- The company's marginal tax rate is 35%, and the firm uses an after-tax MARR of 12%.

11.23 A machine has a first cost of $10,000. End-of-year book values, salvage values, and annual O&M costs are provided over its useful life as follows:

Year End	Book Value	Salvage Value	Operating Costs
1	$8,000	$5,300	$1,500
2	$4,800	$3,900	$2,100
3	$2,880	$2,800	$2,700
4	$1,728	$1,800	$3,400
5	$1,728	$1,400	$4,200
6	$576	$600	$4,900

(a) Determine the economic life of the machine if the MARR is 15% and the marginal tax rate is 40%.
(b) Determine the economic life of the machine if the MARR is changed to 10% and the marginal tax rate remains at 40%.

11.24 Assume, for a particular asset, that

$$I = \$20,000,$$
$$S_n = 12,000 - 2,000n,$$
$$B_n = 20,000 - 2,500n,$$
$$O\&M_n = 3,000 + 1,000(n - 1),$$

and

$$t_m = 0.40,$$

where I = asset purchase price,
  S_n = market value at the end of year n,
  B_n = book value at the end of year n,
  $O\&M_n$ = O&M cost during year n, and
  t_m = marginal tax rate.

(a) Determine the economic service life of the asset if $i = 10\%$.
(b) Determine the economic service life of the asset if $i = 25\%$.
(c) Assuming that $i = 0$, determine the economic service life mathematically.

11.25 Rework Problem 11.8, assuming the following additional information:

- For tax purposes, the entire cost of $15,000 can be depreciated according to a five-year MACRS property class.
- The firm's marginal tax rate is 40%.

11.26 Quintana Electronic Company is considering the purchase of new robot-welding equipment to perform operations currently being performed by less efficient equipment. The new machine's purchase price is $150,000 delivered and installed. A Quintana industrial engineer estimates that the new equipment will produce savings of $30,000 in labor and other direct costs annually, as compared with the present equipment. He estimates the proposed equipment's economic life at 10 years, with a zero salvage value. The present equipment is in good working order and will last, physically, for at least 10 more years. Quintana Company expects to pay income taxes of 40%, and any gains will also be taxed at

40%. Quintana uses a 10% discount rate for analysis performed on an after-tax basis. Depreciation of the new equipment for tax purposes is computed on the basis of the seven-year MACRS property class.

(a) Assuming that the present equipment has zero book value and zero salvage value, should the company buy the proposed equipment?
(b) Assuming that the present equipment is being depreciated at a straight-line rate of 10%, has a book value of $72,000 (a cost of $120,000 and accumulated depreciation of $48,000) and zero net salvage value today, should the company buy the proposed equipment?
(c) Assuming that the present equipment has a book value of $72,000 and a salvage value today of $45,000 and that, if the equipment is retained for 10 more years, its salvage value will be zero, should the company buy the proposed equipment?
(d) Assume that the new equipment will save only $15,000 a year, but that its economic life is expected to be 12 years. If all other conditions are as described in (a), should the company buy the proposed equipment?

11.27 Assume, in Problem 11.26, that Quintana Company decided to purchase the new equipment (hereafter called "Model A"). Two years later, even better equipment (called "Model B") comes onto the market. This equipment makes Model A completely obsolete, with no resale value. The Model B equipment costs $300,000 delivered and installed, but is expected to result in annual savings of $75,000 over the cost of operating the Model A equipment. The economic life of Model B is estimated to be 10 years, with a zero salvage value. (Model B also is classified as a seven-year MACRS property.)

(a) What action should the company take with respect to the potential replacement of Model A with Model B?

(b) If the company decides to purchase the Model B equipment, a mistake must have been made, because good equipment, bought only two years previously, is being scrapped. How did this mistake come about?

Replacement Decisions When the Required Service Period is Long (with Tax Considerations)

11.28 Rework Problem 11.9, assuming the following additional information:

- The current book value of the old machine is $6,248, and the asset has been depreciated according to a seven-year MACRS property class.
- The new asset is also classified as a seven-year MACRS property.
- The company's marginal tax rate is 30%, and the firm uses an after-tax MARR of 8%.

11.29 Rework Problem 11.11, assuming the following additional information:

- The current book value of the old machine is $4,000, and the annual depreciation charge is $800 if the firm decides to keep the old machine for the additional five years.
- The new asset is classified as a seven-year MACRS property.
- The company's marginal tax rate is 40%, and the firm uses an after-tax MARR of 10%.

11.30 Rework Problem 11.13, assuming the following additional information:

- The old switching system has been fully depreciated.
- The new system falls into the five-year MACRS property class.
- The company's marginal tax rate is 40%, and the firm uses an after-tax MARR of 14%.

11.31 Five years ago, a conveyor system was installed in a manufacturing plant at a cost of $35,000. It was estimated that the system, which is still in operating condition, would have a useful life of eight years, with a salvage value of $3,000. If the firm continues to operate the system, the system's estimated market values and operating costs for the next three years are as follows:

Year End	Market Value	Book Value	Operating Costs
0	$11,500	$15,000	
1	$5,200	$11,000	$4,500
2	$3,500	$7,000	$5,300
3	$1,200	$3,000	$6,100

A new system can be installed for $43,500. This system would have an estimated economic life of 10 years, with a salvage value of $3,500. The operating costs for the new system are expected to be $1,500 per year throughout its service life. The firm's MARR is 18%. The system belongs to the seven-year MACRS property class. The firm's marginal tax rate is 35%.

(a) Should the existing system be replaced?
(b) If the decision in (a) is to replace the existing system, when should replacement occur?

11.32 Rework Problem 11.15, assuming the following additional information:

- The old machine has been fully depreciated.
- The new machine will be depreciated under the seven-year MACRS class.
- The marginal tax rate is 35%, and the firm's after-tax MARR is 12%.

11.33 Rework Problem 11.18, assuming the following additional information:

- Option 1: The old sprayer has been fully depreciated. The new sprayer is classified as a seven-year MACRS property.
- Option 2: The larger capacity sprayer is classified as a seven-year MACRS property.

• The company's marginal tax rate is 40%, and the firm uses an after-tax MARR of 12%.

11.34 A six-year old CNC machine that originally cost $8,000 has been fully depreciated, and its current market value is $1,500. If the machine is kept in service for the next five years, its O&M costs and salvage values are estimated as follows:

End of Year	O & M Costs		Salvage Value
	Operation and Repairs	Delays Due to Repairs	
1	$1,300	$600	$1,200
2	$1,500	$800	$1,000
3	$1,700	$1,000	$500
4	$1,900	$1,200	$0
5	$2,000	$1,400	$0

It is suggested that the machine be replaced by a new CNC machine of improved design at a cost of $6,000. It is believed that this purchase will completely eliminate breakdowns, and the resulting combined cost of delays, operation, and repairs will be reduced by $200 a year at each age than is the case with the old machine. Assume a five-year life and a $1,000 terminal salvage value for the challenger. The new machine falls into the five-year MACRS property class. The firm's MARR is 12%, and its marginal tax rate is 30%. Should the old machine be replaced now?

Short Case Studies with Excel

11.35 Chevron Overseas Petroleum, Inc., entered into a 1993 joint-venture agreement with the Republic of Kazakhstan, a former republic of the old Soviet Union, to develop the huge Tengiz oil field. Unfortunately, the climate in the region is harsh, making it difficult to keep oil flowing. The untreated oil comes out of the ground at 114°F. Even though the pipelines are insulated, as the oil travels through the pipeline for processing, hydrate salts begin to precipitate out of the liquid phase as the oil cools. These hydrate salts create a dangerous condition, because they form plugs in the line.

The method for preventing this trap-pressure condition is to inject methanol (MeOH) into the oil stream. This substance keeps the oil flowing and prevents hydrate salts from precipitating out of the liquid phase. The present methanol loading and storage facility is a completely manually controlled system, with no fire protection and a rapidly deteriorating tank that causes leaks. The scope of required repairs and upgrades is extensive. For example, the storage tanks are rusting and are leaking at the riveted joints, the manual control system causes frequent tank overfills, and, as stated previously, there is no fire protection system, as water is not available at this site.

The present storage facility has been in service for five years. Permit requirements necessitate upgrades in order to achieve minimum acceptable standards in Kazakhstan. The upgrades, which will cost $104,000, will extend the life of the current facility to about 10 years. However, upgrades will not completely stop the leaks. The expected spill and leak losses will amount to $5,000 a year. The annual operating costs are expected to be $36,000.

As an alternative, a new methanol storage facility can be designed based on minimum acceptable international oil-industry standards. The new facility, which would cost $325,000, would last about 12 years before a major upgrade would be required. However, it is believed that oil transfer technology will be such that methanol will not be necessary in 10 years; new pipeline heating and insulation systems will make methanol storage and use systems obsolete. With a more closely monitored system that yields lower risk of leaks, spills, and evaporation loss, the expected annual operating cost would be $12,000.

(a) Assume that the storage tanks (the new ones as well as the upgraded ones) will have no salvage values at the end of their useful lives (after considering the removal costs) and that the tanks will be depreciated by the straight-line method according to Kazakhstan's tax law. If Chevron's interest rate is 20% for foreign projects, which option is a better choice?

(b) How would the decision change as you consider the environmental implications of spills (e.g., clean-up costs) and evaporation of product?

11.36 Rivera Industries, a manufacturer of home-heating appliances, is considering the purchase of a used Amada turret punch press, to replace its less advanced present system, which uses four old, small presses. Currently, the four smaller presses are used (in varying sequences, depending on the product) to produce one component of a product until a scheduled time when all machines must retool in order to set up for a different component. Because setup cost is high, production runs of individual components are long. This factor results in large inventory buildups of one component, which are necessary to prevent extended backlogging while other products are being manufactured.

The four presses in use now were purchased six years ago at a price of $100,000. The manufacturing engineer expects that these machines can be used for eight more years, but they will have no market value after that. These presses have been depreciated by the MACRS method (seven-year property). The current book value is $13,387, and the present market value is estimated to be $40,000. The average setup cost, which is determined by the number of required labor hours times the labor rate for the old presses, is $80 per hour, and the number of setups per year expected by the production control department is 200. These conditions yield a yearly setup cost of $16,000. The expected operating and maintenance costs for each

year in the remaining life of this system are estimated as follows:

Year	Setup Costs	O&M Costs
1	$16,000	$15,986
2	$16,000	$16,785
3	$16,000	$17,663
4	$16,000	$18,630
5	$16,000	$19,692
6	$16,000	$20,861
7	$16,000	$22,147
8	$16,000	$23,562

These costs, which were estimated by the manufacturing engineer with the aid of data provided by the vendor, represent a reduction in efficiency and an increase in needed service and repair over time.

The price of the two-year-old Amada turret punch press is $135,000 and would be paid for with cash from the company's capital fund. In addition, the company would incur installation costs totaling $1,200. Also, an expenditure of $12,000 would be required in order to recondition the press to its original condition. The reconditioning would extend the Amada's economic service life to eight years. It would have no salvage value at the end of that time. The Amada would be depreciated under the MACRS with half-year convention as a seven-year property. The average setup cost of the Amada is $15, and the Amada would incur 1,000 setups per year, yielding a yearly setup cost of $15,000. Rivera's accounting department has estimated that at least $26,000, and probably $36,000, per year could be saved by shortening production runs, and thus carrying costs. The operating and maintenance costs of the Amada as estimated by the manufacturing engineer are similar to, but somewhat less than, the O&M costs for the present system:

Year	Setup Costs	O&M Costs
1	$15,000	$11,500
2	$15,000	$11,950
3	$15,000	$12,445
4	$15,000	$12,990
5	$15,000	$13,590
6	$15,000	$14,245
7	$15,000	$14,950
8	$15,000	$15,745

The reduction in the O&M costs is caused by the age difference of the machines and the reduced power requirements of the Amada.

If Rivera Industries delays the replacement of the current four presses for another year, the secondhand Amada machine will no longer be available, and the company will have to buy a brand-new machine at an installed price of $200,450. The expected setup costs would be the same as those for the secondhand machine, but the annual O&M costs would be about 10% lower than the estimated O&M costs for the secondhand machine. The expected economic service life of the brand-new press would be eight years, with no salvage value at the end of that period. The brand-new press also falls into a seven-year MACRS property class.

Rivera's MARR is 12% after taxes, and the marginal income-tax rate is expected to be 40% over the life of the project.

(a) Assuming that the company would need the service of either the Amada press or the current presses for an indefinite period, which option would you recommend?

(b) Assuming that the company would need the press system for only five more years, which option would you recommend?

.12

A Race to the Bottom: With Flawed Analyses, Corps Dredges Ports Nationwide[1]

The Chesapeake and Delaware (C&D) Canal is 35 feet deep and 14 miles long. And according to the Army Corps of Engineers, it flows in two directions at once. At least that's what the agency's environmental studies suggest. In a study that bolstered its $83 million plan to deepen the C&D for the Port of Baltimore, the Corps concluded that the canal's net flow is west to east, which would minimize the project's damage to the Chesapeake Bay. But in a study for its $311 million plan to deepen the Delaware River for the Port of Philadelphia, the same Corps district had concluded that the same canal flows east to west.

The C&D project's economic analysis is as problematic as its water-flow analysis. An ad hoc quartet of tireless Maryland citizen critics has documented at least a dozen mathematical errors, overoptimistic predictions, and other flawed assumptions by the Corps, all exaggerating the benefits to shipping lines or minimizing its costs to taxpayers. The Corps claims that the project's benefits would slightly outweigh the costs; the citizen watchdogs calculate that the costs would be at least 50 times the benefits. In fact, the citizen group claims that the Corps committed a basic math error that boosted the benefit—cost ratio from a failing 0.65 to a passing 1.21.

[1]Michael Grunwald, *The Washington Post*, 09/12/2000, copyright 2000, The Washington Post Co. Used with permission.

Benefit–Cost Analysis

Army Corps studies are supposed to provide impartial evaluations of proposed federal projects, screening out the wasteful and destructive ones and providing an objective seal of approval for the necessary ones. But a review of the proposed canal deepening for Baltimore illustrates how the Corps often justifies projects backed by powerful political interests with questionable technical analyses. If the four citizens hadn't devoted 7,000 hours of their spare time to phone-book-sized studies, and if their congressman hadn't taken an unheard-of stance against a Corps project in his own district, the canal would almost certainly be deeper by now.

Many civil engineers are employed in public-works areas such as water projects, highway construction, and airport construction. One of the most important aspects of any public project is to quantify the benefits of the project in dollar terms. For example, in any airport runway expansion, what is the economic benefit of reducing airport delay? From the airline's point of view, taxiing and arrival delays mean added fuel costs. From the airport's point of view, any delays mean lost revenues in landing and departure fees.

From the public's point of view, delays mean lost earnings, as delays cause people to spend more time on transportation. Comparison of the investment costs of a project with the project's potential benefits, a process known as **benefit–cost analysis**, is an important feature of the economic analysis method.

Up to this point, we have focused our attention on investment decisions in the private sector; the primary objective of these investments was to increase the wealth of corporations. In the public sector, federal, state, and local governments spend hundreds of billions of dollars annually on a wide variety of public activities, such as expanding airport runways. In addition, governments at all levels regulate the behavior of individuals and businesses by influencing the use of enormous quantities of productive resources. How can public decision makers determine whether their decisions, which affect the use of these productive resources, are, in fact, in the best public interest?

Benefit–cost analysis is a decision-making tool used to systematically develop useful information about the desirable and undesirable effects of public projects. In a sense, we may view benefit–cost analysis in the public sector as profitability analysis in the private sector. In other words, benefit–cost analysis attempts to determine whether the social benefits of a proposed public activity outweigh the social costs. Usually, public investment decisions involve a great deal of expenditure, and their benefits are expected to occur over an extended period of time. Examples of benefit–cost analyses include studies of public transportation systems, environmental regulations on noise and pollution, public safety programs, education and training programs, public health programs, flood control systems, water resource development projects, and national defense programs.

The three types of benefit–cost analysis problems are as follows: (1) problems for which the goal is to maximize the benefits for any given set of costs (or budgets), (2) problems for which the goal is to maximize the net benefits when both benefits and costs vary, and (3) problems for which the goal is to minimize costs in order to achieve any given level of benefits (often called cost-effectiveness analysis). These types of decision problems will be considered in this chapter.

12-1 Evaluation of Public Projects

To evaluate public projects designed to accomplish widely differing tasks, we need to measure the benefits or costs in the same units in all projects so that we have a common perspective by which to judge them. In practice, this requirement means expressing both benefits and costs in monetary units, a process that often must be performed without accurate data. In performing benefit–cost analysis, we define **users** as the public and **sponsors** as the government.

The general framework for benefit–cost analysis can be summarized as follows:

1. Identify all users' **benefits** (favorable outcomes) and **disbenefits** (unfavorable outcomes) expected to arise from the project.

2. Quantify, as much as possible, these benefits and disbenefits in dollar terms so that different benefits and the respective costs of attaining them may be compared.

3. Identify the sponsor's costs and quantify them.

4. Determine the equivalent net benefits and net costs at the base period; use a discount rate appropriate for the project.

5. Accept the project if the equivalent users' net benefits exceed the equivalent sponsor's net costs.

We can use benefit–cost analysis to choose among such alternatives as allocating funds for construction of a mass-transit system, an irrigation dam, highways, or an air-traffic control system. If the projects are on the same scale with respect to cost, it is merely a question of choosing the project for where the benefits exceed the costs by the greatest amount. The steps outlined above are for a single (or independent) project evaluation. As is in the case for the internal rate of return criterion, when comparing mutually exclusive alternatives, an incremental benefit–cost ratio must be used. Section 12.2.2 illustrates this important issue in detail.

12.1.1 Valuation of Benefits and Costs

In the abstract, the framework we just developed for benefit–cost analysis is no different from the one we have used throughout this text to evaluate private investment projects. The complications, as we shall discover in practice, arise in trying to identify and assign values to all the benefits and costs of a public project.

12.1.2 Users' Benefits

To begin a benefit–cost analysis, we identify all project benefits and disbenefits to the users, bearing in mind the indirect consequences resulting from the project—the so-called **secondary effects**. For example, construction of a new highway will create new businesses such as gas stations, restaurants, and motels (benefits), but it will also divert some traffic from the old roads, and, as a consequence, some businesses may be lost (disbenefits). Once the benefits and disbenefits are quantified, we define the overall user's benefit B as follows:

$$B = \text{benefits} - \text{disbenefits}.$$

In identifying the user's benefits, we classify each as a **primary benefit**—a benefit directly attributable to the project—or a **secondary benefit**—a benefit indirectly attributable to the project. As an example, the U.S. government at one time was considering building a superconductor research laboratory in Texas. Such a move would bring many scientists and engineers, along with other supporting population members, to the region. Primary national benefits may include the long-term benefits that accrue as a result of various applications of the research to U.S. businesses. Primary regional benefits may include economic benefits created by the research laboratory activities, which would generate many new supporting businesses. The secondary benefits might include the creation of new economic wealth as a consequence of a possible increase in international trade and any increase in the incomes of various regional producers attributable to a growing population.

The reason for making this distinction between primary and secondary benefits is that it may make our analysis more efficient. If primary benefits alone are sufficient to justify project costs, we can save time and effort by not quantifying the secondary benefits.

12.1.3 Sponsor's Costs

We determine the cost to the sponsor by identifying and classifying the expenditures required and any savings (or revenues) to be realized. The sponsor's costs should include both capital investments and annual operating costs. Any sales of products or services that take place on completion of the project will generate some revenues—for example, toll revenues on highways. These revenues reduce the sponsor's costs. Therefore, we calculate the sponsor's costs by combining these cost elements:

Sponsor's costs = capital costs + operating and maintenance costs − revenues.

12.1.4 Social Discount Rate

As we learned in Chapter 9, the selection of an appropriate MARR for evaluating an investment project is a critical issue in the private sector. In public-project analyses, we also need to select an interest rate, called the **social discount rate**, in order to determine equivalent benefits as well as the equivalent costs. Selection of the social discount rate in public-project evaluation is as critical as selection of a MARR in the private sector.

Since present-worth calculations were initiated to evaluate public water resources and related land-use projects in the 1930s, there is a persisting tendency to use relatively low rates of discount as compared with those existing in markets for private assets. During the 1950s and into the 1960s, the rate for water resource projects was 2.63%, which, for much of this period, was even below the yield on long-term government securities. The persistent use of a lower interest rate for water resource projects is a political issue.

In recent years, with the growing interest in performance budgeting and systems analysis that started in the 1960s, government agencies have begun to examine the appropriateness of the discount rate in the public sector in relation to the efficient allocation of resources in the economic system as a whole. Two views of the basis for determining the social discount rate prevail:

1. **Projects without private counterparts:** *The social discount rate should reflect only the prevailing government borrowing rate.* Projects such as dams designed purely for flood control, access roads for noncommercial uses, and reservoirs for community water supply may not have corresponding private counterparts. In such areas of government activity where benefit–cost analysis has

been employed in evaluation, the rate of discount traditionally used has been the cost of government borrowing. In fact, water resource project evaluations follow this view exclusively.

2. **Projects with private counterparts:** *The social discount rate should represent the rate that could have been earned had the funds not been removed from the private sector.* If all public projects were financed by borrowing at the expense of private investment, we may have focused on the opportunity cost of capital in alternative investments in the private sector in order to determine the social discount rate. So in the case of public capital projects similar to projects in the private sector that produce a commodity or a service to be sold on the market (such as electric power) the discount rate employed is the average cost of capital as discussed in Chapter 9. The reasons for using the private rate of return as the opportunity cost of capital for projects similar to those in the private sector are (1) to prevent the public sector from transferring capital from higher yielding to lower yielding investments, and (2) to force public-project evaluators to employ market standards in justifying projects.

The Office of Management and Budget (OMB) holds the second view. Since 1972, the OMB has required that a social discount rate of 10% be used to evaluate federal public projects. Exceptions include water resource projects.

12-2   Benefit–Cost Analysis

An alternative way of expressing the worthiness of a public project is to compare the user's benefits (B) with the sponsor's costs (C) by taking the ratio B/C. In this section, we shall define the benefit–cost (B/C) ratio and explain the relationship between the conventional PW criterion and the B/C ratio.

12.2.1   Definition of Benefit–Cost Ratio

For a given benefit–cost profile, let B and C be the present worths of benefits and costs defined by

$$B = \sum_{n=0}^{N} b_n(1 + i)^{-n} \tag{12.1}$$

$$C = \sum_{n=0}^{N} c_n(1 + i)^{-n}, \tag{12.2}$$

where

b_n = benefit at the end of period n
c_n = expense at the end of period n
N = project life, and
i = sponsor's interest rate (discount rate).

The sponsor's costs (C) consist of the capital expenditure (I) and the annual operating costs (C') accrued in each successive period. (Note the sign convention we use in calculating a benefit–cost ratio: Since we are using a ratio, all benefit and cost flows are expressed in positive units. Recall that in previous equivalent-worth calculations, our sign convention was to explicitly assign "+" for cash inflows and "−" for cash outflows.) Let's assume that a series of initial investments is required during the first K periods, while annual operating and maintenance costs accrue in each following period. Then the equivalent present worth for each component is

$$I = \sum_{n=0}^{K} c_n (1 + i)^{-n} \tag{12.3}$$

and

$$C' = \sum_{n=K+1}^{N} c_n (1 + i)^{-n}, \tag{12.4}$$

and $C = I + C'$.

The B/C ratio[2] is defined as

$$BC(i) = \frac{B}{C} = \frac{B}{I + C'}, \quad \text{where } I + C' > 0. \tag{12.5}$$

If we are to accept a project, $BC(i)$ must be greater than one. Note that the acceptance rule by the B/C-ratio criterion is equivalent to that for the PW criterion, as illustrated in Figure 12.1. Note also that we must express the values of B, C', and I in present-worth equivalents. Alternatively, we can compute these values in terms of annual equivalents and use them in calculating the B/C ratio. The resulting B/C ratio is not affected.

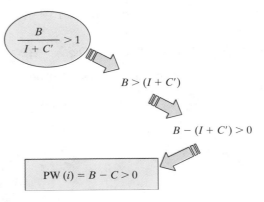

FIGURE 12.1 Relationship between B/C ratio and PW criterion

---

[2]An alternative measure, called the **net B/C ratio**, $B'C(i)$, considers only the initial capital expenditure as cash outlay, and annual net benefits are used:

$$B'C(i) = \frac{B - C}{I}, \quad I > 0.$$

The decision rule has not changed—the ratio must still be greater than one.

Example 12.1 Benefit–Cost Ratio

A public project being considered by a local government has the following estimated benefit–cost profile (see Figure 12.2).

Assume that $i = 10\%$, $N = 5$, and $K = 1$. Compute B, C, I, C', and $BC(10\%)$.

n	b_n	c_n	A_n
0		$10	−$10
1		$10	−$10
2	$20	$5	$15
3	$30	$5	$25
4	$30	$8	$22
5	$20	$8	$12

SOLUTION

We calculate B as follows:

$$B = \$20(P/F, 10\%, 2) + \$30(P/F, 10\%, 3)$$
$$+ \$30(P/F, 10\%, 4) + \$20(P/F, 10\%, 5)$$
$$= \$71.98.$$

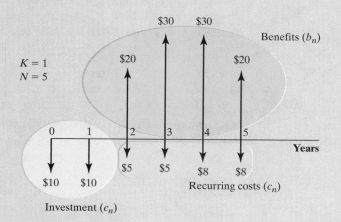

FIGURE 12.2 Classification of a project's cash flow elements into benefits and costs

We calculate C as follows:

$$C = \$10 + \$10(P/F, 10\%, 1) + \$5(P/F, 10\%, 2) + \$5(P/F, 10\%, 3)$$
$$+ \$8(P/F, 10\%, 4) + \$8(P/F, 10\%, 5)$$
$$= \$37.41.$$

We calculate I as follows:

$$I = \$10 + \$10(P/F, 10\%, 1)$$
$$= \$19.09.$$

We calculate C' as follows:

$$C' = C - I$$
$$= \$18.32.$$

Using Eq. (12.5), we can compute the B/C ratio as

$$BC(10\%) = \frac{71.98}{\$19.09 + \$18.32}$$
$$= 1.92 > 1.$$

The B/C ratio exceeds one, so the user's benefits exceed the sponsor's costs.

Comments: Since governments do not tax, depreciation and income taxes are not issues in B–C analysis.

12.2.2 Incremental B/C-Ratio Analysis

Let us now consider how we choose among mutually exclusive public projects. As we explained in Chapter 8, we must use the incremental-investment approach in comparing alternatives based on any relative measure, such as IRR or B/C.

Incremental Analysis Based on BC(i)   To apply incremental analysis, we compute the incremental differences for each term (B, I, and C') and take the B/C ratio based on these differences. To use $BC(i)$ on incremental investment, we may proceed as follows:

1. Eliminate any alternatives with a B/C ratio less than one.
2. Arrange the remaining alternatives in increasing order of the denominator $(I + C')$. Thus, the alternative with the smallest denominator should be the first (j), the alternative with the second smallest (k) should be the second, and so forth.
3. Compute the incremental differences for each term (B, I, and C') for the paired alternatives (j, k) in the list:

$$\Delta B = B_k - B_j;$$
$$\Delta I = I_k - I_j;$$
$$\Delta C' = C'_k - C'_j.$$

4. Compute $BC(i)$ on incremental investment by evaluating

$$BC(i)_{k-j} = \frac{\Delta B}{\Delta I + \Delta C'}.$$

If $BC(i)_{k-j} > 1$, select the k alternative. Otherwise, select the j alternative.

5. Compare the selected alternative with the next one on the list by computing the incremental benefit–cost ratio. Continue the process until you reach the bottom of the list. The alternative selected during the last pairing is the best one.

We may modify the decision procedures when we encounter the following situations:

- If $\Delta I + \Delta C' = 0$, we cannot use the benefit–cost ratio, because this relationship implies that both alternatives require the same initial investment and operating expenditures. In such cases, we simply select the alternative with the largest B value.

- In situations where public projects with unequal service lives are to be compared, but the alternative can be repeated, we may compute all component values (B, C', and I) on an annual basis and use them in incremental analysis.

Example 12.2 Incremental Benefit–Cost Ratios

Consider three investment projects: $A1$, $A2$, and $A3$. Each project has the same service life, and the present worth of each component value (B, I, and C') is computed at 10% as follows:

	Projects		
	A1	**A2**	**A3**
I	$5,000	$20,000	$14,000
B	$12,000	$35,000	$21,000
C'	$4,000	$8,000	$1,000
PW(i)	$3,000	$7,000	$6,000

(a) If all three projects are independent, which projects would be selected, based on $BC(i)$?

(b) If the three projects are mutually exclusive, which project would be the best alternative? Show the sequence of calculations that would be required in order to produce the correct results. Use the *B/C* ratio on incremental investment.

SOLUTION

(a) Since $PW(i)_1$, $PW(i)_2$, and $PW(i)_3$ are positive, all of the projects would be acceptable if they were independent. Also, the $BC(i)$ value for each project is greater than one, so the use of the benefit–cost ratio criterion leads to the same accept–reject conclusion as under the PW criterion:

	A1	A2	A3
$BC(i)$	1.33	1.25	1.40

(b) If these projects are mutually exclusive, we must use the principle of incremental analysis. If we attempt to rank the projects according to the size of the *B/C* ratio, obviously we will observe a different project preference. For example, if we use $BC(i)$ on the total investment, we see that *A3* appears to be the most desirable and *A2* the least desirable project; however, selecting mutually exclusive projects on the basis of *B/C* ratios is incorrect. Certainly, with $PW(i)_2 > PW(i)_3 > PW(i)_1$, project *A2* would be selected under the PW criterion. By computing the incremental *B/C* ratios, we will select a project that is consistent with the PW criterion.

We will first arrange the projects by increasing order of their denominator $(I + C')$ for the $BC(i)$ criterion:

Ranking Base	A1	A3	A2
$I + C'$	$9,000	$15,000	$28,000

We now compare the projects incrementally as follows:

- *A*1 versus *A*3: With the do-nothing alternative, we first drop from consideration any project that has a *B/C* ratio smaller than one. In our example, the *B/C* ratios of all three projects exceed one, so the first incremental comparison is between *A*1 and *A*3:

$$BC(i)_{3-1} = \frac{\$21,000 - \$12,000}{(\$14,000 - \$5,000) + (\$1,000 - \$4,000)}$$
$$= 1.51.$$

Since the ratio is greater than one, we prefer *A*3 over *A*1. Therefore, *A*3 becomes the "current best" alternative.

- *A*3 versus *A*2: Next we must determine whether the incremental benefits to be realized from *A*2 would justify the additional expenditure. Therefore, we need to compare *A*2 and *A*3 as follows:

$$BC(i)_{2-3} = \frac{\$35,000 - \$21,000}{(\$20,000 - \$14,000) + (\$8,000 - \$1,000)}$$
$$= 1.081.$$

The incremental *B/C* ratio again exceeds one, and therefore we prefer *A*2 over *A*3. With no further projects to consider, *A*2 becomes the ultimate choice.

Summary

- **Benefit–cost analysis** is commonly used to evaluate public projects. Several facets unique to public-project analysis are neatly addressed by benefit–cost analysis:

  1. Benefits of a nonmonetary nature can be quantified and factored into the analysis.
  2. A broad range of project users distinct from the sponsor are considered; benefits and disbenefits to *all* these users can (and should) be taken into account.

- Difficulties involved in public-project analysis include the following:

  1. identifying all the users of the project;
  2. identifying all the benefits and disbenefits of the project;
  3. quantifying all the benefits and disbenefits in dollars or some other unit of measure;
  4. selecting an appropriate interest rate at which to discount benefits and costs to a present value.

- The *B/C* ratio is defined as

$$BC(i) = \frac{B}{C} = \frac{B}{I + C'}, \text{ where } I + C' > 0.$$

The decision rule is that, if $BC(i) \geq 1$, the project is acceptable.

Problems

Valuation of Benefits and Costs

12.1 The state of Michigan is considering a bill that would ban the use of road salt on highways and bridges during icy conditions. Road salt is known to be toxic, costly, corrosive, and caustic. Chevron Chemical Company produces a calcium magnesium acetate deicer (CMA), an alternative to road salt, and sells it for $600 a ton as Ice-B-Gon. Road salts, on the other hand, sold for an average of $14 a ton in 2002. Michigan needs about 600,000 tons of road salt each year. (Michigan spent $9.2 million on road salt in 2002.) Chevron estimates that each ton of salt on the road costs $650 in highway corrosion, $525 in rust on vehicles, $150 in corrosion of utility lines, and $100 in damages to water supplies, for a total of $1,425. Salt damage of an unknown cost to vegetation and soil surrounding areas of highways has also occurred. The state of Michigan will ban the use of road salt (at least on expensive steel bridges or near sensitive lakes) if state studies support Chevron's cost claims.

(a) What would be the users' benefits and sponsor's costs if a complete ban on road salt were imposed in Michigan?

(b) How would you go about determining the salt damages (in dollars) to vegetation and soil?

12.2 The Electric Department of the city of Tallahassee, Florida, operates generating and transmission facilities serving approximately 140,000 people in the city and surrounding Leon County. The city has proposed construction of a $300 million 235 MW circulating fluidized bed combustor (CFBC) at Arvah B. Hopkins Station to power a turbine generator currently receiving steam from an existing boiler fueled by gas or oil. Among the advantages associated with the use of CFBC systems are the following:

- A variety of fuels can be burned, including inexpensive low-grade fuels with high ash and high sulfur content.

- The relatively low combustion temperatures inhibit the formation of nitrogen oxides. Acid-gas emissions associated with CFBC units would be expected to be significantly lower than emissions from conventional coal-fueled units.

- The sulfur-removal method, low combustion temperatures, and high-combustion efficiency

characteristic of CFBC units result in solid wastes that are physically and chemically more amenable to land disposal than the solid wastes resulting from conventional coal-burning boilers equipped with flue-gas desulfurization equipment.

Based on the Department of Energy's (DOE) projections of growth and expected market penetration, demonstration of a 235 MW unit could lead to as much as 41,000 MW of CFBC generators being constructed by the year 2010. The proposed project would reduce the city's dependency on oil and gas fuels by converting its largest generating unit to coal-fuel capability. Consequently, substantial reductions of local acid-gas emissions could be realized in comparison with the permitted emissions associated with oil fuel. The city has requested a $50 million cost share from the DOE for the project. Cost sharing under the Clean Coal Technology Program is considered attractive because the DOE cost share would largely offset the risk of using such a new technology. To qualify for the cost-sharing money, the city has to address the following questions for the DOE:

(a) What is the significance of the project at local and national levels?

(b) What items would constitute the users' benefits and disbenefits associated with the project?

(c) What items would constitute the sponsor's costs?

Put yourself in the city engineer's position and respond to these questions.

Benefits and Cost Analyses

12.3 A city government is considering two types of town-dump sanitary systems. Design *A* requires an initial outlay of $400,000, with annual operating and maintenance costs of $50,000 for the next 15 years; design *B* calls for an investment of $300,000, with annual operating and maintenance costs of $80,000 per year for the next 15 years. Fee collections from the residents would be $85,000 per year. The interest rate is 8%, and no salvage value is associated with either system.

(a) Using the benefit–cost ratio ($BC(i)$), which system should be selected?

(b) If a new design (design C), which requires an initial outlay of $350,000 and annual operating and maintenance costs of $65,000, is proposed, would your answer to (a) change?

12.4 The U.S. Government is considering building apartments for government employees working in a foreign country and currently living in locally owned housing. A comparison of two possible buildings indicates the following:

	Building X	Building Y
Original investment by government agencies	$8,000,000	$12,000,000
Estimated annual maintenance costs	$240,000	$180,000
Savings in annual rent now being paid to house employees	$1,960,000	$1,320,000

Assume the salvage or sale value of the apartments to be 60% of the first investment. Use 10% and a 20-year study period to compute the B/C ratio on incremental investment, and make a recommendation as to the best option. (Assume no do-nothing alternative.)

12.5 Three public investment alternatives are available: $A1$, $A2$, and $A3$. Their respective total benefits, costs, and first costs are given in present worth as follows:

Present Worth	Proposals		
	A1	A2	A3
I	$100	$300	$200
B	$400	$700	$500
C'	$100	$200	$150

These alternatives have the same service life. Assuming no do-nothing alternative, which project would you select, based on the benefit–cost ratio on incremental investment ($BC(i)$)?

12.6 A city that operates automobile parking facilities is evaluating a proposal to erect and operate a structure for parking in its downtown area. Three designs for a facility to be built on available sites have been identified as follows (all dollar figures are in thousands):

	Design A	Design B	Design C
Cost of site	$240	$180	$200
Cost of building	$2,200	$700	$1,400
Annual fee collection	$830	$750	$600
Annual maintenance cost	$410	$360	$310
Service life	30 years	30 years	30 years

At the end of the estimated service life, the selected facility would be torn down, and the land would be sold. It is estimated that the proceeds from the resale of the land will be equal to the cost of clearing the site. If the city's interest rate is known to be 10%, which design alternative would be selected, based on the benefit–cost criterion?

12.7 The federal government is planning a hydroelectric project for a river basin. In addition to producing electric power, this project will provide flood control, irrigation, and recreational benefits. The estimated benefits and costs expected to be derived from the three alternatives under consideration are listed as follows:

	Decision Alternatives		
	A	**B**	**C**
Initial cost	$8,000,000	$10,000,000	$15,000,000
Annual benefits:			
Power sales	$1,000,000	$1,200,000	$1,800,000
Flood-control savings	$250,000	$350,000	$500,000
Irrigation benefits	$350,000	$450,000	$600,000
Recreation benefits	$100,000	$200,000	$350,000
O&M costs	$200,000	$250,000	$350,000

The interest rate is 10%, and the life of each project is estimated to be 50 years.

(a) Find the benefit–cost ratio for each alternative.

(b) Select the best alternative, based on $BC(i)$.

12.8 Two different routes are under consideration for a new interstate highway:

	Length of Highway	First Cost	Annual Upkeep Cost
The "long" route	22 miles	$21 million	$140,000
Transmountain shortcut	10 miles	$45 million	$165,000

For either route, the volume of traffic will be 400,000 cars per year. These cars are assumed to operate at $0.25 per mile. Assuming a 40-year life for each road and an interest rate of 10%, determine which route should be selected.

12.9 The government is considering undertaking one out of the four projects $A1$, $A2$, $A3$, and $A4$. These projects are mutually exclusive, and the estimated present worth of their costs and of their benefits are shown in millions of dollars as follows:

Projects	PW of Benefits	PW of Costs
$A1$	$40	$85
$A2$	$150	$110
$A3$	$70	$25
$A4$	$120	$73

All of the projects have the same duration. Assuming no do-nothing alternative, which alternative would you select? Justify your choice by using a benefit–cost ratio on incremental investment. $(BC(i))$.

Short Case Studies

12.10 The City of Portland Sanitation Department is responsible for the collection and disposal of all solid waste within the city limits. The city must collect and dispose of an average of 300 tons of garbage each day. The city is considering ways to improve the current solid-waste collection and disposal system:

- The present collection and disposal system uses Dempster Dumpmaster Frontend Loaders for collection and disposal. Each collecting vehicle has a load capacity of 10 tons, or 24 cubic yards, and dumping is automatic. The incinerator in use was manufactured in 1942. It was designed to incinerate 150 tons every 24 hours. A natural-gas afterburner has been added in an effort to reduce air pollution; however, the incinerator still does not meet state air-pollution requirements. Prison-farm labor is used for the operation of the incinerator. Because the capacity of the incinerator is relatively low, some trash is not incinerated, but is taken to the city landfill instead. The trash landfill is located approximately 11 miles, and the incinerator approximately five, miles, from the center of the city. The mileage and costs in person-hours for delivery to the disposal sites is excessive; a high percentage of empty vehicle miles and person-hours are required because separate methods of disposal are used and the destination sites are remote from the collection areas. The operating cost for the present system is $905,400. This figure includes $624,635 to operate the prison-farm incinerator, $222,928 to operate the existing landfill, and $57,837 to maintain the current incinerator.

• The proposed system calls for a number of portable incinerators, each with 100-ton-per-day capacity for the collection and disposal of refuse waste from three designated areas within the city. Collection vehicles will also be staged at these incineration–disposal sites, with the necessary plant and support facilities for incinerator operation, collection-vehicle fueling and washing, and support building for stores, as well as shower and locker rooms for collection and site crew personnel. The pickup and collection procedure remains essentially the same as in the existing system. The disposal-staging sites, however, are located strategically in the city, based on the volume and location of wastes collected, thus eliminating long hauls and reducing the number of miles the collection vehicles must retravel from pickup to disposal site.

Four variations of the proposed system are being considered, containing one, two, three, and four incinerator-staging areas, respectively. The type of incinerator used in all four variations is a modular prepackaged unit, which can be installed at several sites in the city. Such units more than meet all state and federal standards on their exhaust emissions. The city of Portland needs 24 units, each with a rated capacity of 12.5 tons of garbage per 24 hours. The price per unit is $137,600, which means a capital investment of about $3,302,000. The estimated

plant facilities, such as housing and foundation, are estimated to cost $200,000 per facility. This figure is based on a plan incorporating four incinerator plants strategically located around the city. Each plant would house 8 units and be capable of handling 100 tons of garbage per day. Additional plant features, such as landscaping, are estimated to cost $60,000 per plant.

The annual operating cost of the proposed system would vary according to the type of system configuration. It takes about 1.5 to 1.7 MCF of fuel to incinerate 1 ton of garbage. The conservative 1.7 MCF figure was used for total cost, resulting in a fuel cost of $4.25 per ton of garbage at a cost of $2.50 per MCF. Electric requirements at each plant will be 230kW per day. For a plant operating at full capacity, that requirement means a $0.48-per-ton cost for electricity. Two men can easily operate one plant, but safety factors dictate that there be three operators, at a cost of $7.14 per hour per operator. This translates to a cost of $1.72 per ton. The maintenance cost of each plant is estimated to be $1.19 per ton. Since three plants will require fewer transportation miles, it is necessary to consider the savings accruing from this operating advantage. For example, the configuration with three plant locations will save 6.14 miles per truck per day, on average. At an estimated $0.30-per-mile cost, this would mean that an annual savings of $15,300 is realized when considering minimum trips to the landfill disposer. Labor savings are also realized because of the shorter routes, which permit more pickups during the day. This results in an annual labor savings of $103,500 for three incinerators. The table at the top of the next page summarizes all costs, in thousands of dollars, associated with the present and proposed systems.

A bond will be issued to provide the necessary capital investment at an interest rate of 8%, with a maturity date 20 years in the future.

Costs for Proposed Systems (Unit-thousand)

Item	Present System	Number of Incinerator Sites			
		1	2	3	4
Capital costs					
Incinerators		$3,302	$3,302	$3,302	$3,302
Plant facilities		$600	$900	$1,260	$1,920
Annex buildings		$91	$102	$112	$132
Additional features		$60	$80	$90	$100
Total		$4,053	$4,384	$4,764	$5,454
Annual O&M costs	$905.4	$342	$480	$414	$408
Annual savings					
Pickup transportation		$13.2	$14.7	$15.3	$17.1
Labor		$87.6	$99.3	$103.5	$119.40

The proposed systems are expected to last 20 years, with negligible salvage values. If the current system is to be retained, the annual O&M costs would be expected to increase at an annual rate of 10%. The city will use the bond interest rate as the interest rate for any public-project evaluation.

(a) Determine the operating cost of the current system in terms of dollars per ton of solid waste.

(b) Determine the economics of each solid-waste disposal alternative in terms of dollars per ton of solid waste.

12.11 Because of a rapid growth in population, a small town in Pennsylvania is considering several options to establish a wastewater treatment facility that can handle a wastewater flow of 2MGD (million gallons per day). The town has five options:

- Option 1—No action: This option will lead to continued deterioration of the environment. If growth continues and pollution results, fines imposed (as high as $10,000 per day) would soon exceed construction costs for a new facility.

- Option 2—Land-treatment facility: Provide a system for land treatment of wastewater to be generated over the next 20 years. Out of the five available options this option will require the most land use—800 acres—for treatment of the wastewater. In addition to finding a suitable site, wastewater will have to be pumped a considerable distance out of town will be required. The land cost in the area is $3,000 per acre. The system will use spray irrigation to distribute wastewater over the site. No more than one inch of wastewater can be applied in one week per acre.

- Option 3—Activated sludge-treatment facility: Provide an activated sludge-treatment facility at a site near the planning area. No pumping will be required for this alternative. Only seven acres of land will be needed for construction of the plant, at a cost of $7,000 per acre.

- Option 4—Trickling filter-treatment facility: Provide a trickling filter-treatment facility at the same site selected for the activated sludge-treatment plant of Option 3. The land required will be the same as used for Option 3. Both facilities will provide similar levels of treatment, using different units.

- Option 5—Lagoon treatment system: Utilize a three-cell lagoon system for treatment. The lagoon system requires substantially more land than Options 3 and 4, but less than Option 2. Due to the larger land requirement, this treatment system will have to be located some distance outside of the planning area and will require pumping of the wastewater to reach the site.

The land costs for Options 2–5 are summarized as follows:

Land Cost for Each Option			
Option Number	Land Required (Acres)	Land Cost	Land Value (in 20 Years)
2	800	$2,400,000	$4,334,600
3	7	$49,000	$88,500
4	7	$49,000	$88,500
5	80	$400,000	$722,400

The following tables summarize the capital expenditures and O&M costs, respectively, associated with Options 2–5:

Option Number	Capital Expenditures			
	Equipment	Structure	Pumping	Total
2	$500,000	$700,000	$100,000	$1,300,000
3	$500,000	$2,100,000	$0	$2,600,000
4	$400,000	$2,463,000	$0	$2,863,000
5	$175,000	$1,750,000	$100,000	$2,025,000

Option Number	Annual O&M Costs			
	Energy	Labor	Repairs	Total
2	$200,000	$95,000	$30,000	$325,000
3	$125,000	$65,000	$20,000	$210,000
4	$100,000	$53,000	$15,000	$168,000
5	$50,000	$37,000	$5,000	$92,000

The price of land is assumed to be appreciating at an annual rate of 3%.

The equipment installed for Options 2–5 will require a replacement cycle of 15 years. Its replacement cost will increase at an annual rate of 5% (over the initial cost), and its salvage value at the end of the planning horizon will be 50% of the replacement cost. The entire structure requires replacement after 40 years and will have a salvage value of 60% of the original cost.

The costs of energy and repair will increase at an annual rate of 5% and 2%, respectively. The labor cost will increase at an annual rate of 4%.

(a) If the interest rate (including inflation) is 10%, which option is the most cost effective?

(b) Suppose that a household discharges about 400 gallons of wastewater per day through the facility selected in (a). What should be the monthly assessed bill for this household, assuming the following?

- The analysis period is 120 years.
- Replacement costs for the equipment as well as for the pumping facilities will increase at an annual rate of 5%.
- The replacement cost for the structure will remain constant over the planning period. However, the structure's salvage value will always be 60% of the original cost. (Because the structure has a 40-year replacement cycle, any increase in the future replacement cost will have very little impact on the solution.)
- The equipment's salvage value at the end of its useful life will be 50% of the original cost. For example, the equipment installed for Option 1 will cost $500,000, so its salvage value at the end of 15 years will be $250,000.

- All O&M cost figures are given in today's dollars. For example, the annual energy cost of $20,000 for Option 2 means that the actual energy cost during the first operating year will be $200,000(1.05) = $210,000.

- Option 1 is not considered a viable alternative, as its annual operating cost exceeds $36,500,00.

12.12 The Federal Highway Administration predicts that by the year 2005, Americans will be spending 8.1 billion hours per year in traffic jams. Most traffic experts believe that adding and enlarging highway systems will not alleviate the problem. As a result, current research on traffic management is focusing on three areas: (1) development of computerized dashboard navigational systems, (2) development of roadside sensors and signals that monitor and help manage the flow of traffic, and (3) development of automated steering and speed controls that might allow cars to drive themselves on certain stretches of highway.

In Los Angeles, perhaps the most traffic-congested city in the United States, a Texas Transportation Institute study found that traffic delays cost motorists $8 billion per year. But Los Angeles has already implemented a system of computerized traffic-signal controls that, by some estimates, has reduced travel time by 13.2%, fuel consumption by 12.5%, and pollution by 10%. And between Santa Monica and downtown Los Angeles, testing of an electronic traffic and navigational system—including highway sensors and cars with computerized dashboard maps—is being sponsored by federal, state, and local governments and General Motors Corporation. This test program costs $40 million; to install it throughout Los Angeles could cost $2 billion.

On a national scale, the estimates for implementing "smart" roads and vehicles are even more staggering: It would cost $18 billion to build the highways, $4 billion per year to maintain and operate them, $1 billion for research and development of driver-information aids, and $2.5 billion for vehicle-control devices.

However, advocates say that the rewards far outweigh the costs.

(a) On a national scale, how would you identify the users' benefits and disbenefits for this type of public project?

(b) On a national scale, what would be the sponsor's cost?

(c) Suppose that the users' net benefits grow at 3% per year and the sponsor's costs grow at 4% per year. Assuming a social discount rate of 10%, what would be the B/C ratio over a 20-year study period?

12.13 Atlanta Regional Commission is considering expanding Hartsfield (Atlanta) International Airport. Three options have been proposed to handle the growth expected at Hartsfield:

- Option 1: Build five new runways, a new concourse adjacent to a new eastside terminal, and close-in parking. Cost estimate: $1.5 billion.

- Option 2: Construct a more expensive version of Option 1, putting the new concourse adjacent to the fifth runway. Cost estimate: $1.6 billion.

- Option 3: In addition to the five new runways in Options 1 and 2, build a sixth runway north of the current airport boundaries in the College-Park–East Point–Hapeville corridor side terminal. Cost estimate: $5.9 billion.

A study indicates that a north runway (e.g., the sixth runway in Option 3) would displace more than 3,000 households and 978 businesses. A south runway (Option 2) could dislocate as many as 7,000 households and more than 900 businesses. Under any scenario of airport expansion, there still will be some flight delays. However, some plans project longer delays than others. The following table summarizes some projections for the year 2015 under the no-build scenario and the accompany figure illustrates the three options under consideration:

Dealing with the Delays (Units in Minutes)				
	No Change	Option 1	Option 2	Option 3
Average departure delay	34.80	17.02	19.66	5.71
Average arrival delay	19.97	7.08	6.68	3.91
Average taxiing time	26.04	18.54	19.99	12.80

The committee has six months to study each option before making a final recommendation. What would you recommend?

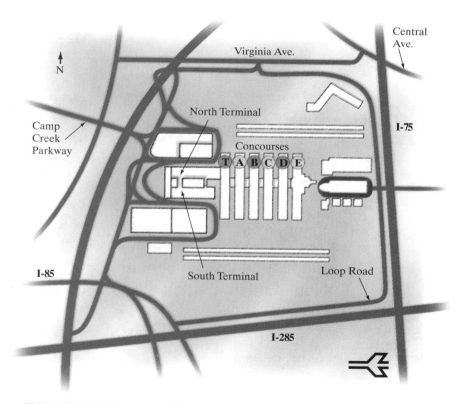

FIGURE 12.3 Proposed future expansion plan for Atlanta Hartsfield Airport

.13

In contrast with Enron's use of complex off-balance-sheet partnerships to hide debt, WorldCom used a very simple accounting shift to reduce its expenses and improve the company's bottom line (net income).

WorldCom's Hidden Loss[1]

The Accounting Principle

The costs of a company's operations—salaries, materials, and the like—are supposed to be treated on a company's books as expenses in the year they are incurred. But, as we have seen, the purchase prices for certain long-lasting, big-ticket items—like buildings or heavy machinery—are treated differently. Rather than forcing companies to recognize such investments all in one year, accounting rules effectively allow them to spread the cost over the years in which the items will be used. The following figure illustrates these two methods of accounting for a simplistic example of a $7 item:

And WorldCom's Numbers

In 2001, WorldCom accounted for more than $3 billion in "line costs" fees it paid to other communications companies to use their networks—as capital expenditures. That way, rather than expensing the entire $3 billion in 2001, the company could spread the costs over a longer period—as long as 40 years. The immediate effect:

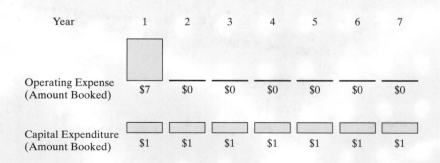

Year	1	2	3	4	5	6	7
Operating Expense (Amount Booked)	$7	$0	$0	$0	$0	$0	$0
Capital Expenditure (Amount Booked)	$1	$1	$1	$1	$1	$1	$1

[1]Source: The New York Times, July 22, 2002.

Understanding Financial Statements

WorldCom showed a profit instead of the loss of about $662 million that it really incurred. The company now says it should have accounted for the costs as operating expenses. The company also shifted nearly $800 million in the first quarter of 2002. The effect of the line-cost shift is shown in the following figure:

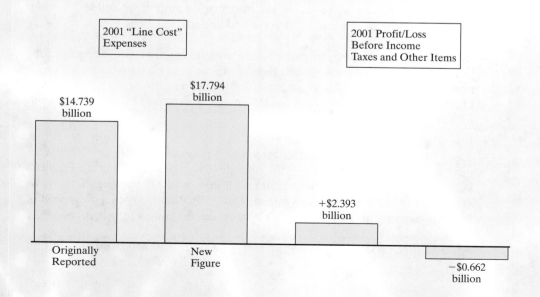

2001 "Line Cost" Expenses

2001 Profit/Loss Before Income Taxes and Other Items

$14.739 billion

$17.794 billion

+$2.393 billion

−$0.662 billion

Originally Reported

New Figure

What are the consequences of these accounting irregularities (fraud)? As the U.S. Securities and Exchange Commission's (SEC) investigation intensified, investors dumped WorldCom's stocks in a hurry, resulting in a significant drop in the company's market value. Eventually, on July 21, 2002, WorldCom had to file for relief under Chapter 11 of the Bankruptcy Code, as its creditors had withdrawn their lines of credit.

Such examples of fraud aside, changes in accounting figures can have an enormous impact on a firm's position. Thus, it is paramount that engineers have a basic understanding of accounting. While knowledge of PW criterion, AE analysis, and the other topics we have covered are essential, understanding how a project impacts the firm's bottom line is also important. In this chapter, we will begin by discussing the characteristics of these financial statements and the factors that comprise each. Our purpose is not to present bookkeeping aspects of accounting, but to acquaint you with financial statements and to give you the basic information you need to make sound engineering economic decisions.

13-1 Accounting: The Basis of Decision Making

We need financial information when we are making business decisions. Virtually all businesses and most individuals keep accounting records to aid them in making decisions. As illustrated in Figure 13.1, accounting is the information system that measures business activities, processes that information into reports, and communicates the results to decision makers. For this reason, we call accounting "the language of business." The better you understand this language, the better you can manage your financial well-being, and the better your financial decisions will be.

Personal financial planning, education expenses, loans, car payments, income taxes, and investments are all based on the information system we call accounting. The use of accounting information is diverse and varied:

- Business managers use accounting information to set goals for their organizations, to evaluate progress toward those goals, and to take corrective actions if necessary. Decisions based on accounting information may include which building or equipment to purchase, how much merchandise inventory to keep on hand, and how much cash to borrow.

- Investors and creditors provide the money a business needs to begin operations. To decide whether to help start a new venture, potential investors evaluate what income they can expect on their investment. This means analyzing the financial statements of the business. Before making a loan, banks determine the borrower's ability to meet scheduled payments. This evaluation includes a projection of future operations and revenue, which is based on accounting information.

An essential product of accounting is a series of financial statements that allow people to make informed decisions. For business use, financial statements are

People make decisions

Business transactions occur

Accounting data for future decisions

Post audit

Company prepares reports to show the results of their operations

FIGURE 13.1 **An illustration of the flow of information in the accounting system**

the documents that report financial information about a business entity to decision makers. They tell us how a business is performing and where it stands financially. These financial statements include the balance sheet, income statement, and statement of cash flows.

13–2 Financial Status for Businesses

All businesses must prepare their financial status. Of the various reports corporations issue to their stockholders, the annual report is by far the most important. The annual report contains basic financial statements as well as management's opinion of the past year's operations and the firm's future prospects. What would managers and investors want to know about a company at the end of the fiscal year (or another fiscal period, such as a quarter)? Managers or investors are likely to ask the following four basic questions:

- What is the company's financial position at the end of the fiscal period?
- How much profit did the company make during the fiscal period?
- How did the company decide to use its profits?
- How much cash did the company generate and spend during the period?

As illustrated in Figure 13.2, the answer to each question is provided by one of the financial statements. The fiscal year (or operating cycle) can be any 12-month term, but is usually January 1 through December 31 of a calendar year.

As mentioned in Section 1.2.1, one of the primary responsibilities of engineers in business is to plan for the acquisition of equipment (capital expenditure) that will enable the firm to design and produce products economically. This task requires estimation of savings and costs associated with the equipment acquisition and the degree of risk associated with project execution. These amounts affect the business' bottom line (profitability), which eventually affects the firm's stock price in the marketplace, as illustrated in Figure 13.3. Therefore, engineers should understand the meanings of various financial statements in order to communicate with upper management the nature of a project's profitability.

For illustration purposes, we use data taken from Dell Computer Corporation, a manufacturer of a wide range of computer systems, including desktops, notebooks,

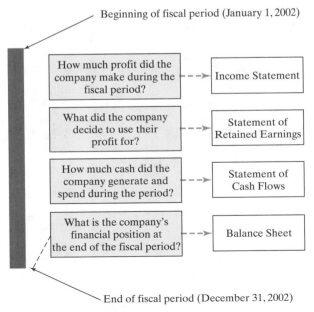

FIGURE 13.2 Information reported on a company's financial statements

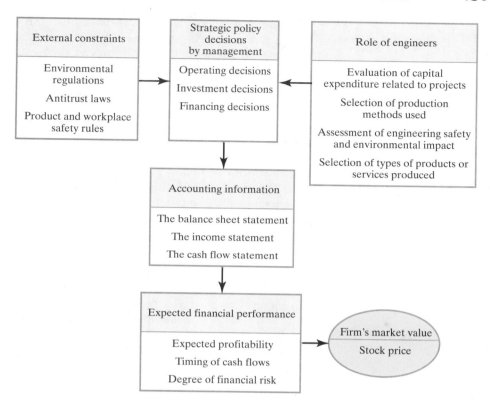

FIGURE 13.3 Summary of major factors affecting stock price

and workstations, to discuss the nature of basic financial statements. In 1984, Michael Dell began his computer business at the University of Texas–Austin, often hiding his IBM PC in his roommate's bathtub when his family came to visit. His dorm-room business officially became Dell Computer Corporation on May 3, 1984. In 2003, Dell became number one in PC sales and the fastest growing firm among all major computer-systems companies worldwide, with 20,800 employees around the globe. Dell's pioneering "direct model" is a simple concept of selling personal computer systems directly to customers. It offers in-person relationships with corporate and institutional customers; telephone and Internet purchasing (the latter now averaging $4.5 million a day); build-to-order computer systems; phone and on-line technical support; and next-day, on-site product service.

The company's revenue for the last four quarters of 2003 totaled $35.40 billion. The increase in net revenues for both fiscal year 2003 and fiscal year 2002 was principally driven by an increase in the number of units sold. Unit shipments grew 21% for fiscal year 2003. In its 2003 annual report, management painted an optimistic picture for the future, stating that Dell will continue to invest in information systems, research, development, and engineering activities; to support its growth; and to provide for new competitive products. Of course, there is no assurance that Dell's revenue will continue to grow at the current annual rate of 14% in the future.

What can individual investors make of all this? Actually, they can make quite a bit. As you will see, investors use the information contained in an annual report to form expectations about future earnings and dividends. Therefore, the annual report is certainly of great interest to investors.

13.2.1 The Balance Sheet

What is a company's financial position at the end of a reporting period? We find the answer to this question in the company's **balance sheet statement**. A company's balance sheet, sometimes called its **statement of financial position**, reports three main categories of items: assets, liabilities, and stockholders' equity. As shown in Table 13.1, the first half of Dell's year-end 2003 and 2002 balance sheets lists the firm's assets, while the remaining portion shows the liabilities and equity, or claims against these assets. The financial statements are based on the most basic tool of accounting, the **accounting equation**. The accounting equation shows the relationship among assets, liabilities, and owners' equity:

$$\text{Assets} = \text{Liabilities} + \text{Stockholders' Equity}.$$

Figure 13.4 illustrates the relationship between assets and liabilities including equity and how these items appear in the balance sheet.

Every business transaction, no matter how simple or how complex, can be expressed in terms of its effect on the accounting equation. Regardless of whether a business grows or contracts, this equality between its assets and the claims against those assets is always maintained. In other words, any change in the amount of total assets is necessarily accompanied by an equal change on the other side of the equation, that is, by an increase or decrease in either the liabilities or the owners' equity.

Assets The dollar amounts shown under the Assets column in Table 13.1 represent how much the company owns at the time of reporting. We list the asset items in the order of their "liquidity," or the length of time it takes to convert them to cash, according to the following three categories:

- **Current assets** can be converted to cash or its equivalent in less than one year. This type of asset generally includes three major accounts:

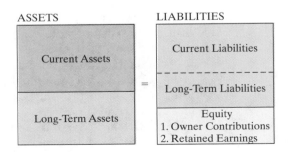

FIGURE 13.4 The four quadrants of the balance sheet

TABLE 13.1 Consolidated Statement of Financial Position				
	31-Jan-03	**1-Feb-02**	**$ Change**	**Percent Change**
Assets (in millions)				
Current assets:				
Cash	$4,232	$3,641	$591	16%
Short-term investments	406	273	133	49%
Accounts receivable, net	2,586	2,269	317	14%
Inventories	306	278	28	10%
Other	1,394	1,416	(22)	(2%)
Total current assets	8,924	7,877	1,047	13%
Property, plant and equipment, net	913	826	87	11%
Long-term investments	5,267	4,373	894	20%
Other noncurrent assets	366	459	(93)	(20%)
Total assets	$15,470	$13,535	$1,935	14%
Liabilities and Stockholders' Equity				
Current liabilities:				
Accounts payable	$5,989	$5,075	$914	18%
Accrued and other	2,944	2,444	500	20%
Total current liabilities	8,933	7,519	1414	19%
Long-term debt	506	520	(14)	(3%)
Other	1,158	802	356	44%
Total liabilities	10,597	8,841	1,756	20%
Stockholders' equity:				
Preferred stock	—	—		
Common stock and capital in excess of $0.01 par value	6,018	5,605	413	7%
Treasury stock	(4,539)	(2,249)	(2,290)	102%
Retained earnings	3,486	1,364	2,122	156%
Other	(92)	(26)	(66)	254%
Total stockholders' equity	4,873	4,694	179	4%
Total liabilities and stockholders' equity	$15,470	$13,535	$1,935	14%

1. The first account is cash and cash equivalents. A firm typically has a cash account at a bank to provide for the funds needed to conduct day-to-day business. Although we state all the assets in terms of dollars, only items labeled as cash represent actual money. Cash-equivalent items include marketable securities.

2. The second account includes *short-term investments*.

3. The third account is *accounts receivable*, which is money that is owed to the firm, but has not yet been received. For example, when Dell receives an order from a retail store, the company will send an invoice along with the shipment to the retailer. Then the unpaid bill immediately falls into the accounts-receivable category. When this bill is paid, it is deducted from the accounts-receivable category and placed into the cash category. Normally, a typical firm will have 30- to 45-day accounts receivable, depending on the frequency of its bills and the payment terms for customers.

4. The fourth account is *inventories*, which show the dollars the company has invested in raw materials, work-in-process, and finished goods available for sale.

- **Fixed assets** are relatively permanent and take time to convert into cash. Fixed assets reflect the amount of money a company has paid for its plant and equipment acquired at some time in the past. The most common fixed assets include the physical investment in the business, such as land, buildings, factory machinery, office equipment, and automobiles. With the exception of land, most fixed assets have a limited useful life. For example, buildings and equipment are used up over a period of years. Each year, a portion of the usefulness of these assets expires, and a portion of their total cost should thus be recognized as a depreciation expense. As stated previously in this book, the term *depreciation* refers to the accounting process for this gradual conversion of fixed assets into expenses. The item "property, plant, and equipment, net" thus represents the current book value of these assets after such depreciation expenses have been deducted.

- **Other assets** are listed at the end. Typical assets in this category include investments made in other companies and intangible assets such as goodwill, copyrights, franchises, and so forth. Goodwill appears on the balance sheet only when an operating business is purchased in its entirety. This item indicates any additional amount paid for the business above the fair market value of the business. (Here, the fair market value is defined as the price that a buyer is willing to pay when the business is offered for sale.)

Liabilities and Stockholders' Equity (Owners' Net Worth) The claims against assets are of two types: liabilities and stockholders' equity. Liabilities refer to money the company owes. Stockholders' equity indicates the portion of the assets of a company that is provided by the investors (owners). Therefore, stockholders' equity is also the liability of a company to its owners. (Recall Figure 13.4, which illustrates the relationship between assets and liabilities including equity.) The different categories of liabilities and stockholders' equity are described as follows:

- **Current liabilities** are those debts that must be paid in the near future (normally within one year). Major current liabilities include accounts and notes payable within a year, as well as accrued expenses (wages, salaries, interest, rent, taxes, etc., owed, but not yet due for payment) and advance payments and deposits from customers.

- **Other liabilities** include *long-term liabilities* such as bonds, mortgages, and long-term notes, which are due and payable more than one year in the future.

- **Stockholders' equity** represents the amount that is available to the stockholders (owners) after all other debts have been paid. It generally consists of preferred and common stock, treasury stock, capital surplus, and retained earnings. Preferred stock is a hybrid between common stock and debt. Such stock promises a fixed dividend (much like a bond's interest payment) but often limited voting rights. In the case of bankruptcy, preferred stockholders receive money after debtholders are paid and before common stockholders. Many firms do not use any preferred stock. The common stockholders' equity, or **net worth**, is a residual and is calculated as follows:

  Assets − Liabilities − Preferred stock = Common stockholders' equity.

  e.g., $15,470 − $10,597 − $0 = 4,873.

- **Common stock** is the aggregate par value of the company's issued stock. Companies rarely issue stocks at a discount (i.e., at an amount below the stated par). Corporations normally set the par value low enough so that, in practice, stock is usually sold at a premium.

- **Paid-in capital** (capital surplus) is the amount of money received from the sale of stock over the par value. Outstanding stock is the number of shares issued that actually is held by the public. If the corporation buys back part of its own issued stock, the value of the repurchased is listed as *treasury stock* on the balance sheet.

- **Retained Earnings** represent the cumulative net income of the firm since its beginning, less the total dividends that have been paid to stockholders. In other words, retained earnings indicate the amount of assets that has been financed by plowing profits back into the business. Therefore, these retained earnings belong to the stockholders.

How to Read Dell's Balance Sheet   With revenue of $35.40 billion for fiscal year 2003, Dell is the world's leading direct-computer-systems company. The $15.47 billion of total assets shown in Table 13.1 were necessary to support the sales of $35.4 billion:

- *Acquisition of Fixed Assets* The net increase in fixed assets is $87 million ($913 million – $826 million). By adding back the depreciation ($211 million) to show the increase in gross fixed assets, we find that Dell acquired $298 million in fixed assets in FY 2003.

- *Debt* Dell had a total long-term debt of $506 million that consists of the several bonds issued in previous years. The interest payments associated with these long-term debts were about $17 million.

- *Equity* Dell had 2.644 billion shares of common stock outstanding. Investors have provided the company with a total capital of $6.018 billion. However, Dell has retained the current as well as previous earnings of $3.486 billion since it was incorporated. These earnings belong to Dell's common stockholders. As end of 2003, the combined net stockholder's equity was $4.873 billion. (This net equity figure includes $4.539 billion worth of Treasury stock.)

- *Share value* Stockholders on average have a total investment of $1.77 per share ($4.873 billion/2.644 billion shares) in the company, known as the stock's book value. In January 2003, the stock traded in the general range of $24 to $28 per share. Note that this market price is quite different from the stock's book value. Many factors affect the market price—most importantly, how investors expect the company to do in the future. Certainly, the company's direct made-to-order business practices have had a major influence on the market value of its stock.

13.2.2 The Income Statement

The second financial report is the **income statement**, which indicates whether the company is making or losing money during a stated *period*. Most businesses prepare quarterly and monthly income statements in addition to annual ones. For Dell's income statement, the accounting period begins on February 1 and ends on January 31 of the following year. Table 13.2 gives the 2003, 2002, and 2001, income statements for Dell.

Net Income   Typical elements that are itemized in the income statement are as follows:

- The **net revenue** (or **net sales**) figure represents the gross sales less any sales return and allowances.

- The expenses and costs of doing business are listed on the next several lines as deductions from the revenues. The largest expense for a typical manufacturing firm is its production expense for making a product (such as labor, materials, and overhead), called **cost of revenue** (or **cost of goods sold**).

- Net revenue less the cost of revenue indicates the **gross margin**.

- Next, we subtract any other **operating expenses** from operating income. These other operating expenses are items such as interest, lease, selling, research and development (R&D), and administration expenses. This operation results in the operating income period.

TABLE 13.2 Income Statement for Dell Corporation
(in millions, except per-share amount)

	Fiscal Year Ended		
	31-Jan-03	**1-Feb-02**	**2-Feb-01**
Net revenue	$35,404	$31,168	$31,888
Cost of revenue	29,055	25,661	25,445
Gross margin	6,349	5,507	6,443
Operating expenses:			
Selling, general, and administrative	3,050	2,784	3,193
Research, development, and engineering	455	452	482
Special charges	0	482	105
Total operating expenses	3,505	3,718	3,780
Operating income	2,844	1,789	2,663
Investment and other income	183	58	531
Income before income taxes	3,027	1,731	3,194
Provision for income taxes	905	485	958
Cumulative effect of change in accounting principle, net		0	59
Net income	2,122	1,246	2,177
Earnings per common share:			
Basic	0.82	0.48	0.87
Diluted	0.80	0.46	0.81
Weighted-average shares outstanding:			
Basic	2,584	2,602	2,582
Diluted	2,644	2,726	2,746

- If the company generated **other income** from investments or any nonoperating activities, this item will be a part of income subject to income taxes as well.

- Finally, we determine the **net income** (or net profit) by subtracting the income taxes from the taxable income. This net income is also commonly known as the *accounting income*. Table 13.3 illustrates calculation of the gross

TABLE 13.3 Understanding Operating Margin and Net Margin

Dell Computer Corporation
Statement of Operations (Year Ended January 31 2003)

Sales	$35,404	100.00%
Less: Cost of Goods Sold	$29,055	82.07%
Gross Profit (Gross Margin)	$6,349	17.93%
Less: Operating expenses	$3,505	9.90%
Operating Profit (Operating Margin)	$2,844	8.03%
Other Income	$183	0.52%
Net Income before Taxes (NIBT)	$3,027	8.55%
Less: Taxes	$905	2.56%
Net Income (Net Margin)	$2,122	6.00%

margin, operating margin, and net margin, which are expressed as percentages of the total sales. Dell's net margin is about 6%, meaning that for every dollar of sale, Dell is making 6 cents of net profit.

Earnings per Share   Another important piece of financial information provided in the income statement is the **earnings per share** (EPS) figure. In simple situations, we compute this quantity by dividing the available earnings to common stockholders by the number of shares of common stock outstanding. Stockholders and potential investors want to know what their relative share of profits is, not just the total dollar amount. Presentation of profits on a per-share basis allows stockholders to relate earnings to what they paid for a share of stock. Naturally, companies want to report a higher EPS to their investors as a means of summarizing how well they managed their businesses for the benefit of their owners. Dell earned $0.80 per share in 2003, up from $0.46 in 2002, but it paid no dividend.

Retained Earnings   As a supplement to the income statement, corporations also report their retained earnings during the accounting period. When a corporation makes some profits, it has to decide what to do with these profits. The corporation may decide to pay out some of the profits as dividends to its stockholders. Alternatively, it may retain the remaining profits in the business in order to finance expansion or support other business activities.

When the corporation declares dividends, preferred stock has priority over common stock in regards to the receipt of dividends. Preferred stock pays a stated dividend, much like the interest payment on bonds. The dividend is not a legal liability until the board of directors has declared it. However, many corporations view the dividend payments to preferred stockholders as a liability. Therefore, the term "available earnings for common stockholders" reflects the net earnings of the corporation less the preferred-stock dividends. When preferred-and common-stock dividends are subtracted from net income, the remainder is retained earnings (profits) for the year. As mentioned previously, these retained earnings are reinvested into the business.

Understanding Dell's Income Statement   Net sales were $35.404 billion in 2003, compared with $31.168 billion in 2002, a gain of 13.60%. Profits from operations (operating income) rose 58.97% to 2.844 billion, and net income was up 70.30% to $2.122 billion. We can infer the following:

- *Dividends* Dell issued no preferred stock, so there is no required cash dividend. Therefore, the net income of $2.122 billion belongs to the common stockholders.

- *EPS* Earnings per common share climbed at a faster pace to $0.80, an increase of 73.91%. Dell could have retained this income fully for reinvestment in the firm or paid it as dividends to its common stockholders. Instead, Dell repurchased and retired 50 million shares of common stock for $2.29 billion. We can see that Dell had earnings available to common stockholders of $2.122 billion. The beginning balance of the retained earnings was $1.364 billion. Therefore, the total retained earnings grew to $3.486 billion, and of this amount, Dell paid out $2.29 billion to repurchase 50 million of its own shares.

13.2.3   **The Cash Flow Statement**

The income statement explained in the previous section only indicates whether the company was making or losing money during the reporting period. Therefore, the emphasis was on determining the net income (profits) of the firm, mainly for the operating activities. However, the income statement ignores two other important business activities for the period: financing and investing activities. Therefore, we need another financial statement—the **cash flow statement**—that details how the company generated cash and how the company used its cash during the reporting period. This statement is concerned with how the company actually used its cash in its period, thus explaining how the firm went from the level of cash in its accounts reported at the start of the year to the level of cash it had at the end of the year.

Sources and Uses of Cash   The difference between the sources (inflows) and uses (outflows) of cash represents the net cash flow during the reporting period. This is a very important piece of information, because investors determine the value of an asset (or a whole firm) by the cash flows it generates. Figure 13.5 illustrates how a firm generates cash flows and summarizes the sources and uses of cash during its business cycle.

Certainly, a firm's net income is important, but cash flows are even more important, because we need cash to pay dividends and to purchase the assets required for continuing operations. As we mentioned previously, the goal of the firm should be to maximize the price of its stock. Since the value of any asset depends on the cash flows produced by the asset, managers want to maximize cash flows available to its investors over the long run. Therefore, we should make investment

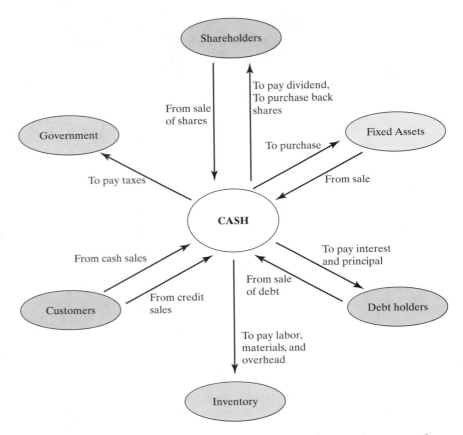

FIGURE 13.5 The cash flow cycle in a typical manufacturing firm

decisions based on cash flows rather than profits. For such investment decisions, it is necessary to convert profits (as determined in the income statement) to cash flows. Table 13.4 is Dell's statement of cash flows as it would appear in the company's annual report.

Reporting Format In preparing a cash flow statement such as in Table 13.4, many companies identify the sources and uses of cash according to the types of business activities. There are three types of activities:

- **Operating activities:** We start with the net change in operating cash flows from the income statement. Here, operating cash flows represent those cash flows related to the production and sales of goods or services. All noncash expenses are added back to net income (or after-tax profits). For example, an expense such as depreciation is only an accounting expense (bookkeeping entry). While we may charge such items against current income as an expense, they do not involve an actual cash outflow. The actual cash flow may have occurred when

TABLE 13.4 Cash Flow Statement for Dell Computer Corporation		
	Fiscal Year Ended	
(in millions)	**31- Jan-03**	**02-Feb-02**
Cash flows from operating activities:		
Net income	$2,122	$ 1,246
Depreciation and amortization	$211	$487
Changes in working capital	$1,210	$826
Changes in non-current assets and liabilities	$212	$62
Special charges and other adjustments	$(217)	$937
Net cash provided by operating activities	$3,538	$3,797
Cash flows from investing activities:		
Marketable securities:		
Purchases	$(8,736)	$(5,382)
Sales	$7,660	$3,425
Capital expenditures	$305	$303
Net cash used in investing activities	$1,381	$2,260
Cash flows from financing activities:		
Purchase of common stock	$(2,290)	$(3,000)
Issuance of common stock under employee plans and other	$265	$298
Net cash used in financing activities	$(2,025)	$(2,702)
Effect of exchange rate changes on cash	$459	$(104)
Net increase in cash	$591	$(1,269)
Cash at beginning of period	$3,641	$4,910
Cash at end of period	$4,232	$3,641

the asset was purchased. Any adjustments in **working-capital** terms will also be listed here. Once again, working capital is defined as the difference between the current assets and the current liabilities. Further, we can determine the net change in **working capital requirement** by the difference between "change in current assets" and "change in current liabilities." If this net change being positive, the working capital requirement appears as *uses* of cash in the cash flow statement.

- **Investing activities:** After determining the operating cash flows, we consider any cash flow transactions related to investment activities. Investment activities include transactions such as purchasing new fixed assets (cash out-flow), reselling old equipment (cash inflow), and buying and selling financial assets.

- **Financing activities:** Finally, we detail cash transactions related to financing any capital used in business. For example, the company could borrow or sell more stock, resulting in cash inflows. Paying off existing debt would result in cash outflows.

By summarizing cash inflows and outflows from these three types of activities for a given accounting period, we obtain the net changes in cash flow position of the company.

How to Read Dell's Cash Flow Statement As shown in Table 13.4, Dell's cash flow from operations in fiscal year 2003 amounted to $591. Note that this amount is significantly less than the $2.122 billion earned during the reporting period. Where did the rest of the money go? Basically, we are trying to explain the sources and uses of funds in the three cash flow areas, as depicted in Figure 13.6.

The main reason for the difference is the accrual-basis accounting principle used by the Dell Corporation. In **accrual-basis accounting,** an accountant recognizes the impact of a business event as it occurs. When the business performs a service, makes a sale, or incurs an expense, the accountant enters the transaction into the books, no matter whether cash has been received or paid. For example, the increase in accounts receivable during 2003 ($2,586 million $-$$2,269 million $=$ $317 million) represents the amount of total sales on credit. Since this figure was included in the total sales in determining the net income, we need to subtract this figure in order to determine the true cash position. After adjustments, the net cash provided from operating activities is $3.538 billion.

From the investment activities, there was an investment inflow of $305 million in sales of business assets. Dell sold $7.66 billion worth of stocks and bonds during the period and reinvested $8.736 billion in various financial securities. The net cash flow used from these investing activities amounted to ($1,381) million, which means that there was an outflow.

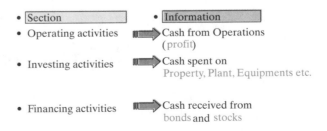

FIGURE 13.6 Explaining the cash flow statement by the type of cash flow activities

Financing decision produced a net outflow of $2.025 billion, including a repurchase of Dell's own shares. (This repurchase of Dell's own stock is equivalent to Dell investing its idle cash from operation in the stock market. Dell could have bought another company's stock, such as that of IBM or Microsoft, with the money. However, it liked its own stock better than any other stocks on the market, so it ended up buying that instead.)

Finally, there was the effect of exchange-rate changes on cash for foreign sales. This amounts to a net increase of $459 million.

Together, the three types of activities generated a total cash flow of $591 million. With the initial cash balance of $3.641 billion, the ending cash balance now increases to $4.232 billion. This same amount denotes the change in Dell's cash position as shown in the cash accounts on the balance sheet.

13-3   Using Ratios to Make Business Decisions

Financial statements tell us what has happened during a particular period of time. In that sense, financial statements are essentially historical documents. However, most users of financial statements are concerned about what will happen in the future. For example,

- stockholders are concerned with future earnings and dividends,

- creditors are concerned with the company's ability to repay its future debts,

- managers are concerned with the company's ability to finance future expansion, and

- engineers are concerned with planning actions that will influence the future course of business events.

Despite the fact that financial statements are historical documents, they can still provide valuable information that addresses all of these concerns. An important part of financial analysis is the calculation and interpretation of various financial ratios, which provide insight into a firm's future status. In this section, we consider some of the widely used ratios that analysts use in attempting to predict the future course of events in business organizations. We may group these ratios in five categories (debt management, liquidity, asset management, profitability, and market trend), as outlined in Figure 13.7. In all of the upcoming financial-ratio calculations, we will use the 2003 financial statements for Dell Computer Corporation as summarized in Tables, 13.1, 13.2, and 13.4.

We often compare a company's financial ratios with industry average figures; however, we should note at this point that an industry average is not an absolute number that all firms should strive to maintain. In fact, some very well-managed firms will be above the average, while other good firms will be below it. However, if a firm's ratios are quite different from the average for its industry, we should examine the reason that this variance occurs.

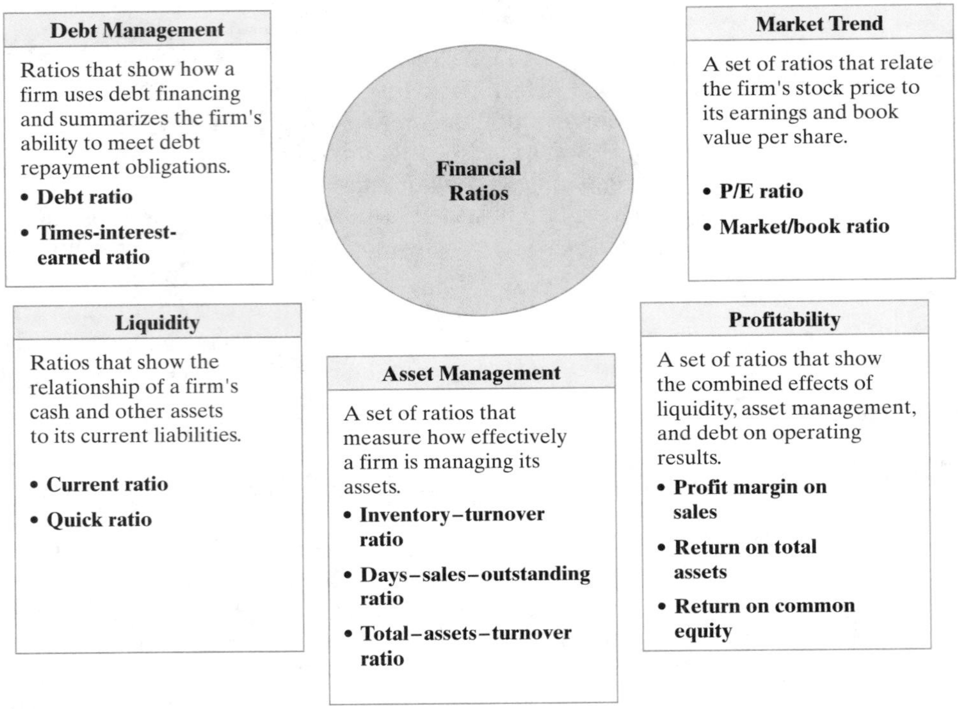

FIGURE 13.7 Types of financial ratios used in evaluating a firm's financial health

13.3.1 Debt Management Analysis

All businesses need assets in order to operate. To acquire assets, a firm must raise capital. When the firm finances its long-term needs externally, it may obtain funds from the capital markets. Capital comes in two forms, **debt** and **equity**. Debt capital refers to borrowed capital from financial institutions and bond market. Equity capital refers to capital obtained from the owners of the company.

The basic methods of debt financing include bank loans and bond sales. For example, say that a firm needs $10,000 to purchase a computer. In this situation, the firm could borrow the money from a bank and repay the loan and the specified interest in a few years, an approach known as a *short-term debt financing*. Now suppose that the firm needs $100 million for a construction project. It would normally be very expensive (or require a substantial amount of mortgage) to borrow the money directly from a bank. In this situation, the firm would go public in order to borrow money on a long-term basis. When investors lend capital to a company and the company agrees to repay the loan at an agreed interest rate, the investor is a creditor of the corporation. The document that records the nature of the arrangement between the issuing company and the investor is called a **bond**. Raising capital through issuing bonds is called *long-term debt financing*.

Similarly, there are different types of equity capital. For example, the equity of a proprietorship represents the money provided by the owner. For a corporation, equity capital comes in two forms: *preferred* and *common stock*. Investors provide capital to a corporation, and the company agrees to provide the investor with fractional ownership in the corporation.

Since a company must pay its creditors on time and in full to remain solvent and out of bankruptcy, one primary concern of financial analysis is to determine how able a firm is to cover its required debt payments. To do so, we first examine the extent to which a company uses debt financing (or financial leverage) in business operation as follows:

- check the balance sheet to determine the extent to which borrowed funds have been used to finance assets, and
- review the income statement to see the extent to which fixed charges (interests) are covered by operating profits.

Two essential indicators of a business' ability to pay its long-term liabilities are the *debt ratio* and the *times-interest-earned ratio*.

Debt Ratio   The relationship between total liabilities and total assets, generally called the **debt ratio**, tells us the proportion of the company's assets that it has financed with debt:

$$\text{Debt ratio} = \frac{\text{total debt}}{\text{total assets}}.$$

For example, Dell's debt ratio for 2003 can be calculated as follows:

$$\text{Debt ratio} = \frac{\$10,597}{\$15,470} = 68.50\%.$$

Total debt includes both current liabilities and long-term debt. If the debt ratio is one, then the company has used debt to finance all of its assets. As of January 31, 2003, Dell's debt ratio was 68.50%; this means that its creditors have supplied close to 69% of the firm's total financing. Certainly, most creditors prefer low debt ratios, because the lower the ratio, the greater the cushion against creditors' losses in case of liquidation. If a company seeking financing already has large liabilities, then additional debt payments may be too much for the business to handle. For this highly leveraged company, creditors generally charge higher interest rates on new borrowing in order to help protect themselves.

Times-Interest-Earned Ratio   The most common measure of the ability of a company's operations to provide protection to the long-term creditor is the **times-interest-earned ratio**. We find this ratio by dividing earnings before interest and income taxes (EBIT) by the yearly interest charges that must be met. For example, Dell issued $506 million worth of senior notes and long-term bonds with a combined interest rate of 6.8%. This results in $17 million in interest expenses in the year 2003, so we calculate the following:

$$\text{Times-interest-earned ratio} = \frac{\text{EBIT}}{\text{Interest expense}}$$

$$= \frac{\$3,027 \text{ million } + \$17 \text{ million}}{\$17 \text{ million}}$$

$$= 179 \text{ times.}$$

The ratio measures the extent to which operating income can decline before the firm is unable to meet its annual interest costs. Failure to meet this obligation can bring legal action by the firm's creditors, possibly resulting in bankruptcy. Note that we use the earnings before interest and income taxes, rather than net income, in the numerator. Because Dell must pay interest with pretax dollars, Dell's ability to pay current interest is not affected by income taxes. Only those earnings remaining after all interest charges have been incurred are subject to income taxes. For Dell, the times-interest-earned ratio for 2003 is 179 times. This ratio is notably higher compared with that of the industry average, 157 times, during the same operating period.

13.3.2 Liquidity Analysis

Dell's short-term suppliers and creditors are also concerned with the level of liabilities as well. Short-term creditors want to be repaid on time. Therefore, they focus on Dell's cash flows and on its working capital, as these quantities are Dell's primary sources of cash in the near future. The excess of current assets over current liabilities is known as **working capital**. This figure indicates the extent to which current assets can be converted to cash in order to meet current obligations. Therefore, we view a firm's net working capital as a measure of its *liquidity position*. In general, the larger the working capital, the more able the business is to pay its debts.

Current Ratio   We calculate the current ratio by dividing current assets by current liabilities:

$$\text{Current ratio} = \frac{\text{current assets}}{\text{current liabilities}}.$$

For example, Dell's current ratio in 2003 can be calculated as follows:

$$\text{Current ratio} = \frac{\$8,924}{\$8,933} = 1.00 \text{ times.}$$

If a company is getting into financial difficulty, it begins paying its bills (accounts payable) more slowly, borrowing from its bank, and so on. If current liabilities are rising faster than current assets, the current ratio will fall, and this could spell trouble. What is an acceptable current ratio? The answer depends on the nature of the industry. The general rule of thumb calls for a current ratio of 2 to 1. This rule, of course, is subject to many exceptions, depending heavily on the composition of the assets involved.

Quick (Acid Test) Ratio   The quick ratio tells us whether the company could pay all its current liabilities if they came due immediately. We calculate the quick ratio by deducting inventories from current assets and then dividing the remainder by current liabilities:

$$\text{Quick ratio} = \frac{\text{current assets} - \text{inventories}}{\text{current liabilities}}.$$

For example, Dell's quick ratio in 2003 can be calculated as follows:

$$\text{Quick ratio} = \frac{\$8,924 - \$306}{\$8,933} = 0.96 \text{ times}.$$

The quick ratio measures how well a company can meet its obligations without having to liquidate or depend too heavily on selling its inventory. Inventories are typically the least liquid of a firm's current assets; hence, they are the assets on which losses are most likely to occur in case of liquidation. Although Dell's current ratio for 2003—1—may look below average for its industry (1.5), its liquidity position is relatively strong, as it carried a very little amount of inventory in its current assets (only $306 million out of $8.924 billion of current assets).

13.3.3   Asset Management Analysis

The ability to sell inventory and collect accounts receivable is fundamental to business success. Therefore, the third group of ratios measures how effectively the firm is managing its assets. We will review three ratios related to a firm's asset management: (1) inventory-turnover ratio, (2) days-sale-outstanding ratio, and (3) total-asset-turnover ratio. The purpose of these ratios is to answer the following question: Does the total amount of each type of asset as reported on the balance sheet seem reasonable in view of current and projected sales levels? Any asset acquisition requires the use of funds. If a firm has an excess of assets, its cost of capital will be too high; as a result, its profits will be depressed. On the other hand, if assets are too low, the firm is likely to lose profitable sales.

Inventory-Turnover Ratio   The inventory-turnover ratio measures how many times a company has sold and replaced its inventory during the year. We compute the ratio by dividing sales by the average level of inventories on hand. We compute the average inventory figure by taking the average of the beginning and ending inventory figures. Since Dell has a beginning inventory figure of $278 million and an ending inventory figure of $306 million, its average inventory for the year would be $292 million. Then we compute Dell's inventory-turnover ratio for 2003 as follows:

$$\text{Inventory-turnover ratio} = \frac{\text{sales}}{\text{average inventory balance}}$$
$$= \frac{\$35,404}{\$292} = 121.24 \text{ times}.$$

As a rough approximation, Dell was able to sell and restock its inventory 121 times in 2003. Dell's turnover of 121 times is much faster than that of its industry average, 90.9 times, during the same operating period. This result suggests that Dell's competitors are holding excessive stocks of inventory; excess stocks are, of course, unproductive, and they represent an investment with a low or zero rate of return.

Days-Sales-Outstanding (Accounts Receivable Turnover) Ratio The days-sales-outstanding (DSO) ratio is a rough measure of how many times a company's accounts receivable have been turned into cash during the year. We determine the ratio, also called the **average collection period**, by dividing accounts receivable by average sales per day. In other words, this ratio indicates the average length of time the firm must wait after making a sale before receiving cash. For Dell in 2003, we have

$$
\begin{aligned}
\text{DSO} &= \frac{\text{receivables}}{\text{average sales per day}} = \frac{\text{receivables}}{\text{annual sales}/365} \\
&= \frac{\$2{,}586}{\$35{,}404/365} = \frac{\$2{,}586}{\$97.0} \\
&= 26.66 \text{ days.}
\end{aligned}
$$

Thus, for Dell, on average, it takes about 27 days to collect on a credit sale. During the same period, Compaq's[2] average collection period was 49 days. Whether Dell's average of 27 days taken to collect on an account is good or bad depends on the credit terms Dell is offering its customers. If credit terms are 30 days, we can say that Dell's customers, on average, are paying their bills on time. In order to improve their working-capital position, most customers tend to withhold payment for as long as the credit terms will allow and may even go over by a few days. The long collection period may signal that customers are in financial trouble or the company has poor credit management.

Total-Assets-Turnover Ratio The total-assets-turnover ratio measures how effectively a firm uses its total assets in generating its revenues. It is the ratio of sales to all of the firm's assets. For Dell in 2003,

$$
\begin{aligned}
\text{Total-assets-turnover ratio} &= \frac{\text{sales}}{\text{total assets}} \\
&= \frac{\$35{,}404}{\$15{,}470} \\
&= 2.29 \text{ times.}
\end{aligned}
$$

Dell's ratio of 2.29 times, when compared with the industry average ratio of 1.41 times, is almost 62% faster, indicating that Dell is using its total assets about 62% more intensively than its peers. In fact, Dell's total investment in plant and equipment is about one-fifth of Compaq's. If we view Dell's ratio as the industry average, we can say that Compaq has too much investment in inventory, plant, and equipment compared with the size of sales.

---

[2]Hewlett-Packard (HP) completed its merger transaction involving Compaq Computer Corporation on May 3, 2002.

13.3.4 **Profitability Analysis**

One of the most important goals for any business is to earn a profit. The ratios examined thus far provide useful clues as to the effectiveness of a firm's operations, but profitability ratios show the combined effects of liquidity, asset management, and debt on operating results. Therefore, ratios that measure profitability play a large role in decision making.

Profit Margin on Sales   We calculate the profit margin on sales by dividing net income by sales. This ratio indicates the profit per dollar of sales. For Dell in 2003,

$$\text{Profit margin on sales} = \frac{\text{net income available to common stockholders}}{\text{sales}}$$

$$= \frac{\$2,122}{\$35,404} = 6.00\%.$$

Thus, Dell's profit margin is equivalent to 6 cents for each sales dollar generated. Dell's profit margin is greater than Compaq's profit margin of 1.48%. This difference indicates that, although Compaq's sales were about 50% more than Dell's during the same operating period, Compaq's operation is less efficient than Dell's operation. Compaq's low profit margin is also a result of its heavy use of debt and the very high volume of inventory that it carries. Recall that net income is income after taxes. Thus, if two firms have identical operations in the sense that their sales, operating costs, and earnings before income tax are the same, but one company uses more debt than the other, then the company with more debt will have higher interest charges. Those interest charges will pull net income down, and because sales are the same, the result will be a relatively low profit margin for the indebted company.

Return on Total Assets   The return on total assets (ROA), or simply return on assets, measures a company's success in using its assets to earn a profit. The ratio of net income to total assets measures the return on total assets after interest and taxes. For Dell in 2003,

$$\text{Return on total assets} = \frac{\text{net income} + \text{interest expense}(1 - \text{tax rate})}{\text{average total assets}}$$

$$= \frac{\$2.122 + \$17(1 - 0.30)}{(\$15,470 + \$13,535)/2} = 14.71\%.$$

Adding a portion of interest expenses back to net income results in an adjusted earnings figure that shows what earnings would have been if the assets had been acquired solely through equity. (Note that Dell's effective tax rate is 30% in year 2003.) With this adjustment, we may be able to compare the return on total assets for companies with differing amounts of debt. Again, Dell's 14.71% return on assets is well above the industry average of 4.4%. This high return results from (1) the company's high basic earning power and (2) its low use of debt, both of which cause its net income to be relatively high.

Return on Common Equity   Another popular measure of profitability is the rate of return on common equity (ROE). This ratio shows the relationship between net income and common stockholders' investment in the company. That is, it answers the question, how much income is earned for every $1 invested by common shareholders? To compute this ratio, we first subtract preferred dividends from net income; the result is known as "net income available to common stockholders." We then divide this net income available to common stockholders by the average common (stockholders) equity during the year. We compute average common equity by using the beginning and ending balances. At the beginning of fiscal year 2003, Dell's common equity balance was $4.694 billion, and its ending balance was $4.873 billion. The average balance is then simply $4.7835 billion. So,

$$\text{Return on common equity} = \frac{\text{net income available to common stockholders}}{\text{average common equity}}$$

$$= \frac{\$2,122}{(\$4,694 + \$4,873)/2}$$

$$= \frac{\$2,122}{\$4,783.5} = 44.36\%.$$

The rate of return on common equity for Dell was 44.36% during 2003. During the same period, Compaq's return on common equity amounted to 8.85%, which is considered a poor performance in the computer industry in general. Figure 13.8 illustrates how the debt-to-equity ratio (total debt over total equity) (distinct from the debt ratio) impacts the return on equity.

13.3.5 Market-Value Analysis

When purchasing a company's stock, what would be your primary factors in valuing that stock? In general, investors purchase stock to earn a return on their investment. This return consists of two parts: (1) gains (or losses) from selling the stock at a price that is higher (or lower) than the purchase price and (2) dividends, the periodic distributions of profits to stockholders. The market-value ratios, such as price–earnings ratio and market–book ratio, relate the firm's stock price to its earnings and book

- This is an example of a healthy company that might not have a spectacular ROE because there is so much equity in the company.

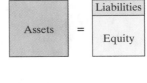

- This an example of a highly leveraged company that might have a spectacular ROE because the owners have put so little of their own resources into the company.

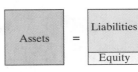

FIGURE 13.8 How the debt-to-equity ratio impacts the return on equity

value per share. These ratios give management an indication of what investors think of the company's past performance and future prospects. If a firm's asset management and debt management are sound and its profit is rising, then its market-value ratios and stock price will be high.

Price/Earnings Ratio   The price/earnings (P/E) ratio shows how much investors are willing to pay per dollar of reported profits. Dell's stock sold for $25 in early February of 2003, so with an EPS of $0.80, Dell's P/E ratio is 31.25:

$$\text{P/E ratio} = \frac{\text{price per share}}{\text{earnings per share}} = \frac{\$25}{\$0.80} = 31.25.$$

That is, the stock was selling for about 31.25 times its current earnings per share. In general, P/E ratios are higher for firms with high growth prospects, with all other things held constant, but lower for firms with lower expected earnings. Dell's expected annual increase in operating earnings is 30% over the next three to five years. Since Dell's expected growth is greater than 15%, the average for the computer industry, we may infer that investors value Dell's stock more highly than most other stocks in the industry. However, all stocks with high P/E ratios will also carry high risk whenever the expected growths do not materialize. Any slight earnings disappointment tends to punish the market price significantly.

Book Value per Share   Another ratio frequently used in assessing the well-being of common stockholders is book value per share. The book value per share measures the amount that would be distributed to holders of each share of common stock if all assets were sold at their balance-sheet carrying amounts and if all creditors were paid off. We compute the book value per share for Dell's common stock in 2003 as follows:

$$\text{Book value per share} = \frac{\text{total stockholders' equity} - \text{preferred stock}}{\text{shares outstanding}}$$

$$= \frac{\$4,873}{\$2,644} = \$1.84.$$

If we compare this book value with the market price of $31 at the time of publication, then we may say that the stock appears to be overpriced. Once again, market prices reflect expectations about future earnings and dividends, whereas book value largely reflects the results of events that occurred in the past. Therefore, the market value of a stock tends to exceed its book value. Table 13.5 summarizes the financial ratios for Dell Computer Corporation in reference to industry and the S & P averages.

13.3.6   Limitations of Financial Ratios in Business Decisions

Business decisions are made in a world of uncertainty. As useful as ratios are, they have limitations. We can draw an analogy between the use of financial ratios in decision making and a physician's use of a thermometer. A reading of 102°F indicates that something is wrong with the patient, but the temperature alone does not

TABLE 13.5 Comparisons of Key Financial Ratios for Dell Computer Corporation with the Industry Average and the S&P 500

Category	Financial Ratio	Dell	Industry	S&P 500
Debt Management	Debt ratio	0.68	0.56	0.48
	Time-interest-earned	179	157	2.4
Liquidity	Current ratio	1.0	1.4	1.5
	Quick ratio	0.96	1.2	1.1
Asset Management	Inventory–turnover ratio	121.24	90.9	26
	Days-sales-outstanding ratio	26.66	14.5	5.8
	Total-assets turnover ratio	2.29	2.0	0.3
Profitability	Profit margin	6.0%	3.6%	4.1%
	Return on total assets	14.71%	4.4%	2.0%
	Return on common equity	44.36%	9.3%	12.1%
Market Trend	P/E ratio	31.25	56.7	37.0
	Book value/share	$1.84	$2.91	$10.67

S&P 500: A list of 500 large U.S. companies listed by Standard & Poors.

indicate what the problem is or how to cure it. In other words, ratio analysis is useful, but analysts should be aware of ever-changing market conditions and make adjustments as necessary. It is also difficult to generalize about whether a particular ratio is "good" or "bad." For example, a high current ratio may indicate a strong liquidity position, which is good, but holding too much cash in a bank account (which will increase the current ratio) may not be the best use of funds. In addition, ratio analysis based on any one year may not represent the true business condition. It is important to analyze trends in various financial ratios as well as their absolute levels, for trends give clues as to whether the financial situation is likely to improve or to deteriorate. To do a **trend analysis**, one simply plots a ratio over time. As a typical engineering student, your judgment in interpreting a set of financial ratios is understandably weak at this point, but it will improve as you encounter many facets of business decisions in the real world. Again, accounting is a language of business, and as you speak it more often, it will provide useful insights into a firm's operations.

Summary

- The primary purposes of this chapter were (1) to describe basic financial statements, (2) to present some background information on cash flows and corporate profitability, and (3) to discuss techniques used by investors and managers to analyze financial statements.

- Before making any major business decisions, it is important to understand the impact of the decisions on the firm's financial statements.

- The three basic financial statements contained in a company's annual report are the balance sheet, the income statement, and the statement of cash flows.

Investors use the information provided in these statements to form expectations about future levels of earnings and dividends and about the firm's riskiness.

- A firm's balance sheet shows a snapshot of the firm's financial position at a particular point in time.

- A firm's income statement reports the results of operations over a period of time, and it shows earnings per share as its "bottom line."

- A firm's statement of cash flows reports the impact of operating, investing, and financing activities on cash flows over an accounting period.

- The purpose of calculating a set of financial ratios is twofold: (1) to examine the relative strengths and weaknesses of a company as compared with other companies in the same industry and (2) to show whether the company's financial position has been improving or deteriorating over time.

- Liquidity ratios show the relationship of a firm's current assets to its current liabilities, and thus its ability to meet maturing debts. Two commonly used liquidity ratios are the current ratio and the quick (acid test) ratio.

- Asset management ratios measure how effectively a firm is managing its assets. Some of the major ratios include the inventory-turnover ratio, the collection period, and the total-assets-turnover ratio.

- Debt management ratios reveal (1) the extent to which a firm is financed with debt and (2) the firm's likelihood of defaulting on its debt obligations. In this category, we may include the debt ratio and the times-interest-earned ratio.

- Profitability ratios show the combined effects of liquidity, asset management, and debt management policies on operating results. Profitability ratios include the profit margin on sales, the return on total assets, and the return on common equity.

- Market-value ratios relate a firm's stock price to its earnings and book value per share and give management an indication of what investors think of the company's past performance and future prospects. Market value ratios include the price/earnings ratio and the market/book ratio.

- Trend analysis, where one plots a ratio over time, is important, because it reveals whether a firm's ratios are improving or deteriorating over time.

Problems

Financial Statements

13.1 Definitional problems: Listed as follows are eight terms that relate to financial statements:

1. Balance sheet statement
2. Income statement
3. Cash flow statement
4. Operating activities
5. Investment activities
6. Financing activities
7. Treasury account
8. Capital account

Choose the term from the list that most appropriately completes each of the following statements:

- As an outside investor, you would view a firm's _____ as the most important financial report for gauging the quality of earnings.

● Retained earnings as reported in the _____ represent income earned by the firm in past years that has not been paid out as dividends.

● The _____ is designed to show how a firm's operations have affected its cash position by providing actual net cash flows into or out of the firm during some specified period.

● Typically, a firm's cash flow statement is categorized into three activities: _____, _____, and _____.

● When you issue stock, the money raised beyond the par value is shown in the _____ in the balance-sheet statement.

13.2 Definitional problems: Listed as follows are 11 terms that relate to ratio analysis:

1. Book value per share
2. Inventory turnover
3. Debt-to-equity ratio
4. Average collection period
5. Average sales period
6. Return on common equity
7. Earnings per share
8. Price/earnings ratio
9. Return on total assets
10. Current ratio
11. Accounts-receivable turnover

Choose the financial ratio or term from the list that most appropriately completes each of the following statements:

● The _____ tends to have an effect on the market price per share, as reflected in the price/earnings ratio.

● The _____ indicates whether a stock is relatively cheap or relatively expensive in relation to current earnings.

● The _____ measures the amount that would be distributed to holders of common stock if all assets were sold at their balance-sheet carrying amount and if all creditors were paid off.

● The _____ is a rough measure of how many times a company's accounts receivable has been turned into cash during the year.

● The _____ is a measure of the amount of assets being provided by creditors for each dollar of assets being provided by the stockholders.

● The _____ measures how well management has employed assets.

13.3 Consider the following balance-sheet entries for Delta Corporation:

**Balance Sheet Statement
as of December 31, 2003**

Assets:	
Cash	$150,000
Marketable securities	$200,000
Accounts receivable	$150,000
Inventories	$50,000
Prepaid taxes and insurance	$30,000
Manufacturing plant at cost	$600,000
Less accumulated depreciation	$100,000
Net fixed assets	$500,000
Goodwill	$20,000
Liabilities and Shareholders' Equity:	
Notes payable	$50,000
Accounts payable	$100,000
Income taxes payable	$80,000
Long-term mortgage bonds	$400,000
Preferred stock, 6%, $100 par value (1,000 shares)	$100,000
Common stock, $15 par value (10,000 shares)	$150,000
Capital surplus	$150,000
Retained earnings	$70,000

(a) Compute the following for the firm:

Current assets: $ _____
Current liabilities: $ _____
Working capital: $ _____
Shareholders' equity: $ _____

(b) If the firm had a net income after taxes of $500,000, what are the earnings per share?

(c) When the firm issued its common stock, what was the market price of the stock per share?

13.4 A chemical-processing firm is planning on adding a second polyethylene plant at another location. The financial information for the first project year is provided as follows:

Sales	$1,500,000
Manufacturing costs	
Direct materials	$150,000
Direct labor	$200,000
Overhead	$100,000
Depreciation	$200,000
Operating expenses	$150,000
Equipment purchase	$400,000
Borrowing to finance equipment	$200,000
Increase in inventories	$100,000
Decrease in accounts receivable	$20,000
Increase in wages payable	$30,000
Decrease in notes payable	$40,000
Income taxes	$272,000
Interest payment on financing	$20,000

(a) Compute the working-capital requirement during this project period.

(b) What is the taxable income during this project period?

(c) What is the net income during this project period?

(d) Compute the net cash flow from this project during the first year.

Financial-Ratio Analysis

13.5 Consider the following financial statements for Bay Network Corporation.

Unit (000)	Dec. 2002	Dec. 2001
Balance Sheet Summary		
Cash	158,043	20,098
Securities	285,116	0
Receivables	24,582	8,056
Allowances	632	0
Inventory	0	0
Current Assets	377,833	28,834
Property and		
Equipment, Net	20,588	10,569
Depreciation	8,172	2,867
Total Assets	513,378	36,671
Current Liabilities	55,663	14,402
Bonds	0	0
Preferred		
Mandatory	0	0
Preferred Stock	0	0
Common Stock	2	1
Other Stockholders'		
Equity	457,713	17,064
Total Liabilities		
and Equity	513,378	36,671
Income Statement Summary		
Total Revenues	102,606	3,807
Cost of Sales	45,272	4,416
Other Expenses	71,954	31,661
Loss Provision	0	0
Interest Income	8,011	1,301
Income Pretax	−6,609	−69
Income Tax	2,425	2
Income Continuing	−9,034	−30,971
Discontinued	0	0
Extraordinary	0	0
Changes	0	0
Net Income	**−9,034**	**−30,971**
EPS Primary	**$ −0.1**	**$ −0.80**
EPS Diluted	$−0.10	$−0.80
	$−0.05	$−0.40
	(split adjusted)	(split adjusted)

The closing stock price for Bay Network was $12.45 on December 31, 2002. Based on the financial data presented, compute the following financial ratios, and make an informed analysis of Bay's financial health:

(a) Debt ratio

(b) Times-interest-earned ratio

(c) Current ratio

(d) Quick (acid test) ratio

(e) Inventory-turnover ratio

(f) Days-sales-outstanding ratio

(g) Total-assets-turnover ratio

(h) Profit margin on sales

(i) Return on total assets

(j) Return on common equity

(k) Price/earnings ratio

(l) Book value per share

13.6 The following financial statements summarize the financial conditions for Flectronics, Inc., an electronics outsourcing contractor, for fiscal year 2002:

US $ (000)	Aug. 2002	Aug. 2001
Balance Sheet Summary		
Cash	1,325,637	225,228
Securities	362,769	83,576
Receivables	1,123,901	674,193
Allowances	5,580	−3,999
Inventory	1,080,083	788,519
Current Assets	3,994,084	1,887,558
Property and Equipment, Net	1,186,885	859,831
Depreciation	533,311	−411,792
Total Assets	4,834,696	2,410,568
Current Liabilities	1,113,186	840,834
Bonds	922,653	385,519
Preferred Mandatory	0	0
Preferred Stock	0	0
Common Stock	271	117
Other Stockholders' Equity	2,792,820	1,181,209
Total Liabilities and Equity	4,834,696	2,410,568
Income Statement Summary		
Total Revenues	8,391,409	5,288,294
Cost of Sales	7,614,589	4,749,988
Other Expenses	335,808	237,063
Loss Provision	2,143	2,254

US $ (000)	Aug. 2002	Aug. 2001
Interest Expense	36,479	24,759
Income Pretax	432,342	298,983
Income Tax	138,407	100,159
Income Continuing	293,935	198,159
Discontinued	0	0
Extraordinary	0	0
Changes	0	0
Net Income	**293,935**	**198,159**
EPS Primary	**$1.19**	**$1.72**
EPS Diluted	$1.13	$1.65

Unlike Bay Network Corporation in Problem 13.5, the company made profits for several years. Compute the following financial ratios, and interpret the firm's financial health during fiscal year 2002:

(a) Debt ratio

(b) Times-interest-earned ratio

(c) Current ratio

(d) Quick (acid test) ratio

(e) Inventory-turnover ratio

(f) Days-sales-outstanding ratio

(g) Total-assets-turnover ratio

(h) Profit margin on sales

(i) Return on total assets

(j) Return on common equity

(k) Price/earnings ratio, with a share price of $65.

(l) Book value per share

13.7 J. C. Olson & Co. had earnings per share of $8 in year 2003, and it paid a $4 dividend. Book value per share at year end was $80. During the same period, the total retained earnings increased by $24 million. Olson has no preferred stock, and no new common stock was issued during the year. If Olson's year-end debt (which equals its total liabilities) was $240 million, what was the company's year-end debt ratio?

Short Case Study with Excel

13.8 Consider the Coca-Cola Company and Pepsi-Cola Company. Both companies compete with each other in the soft-drink sector. Get the most recent annual report for each company, and answer the given questions. [*Note*: You can visit the firms' websites to download their annual reports. (Look for "Investor' Relations.")]

(a) Based on the most recent financial statements, comment on each company's financial performance in the following areas:

- Asset management
- Liquidity
- Debt management
- Profitability
- Market value

(b) Check the current stock prices for both companies. The stock ticker symbols are KO for Coca-Cola and PEP for Pepsi. Based on your analysis in part (a), in which company would you invest your money on and why?

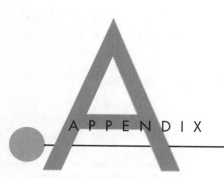

APPENDIX

Electronic Spreadsheets:

A Summary of Built-In

Financial Functions

Some financial functions are available in electronic spreadsheets, namely, Excel and Lotus 1-2-3. As you will see, the keystrokes and choices are similar for both.

A-1   Nominal versus Effective Interest Rates

A conversion from a nominal interest rate to an effective interest rate (or vice versa) is easily obtained with Excel.

Function Description	Excel	Lotus 1-2-3
Effective interest rate	=EFFECT(**nominal_rate, npery**)	
Nominal interest rate	=NOMINAL(**effect_rate, npery**)	

Effect_rate is the effective interest rate; *Npery* is the number of compounding periods per year; *Nominal_rate* is the nominal interest rate.

A-2   Single-Sum Compounding Functions

Single-sum compounding functions deal with either the single-payment compound-amount factor or the present-worth factor. With these functions, the future amount when the present single sum is given, the present amount when its future amount is specified, or the unknown interest (growth) rate and number of interest periods can be computed.

Function Description	Excel	Lotus 1-2-3
1. Calculating the future worth of a single payment	$=FV(i, N, 0, P)$	
2. Calculating the present worth of a single payment	$=PV(i, N, 0, F)$	
3. Calculating an unknown number of payment periods (N)	$=NPER(i, 0, P, F)$	$@CTERM(i, F, P)$
4. Calculating an unknown interest rate	$=RATE(N, 0, P, F, 0, \text{guess})$	$@RATE(F, P, N)$

i is the interest rate; N is the number of the interest periods; F is the future worth specified at the end of the period N; P is the present worth at period 0; *Guess* is your estimate of the interest rate.

A-3   Annuity Functions

Annuity functions provide a full range capability to calculate the future value of an annuity, the present value of an annuity, and the interest rate used in an annuity payment. If a cash payment is made at the end of a period, it is an **ordinary annuity**. If a cash payment is made at the beginning of a period, it is known as an **annuity due**. When using annuity functions, you can specify the choice of annuity by setting type parameters; if type = 0, or is omitted, it is an ordinary annuity. If type = 1, it is an annuity due. In annuity functions, the parameters in boldface must be specified by the user.

Function Description	Excel	Lotus 1-2-3
1. Calculating the number of payment periods (N) in an annuity	=NPER(**i, A, 0, F,** type)	@TERM(**A, i, F**)
2. Calculating the number of payment periods (N) of an annuity when its present worth is specified	=NPER(**i, A, P, F,** type)	@NPER(**A, i, F,** type, **P**)
3. Calculating the future worth of an annuity (equal-payment series compound-amount factor)	=FV(**i, N, A**)	@FV(**A, i, N**)
4. Calculating the future worth of an annuity when its present worth is specified	=FV(**i, N, A, P,** type)	@FVAL(**A, i, N,** type, **P**)
5. Calculating the periodic equal payments of an annuity (capital recovery factor)	=PMT(**i, N, P**)	@PMT(**P, i, N**)
6. Calculating the periodic equal payment of an annuity when its future worth is specified	=PMT(**i, N, P, F,** type)	@PAYMT(**P, i, N,** type, **F**)
7. Calculating the present worth of an annuity (equal-payment series present-worth factor)	=PV(**i, N, A**)	@PV(**A, i, N**)
8. Calculating the present worth of an annuity when its optional future worth is specified	=PV(**i, N, A, F,** type)	@PVAL(**A, i, N,** type, **F**)
9. Calculating the interest rate used in an annuity	=RATE(**N, A, P, F,** type, guess)	@IRATE(**N, A, P,** type, **F**, guess)

Rate (i) is the interest rate per period; *Per (n)* is the period for which you want to find the interest and must be in the range 1 to **nper**; *Pmt (A)* is the payment made each period and cannot change over the life of the annuity; *Nper (N)* is the total number of payment periods in an annuity; *Pv (P)* is the present value, or the lump-sum (starting) amount that a series of future payments is worth right now; *Fv (F)* is the future value, or a cash balance you want to attain after the last payment is made. If *fv* is omitted, it is assumed to be 0 (the future value of a loan, for example, is 0); *Type* is the number 0 or 1 and indicates when payments are due. If type is omitted, it is assumed to be 0.

Set *type* equal to	If payments are due
0	At the end of the period
1	At the beginning of the period

Guess is your estimate of the interest rate.

A–4 Loan Analysis Functions

When you need to compute the monthly, interest, and principal payments, several commands are available to facilitate a typical loan analysis.

Function Description	Excel	Lotus 1-2-3
1. Calculating the periodic	=PMT(*i, N, P,* type)	@PMT(*P, i, N*) loan payment size (*A*)
		@PAYMT(*P, i, N,* type, *F*)
2. Calculating the portion of a loan interest payment for a given period *n*	=IPMT(*i, n, N, P, F,* type)	@IPAYMT(*P, i, N, n*)
3. Calculating the cumulative interest payment between two interval periods	=CUMIMPT(*i, N, P, start_period, end_period,* type)	@IPAYMT(*P, i, N, n, start, end,* type)
4. Calculating the portion of a loan principal payment for a given period *n*	=PPMT(*i, n, N, P, F,* type)	@PPAYMT(*P, i, N, n*)
5. Calculating the cumulative principal payment between two interval periods	=CUMPRINC(*i, N, P, start_period, end_period,* type)	@PPAYMT(*P, i, N, n, start, end,* type)

Rate (*i*) is the interest rate; *Nper* (*N*) is the total number of payment periods; *Pv* (*P*) is the present value; *Start_period* is the first period in the calculation. Payment periods are numbered beginning with 1; *End_period* is the last period in the calculation; *Type* is the timing of the payment.

Type	Timing
0	Payment at the end of the period
1	Payment at the beginning of the period

A–5 Bond Functions

Several financial functions are available to evaluate investments in bonds. In particular, the yield calculation at bond maturity and bond-pricing decision can be easily made.

Function Description	Excel	Lotus 1-2-3
1. Calculating accrued interest	=ACCRINT(*issue, first_interest, maturity, coupon,* par, frequency, basis)	@ACCRUED (*settlement, settlement, rate,* par, frequency, basis)
2. Calculating bond price	=PRICE (*settlement, maturity,* redemption, frequency, basis)	@ACCRUED (*settlement, rate, yield, maturity, coupon, yield,* frequency, basis)

Function Description	Excel	Lotus 1-2-3
3. Calculating the maturity yield	=YIELD(*settlement, maturity, rate, price,* redemption, frequency, basis)	@YIELD (*settlement, maturity, coupon,price,* frequency, basis)

Settlement is the security's settlement date, expressed as a serial date number. Use the *NOW* command to convert a date to a serial number; *Maturity* is the security's maturity date, expressed as a serial date number; *First_interest* is the security's first interest date, expressed as a serial date number; *Coupon* is the security's annual coupon rate; *Issue* is a number representing the issue date; *Par* is the security's par value. If you omit par, ACCRINT uses $1,000; *Price* is the security's price per $100 face value; *Redemption* is the security's redemption value per $100 face value; *Frequency* is the number of coupon payments per year. For annual payments, frequency = 1; for semi-annual payment, frequency = 2; *Yield* is the security's annual yield; *Basis* (or calendar) is the type of day count basis to use:

Basis (or Calendar)	Day Count Basis
0 or omitted	US(NASD) 30/360
1	Actual/Actual
2	Actual/360
3	Actual/365
4	European 30/360

Firstcpn is a number representing the first coupon date.

A-6 Project Evaluation Tools

Several measures of investment worth are available to calculate the NPW, IRR, and annual equivalent of a project's cash flow series.

Function Description	Excel	Lotus 1-2-3
1. Net present value calculation	=NPV(*i, range*)	@NPV(*i, range*)
2. Rate of return calculation	=IRR(*range, guess*)	@IRR(*guess, range*)
3. Annual equivalent calculation	=PMT(*i, N, NPV (i, range),* type)	@PMT(NPV(*i,range*), *i, N*)

i is the minimum attractive rate of return (MARR); *Range* is the cell address where the cash flow streams are stored; *Guess* is the estimated interest rate in solving IRR.

A-7 Depreciation Functions

Function Description	Excel	Lotus 1-2-3
1. Straight-line method	=SLN(*cost, salvage, life*)	@SLN(*cost, salvage, life*)
2. Double declining-balance method: Calculates 200% declining-balance depreciation	=DDB(*cost, salvage, life, period,* factor)	@DDB(*cost, salvage, life, period*)

Function Description	Excel	Lotus 1-2-3
3. Variable declining balance method: Calculates the depreciation by using the variable-rate declining-balance method	=VDB(*cost, salvage, life, life_start, end_period, factor,* no_switch)	@VDB(*cost, salvage, life, start_period, end_period,* depreciation_percent, switch)
4. Sum of the years' digits' method: Calculates the sum of the years' digits depreciation	=SYD(*cost, salvage, life, period*)	@SYD(*cost, salvage, life, period*)

Cost (I) is the initial cost (cost basis) of the asset; *Salvage (S)* is the value at the end of the depreciable life; *Life (N)* is the number of periods over which the asset is being depreciated (known as tax life or depreciable life); *Period (n)* is the period for which you want to calculate the depreciation; *Factor (or depreciation percent)* is the rate at which the balance declines. If factor is omitted, it is assumed to be 2 (double-declining balance method). For a 150% declining-balance method, enter factor = 1.5; *Start_period* is the starting period for which you want to calculate the depreciation; *End_period* is the ending period for which you want to calculate the depreciation; *No_switch* is a logical value specifying whether to switch to straight-line depreciation even when depreciation is greater than the declining-balance method. *If no_switch* (or switch for Lotus) is TRUE, it does not switch to the straight-line method even when the depreciation is greater than the declining balance calculation; *If no_switch* is FALSE or omitted, it switches to straight-line depreciation when the depreciation is greater than the declining-balance calculation.

Interest Factors

for Discrete Compounding

0.25%

	Single Payment		Equal Payment Series				Gradient Series		
N	**Compound Amount Factor** (F/P,i,N)	**Present Worth Factor** (P/F,i,N)	**Compound Amount Factor** (F/A,i,N)	**Sinking Fund Factor** (A/F,i,N)	**Present Worth Factor** (P/A,i,N)	**Capital Recovery Factor** (A/P,i,N)	**Gradient Uniform Series** (A/G,i,N)	**Gradient Present Worth** (P/G,i,N)	**N**
1	1.0025	0.9975	1.0000	1.0000	0.9975	1.0025	0.0000	0.0000	1
2	1.0050	0.9950	2.0025	0.4994	1.9925	0.5019	0.4994	0.9950	2
3	1.0075	0.9925	3.0075	0.3325	2.9851	0.3350	0.9983	2.9801	3
4	1.0100	0.9901	4.0150	0.2491	3.9751	0.2516	1.4969	5.9503	4
5	1.0126	0.9876	5.0251	0.1990	4.9627	0.2015	1.9950	9.9007	5
6	1.0151	0.9851	6.0376	0.1656	5.9478	0.1681	2.4927	14.8263	6
7	1.0176	0.9827	7.0527	0.1418	6.9305	0.1443	2.9900	20.7223	7
8	1.0202	0.9802	8.0704	0.1239	7.9107	0.1264	3.4869	27.5839	8
9	1.0227	0.9778	9.0905	0.1100	8.8885	0.1125	3.9834	35.4061	9
10	1.0253	0.9753	10.1133	0.0989	9.8639	0.1014	4.4794	44.1842	10
11	1.0278	0.9729	11.1385	0.0898	10.8368	0.0923	4.9750	53.9133	11
12	1.0304	0.9705	12.1664	0.0822	11.8073	0.0847	5.4702	64.5886	12
13	1.0330	0.9681	13.1968	0.0758	12.7753	0.0783	5.9650	76.2053	13
14	1.0356	0.9656	14.2298	0.0703	13.7410	0.0728	6.4594	88.7587	14
15	1.0382	0.9632	15.2654	0.0655	14.7042	0.0680	6.9534	102.2441	15
16	1.0408	0.9608	16.3035	0.0613	15.6650	0.0638	7.4469	116.6567	16
17	1.0434	0.9584	17.3443	0.0577	16.6235	0.0602	7.9401	131.9917	17
18	1.0460	0.9561	18.3876	0.0544	17.5795	0.0569	8.4328	148.2446	18
19	1.0486	0.9537	19.4336	0.0515	18.5332	0.0540	8.9251	165.4106	19
20	1.0512	0.9513	20.4822	0.0488	19.4845	0.0513	9.4170	183.4851	20
21	1.0538	0.9489	21.5334	0.0464	20.4334	0.0489	9.9085	202.4634	21
22	1.0565	0.9466	22.5872	0.0443	21.3800	0.0468	10.3995	222.3410	22
23	1.0591	0.9442	23.6437	0.0423	22.3241	0.0448	10.8901	243.1131	23
24	1.0618	0.9418	24.7028	0.0405	23.2660	0.0430	11.3804	264.7753	24
25	1.0644	0.9395	25.7646	0.0388	24.2055	0.0413	11.8702	287.3230	25
26	1.0671	0.9371	26.8290	0.0373	25.1426	0.0398	12.3596	310.7516	26
27	1.0697	0.9348	27.8961	0.0358	26.0774	0.0383	12.8485	335.0566	27
28	1.0724	0.9325	28.9658	0.0345	27.0099	0.0370	13.3371	360.2334	28
29	1.0751	0.9301	30.0382	0.0333	27.9400	0.0358	13.8252	386.2776	29
30	1.0778	0.9278	31.1133	0.0321	28.8679	0.0346	14.3130	413.1847	30
31	1.0805	0.9255	32.1911	0.0311	29.7934	0.0336	14.8003	440.9502	31
32	1.0832	0.9232	33.2716	0.0301	30.7166	0.0326	15.2872	469.5696	32
33	1.0859	0.9209	34.3547	0.0291	31.6375	0.0316	15.7736	499.0386	33
34	1.0886	0.9186	35.4406	0.0282	32.5561	0.0307	16.2597	529.3528	34
35	1.0913	0.9163	36.5292	0.0274	33.4724	0.0299	16.7454	560.5076	35
36	1.0941	0.9140	37.6206	0.0266	34.3865	0.0291	17.2306	592.4988	36
40	1.1050	0.9050	42.0132	0.0238	38.0199	0.0263	19.1673	728.7399	40
48	1.1273	0.8871	50.9312	0.0196	45.1787	0.0221	23.0209	1040.0552	48
50	1.1330	0.8826	53.1887	0.0188	46.9462	0.0213	23.9802	1125.7767	50
60	1.1616	0.8609	64.6467	0.0155	55.6524	0.0180	28.7514	1600.0845	60
72	1.1969	0.8355	78.7794	0.0127	65.8169	0.0152	34.4221	2265.5569	72
80	1.2211	0.8189	88.4392	0.0113	72.4260	0.0138	38.1694	2764.4568	80
84	1.2334	0.8108	93.3419	0.0107	75.6813	0.0132	40.0331	3029.7592	84
90	1.2520	0.7987	100.7885	0.0099	80.5038	0.0124	42.8162	3446.8700	90
96	1.2709	0.7869	108.3474	0.0092	85.2546	0.0117	45.5844	3886.2832	96
100	1.2836	0.7790	113.4500	0.0088	88.3825	0.0113	47.4216	4191.2417	100
108	1.3095	0.7636	123.8093	0.0081	94.5453	0.0106	51.0762	4829.0125	108
120	1.3494	0.7411	139.7414	0.0072	103.5618	0.0097	56.5084	5852.1116	120
240	1.8208	0.5492	328.3020	0.0030	180.3109	0.0055	107.5863	19398.9852	240
360	2.4568	0.4070	582.7369	0.0017	237.1894	0.0042	152.8902	36263.9299	360

0.50%

	Single Payment		Equal Payment Series				Gradient Series		
	Compound Amount Factor	Present Worth Factor	Compound Amount Factor	Sinking Fund Factor	Present Worth Factor	Capital Recovery Factor	Gradient Uniform Series	Gradient Present Worth	
N	(F/P,i,N)	(P/F,i,N)	(F/A,i,N)	(A/F,i,N)	(P/A,i,N)	(A/P,i,N)	(A/G,i,N)	(P/G,i,N)	N
1	1.0050	0.9950	1.0000	1.0000	0.9950	1.0050	0.0000	0.0000	1
2	1.0100	0.9901	2.0050	0.4988	1.9851	0.5038	0.4988	0.9901	2
3	1.0151	0.9851	3.0150	0.3317	2.9702	0.3367	0.9967	2.9604	3
4	1.0202	0.9802	4.0301	0.2481	3.9505	0.2531	1.4938	5.9011	4
5	1.0253	0.9754	5.0503	0.1980	4.9259	0.2030	1.9900	9.8026	5
6	1.0304	0.9705	6.0755	0.1646	5.8964	0.1696	2.4855	14.6552	6
7	1.0355	0.9657	7.1059	0.1407	6.8621	0.1457	2.9801	20.4493	7
8	1.0407	0.9609	8.1414	0.1228	7.8230	0.1278	3.4738	27.1755	8
9	1.0459	0.9561	9.1821	0.1089	8.7791	0.1139	3.9668	34.8244	9
10	1.0511	0.9513	10.2280	0.0978	9.7304	0.1028	4.4589	43.3865	10
11	1.0564	0.9466	11.2792	0.0887	10.6770	0.0937	4.9501	52.8526	11
12	1.0617	0.9419	12.3356	0.0811	11.6189	0.0861	5.4406	63.2136	12
13	1.0670	0.9372	13.3972	0.0746	12.5562	0.0796	5.9302	74.4602	13
14	1.0723	0.9326	14.4642	0.0691	13.4887	0.0741	6.4190	86.5835	14
15	1.0777	0.9279	15.5365	0.0644	14.4166	0.0694	6.9069	99.5743	15
16	1.0831	0.9233	16.6142	0.0602	15.3399	0.0652	7.3940	113.4238	16
17	1.0885	0.9187	17.6973	0.0565	16.2586	0.0615	7.8803	128.1231	17
18	1.0939	0.9141	18.7858	0.0532	17.1728	0.0582	8.3658	143.6634	18
19	1.0994	0.9096	19.8797	0.0503	18.0824	0.0553	8.8504	160.0360	19
20	1.1049	0.9051	20.9791	0.0477	18.9874	0.0527	9.3342	177.2322	20
21	1.1104	0.9006	22.0840	0.0453	19.8880	0.0503	9.8172	195.2434	21
22	1.1160	0.8961	23.1944	0.0431	20.7841	0.0481	10.2993	214.0611	22
23	1.1216	0.8916	24.3104	0.0411	21.6757	0.0461	10.7806	233.6768	23
24	1.1272	0.8872	25.4320	0.0393	22.5629	0.0443	11.2611	254.0820	24
25	1.1328	0.8828	26.5591	0.0377	23.4456	0.0427	11.7407	275.2686	25
26	1.1385	0.8784	27.6919	0.0361	24.3240	0.0411	12.2195	297.2281	26
27	1.1442	0.8740	28.8304	0.0347	25.1980	0.0397	12.6975	319.9523	27
28	1.1499	0.8697	29.9745	0.0334	26.0677	0.0384	13.1747	343.4332	28
29	1.1556	0.8653	31.1244	0.0321	26.9330	0.0371	13.6510	367.6625	29
30	1.1614	0.8610	32.2800	0.0310	27.7941	0.0360	14.1265	392.6324	30
31	1.1672	0.8567	33.4414	0.0299	28.6508	0.0349	14.6012	418.3348	31
32	1.1730	0.8525	34.6086	0.0289	29.5033	0.0339	15.0750	444.7618	32
33	1.1789	0.8482	35.7817	0.0279	30.3515	0.0329	15.5480	471.9055	33
34	1.1848	0.8440	36.9606	0.0271	31.1955	0.0321	16.0202	499.7583	34
35	1.1907	0.8398	38.1454	0.0262	32.0354	0.0312	16.4915	528.3123	35
36	1.1967	0.8356	39.3361	0.0254	32.8710	0.0304	16.9621	557.5598	36
40	1.2208	0.8191	44.1588	0.0226	36.1722	0.0276	18.8359	681.3347	40
48	1.2705	0.7871	54.0978	0.0185	42.5803	0.0235	22.5437	959.9188	48
50	1.2832	0.7793	56.6452	0.0177	44.1428	0.0227	23.4624	1035.6966	50
60	1.3489	0.7414	69.7700	0.0143	51.7256	0.0193	28.0064	1448.6458	60
72	1.4320	0.6983	86.4089	0.0116	60.3395	0.0166	33.3504	2012.3478	72
80	1.4903	0.6710	98.0677	0.0102	65.8023	0.0152	36.8474	2424.6455	80
84	1.5204	0.6577	104.0739	0.0096	68.4530	0.0146	38.5763	2640.6641	84
90	1.5666	0.6383	113.3109	0.0088	72.3313	0.0138	41.1451	2976.0769	90
96	1.6141	0.6195	122.8285	0.0081	76.0952	0.0131	43.6845	3324.1846	96
100	1.6467	0.6073	129.3337	0.0077	78.5426	0.0127	45.3613	3562.7934	100
108	1.7137	0.5835	142.7399	0.0070	83.2934	0.0120	48.6758	4054.3747	108
120	1.8194	0.5496	163.8793	0.0061	90.0735	0.0111	53.5508	4823.5051	120
240	3.3102	0.3021	462.0409	0.0022	139.5808	0.0072	96.1131	13415.5395	240
360	6.0226	0.1660	1004.5150	0.0010	166.7916	0.0060	128.3236	21403.3041	360

0.75%

	Single Payment		Equal Payment Series				Gradient Series		
N	Compound Amount Factor (F/P,i,N)	Present Worth Factor (P/F,i,N)	Compound Amount Factor (F/A,i,N)	Sinking Fund Factor (A/F,i,N)	Present Worth Factor (P/A,i,N)	Capital Recovery Factor (A/P,i,N)	Gradient Uniform Series (A/G,i,N)	Gradient Present Worth (P/G,i,N)	N
1	1.0075	0.9926	1.0000	1.0000	0.9926	1.0075	0.0000	0.0000	1
2	1.0151	0.9852	2.0075	0.4981	1.9777	0.5056	0.4981	0.9852	2
3	1.0227	0.9778	3.0226	0.3308	2.9556	0.3383	0.9950	2.9408	3
4	1.0303	0.9706	4.0452	0.2472	3.9261	0.2547	1.4907	5.8525	4
5	1.0381	0.9633	5.0756	0.1970	4.8894	0.2045	1.9851	9.7058	5
6	1.0459	0.9562	6.1136	0.1636	5.8456	0.1711	2.4782	14.4866	6
7	1.0537	0.9490	7.1595	0.1397	6.7946	0.1472	2.9701	20.1808	7
8	1.0616	0.9420	8.2132	0.1218	7.7366	0.1293	3.4608	26.7747	8
9	1.0696	0.9350	9.2748	0.1078	8.6716	0.1153	3.9502	34.2544	9
10	1.0776	0.9280	10.3443	0.0967	9.5996	0.1042	4.4384	42.6064	10
11	1.0857	0.9211	11.4219	0.0876	10.5207	0.0951	4.9253	51.8174	11
12	1.0938	0.9142	12.5076	0.0800	11.4349	0.0875	5.4110	61.8740	12
13	1.1020	0.9074	13.6014	0.0735	12.3423	0.0810	5.8954	72.7632	13
14	1.1103	0.9007	14.7034	0.0680	13.2430	0.0755	6.3786	84.4720	14
15	1.1186	0.8940	15.8137	0.0632	14.1370	0.0707	6.8606	96.9876	15
16	1.1270	0.8873	16.9323	0.0591	15.0243	0.0666	7.3413	110.2973	16
17	1.1354	0.8807	18.0593	0.0554	15.9050	0.0629	7.8207	124.3887	17
18	1.1440	0.8742	19.1947	0.0521	16.7792	0.0596	8.2989	139.2494	18
19	1.1525	0.8676	20.3387	0.0492	17.6468	0.0567	8.7759	154.8671	19
20	1.1612	0.8612	21.4912	0.0465	18.5080	0.0540	9.2516	171.2297	20
21	1.1699	0.8548	22.6524	0.0441	19.3628	0.0516	9.7261	188.3253	21
22	1.1787	0.8484	23.8223	0.0420	20.2112	0.0495	10.1994	206.1420	22
23	1.1875	0.8421	25.0010	0.0400	21.0533	0.0475	10.6714	224.6682	23
24	1.1964	0.8358	26.1885	0.0382	21.8891	0.0457	11.1422	243.8923	24
25	1.2054	0.8296	27.3849	0.0365	22.7188	0.0440	11.6117	263.8029	25
26	1.2144	0.8234	28.5903	0.0350	23.5422	0.0425	12.0800	284.3888	26
27	1.2235	0.8173	29.8047	0.0336	24.3595	0.0411	12.5470	305.6387	27
28	1.2327	0.8112	31.0282	0.0322	25.1707	0.0397	13.0128	327.5416	28
29	1.2420	0.8052	32.2609	0.0310	25.9759	0.0385	13.4774	350.0867	29
30	1.2513	0.7992	33.5029	0.0298	26.7751	0.0373	13.9407	373.2631	30
31	1.2607	0.7932	34.7542	0.0288	27.5683	0.0363	14.4028	397.0602	31
32	1.2701	0.7873	36.0148	0.0278	28.3557	0.0353	14.8636	421.4675	32
33	1.2796	0.7815	37.2849	0.0268	29.1371	0.0343	15.3232	446.4746	33
34	1.2892	0.7757	38.5646	0.0259	29.9128	0.0334	15.7816	472.0712	34
35	1.2989	0.7699	39.8538	0.0251	30.6827	0.0326	16.2387	498.2471	35
36	1.3086	0.7641	41.1527	0.0243	31.4468	0.0318	16.6946	524.9924	36
40	1.3483	0.7416	46.4465	0.0215	34.4469	0.0290	18.5058	637.4693	40
48	1.4314	0.6986	57.5207	0.0174	40.1848	0.0249	22.0691	886.8404	48
50	1.4530	0.6883	60.3943	0.0166	41.5664	0.0241	22.9476	953.8486	50
60	1.5657	0.6387	75.4241	0.0133	48.1734	0.0208	27.2665	1313.5189	60
72	1.7126	0.5839	95.0070	0.0105	55.4768	0.0180	32.2882	1791.2463	72
80	1.8180	0.5500	109.0725	0.0092	59.9944	0.0167	35.5391	2132.1472	80
84	1.8732	0.5338	116.4269	0.0086	62.1540	0.0161	37.1357	2308.1283	84
90	1.9591	0.5104	127.8790	0.0078	65.2746	0.0153	39.4946	2577.9961	90
96	2.0489	0.4881	139.8562	0.0072	68.2584	0.0147	41.8107	2853.9352	96
100	2.1111	0.4737	148.1445	0.0068	70.1746	0.0143	43.3311	3040.7453	100
108	2.2411	0.4462	165.4832	0.0060	73.8394	0.0135	46.3154	3419.9041	108
120	2.4514	0.4079	193.5143	0.0052	78.9417	0.0127	50.6521	3998.5621	120
240	6.0092	0.1664	667.8869	0.0015	111.1450	0.0090	85.4210	9494.1162	240
360	14.7306	0.0679	1830.7435	0.0005	124.2819	0.0080	107.1145	13312.3871	3600

1.0%

	Single Payment		Equal Payment Series				Gradient Series		
N	Compound Amount Factor $(F/P,i,N)$	Present Worth Factor $(P/F,i,N)$	Compound Amount Factor $(F/A,i,N)$	Sinking Fund Factor $(A/F,i,N)$	Present Worth Factor $(P/A,i,N)$	Capital Recovery Factor $(A/P,i,N)$	Gradient Uniform Series $(A/G,i,N)$	Gradient Present Worth $(P/G,i,N)$	**N**
1	1.0100	0.9901	1.0000	1.0000	0.9901	1.0100	0.0000	0.0000	1
2	1.0201	0.9803	2.0100	0.4975	1.9704	0.5075	0.4975	0.9803	2
3	1.0303	0.9706	3.0301	0.3300	2.9410	0.3400	0.9934	2.9215	3
4	1.0406	0.9610	4.0604	0.2463	3.9020	0.2563	1.4876	5.8044	4
5	1.0510	0.9515	5.1010	0.1960	4.8534	0.2060	1.9801	9.6103	5
6	1.0615	0.9420	6.1520	0.1625	5.7955	0.1725	2.4710	14.3205	6
7	1.0721	0.9327	7.2135	0.1386	6.7282	0.1486	2.9602	19.9168	7
8	1.0829	0.9235	8.2857	0.1207	7.6517	0.1307	3.4478	26.3812	8
9	1.0937	0.9143	9.3685	0.1067	8.5660	0.1167	3.9337	33.6959	9
10	1.1046	0.9053	10.4622	0.0956	9.4713	0.1056	4.4179	41.8435	10
11	1.1157	0.8963	11.5668	0.0865	10.3676	0.0965	4.9005	50.8067	11
12	1.1268	0.8874	12.6825	0.0788	11.2551	0.0888	5.3815	60.5687	12
13	1.1381	0.8787	13.8093	0.0724	12.1337	0.0824	5.8607	71.1126	13
14	1.1495	0.8700	14.9474	0.0669	13.0037	0.0769	6.3384	82.4221	14
15	1.1610	0.8613	16.0969	0.0621	13.8651	0.0721	6.8143	94.4810	15
16	1.1726	0.8528	17.2579	0.0579	14.7179	0.0679	7.2886	107.2734	16
17	1.1843	0.8444	18.4304	0.0543	15.5623	0.0643	7.7613	120.7834	17
18	1.1961	0.8360	19.6147	0.0510	16.3983	0.0610	8.2323	134.9957	18
19	1.2081	0.8277	20.8109	0.0481	17.2260	0.0581	8.7017	149.8950	19
20	1.2202	0.8195	22.0190	0.0454	18.0456	0.0554	9.1694	165.4664	20
21	1.2324	0.8114	23.2392	0.0430	18.8570	0.0530	9.6354	181.6950	21
22	1.2447	0.8034	24.4716	0.0409	19.6604	0.0509	10.0998	198.5663	22
23	1.2572	0.7954	25.7163	0.0389	20.4558	0.0489	10.5626	216.0660	23
24	1.2697	0.7876	26.9735	0.0371	21.2434	0.0471	11.0237	234.1800	24
25	1.2824	0.7798	28.2432	0.0354	22.0232	0.0454	11.4831	252.8945	25
26	1.2953	0.7720	29.5256	0.0339	22.7952	0.0439	11.9409	272.1957	26
27	1.3082	0.7644	30.8209	0.0324	23.5596	0.0424	12.3971	292.0702	27
28	1.3213	0.7568	32.1291	0.0311	24.3164	0.0411	12.8516	312.5047	28
29	1.3345	0.7493	33.4504	0.0299	25.0658	0.0399	13.3044	333.4863	29
30	1.3478	0.7419	34.7849	0.0287	25.8077	0.0387	13.7557	355.0021	30
31	1.3613	0.7346	36.1327	0.0277	26.5423	0.0377	14.2052	377.0394	31
32	1.3749	0.7273	37.4941	0.0267	27.2696	0.0367	14.6532	399.5858	32
33	1.3887	0.7201	38.8690	0.0257	27.9897	0.0357	15.0995	422.6291	33
34	1.4026	0.7130	40.2577	0.0248	28.7027	0.0348	15.5441	446.1572	34
35	1.4166	0.7059	41.6603	0.0240	29.4086	0.0340	15.9871	470.1583	35
36	1.4308	0.6989	43.0769	0.0232	30.1075	0.0332	16.4285	494.6207	36
40	1.4889	0.6717	48.8864	0.0205	32.8347	0.0305	18.1776	596.8561	40
48	1.6122	0.6203	61.2226	0.0163	37.9740	0.0263	21.5976	820.1460	48
50	1.6446	0.6080	64.4632	0.0155	39.1961	0.0255	22.4363	879.4176	50
60	1.8167	0.5504	81.6697	0.0122	44.9550	0.0222	26.5333	1192.8061	60
72	2.0471	0.4885	104.7099	0.0096	51.1504	0.0196	31.2386	1597.8673	72
80	2.2167	0.4511	121.6715	0.0082	54.8882	0.0182	34.2492	1879.8771	80
84	2.3067	0.4335	130.6723	0.0077	56.6485	0.0177	35.7170	2023.3153	84
90	2.4486	0.4084	144.8633	0.0069	59.1609	0.0169	37.8724	2240.5675	90
96	2.5993	0.3847	159.9273	0.0063	61.5277	0.0163	39.9727	2459.4298	96
100	2.7048	0.3697	170.4814	0.0059	63.0289	0.0159	41.3426	2605.7758	100
108	2.9289	0.3414	192.8926	0.0052	65.8578	0.0152	44.0103	2898.4203	108
120	3.3004	0.3030	230.0387	0.0043	69.7005	0.0143	47.8349	3334.1148	120
240	10.8926	0.0918	989.2554	0.0010	90.8194	0.0110	75.7393	6878.6016	240
360	35.9496	0.0278	3494.9641	0.0003	97.2183	0.0103	89.6995	8720.4323	360

1.25%

	Single Payment		Equal Payment Series				Gradient Series		
	Compound Amount Factor	Present Worth Factor	Compound Amount Factor	Sinking Fund Factor	Present Worth Factor	Capital Recovery Factor	Gradient Uniform Series	Gradient Present Worth	
N	(F/P,i,N)	(P/F,i,N)	(F/A,i,N)	(A/F,i,N)	(P/A,i,N)	(A/P,i,N)	(A/G,i,N)	(P/G,i,N)	N
1	1.0125	0.9877	1.0000	1.0000	0.9877	1.0125	0.0000	0.0000	1
2	1.0252	0.9755	2.0125	0.4969	1.9631	0.5094	0.4969	0.9755	2
3	1.0380	0.9634	3.0377	0.3292	2.9265	0.3417	0.9917	2.9023	3
4	1.0509	0.9515	4.0756	0.2454	3.8781	0.2579	1.4845	5.7569	4
5	1.0641	0.9398	5.1266	0.1951	4.8178	0.2076	1.9752	9.5160	5
6	1.0774	0.9282	6.1907	0.1615	5.7460	0.1740	2.4638	14.1569	6
7	1.0909	0.9167	7.2680	0.1376	6.6627	0.1501	2.9503	19.6571	7
8	1.1045	0.9054	8.3589	0.1196	7.5681	0.1321	3.4348	25.9949	8
9	1.1183	0.8942	9.4634	0.1057	8.4623	0.1182	3.9172	33.1487	9
10	1.1323	0.8832	10.5817	0.0945	9.3455	0.1070	4.3975	41.0973	10
11	1.1464	0.8723	11.7139	0.0854	10.2178	0.0979	4.8758	49.8201	11
12	1.1608	0.8615	12.8604	0.0778	11.0793	0.0903	5.3520	59.2967	12
13	1.1753	0.8509	14.0211	0.0713	11.9302	0.0838	5.8262	69.5072	13
14	1.1900	0.8404	15.1964	0.0658	12.7706	0.0783	6.2982	80.4320	14
15	1.2048	0.8300	16.3863	0.0610	13.6005	0.0735	6.7682	92.0519	15
16	1.2199	0.8197	17.5912	0.0568	14.4203	0.0693	7.2362	104.3481	16
17	1.2351	0.8096	18.8111	0.0532	15.2299	0.0657	7.7021	117.3021	17
18	1.2506	0.7996	20.0462	0.0499	16.0295	0.0624	8.1659	130.8958	18
19	1.2662	0.7898	21.2968	0.0470	16.8193	0.0595	8.6277	145.1115	19
20	1.2820	0.7800	22.5630	0.0443	17.5993	0.0568	9.0874	159.9316	20
21	1.2981	0.7704	23.8450	0.0419	18.3697	0.0544	9.5450	175.3392	21
22	1.3143	0.7609	25.1431	0.0398	19.1306	0.0523	10.0006	191.3174	22
23	1.3307	0.7515	26.4574	0.0378	19.8820	0.0503	10.4542	207.8499	23
24	1.3474	0.7422	27.7881	0.0360	20.6242	0.0485	10.9056	224.9204	24
25	1.3642	0.7330	29.1354	0.0343	21.3573	0.0468	11.3551	242.5132	25
26	1.3812	0.7240	30.4996	0.0328	22.0813	0.0453	11.8024	260.6128	26
27	1.3985	0.7150	31.8809	0.0314	22.7963	0.0439	12.2478	279.2040	27
28	1.4160	0.7062	33.2794	0.0300	23.5025	0.0425	12.6911	298.2719	28
29	1.4337	0.6975	34.6954	0.0288	24.2000	0.0413	13.1323	317.8019	29
30	1.4516	0.6889	36.1291	0.0277	24.8889	0.0402	13.5715	337.7797	30
31	1.4698	0.6804	37.5807	0.0266	25.5693	0.0391	14.0086	358.1912	31
32	1.4881	0.6720	39.0504	0.0256	26.2413	0.0381	14.4438	379.0227	32
33	1.5067	0.6637	40.5386	0.0247	26.9050	0.0372	14.8768	400.2607	33
34	1.5256	0.6555	42.0453	0.0238	27.5605	0.0363	15.3079	421.8920	34
35	1.5446	0.6474	43.5709	0.0230	28.2079	0.0355	15.7369	443.9037	35
36	1.5639	0.6394	45.1155	0.0222	28.8473	0.0347	16.1639	466.2830	36
40	1.6436	0.6084	51.4896	0.0194	31.3269	0.0319	17.8515	559.2320	40
48	1.8154	0.5509	65.2284	0.0153	35.9315	0.0278	21.1299	759.2296	48
50	1.8610	0.5373	68.8818	0.0145	37.0129	0.0270	21.9295	811.6738	50
60	2.1072	0.4746	88.5745	0.0113	42.0346	0.0238	25.8083	1084.8429	60
72	2.4459	0.4088	115.6736	0.0086	47.2925	0.0211	30.2047	1428.4561	72
80	2.7015	0.3702	136.1188	0.0073	50.3867	0.0198	32.9822	1661.8651	80
84	2.8391	0.3522	147.1290	0.0068	51.8222	0.0193	34.3258	1778.8384	84
90	3.0588	0.3269	164.7050	0.0061	53.8461	0.0186	36.2855	1953.8303	90
96	3.2955	0.3034	183.6411	0.0054	55.7246	0.0179	38.1793	2127.5244	96
100	3.4634	0.2887	197.0723	0.0051	56.9013	0.0176	39.4058	2242.2411	100
108	3.8253	0.2614	226.0226	0.0044	59.0865	0.0169	41.7737	2468.2636	108
120	4.4402	0.2252	275.2171	0.0036	61.9828	0.0161	45.1184	2796.5694	120
240	19.7155	0.0507	1497.2395	0.0007	75.9423	0.0132	67.1764	5101.5288	240
360	87.5410	0.0114	6923.2796	0.0001	79.0861	0.0126	75.8401	5997.9027	360

	Single Payment		Equal Payment Series				Gradient Series		
N	Compound Amount Factor (F/P,i,N)	Present Worth Factor (P/F,i,N)	Compound Amount Factor (F/A,i,N)	Sinking Fund Factor (A/F,i,N)	Present Worth Factor (P/A,i,N)	Capital Recovery Factor (A/P,i,N)	Gradient Uniform Series (A/G,i,N)	Gradient Present Worth (P/G,i,N)	N
1	1.0150	0.9852	1.0000	1.0000	0.9852	1.0150	0.0000	0.0000	1
2	1.0302	0.9707	2.0150	0.4963	1.9559	0.5113	0.4963	0.9707	2
3	1.0457	0.9563	3.0452	0.3284	2.9122	0.3434	0.9901	2.8833	3
4	1.0614	0.9422	4.0909	0.2444	3.8544	0.2594	1.4814	5.7098	4
5	1.0773	0.9283	5.1523	0.1941	4.7826	0.2091	1.9702	9.4229	5
6	1.0934	0.9145	6.2296	0.1605	5.6972	0.1755	2.4566	13.9956	6
7	1.1098	0.9010	7.3230	0.1366	6.5982	0.1516	2.9405	19.4018	7
8	1.1265	0.8877	8.4328	0.1186	7.4859	0.1336	3.4219	25.6157	8
9	1.1434	0.8746	9.5593	0.1046	8.3605	0.1196	3.9008	32.6125	9
10	1.1605	0.8617	10.7027	0.0934	9.2222	0.1084	4.3772	40.3675	10
11	1.1779	0.8489	11.8633	0.0843	10.0711	0.0993	4.8512	48.8568	11
12	1.1956	0.8364	13.0412	0.0767	10.9075	0.0917	5.3227	58.0571	12
13	1.2136	0.8240	14.2368	0.0702	11.7315	0.0852	5.7917	67.9454	13
14	1.2318	0.8118	15.4504	0.0647	12.5434	0.0797	6.2582	78.4994	14
15	1.2502	0.7999	16.6821	0.0599	13.3432	0.0749	6.7223	89.6974	15
16	1.2690	0.7880	17.9324	0.0558	14.1313	0.0708	7.1839	101.5178	16
17	1.2880	0.7764	19.2014	0.0521	14.9076	0.0671	7.6431	113.9400	17
18	1.3073	0.7649	20.4894	0.0488	15.6726	0.0638	8.0997	126.9435	18
19	1.3270	0.7536	21.7967	0.0459	16.4262	0.0609	8.5539	140.5084	19
20	1.3469	0.7425	23.1237	0.0432	17.1686	0.0582	9.0057	154.6154	20
21	1.3671	0.7315	24.4705	0.0409	17.9001	0.0559	9.4550	169.2453	21
22	1.3876	0.7207	25.8376	0.0387	18.6208	0.0537	9.9018	184.3798	22
23	1.4084	0.7100	27.2251	0.0367	19.3309	0.0517	10.3462	200.0006	23
24	1.4295	0.6995	28.6335	0.0349	20.0304	0.0499	10.7881	216.0901	24
25	1.4509	0.6892	30.0630	0.0333	20.7196	0.0483	11.2276	232.6310	25
26	1.4727	0.6790	31.5140	0.0317	21.3986	0.0467	11.6646	249.6065	26
27	1.4948	0.6690	32.9867	0.0303	22.0676	0.0453	12.0992	267.0002	27
28	1.5172	0.6591	34.4815	0.0290	22.7267	0.0440	12.5313	284.7958	28
29	1.5400	0.6494	35.9987	0.0278	23.3761	0.0428	12.9610	302.9779	29
30	1.5631	0.6398	37.5387	0.0266	24.0158	0.0416	13.3883	321.5310	30
31	1.5865	0.6303	39.1018	0.0256	24.6461	0.0406	13.8131	340.4402	31
32	1.6103	0.6210	40.6883	0.0246	25.2671	0.0396	14.2355	359.6910	32
33	1.6345	0.6118	42.2986	0.0236	25.8790	0.0386	14.6555	379.2691	33
34	1.6590	0.6028	43.9331	0.0228	26.4817	0.0378	15.0731	399.1607	34
35	1.6839	0.5939	45.5921	0.0219	27.0756	0.0369	15.4882	419.3521	35
36	1.7091	0.5851	47.2760	0.0212	27.6607	0.0362	15.9009	439.8303	36
40	1.8140	0.5513	54.2679	0.0184	29.9158	0.0334	17.5277	524.3568	40
48	2.0435	0.4894	69.5652	0.0144	34.0426	0.0294	20.6667	703.5462	48
50	2.1052	0.4750	73.6828	0.0136	34.9997	0.0286	21.4277	749.9636	50
60	2.4432	0.4093	96.2147	0.0104	39.3803	0.0254	25.0930	988.1674	60
72	2.9212	0.3423	128.0772	0.0078	43.8447	0.0228	29.1893	1279.7938	72
80	3.2907	0.3039	152.7109	0.0065	46.4073	0.0215	31.7423	1473.0741	80
84	3.4926	0.2863	166.1726	0.0060	47.5786	0.0210	32.9668	1568.5140	84
90	3.8189	0.2619	187.9299	0.0053	49.2099	0.0203	34.7399	1709.5439	90
96	4.1758	0.2395	211.7202	0.0047	50.7017	0.0197	36.4381	1847.4725	96
100	4.4320	0.2256	228.8030	0.0044	51.6247	0.0194	37.5295	1937.4506	100
108	4.9927	0.2003	266.1778	0.0038	53.3137	0.0188	39.6171	2112.1348	108
120	5.9693	0.1675	331.2882	0.0030	55.4985	0.0180	42.5185	2359.7114	120
240	35.6328	0.0281	2308.8544	0.0004	64.7957	0.0154	59.7368	3870.6912	240
360	212.7038	0.0047	14113.5854	0.0001	66.3532	0.0151	64.9662	4310.7165	360

1.75%

	Single Payment		Equal Payment Series				Gradient Series		
N	Compound Amount Factor (F/P,i,N)	Present Worth Factor (P/F,i,N)	Compound Amount Factor (F/A,i,N)	Sinking Fund Factor (A/F,i,N)	Present Worth Factor (P/A,i,N)	Capital Recovery Factor (A/P,i,N)	Gradient Uniform Series (A/G,i,N)	Gradient Present Worth (P/G,i,N)	N
1	1.0175	0.9828	1.0000	1.0000	0.9828	1.0175	0.0000	0.0000	1
2	1.0353	0.9659	2.0175	0.4957	1.9487	0.5132	0.4957	0.9659	2
3	1.0534	0.9493	3.0528	0.3276	2.8980	0.3451	0.9884	2.8645	3
4	1.0719	0.9330	4.1062	0.2435	3.8309	0.2610	1.4783	5.6633	4
5	1.0906	0.9169	5.1781	0.1931	4.7479	0.2106	1.9653	9.3310	5
6	1.1097	0.9011	6.2687	0.1595	5.6490	0.1770	2.4494	13.8367	6
7	1.1291	0.8856	7.3784	0.1355	6.5346	0.1530	2.9306	19.1506	7
8	1.1489	0.8704	8.5075	0.1175	7.4051	0.1350	3.4089	25.2435	8
9	1.1690	0.8554	9.6564	0.1036	8.2605	0.1211	3.8844	32.0870	9
10	1.1894	0.8407	10.8254	0.0924	9.1012	0.1099	4.3569	39.6535	10
11	1.2103	0.8263	12.0148	0.0832	9.9275	0.1007	4.8266	47.9162	11
12	1.2314	0.8121	13.2251	0.0756	10.7395	0.0931	5.2934	56.8489	12
13	1.2530	0.7981	14.4565	0.0692	11.5376	0.0867	5.7573	66.4260	13
14	1.2749	0.7844	15.7095	0.0637	12.3220	0.0812	6.2184	76.6227	14
15	1.2972	0.7709	16.9844	0.0589	13.0929	0.0764	6.6765	87.4149	15
16	1.3199	0.7576	18.2817	0.0547	13.8505	0.0722	7.1318	98.7792	16
17	1.3430	0.7446	19.6016	0.0510	14.5951	0.0685	7.5842	110.6926	17
18	1.3665	0.7318	20.9446	0.0477	15.3269	0.0652	8.0338	123.1328	18
19	1.3904	0.7192	22.3112	0.0448	16.0461	0.0623	8.4805	136.0783	19
20	1.4148	0.7068	23.7016	0.0422	16.7529	0.0597	8.9243	149.5080	20
21	1.4395	0.6947	25.1164	0.0398	17.4475	0.0573	9.3653	163.4013	21
22	1.4647	0.6827	26.5559	0.0377	18.1303	0.0552	9.8034	177.7385	22
23	1.4904	0.6710	28.0207	0.0357	18.8012	0.0532	10.2387	192.5000	23
24	1.5164	0.6594	29.5110	0.0339	19.4607	0.0514	10.6711	207.6671	24
25	1.5430	0.6481	31.0275	0.0322	20.1088	0.0497	11.1007	223.2214	25
26	1.5700	0.6369	32.5704	0.0307	20.7457	0.0482	11.5274	239.1451	26
27	1.5975	0.6260	34.1404	0.0293	21.3717	0.0468	11.9513	255.4210	27
28	1.6254	0.6152	35.7379	0.0280	21.9870	0.0455	12.3724	272.0321	28
29	1.6539	0.6046	37.3633	0.0268	22.5916	0.0443	12.7907	288.9623	29
30	1.6828	0.5942	39.0172	0.0256	23.1858	0.0431	13.2061	306.1954	30
31	1.7122	0.5840	40.7000	0.0246	23.7699	0.0421	13.6188	323.7163	31
32	1.7422	0.5740	42.4122	0.0236	24.3439	0.0411	14.0286	341.5097	32
33	1.7727	0.5641	44.1544	0.0226	24.9080	0.0401	14.4356	359.5613	33
34	1.8037	0.5544	45.9271	0.0218	25.4624	0.0393	14.8398	377.8567	34
35	1.8353	0.5449	47.7308	0.0210	26.0073	0.0385	15.2412	396.3824	35
36	1.8674	0.5355	49.5661	0.0202	26.5428	0.0377	15.6399	415.1250	36
40	2.0016	0.4996	57.2341	0.0175	28.5942	0.0350	17.2066	492.0109	40
48	2.2996	0.4349	74.2628	0.0135	32.2938	0.0310	20.2084	652.6054	48
50	2.3808	0.4200	78.9022	0.0127	33.1412	0.0302	20.9317	693.7010	50
60	2.8318	0.3531	104.6752	0.0096	36.9640	0.0271	24.3885	901.4954	60
72	3.4872	0.2868	142.1263	0.0070	40.7564	0.0245	28.1948	1149.1181	72
80	4.0064	0.2496	171.7938	0.0058	42.8799	0.0233	30.5329	1309.2482	80
84	4.2943	0.2329	188.2450	0.0053	43.8361	0.0228	31.6442	1387.1584	84
90	4.7654	0.2098	215.1646	0.0046	45.1516	0.0221	33.2409	1500.8798	90
96	5.2882	0.1891	245.0374	0.0041	46.3370	0.0216	34.7556	1610.4716	96
100	5.6682	0.1764	266.7518	0.0037	47.0615	0.0212	35.7211	1681.0886	100
108	6.5120	0.1536	314.9738	0.0032	48.3679	0.0207	37.5494	1816.1852	108
120	8.0192	0.1247	401.0962	0.0025	50.0171	0.0200	40.0469	2003.0269	120
240	64.3073	0.0156	3617.5602	0.0003	56.2543	0.0178	53.3518	3001.2678	240
360	515.6921	0.0019	29410.9747	0.0000	57.0320	0.0175	56.4434	3219.0833	360

2.0%

	Single Payment		Equal Payment Series				Gradient Series		
	Compound Amount Factor	Present Worth Factor	Compound Amount Factor	Sinking Fund Factor	Present Worth Factor	Capital Recovery Factor	Gradient Uniform Series	Gradient Present Worth	
N	(F/P,i,N)	(P/F,i,N)	(F/A,i,N)	(A/F,i,N)	(P/A,i,N)	(A/P,i,N)	(A/G,i,N)	(P/G,i,N)	N
1	1.0200	0.9804	1.0000	1.0000	0.9804	1.0200	0.0000	0.0000	1
2	1.0404	0.9612	2.0200	0.4950	1.9416	0.5150	0.4950	0.9612	2
3	1.0612	0.9423	3.0604	0.3268	2.8839	0.3468	0.9868	2.8458	3
4	1.0824	0.9238	4.1216	0.2426	3.8077	0.2626	1.4752	5.6173	4
5	1.1041	0.9057	5.2040	0.1922	4.7135	0.2122	1.9604	9.2403	5
6	1.1262	0.8880	6.3081	0.1585	5.6014	0.1785	2.4423	13.6801	6
7	1.1487	0.8706	7.4343	0.1345	6.4720	0.1545	2.9208	18.9035	7
8	1.1717	0.8535	8.5830	0.1165	7.3255	0.1365	3.3961	24.8779	8
9	1.1951	0.8368	9.7546	0.1025	8.1622	0.1225	3.8681	31.5720	9
10	1.2190	0.8203	10.9497	0.0913	8.9826	0.1113	4.3367	38.9551	10
11	1.2434	0.8043	12.1687	0.0822	9.7868	0.1022	4.8021	46.9977	11
12	1.2682	0.7885	13.4121	0.0746	10.5753	0.0946	5.2642	55.6712	12
13	1.2936	0.7730	14.6803	0.0681	11.3484	0.0881	5.7231	64.9475	13
14	1.3195	0.7579	15.9739	0.0626	12.1062	0.0826	6.1786	74.7999	14
15	1.3459	0.7430	17.2934	0.0578	12.8493	0.0778	6.6309	85.2021	15
16	1.3728	0.7284	18.6393	0.0537	13.5777	0.0737	7.0799	96.1288	16
17	1.4002	0.7142	20.0121	0.0500	14.2919	0.0700	7.5256	107.5554	17
18	1.4282	0.7002	21.4123	0.0467	14.9920	0.0667	7.9681	119.4581	18
19	1.4568	0.6864	22.8406	0.0438	15.6785	0.0638	8.4073	131.8139	19
20	1.4859	0.6730	24.2974	0.0412	16.3514	0.0612	8.8433	144.6003	20
21	1.5157	0.6598	25.7833	0.0388	17.0112	0.0588	9.2760	157.7959	21
22	1.5460	0.6468	27.2990	0.0366	17.6580	0.0566	9.7055	171.3795	22
23	1.5769	0.6342	28.8450	0.0347	18.2922	0.0547	10.1317	185.3309	23
24	1.6084	0.6217	30.4219	0.0329	18.9139	0.0529	10.5547	199.6305	24
25	1.6406	0.6095	32.0303	0.0312	19.5235	0.0512	10.9745	214.2592	25
26	1.6734	0.5976	33.6709	0.0297	20.1210	0.0497	11.3910	229.1987	26
27	1.7069	0.5859	35.3443	0.0283	20.7069	0.0483	11.8043	244.4311	27
28	1.7410	0.5744	37.0512	0.0270	21.2813	0.0470	12.2145	259.9392	28
29	1.7758	0.5631	38.7922	0.0258	21.8444	0.0458	12.6214	275.7064	29
30	1.8114	0.5521	40.5681	0.0246	22.3965	0.0446	13.0251	291.7164	30
31	1.8476	0.5412	42.3794	0.0236	22.9377	0.0436	13.4257	307.9538	31
32	1.8845	0.5306	44.2270	0.0226	23.4683	0.0426	13.8230	324.4035	32
33	1.9222	0.5202	46.1116	0.0217	23.9886	0.0417	14.2172	341.0508	33
34	1.9607	0.5100	48.0338	0.0208	24.4986	0.0408	14.6083	357.8817	34
35	1.9999	0.5000	49.9945	0.0200	24.9986	0.0400	14.9961	374.8826	35
36	2.0399	0.4902	51.9944	0.0192	25.4888	0.0392	15.3809	392.0405	36
40	2.2080	0.4529	60.4020	0.0166	27.3555	0.0366	16.8885	461.9931	40
48	2.5871	0.3865	79.3535	0.0126	30.6731	0.0326	19.7556	605.9657	48
50	2.6916	0.3715	84.5794	0.0118	31.4236	0.0318	20.4420	642.3606	50
60	3.2810	0.3048	114.0515	0.0088	34.7609	0.0288	23.6961	823.6975	60
72	4.1611	0.2403	158.0570	0.0063	37.9841	0.0263	27.2234	1034.0557	72
80	4.8754	0.2051	193.7720	0.0052	39.7445	0.0252	29.3572	1166.7868	80
84	5.2773	0.1895	213.8666	0.0047	40.5255	0.0247	30.3616	1230.4191	84
90	5.9431	0.1683	247.1567	0.0040	41.5869	0.0240	31.7929	1322.1701	90
96	6.6929	0.1494	284.6467	0.0035	42.5294	0.0235	33.1370	1409.2973	96
100	7.2446	0.1380	312.2323	0.0032	43.0984	0.0232	33.9863	1464.7527	100
108	8.4883	0.1178	374.4129	0.0027	44.1095	0.0227	35.5774	1569.3025	108
120	10.7652	0.0929	488.2582	0.0020	45.3554	0.0220	37.7114	1710.4160	120
240	115.8887	0.0086	5744.4368	0.0002	49.5686	0.0202	47.9110	2374.8800	240
360	1247.5611	0.0008	62328.0564	0.0000	49.9599	0.0200	49.7112	2483.5679	360

3.0%

	Single Payment		Equal Payment Series				Gradient Series		
N	Compound Amount Factor (F/P, i, N)	Present Worth Factor (P/F, i, N)	Compound Amount Factor (F/A, i, N)	Sinking Fund Factor (A/F, i, N)	Present Worth Factor (P/A, i, N)	Capital Recovery Factor (A/P, i, N)	Gradient Uniform Series (A/G, i, N)	Gradient Present Worth (P/G, i, N)	N
1	1.0300	0.9709	1.0000	1.0000	0.9709	1.0300	0.0000	0.0000	1
2	1.0609	0.9426	2.0300	0.4926	1.9135	0.5226	0.4926	0.9426	2
3	1.0927	0.9151	3.0909	0.3235	2.8286	0.3535	0.9803	2.7729	3
4	1.1255	0.8885	4.1836	0.2390	3.7171	0.2690	1.4631	5.4383	4
5	1.1593	0.8626	5.3091	0.1884	4.5797	0.2184	1.9409	8.8888	5
6	1.1941	0.8375	6.4684	0.1546	5.4172	0.1846	2.4138	13.0762	6
7	1.2299	0.8131	7.6625	0.1305	6.2303	0.1605	2.8819	17.9547	7
8	1.2668	0.7894	8.8923	0.1125	7.0197	0.1425	3.3450	23.4806	8
9	1.3048	0.7664	10.1591	0.0984	7.7861	0.1284	3.8032	29.6119	9
10	1.3439	0.7441	11.4639	0.0872	8.5302	0.1172	4.2565	36.3088	10
11	1.3842	0.7224	12.8078	0.0781	9.2526	0.1081	4.7049	43.5330	11
12	1.4258	0.7014	14.1920	0.0705	9.9540	0.1005	5.1485	51.2482	12
13	1.4685	0.6810	15.6178	0.0640	10.6350	0.0940	5.5872	59.4196	13
14	1.5126	0.6611	17.0863	0.0585	11.2961	0.0885	6.0210	68.0141	14
15	1.5580	0.6419	18.5989	0.0538	11.9379	0.0838	6.4500	77.0002	15
16	1.6047	0.6232	20.1569	0.0496	12.5611	0.0796	6.8742	86.3477	16
17	1.6528	0.6050	21.7616	0.0460	13.1661	0.0760	7.2936	96.0280	17
18	1.7024	0.5874	23.4144	0.0427	13.7535	0.0727	7.7081	106.0137	18
19	1.7535	0.5703	25.1169	0.0398	14.3238	0.0698	8.1179	116.2788	19
20	1.8061	0.5537	26.8704	0.0372	14.8775	0.0672	8.5229	126.7987	20
21	1.8603	0.5375	28.6765	0.0349	15.4150	0.0649	8.9231	137.5496	21
22	1.9161	0.5219	30.5368	0.0327	15.9396	0.0627	9.3186	148.5094	22
23	1.9736	0.5067	32.4529	0.0308	16.4436	0.0608	9.7093	159.6566	23
24	2.0328	0.4919	34.4265	0.0290	16.9355	0.0590	10.0954	170.9711	24
25	2.0938	0.4776	36.4593	0.0274	17.4131	0.0574	10.4768	182.4336	25
26	2.1566	0.4637	38.5530	0.0259	17.8768	0.0559	10.8535	194.0260	26
27	2.2213	0.4502	40.7096	0.0246	18.3270	0.0546	11.2255	205.7309	27
28	2.2879	0.4371	42.9309	0.0233	18.7641	0.0533	11.5930	217.5320	28
29	2.3566	0.4243	45.2189	0.0221	19.1885	0.0521	11.9558	229.4137	29
30	2.4273	0.4120	47.5754	0.0210	19.6004	0.0510	12.3141	241.3613	30
31	2.5001	0.4000	50.0027	0.0200	20.0004	0.0500	12.6678	253.3609	31
32	2.5751	0.3883	52.5028	0.0190	20.3888	0.0490	13.0169	265.3993	32
33	2.6523	0.3770	55.0778	0.0182	20.7658	0.0482	13.3616	277.4642	33
34	2.7319	0.3660	57.7302	0.0173	21.1318	0.0473	13.7018	289.5437	34
35	2.8139	0.3554	60.4621	0.0165	21.4872	0.0465	14.0375	301.6267	35
40	3.2620	0.3066	75.4013	0.0133	23.1148	0.0433	15.6502	361.7499	40
45	3.7816	0.2644	92.7199	0.0108	24.5187	0.0408	17.1556	420.6325	45
50	4.3839	0.2281	112.7969	0.0089	25.7298	0.0389	18.5575	477.4803	50
55	5.0821	0.1968	136.0716	0.0073	26.7744	0.0373	19.8600	531.7411	55
60	5.8916	0.1697	163.0534	0.0061	27.6756	0.0361	21.0674	583.0526	60
65	6.8300	0.1464	194.3328	0.0051	28.4529	0.0351	22.1841	631.2010	65
70	7.9178	0.1263	230.5941	0.0043	29.1234	0.0343	23.2145	676.0869	70
75	9.1789	0.1089	272.6309	0.0037	29.7018	0.0337	24.1634	717.6978	75
80	10.6409	0.0940	321.3630	0.0031	30.2008	0.0331	25.0353	756.0865	80
85	12.3357	0.0811	377.8570	0.0026	30.6312	0.0326	25.8349	791.3529	85
90	14.3005	0.0699	443.3489	0.0023	31.0024	0.0323	26.5667	823.6302	90
95	16.5782	0.0603	519.2720	0.0019	31.3227	0.0319	27.2351	853.0742	95
100	19.2186	0.0520	607.2877	0.0016	31.5989	0.0316	27.8444	879.8540	100

4.0%

	Single Payment		Equal Payment Series				Gradient Series		
	Compound Amount Factor	Present Worth Factor	Compound Amount Factor	Sinking Fund Factor	Present Worth Factor	Capital Recovery Factor	Gradient Uniform Series	Gradient Present Worth	
N	(F/P, i, N)	(P/F, i, N)	(F/A, i, N)	(A/F, i, N)	(P/A, i, N)	(A/P, i, N)	(A/G, i, N)	(P/G, i, N)	N
1	1.0400	0.9615	1.0000	1.0000	0.9615	1.0400	0.0000	0.0000	1
2	1.0816	0.9246	2.0400	0.4902	1.8861	0.5302	0.4902	0.9246	2
3	1.1249	0.8890	3.1216	0.3203	2.7751	0.3603	0.9739	2.7025	3
4	1.1699	0.8548	4.2465	0.2355	3.6299	0.2755	1.4510	5.2670	4
5	1.2167	0.8219	5.4163	0.1846	4.4518	0.2246	1.9216	8.5547	5
6	1.2653	0.7903	6.6330	0.1508	5.2421	0.1908	2.3857	12.5062	6
7	1.3159	0.7599	7.8983	0.1266	6.0021	0.1666	2.8433	17.0657	7
8	1.3686	0.7307	9.2142	0.1085	6.7327	0.1485	3.2944	22.1806	8
9	1.4233	0.7026	10.5828	0.0945	7.4353	0.1345	3.7391	27.8013	9
10	1.4802	0.6756	12.0061	0.0833	8.1109	0.1233	4.1773	33.8814	10
11	1.5395	0.6496	13.4864	0.0741	8.7605	0.1141	4.6090	40.3772	11
12	1.6010	0.6246	15.0258	0.0666	9.3851	0.1066	5.0343	47.2477	12
13	1.6651	0.6006	16.6268	0.0601	9.9856	0.1001	5.4533	54.4546	13
14	1.7317	0.5775	18.2919	0.0547	10.5631	0.0947	5.8659	61.9618	14
15	1.8009	0.5553	20.0236	0.0499	11.1184	0.0899	6.2721	69.7355	15
16	1.8730	0.5339	21.8245	0.0458	11.6523	0.0858	6.6720	77.7441	16
17	1.9479	0.5134	23.6975	0.0422	12.1657	0.0822	7.0656	85.9581	17
18	2.0258	0.4936	25.6454	0.0390	12.6593	0.0790	7.4530	94.3498	18
19	2.1068	0.4746	27.6712	0.0361	13.1339	0.0761	7.8342	102.8933	19
20	2.1911	0.4564	29.7781	0.0336	13.5903	0.0736	8.2091	111.5647	20
21	2.2788	0.4388	31.9692	0.0313	14.0292	0.0713	8.5779	120.3414	21
22	2.3699	0.4220	34.2480	0.0292	14.4511	0.0692	8.9407	129.2024	22
23	2.4647	0.4057	36.6179	0.0273	14.8568	0.0673	9.2973	138.1284	23
24	2.5633	0.3901	39.0826	0.0256	15.2470	0.0656	9.6479	147.1012	24
25	2.6658	0.3751	41.6459	0.0240	15.6221	0.0640	9.9925	156.1040	25
26	2.7725	0.3607	44.3117	0.0226	15.9828	0.0626	10.3312	165.1212	26
27	2.8834	0.3468	47.0842	0.0212	16.3296	0.0612	10.6640	174.1385	27
28	2.9987	0.3335	49.9676	0.0200	16.6631	0.0600	10.9909	183.1424	28
29	3.1187	0.3207	52.9663	0.0189	16.9837	0.0589	11.3120	192.1206	29
30	3.2434	0.3083	56.0849	0.0178	17.2920	0.0578	11.6274	201.0618	30
31	3.3731	0.2965	59.3283	0.0169	17.5885	0.0569	11.9371	209.9556	31
32	3.5081	0.2851	62.7015	0.0159	17.8736	0.0559	12.2411	218.7924	32
33	3.6484	0.2741	66.2095	0.0151	18.1476	0.0551	12.5396	227.5634	33
34	3.7943	0.2636	69.8579	0.0143	18.4112	0.0543	12.8324	236.2607	34
35	3.9461	0.2534	73.6522	0.0136	18.6646	0.0536	13.1198	244.8768	35
40	4.8010	0.2083	95.0255	0.0105	19.7928	0.0505	14.4765	286.5303	40
45	5.8412	0.1712	121.0294	0.0083	20.7200	0.0483	15.7047	325.4028	45
50	7.1067	0.1407	152.6671	0.0066	21.4822	0.0466	16.8122	361.1638	50
55	8.6464	0.1157	191.1592	0.0052	22.1086	0.0452	17.8070	393.6890	55
60	10.5196	0.0951	237.9907	0.0042	22.6235	0.0442	18.6972	422.9966	60
65	12.7987	0.0781	294.9684	0.0034	23.0467	0.0434	19.4909	449.2014	65
70	15.5716	0.0642	364.2905	0.0027	23.3945	0.0427	20.1961	472.4789	70
75	18.9453	0.0528	448.6314	0.0022	23.6804	0.0422	20.8206	493.0408	75
80	23.0498	0.0434	551.2450	0.0018	23.9154	0.0418	21.3718	511.1161	80
85	28.0436	0.0357	676.0901	0.0015	24.1085	0.0415	21.8569	526.9384	85
90	34.1193	0.0293	827.9833	0.0012	24.2673	0.0412	22.2826	540.7369	90
95	41.5114	0.0241	1012.7846	0.0010	24.3978	0.0410	22.6550	552.7307	95
100	50.5049	0.0198	1237.6237	0.0008	24.5050	0.0408	22.9800	563.1249	100

5.0%

	Single Payment		Equal Payment Series				Gradient Series		
	Compound Amount Factor	Present Worth Factor	Compound Amount Factor	Sinking Fund Factor	Present Worth Factor	Capital Recovery Factor	Gradient Uniform Series	Gradient Present Worth	
N	(F/P, i, N)	(P/F, i, N)	(F/A, i, N)	(A/F, i, N)	(P/A, i, N)	(A/P, i, N)	(A/G, i, N)	(P/G, i, N)	N
1	1.0500	0.9524	1.0000	1.0000	0.9524	1.0500	0.0000	0.0000	1
2	1.1025	0.9070	2.0500	0.4878	1.8594	0.5378	0.4878	0.9070	2
3	1.1576	0.8638	3.1525	0.3172	2.7232	0.3672	0.9675	2.6347	3
4	1.2155	0.8227	4.3101	0.2320	3.5460	0.2820	1.4391	5.1028	4
5	1.2763	0.7835	5.5256	0.1810	4.3295	0.2310	1.9025	8.2369	5
6	1.3401	0.7462	6.8019	0.1470	5.0757	0.1970	2.3579	11.9680	6
7	1.4071	0.7107	8.1420	0.1228	5.7864	0.1728	2.8052	16.2321	7
8	1.4775	0.6768	9.5491	0.1047	6.4632	0.1547	3.2445	20.9700	8
9	1.5513	0.6446	11.0266	0.0907	7.1078	0.1407	3.6758	26.1268	9
10	1.6289	0.6139	12.5779	0.0795	7.7217	0.1295	4.0991	31.6520	10
11	1.7103	0.5847	14.2068	0.0704	8.3064	0.1204	4.5144	37.4988	11
12	1.7959	0.5568	15.9171	0.0628	8.8633	0.1128	4.9219	43.6241	12
13	1.8856	0.5303	17.7130	0.0565	9.3936	0.1065	5.3215	49.9879	13
14	1.9799	0.5051	19.5986	0.0510	9.8986	0.1010	5.7133	56.5538	14
15	2.0789	0.4810	21.5786	0.0463	10.3797	0.0963	6.0973	63.2880	15
16	2.1829	0.4581	23.6575	0.0423	10.8378	0.0923	6.4736	70.1597	16
17	2.2920	0.4363	25.8404	0.0387	11.2741	0.0887	6.8423	77.1405	17
18	2.4066	0.4155	28.1324	0.0355	11.6896	0.0855	7.2034	84.2043	18
19	2.5270	0.3957	30.5390	0.0327	12.0853	0.0827	7.5569	91.3275	19
20	2.6533	0.3769	33.0660	0.0302	12.4622	0.0802	7.9030	98.4884	20
21	2.7860	0.3589	35.7193	0.0280	12.8212	0.0780	8.2416	105.6673	21
22	2.9253	0.3418	38.5052	0.0260	13.1630	0.0760	8.5730	112.8461	22
23	3.0715	0.3256	41.4305	0.0241	13.4886	0.0741	8.8971	120.0087	23
24	3.2251	0.3101	44.5020	0.0225	13.7986	0.0725	9.2140	127.1402	24
25	3.3864	0.2953	47.7271	0.0210	14.0939	0.0710	9.5238	134.2275	25
26	3.5557	0.2812	51.1135	0.0196	14.3752	0.0696	9.8266	141.2585	26
27	3.7335	0.2678	54.6691	0.0183	14.6430	0.0683	10.1224	148.2226	27
28	3.9201	0.2551	58.4026	0.0171	14.8981	0.0671	10.4114	155.1101	28
29	4.1161	0.2429	62.3227	0.0160	15.1411	0.0660	10.6936	161.9126	29
30	4.3219	0.2314	66.4388	0.0151	15.3725	0.0651	10.9691	168.6226	30
31	4.5380	0.2204	70.7608	0.0141	15.5928	0.0641	11.2381	175.2333	31
32	4.7649	0.2099	75.2988	0.0133	15.8027	0.0633	11.5005	181.7392	32
33	5.0032	0.1999	80.0638	0.0125	16.0025	0.0625	11.7566	188.1351	33
34	5.2533	0.1904	85.0670	0.0118	16.1929	0.0618	12.0063	194.4168	34
35	5.5160	0.1813	90.3203	0.0111	16.3742	0.0611	12.2498	200.5807	35
40	7.0400	0.1420	120.7998	0.0083	17.1591	0.0583	13.3775	229.5452	40
45	8.9850	0.1113	159.7002	0.0063	17.7741	0.0563	14.3644	255.3145	45
50	11.4674	0.0872	209.3480	0.0048	18.2559	0.0548	15.2233	277.9148	50
55	14.6356	0.0683	272.7126	0.0037	18.6335	0.0537	15.9664	297.5104	55
60	18.6792	0.0535	353.5837	0.0028	18.9293	0.0528	16.6062	314.3432	60
65	23.8399	0.0419	456.7980	0.0022	19.1611	0.0522	17.1541	328.6910	65
70	30.4264	0.0329	588.5285	0.0017	19.3427	0.0517	17.6212	340.8409	70
75	38.8327	0.0258	756.6537	0.0013	19.4850	0.0513	18.0176	351.0721	75
80	49.5614	0.0202	971.2288	0.0010	19.5965	0.0510	18.3526	359.6460	80
85	63.2544	0.0158	1245.0871	0.0008	19.6838	0.0508	18.6346	366.8007	85
90	80.7304	0.0124	1594.6073	0.0006	19.7523	0.0506	18.8712	372.7488	90
95	103.0347	0.0097	2040.6935	0.0005	19.8059	0.0505	19.0689	377.6774	95
100	131.5013	0.0076	2610.0252	0.0004	19.8479	0.0504	19.2337	381.7492	100

6.0%

	Single Payment		Equal Payment Series				Gradient Series		
	Compound Amount Factor	Present Worth Factor	Compound Amount Factor	Sinking Fund Factor	Present Worth Factor	Capital Recovery Factor	Gradient Uniform Series	Gradient Present Worth	
N	(F/P,i,N)	(P/F,i,N)	(F/A,i,N)	(A/F,i,N)	(P/A,i,N)	(A/P,i,N)	(A/G,i,N)	(P/G,i,N)	N
1	1.0600	0.9434	1.0000	1.0000	0.9434	1.0600	0.0000	0.0000	1
2	1.1236	0.8900	2.0600	0.4854	1.8334	0.5454	0.4854	0.8900	2
3	1.1910	0.8396	3.1836	0.3141	2.6730	0.3741	0.9612	2.5692	3
4	1.2625	0.7921	4.3746	0.2286	3.4651	0.2886	1.4272	4.9455	4
5	1.3382	0.7473	5.6371	0.1774	4.2124	0.2374	1.8836	7.9345	5
6	1.4185	0.7050	6.9753	0.1434	4.9173	0.2034	2.3304	11.4594	6
7	1.5036	0.6651	8.3938	0.1191	5.5824	0.1791	2.7676	15.4497	7
8	1.5938	0.6274	9.8975	0.1010	6.2098	0.1610	3.1952	19.8416	8
9	1.6895	0.5919	11.4913	0.0870	6.8017	0.1470	3.6133	24.5768	9
10	1.7908	0.5584	13.1808	0.0759	7.3601	0.1359	4.0220	29.6023	10
11	1.8983	0.5268	14.9716	0.0668	7.8869	0.1268	4.4213	34.8702	11
12	2.0122	0.4970	16.8699	0.0593	8.3838	0.1193	4.8113	40.3369	12
13	2.1329	0.4688	18.8821	0.0530	8.8527	0.1130	5.1920	45.9629	13
14	2.2609	0.4423	21.0151	0.0476	9.2950	0.1076	5.5635	51.7128	14
15	2.3966	0.4173	23.2760	0.0430	9.7122	0.1030	5.9260	57.5546	15
16	2.5404	0.3936	25.6725	0.0390	10.1059	0.0990	6.2794	63.4592	16
17	2.6928	0.3714	28.2129	0.0354	10.4773	0.0954	6.6240	69.4011	17
18	2.8543	0.3503	30.9057	0.0324	10.8276	0.0924	6.9597	75.3569	18
19	3.0256	0.3305	33.7600	0.0296	11.1581	0.0896	7.2867	81.3062	19
20	3.2071	0.3118	36.7856	0.0272	11.4699	0.0872	7.6051	87.2304	20
21	3.3996	0.2942	39.9927	0.0250	11.7641	0.0850	7.9151	93.1136	21
22	3.6035	0.2775	43.3923	0.0230	12.0416	0.0830	8.2166	98.9412	22
23	3.8197	0.2618	46.9958	0.0213	12.3034	0.0813	8.5099	104.7007	23
24	4.0489	0.2470	50.8156	0.0197	12.5504	0.0797	8.7951	110.3812	24
25	4.2919	0.2330	54.8645	0.0182	12.7834	0.0782	9.0722	115.9732	25
26	4.5494	0.2198	59.1564	0.0169	13.0032	0.0769	9.3414	121.4684	26
27	4.8223	0.2074	63.7058	0.0157	13.2105	0.0757	9.6029	126.8600	27
28	5.1117	0.1956	68.5281	0.0146	13.4062	0.0746	9.8568	132.1420	28
29	5.4184	0.1846	73.6398	0.0136	13.5907	0.0736	10.1032	137.3096	29
30	5.7435	0.1741	79.0582	0.0126	13.7648	0.0726	10.3422	142.3588	30
31	6.0881	0.1643	84.8017	0.0118	13.9291	0.0718	10.5740	147.2864	31
32	6.4534	0.1550	90.8898	0.0110	14.0840	0.0710	10.7988	152.0901	32
33	6.8406	0.1462	97.3432	0.0103	14.2302	0.0703	11.0166	156.7681	33
34	7.2510	0.1379	104.1838	0.0096	14.3681	0.0696	11.2276	161.3192	34
35	7.6861	0.1301	111.4348	0.0090	14.4982	0.0690	11.4319	165.7427	35
40	10.2857	0.0972	154.7620	0.0065	15.0463	0.0665	12.3590	185.9568	40
45	13.7646	0.0727	212.7435	0.0047	15.4558	0.0647	13.1413	203.1096	45
50	18.4202	0.0543	290.3359	0.0034	15.7619	0.0634	13.7964	217.4574	50
55	24.6503	0.0406	394.1720	0.0025	15.9905	0.0625	14.3411	229.3222	55
60	32.9877	0.0303	533.1282	0.0019	16.1614	0.0619	14.7909	239.0428	60
65	44.1450	0.0227	719.0829	0.0014	16.2891	0.0614	15.1601	246.9450	65
70	59.0759	0.0169	967.9322	0.0010	16.3845	0.0610	15.4613	253.3271	70
75	79.0569	0.0126	1300.9487	0.0008	16.4558	0.0608	15.7058	258.4527	75
80	105.7960	0.0095	1746.5999	0.0006	16.5091	0.0606	15.9033	262.5493	80
85	141.5789	0.0071	2342.9817	0.0004	16.5489	0.0604	16.0620	265.8096	85
90	189.4645	0.0053	3141.0752	0.0003	16.5787	0.0603	16.1891	268.3946	90
95	253.5463	0.0039	4209.1042	0.0002	16.6009	0.0602	16.2905	270.4375	95
100	339.3021	0.0029	5638.3681	0.0002	16.6175	0.0602	16.3711	272.0471	100

7.0%

	Single Payment		Equal Payment Series				Gradient Series		
	Compound Amount Factor	Present Worth Factor	Compound Amount Factor	Sinking Fund Factor	Present Worth Factor	Capital Recovery Factor	Gradient Uniform Series	Gradient Present Worth	
N	(F/P,i,N)	(P/F,i,N)	(F/A,i,N)	(A/F,i,N)	(P/A,i,N)	(A/P,i,N)	(A/G,i,N)	(P/G,i,N)	N
1	1.0700	0.9346	1.0000	1.0000	0.9346	1.0700	0.0000	0.0000	1
2	1.1449	0.8734	2.0700	0.4831	1.8080	0.5531	0.4831	0.8734	2
3	1.2250	0.8163	3.2149	0.3111	2.6243	0.3811	0.9549	2.5060	3
4	1.3108	0.7629	4.4399	0.2252	3.3872	0.2952	1.4155	4.7947	4
5	1.4026	0.7130	5.7507	0.1739	4.1002	0.2439	1.8650	7.6467	5
6	1.5007	0.6663	7.1533	0.1398	4.7665	0.2098	2.3032	10.9784	6
7	1.6058	0.6227	8.6540	0.1156	5.3893	0.1856	2.7304	14.7149	7
8	1.7182	0.5820	10.2598	0.0975	5.9713	0.1675	3.1465	18.7889	8
9	1.8385	0.5439	11.9780	0.0835	6.5152	0.1535	3.5517	23.1404	9
10	1.9672	0.5083	13.8164	0.0724	7.0236	0.1424	3.9461	27.7156	10
11	2.1049	0.4751	15.7836	0.0634	7.4987	0.1334	4.3296	32.4665	11
12	2.2522	0.4440	17.8885	0.0559	7.9427	0.1259	4.7025	37.3506	12
13	2.4098	0.4150	20.1406	0.0497	8.3577	0.1197	5.0648	42.3302	13
14	2.5785	0.3878	22.5505	0.0443	8.7455	0.1143	5.4167	47.3718	14
15	2.7590	0.3624	25.1290	0.0398	9.1079	0.1098	5.7583	52.4461	15
16	2.9522	0.3387	27.8881	0.0359	9.4466	0.1059	6.0897	57.5271	16
17	3.1588	0.3166	30.8402	0.0324	9.7632	0.1024	6.4110	62.5923	17
18	3.3799	0.2959	33.9990	0.0294	10.0591	0.0994	6.7225	67.6219	18
19	3.6165	0.2765	37.3790	0.0268	10.3356	0.0968	7.0242	72.5991	19
20	3.8697	0.2584	40.9955	0.0244	10.5940	0.0944	7.3163	77.5091	20
21	4.1406	0.2415	44.8652	0.0223	10.8355	0.0923	7.5990	82.3393	21
22	4.4304	0.2257	49.0057	0.0204	11.0612	0.0904	7.8725	87.0793	22
23	4.7405	0.2109	53.4361	0.0187	11.2722	0.0887	8.1369	91.7201	23
24	5.0724	0.1971	58.1767	0.0172	11.4693	0.0872	8.3923	96.2545	24
25	5.4274	0.1842	63.2490	0.0158	11.6536	0.0858	8.6391	100.6765	25
26	5.8074	0.1722	68.6765	0.0146	11.8258	0.0846	8.8773	104.9814	26
27	6.2139	0.1609	74.4838	0.0134	11.9867	0.0834	9.1072	109.1656	27
28	6.6488	0.1504	80.6977	0.0124	12.1371	0.9824	9.3289	113.2264	28
29	7.1143	0.1406	87.3465	0.0114	12.2777	0.0814	9.5427	117.1622	29
30	7.6123	0.1314	94.4608	0.0106	12.4090	0.0806	9.7487	120.9718	30
31	8.1451	0.1228	102.0730	0.0098	12.5318	0.0798	9.9471	124.6550	31
32	8.7153	0.1147	110.2182	0.0091	12.6466	0.0791	10.1381	128.2120	32
33	9.3253	0.1072	118.9334	0.0084	12.7538	0.0784	10.3219	131.6435	33
34	9.9781	0.1002	128.2588	0.0078	12.8540	0.0778	10.4987	134.9507	34
35	10.6766	0.0937	138.2369	0.0072	12.9477	0.0772	10.6687	138.1353	35
40	14.9745	0.0668	199.6351	0.0050	13.3317	0.0750	11.4233	152.2928	40
45	21.0025	0.0476	285.7493	0.0035	13.6055	0.0735	12.0360	163.7559	45
50	29.4570	0.0339	406.5289	0.0025	13.8007	0.0725	12.5287	172.9051	50
55	41.3150	0.0242	575.9286	0.0017	13.9399	0.0717	12.9215	180.1243	55
60	57.9464	0.0173	813.5204	0.0012	14.0392	0.0712	13.2321	185.7677	60
65	81.2729	0.0123	1146.7552	0.0009	14.1099	0.0709	13.4760	190.1452	65
70	113.9894	0.0088	1614.1342	0.0006	14.1604	0.0706	13.6662	193.5185	70
75	159.8760	0.0063	2269.6574	0.0004	14.1964	0.0704	13.8136	196.1035	75
80	224.2344	0.0045	3189.0627	0.0003	14.2220	0.0703	13.9273	198.0748	80
85	314.5003	0.0032	4478.5761	0.0002	14.2403	0.0702	14.0146	199.5717	85
90	441.1030	0.0023	6287.1854	0.0002	14.2533	0.0702	14.0812	200.7042	90
95	618.6697	0.0016	8823.8535	0.0001	14.2626	0.0701	14.1319	201.5581	95
100	867.7163	0.0012	12381.6618	0.0001	14.2693	0.0701	14.1703	202.2001	100

8.0%

	Single Payment		Equal Payment Series				Gradient Series		
	Compound Amount Factor	Present Worth Factor	Compound Amount Factor	Sinking Fund Factor	Present Worth Factor	Capital Recovery Factor	Gradient Uniform Series	Gradient Present Worth	
N	(F/P,i,N)	(P/F,i,N)	(F/A,i,N)	(A/F,i,N)	(P/A,i,N)	(A/P,i,N)	(A/G,i,N)	(P/G,i,N)	N
1	1.0800	0.9259	1.0000	1.0000	0.9259	1.0800	0.0000	0.0000	1
2	1.1664	0.8573	2.0800	0.4808	1.7833	0.5608	0.4808	0.8573	2
3	1.2597	0.7938	3.2464	0.3080	2.5771	0.3880	0.9487	2.4450	3
4	1.3605	0.7350	4.5061	0.2219	3.3121	0.3019	1.4040	4.6501	4
5	1.4693	0.6806	5.8666	0.1705	3.9927	0.2505	1.8465	7.3724	5
6	1.5869	0.6302	7.3359	0.1363	4.6229	0.2163	2.2763	10.5233	6
7	1.7138	0.5835	8.9228	0.1121	5.2064	0.1921	2.6937	14.0242	7
8	1.8509	0.5403	10.6366	0.0940	5.7466	0.1740	3.0985	17.8061	8
9	1.9990	0.5002	12.4876	0.0801	6.2469	0.1601	3.4910	21.8081	9
10	2.1589	0.4632	14.4866	0.0690	6.7101	0.1490	3.8713	25.9768	10
11	2.3316	0.4289	16.6455	0.0601	7.1390	0.1401	4.2395	30.2657	11
12	2.5182	0.3971	18.9771	0.0527	7.5361	0.1327	4.5957	34.6339	12
13	2.7196	0.3677	21.4953	0.0465	7.9038	0.1265	4.9402	39.0463	13
14	2.9372	0.3405	24.2149	0.0413	8.2442	0.1213	5.2731	43.4723	14
15	3.1722	0.3152	27.1521	0.0368	8.5595	0.1168	5.5945	47.8857	15
16	3.4259	0.2919	30.3243	0.0330	8.8514	0.1130	5.9046	52.2640	16
17	3.7000	0.2703	33.7502	0.0296	9.1216	0.1096	6.2037	56.5883	17
18	3.9960	0.2502	37.4502	0.0267	9.3719	0.1067	6.4920	60.8426	18
19	4.3157	0.2317	41.4463	0.0241	9.6036	0.1041	6.7697	65.0134	19
20	4.6610	0.2145	45.7620	0.0219	9.8181	0.1019	7.0369	69.0898	20
21	5.0338	0.1987	50.4229	0.0198	10.0168	0.0998	7.2940	73.0629	21
22	5.4365	0.1839	55.4568	0.0180	10.2007	0.0980	7.5412	76.9257	22
23	5.8715	0.1703	60.8933	0.0164	10.3711	0.0964	7.7786	80.6726	23
24	6.3412	0.1577	66.7648	0.0150	10.5288	0.0950	8.0066	84.2997	24
25	6.8485	0.1460	73.1059	0.0137	10.6748	0.0937	8.2254	87.8041	25
26	7.3964	0.1352	79.9544	0.0125	10.8100	0.0925	8.4352	91.1842	26
27	7.9881	0.1252	87.3508	0.0114	10.9352	0.0914	8.6363	94.4390	27
78	8.6271	0.1159	95.3388	0.0105	11.0511	0.0905	8.8289	97.5687	28
29	9.3173	0.1073	103.9659	0.0096	11.1584	0.0896	9.0133	100.5738	29
30	10.0627	0.0994	113.2832	0.0088	11.2578	0.0888	9.1897	103.4558	30
31	10.8677	0.0920	123.3459	0.0081	11.3498	0.0881	9.3584	106.2163	31
32	11.7371	0.0852	134.2135	0.0075	11.4350	0.0875	9.5197	108.8575	32
33	12.6760	0.0789	145.9506	0.0069	11.5139	0.0869	9.6737	111.3819	33
34	13.6901	0.0730	158.6267	0.0063	11.5869	0.0863	9.8208	113.7924	34
35	14.7853	0.0676	172.3168	0.0058	11.6546	0.0858	9.9611	116.0920	35
40	21.7245	0.0460	259.0565	0.0039	11.9246	0.0839	10.5699	126.0422	40
45	31.9204	0.0313	386.5056	0.0026	12.1084	0.0826	11.0447	133.7331	45
50	46.9016	0.0213	573.7702	0.0017	12.2335	0.0817	11.4107	139.5928	50
55	68.9139	0.0145	848.9232	0.0012	12.3186	0.0812	11.6902	144.0065	55
60	101.2571	0.0099	1253.2133	0.0008	12.3766	0.0808	11.9015	147.3000	60
65	148.7798	0.0067	1847.2481	0.0005	12.4160	0.0805	12.0602	149.7387	65
70	218.6064	0.0046	2720.0801	0.0004	12.4428	0.0804	12.1783	151.5326	70
75	321.2045	0.0031	4002.5566	0.0002	12.4611	0.0802	12.2658	152.8448	75
80	471.9548	0.0021	5886.9354	0.0002	12.4735	0.0802	12.3301	153.8001	80
85	693.4565	0.0014	8655.7061	0.0001	12.4820	0.0801	12.3772	154.4925	85
90	1018.9151	0.0010	12723.9386	0.0001	12.4877	0.0801	12.4116	154.9925	90
95	1497.1205	0.0007	18701.5069	0.0001	12.4917	0.0801	12.4365	155.3524	95
100	2199.7613	0.0005	27484.5157	0.0000	12.4943	0.0800	12.4545	155.6107	100

9.0%

	Single Payment		Equal Payment Series				Gradient Series		
	Compound Amount Factor	Present Worth Factor	Compound Amount Factor	Sinking Fund Factor	Present Worth Factor	Capital Recovery Factor	Gradient Uniform Series	Gradient Present Worth	
N	(F/P,i,N)	(P/F,i,N)	(F/A,i,N)	(A/F,i,N)	(P/A,i,N)	(A/P,i,N)	(A/G,i,N)	(P/G,i,N)	N
1	1.0900	0.9174	1.0000	1.0000	0.9174	1.0900	0.0000	0.0000	1
2	1.1881	0.8417	2.0900	0.4785	1.7591	0.5685	0.4785	0.8417	2
3	1.2950	0.7722	3.2781	0.3051	2.5313	0.3951	0.9426	2.3860	3
4	1.4116	0.7084	4.5731	0.2187	3.2397	0.3087	1.3925	4.5113	4
5	1.5386	0.6499	5.9847	0.1671	3.8897	0.2571	1.8282	7.1110	5
6	1.6771	0.5963	7.5233	0.1329	4.4859	0.2229	2.2498	10.0924	6
7	1.8280	0.5470	9.2004	0.1087	5.0330	0.1987	2.6574	13.3746	7
8	1.9926	0.5019	11.0285	0.0907	5.5348	0.1807	3.0512	16.8877	8
9	2.1719	0.4604	13.0210	0.0768	5.9952	0.1668	3.4312	20.5711	9
10	2.3674	0.4224	15.1929	0.0658	6.4177	0.1558	3.7978	24.3728	10
11	2.5804	0.3875	17.5603	0.0569	6.8052	0.1469	4.1510	28.2481	11
12	2.8127	0.3555	20.1407	0.0497	7.1607	0.1397	4.4910	32.1590	12
13	3.0658	0.3262	22.9534	0.0436	7.4869	0.1336	4.8182	36.0731	13
14	3.3417	0.2992	26.0192	0.0384	7.7862	0.1284	5.1326	39.9633	14
15	3.6425	0.2745	29.3609	0.0341	8.0607	0.1241	5.4346	43.8069	15
16	3.9703	0.2519	33.0034	0.0303	8.3126	0.1203	5.7245	47.5849	16
17	4.3276	0.2311	36.9737	0.0270	8.5436	0.1170	6.0024	51.2821	17
18	4.7171	0.2120	41.3013	0.0242	8.7556	0.1142	6.2687	54.8860	18
19	5.1417	0.1945	46.0185	0.0217	8.9501	0.1117	6.5236	58.3868	19
20	5.6044	0.1784	51.1601	0.0195	9.1285	0.1095	6.7674	61.7770	20
21	6.1088	0.1637	56.7645	0.0176	9.2922	0.1076	7.Z0006	65.0509	21
22	6.6586	0.1502	62.8733	0.0159	9.4424	0.1059	7.2232	68.2048	22
23	7.2579	0.1378	69.5319	0.0144	9.5802	0.1044	7.4357	71.2359	23
24	7.9111	0.1264	76.7898	0.0130	9.7066	0.1030	7.6384	74.1433	24
25	8.6231	0.1160	84.7009	0.0118	9.8226	0.1018	7.8316	76.9265	25
26	9.3992	0.1064	93.3240	0.0107	9.9290	0.1007	8.0156	79.5863	26
27	10.2451	0.0976	102.7231	0.0097	10.0266	0.0997	8.1906	82.1241	27
28	11.1671	0.0895	112.9682	0.0089	10.1161	0.0989	8.3571	84.5419	28
29	12.1722	0.0822	124.1354	0.0081	10.1983	0.0981	8.5154	86.8422	29
30	13.2677	0.0754	136.3075	0.0073	10.2737	0.0973	8.6657	89.0280	30
31	14.4618	0.0691	149.5752	0.0067	10.3428	0.0967	8.8083	91.1024	31
32	15.7633	0.0634	164.0370	0.0061	10.4062	0.0961	8.9436	93.0690	32
33	17.1820	0.0582	179.8003	0.0056	10.4644	0.0956	9.0718	94.9314	33
34	18.7284	0.0534	196.9823	0.0051	10.5178	0.0951	9.1933	96.6935	34
35	20.4140	0.0490	215.7108	0.0046	10.5668	0.0946	9.3083	98.3590	35
40	31.4094	0.0318	337.8824	0.0030	10.7574	0.0930	9.7957	105.3762	40
45	48.3273	0.0207	525.8587	0.0019	10.8812	0.0919	10.1603	110.5561	45
50	74.3575	0.0134	815.0836	0.0012	10.9617	0.0912	10.4295	114.3251	50
55	114.4083	0.0087	1260.0918	0.0008	11.0140	0.0908	10.6261	117.0362	55
60	176.0313	0.0057	1944.7921	0.0005	11.0480	0.0905	10.7683	118.9683	60
65	270.8460	0.0037	2998.2885	0.0003	11.0701	0.0903	10.8702	120.3344	65
70	416.7301	0.0024	4619.2232	0.0002	11.0844	0.0902	10.9427	121.2942	70
75	641.1909	0.0016	7113.2321	0.0001	11.0938	0.0901	10.9940	121.9646	75
80	986.5517	0.0010	10950.5741	0.0001	11.0998	0.0901	11.0299	122.4306	80
85	1517.9320	0.0007	16854.8003	0.0001	11.1038	0.0901	11.0551	122.7533	85
90	2335.5266	0.0004	25939.1842	0.0000	11.1064	0.0900	11.0726	122.9758	90
95	3593.4971	0.0003	39916.6350	0.0000	11.1080	0.0900	11.0847	123.1287	95
100	5529.0408	0.0002	61422.6755	0.0000	11.1091	0.0900	11.0930	123.2335	100

10.0%

	Single Payment		Equal Payment Series				Gradient Series		
N	Compound Amount Factor (F/P,i,N)	Present Worth Factor (P/F,i,N)	Compound Amount Factor (F/A,i,N)	Sinking Fund Factor (A/F,i,N)	Present Worth Factor (P/A,i,N)	Capital Recovery Factor (A/P,i,N)	Gradient Uniform Series (A/G,i,N)	Gradient Present Worth (P/G,i,N)	N
1	1.1000	0.9091	1.0000	1.0000	0.9091	1.1000	0.0000	0.0000	1
2	1.2100	0.8264	2.1000	0.4762	1.7355	0.5762	0.4762	0.8264	2
3	1.3310	0.7513	3.3100	0.3021	2.4869	0.4021	0.9366	2.3291	3
4	1.4641	0.6830	4.6410	0.2155	3.1699	0.3155	1.3812	4.3781	4
5	1.6105	0.6209	6.1051	0.1638	3.7908	0.2638	1.8101	6.8618	5
6	1.7716	0.5645	7.7156	0.1296	4.3553	0.2296	2.2236	9.6842	6
7	1.9487	0.5132	9.4872	0.1054	4.8684	0.2054	2.6216	12.7631	7
8	2.1436	0.4665	11.4359	0.0874	5.3349	0.1874	3.0045	16.0287	8
9	2.3579	0.4241	13.5795	0.0736	5.7590	0.1736	3.3724	19.4215	9
10	2.5937	0.3855	15.9374	0.0627	6.1446	0.1627	3.7255	22.8913	10
11	2.8531	0.3505	18.5312	0.0540	6.4951	0.1540	4.0641	26.3963	11
12	3.1384	0.3186	21.3843	0.0468	6.8137	0.1468	4.3884	29.9012	12
13	3.4523	0.2897	24.5227	0.0408	7.1034	0.1408	4.6988	33.3772	13
14	3.7975	0.2633	27.9750	0.0357	7.3667	0.1357	4.9955	36.8005	14
15	4.1772	0.2394	31.7725	0.0315	7.6061	0.1315	5.2789	40.1520	15
16	4.5950	0.2176	35.9497	0.0278	7.8237	0.1278	5.5493	43.4164	16
17	5.0545	0.1978	40.5447	0.0247	8.0216	0.1247	5.8071	46.5819	17
18	5.5599	0.1799	45.5992	0.0219	8.2014	0.1219	6.0526	49.6395	18
19	6.1159	0.1635	51.1591	0.0195	8.3649	0.1195	6.2861	52.5827	19
20	6.7275	0.1486	57.2750	0.0175	8.5136	0.1175	6.5081	55.4069	20
21	7.4002	0.1351	64.0025	0.0156	8.6487	0.1156	6.7189	58.1095	21
22	8.1403	0.1228	71.4027	0.0140	8.7715	0.1140	6.9189	60.6893	22
23	8.9543	0.1117	79.5430	0.0126	8.8832	0.1126	7.1085	63.1462	23
24	9.8497	0.1015	88.4973	0.0113	8.9847	0.1113	7.2881	65.4813	24
25	10.8347	0.0923	98.3471	0.0102	9.0770	0.1102	7.4580	67.6964	25
26	11.9182	0.0839	109.1818	0.0092	9.1609	0.1092	7.6186	69.7940	26
27	13.1100	0.0763	121.0999	0.0083	9.2372	0.1083	7.7704	71.7773	27
28	14.4210	0.0693	134.2099	0.0075	9.3066	0.1075	7.9137	73.6495	28
29	15.8631	0.0630	148.6309	0.0067	9.3696	0.1067	8.0489	75.4146	29
30	17.4494	0.0573	164.4940	0.0061	9.4269	0.1061	8.1762	77.0766	30
31	19.1943	0.0521	181.9434	0.0055	9.4790	0.1055	8.2962	78.6395	31
32	21.1138	0.0474	201.1378	0.0050	9.5264	0.1050	8.4091	80.1078	32
33	23.2252	0.0431	222.2515	0.0045	9.5694	0.1045	8.5152	81.4856	33
34	25.5477	0.0391	245.4767	0.0041	9.6086	0.1041	8.6149	82.7773	34
35	28.1024	0.0356	271.0244	0.0037	9.6442	0.1037	8.7086	83.9872	35
40	45.2593	0.0221	442.5926	0.0023	9.7791	0.1023	9.0962	88.9525	40
45	72.8905	0.0137	718.9048	0.0014	9.8628	0.1014	9.3740	92.4544	45
50	117.3909	0.0085	1163.9085	0.0009	9.9148	0.1009	9.5704	94.8889	50
55	189.0591	0.0053	1880.5914	0.0005	9.9471	0.1005	9.7075	96.5619	55
60	304.4816	0.0033	3034.8164	0.0003	9.9672	0.1003	9.8023	97.7010	60
65	490.3707	0.0020	4893.7073	0.0002	9.9796	0.1002	9.8672	98.4705	65
70	789.7470	0.0013	7887.4696	0.0001	9.9873	0.1001	9.9113	98.9870	70
75	1271.8954	0.0008	12708.9537	0.0001	9.9921	0.1001	9.9410	99.3317	75
80	2048.4002	0.0005	20474.0021	0.0000	9.9951	0.1000	9.9609	99.5606	80
85	3298.9690	0.0003	32979.6903	0.0000	9.9970	0.1000	9.9742	99.7120	85
90	5313.0226	0.0002	53120.2261	0.0000	9.9981	0.1000	9.9831	99.8118	90
95	8556.6760	0.0001	85556.7605	0.0000	9.9988	0.1000	9.9889	99.8773	95
100	13780.6123	0.0001	137796.1234	0.0000	9.9993	0.1000	9.9927	99.9202	100

11.0%

	Single Payment		Equal Payment Series				Gradient Series		
N	Compound Amount Factor (F/P,i,N)	Present Worth Factor (P/F,i,N)	Compound Amount Factor (F/A,i,N)	Sinking Fund Factor (A/F,i,N)	Present Worth Factor (P/A,i,N)	Capital Recovery Factor (A/P,i,N)	Gradient Uniform Series (A/G,i,N)	Gradient Present Worth (P/G,i,N)	N
1	1.1100	0.9009	1.0000	1.0000	0.9009	1.1100	0.0000	0.0000	1
2	1.2321	0.8116	2.1100	0.4739	1.7125	0.5839	0.4739	0.8116	2
3	1.3676	0.7312	3.3421	0.2992	2.4437	0.4092	0.9306	2.2740	3
4	1.5181	0.6587	4.7097	0.2123	3.1024	0.3223	1.3700	4.2502	4
5	1.6851	0.5935	6.2278	0.1606	3.6959	0.2706	1.7923	6.6240	5
6	1.8704	0.5346	7.9129	0.1264	4.2305	0.2364	2.1976	9.2972	6
7	2.0762	0.4817	9.7833	0.1022	4.7122	0.2122	2.5863	12.1872	7
8	2.3045	0.4339	11.8594	0.0843	5.1461	0.1943	2.9585	15.2246	8
9	2.5580	0.3909	14.1640	0.0706	5.5370	0.1806	3.3144	18.3520	9
10	2.8394	0.3522	16.7220	0.0598	5.8892	0.1698	3.6544	21.5217	10
11	3.1518	0.3173	19.5614	0.0511	6.2065	0.1611	3.9788	24.6945	11
12	3.4985	0.2858	22.7132	0.0440	6.4924	0.1540	4.2879	27.8388	12
13	3.8833	0.2575	26.2116	0.0382	6.7499	0.1482	4.5822	30.9290	13
14	4.3104	0.2320	30.0949	0.0332	6.9819	0.1432	4.8619	33.9449	14
15	4.7846	0.2090	34.4054	0.0291	7.1909	0.1391	5.1275	36.8709	15
16	5.3109	0.1883	39.1899	0.0255	7.3792	0.1355	5.3794	39.6953	16
17	5.8951	0.1696	44.5008	0.0225	7.5488	0.1325	5.6180	42.4095	17
18	6.5436	0.1528	50.3959	0.0198	7.7016	0.1298	5.8439	45.0074	18
19	7.2633	0.1377	56.9395	0.0176	7.8393	0.1276	6.0574	47.4856	19
20	8.0623	0.1240	64.2028	0.0156	7.9633	0.1256	6.2590	49.8423	20
21	8.9492	0.1117	72.2651	0.0138	8.0751	0.1238	6.4491	52.0771	21
22	9.9336	0.1007	81.2143	0.0123	8.1757	0.1223	6.6283	54.1912	22
23	11.0263	0.0907	91.1479	0.0110	8.2664	0.1210	6.7969	56.1864	23
24	12.2392	0.0817	102.1742	0.0098	8.3481	0.1198	6.9555	58.0656	24
25	13.5855	0.0736	114.4133	0.0087	8.4217	0.1187	7.1045	59.8322	25
26	15.0799	0.0663	127.9988	0.0078	8.4881	0.1178	7.2443	61.4900	26
27	16.7386	0.0597	143.0786	0.0070	8.5478	0.1170	7.3754	63.0433	27
28	18.5799	0.0538	159.8173	0.0063	8.6016	0.1163	7.4982	64.4965	28
29	20.6237	0.0485	178.3972	0.0056	8.6501	0.1156	7.6131	65.8542	29
30	22.8923	0.0437	199.0209	0.0050	8.6938	0.1150	7.7206	67.1210	30
31	25.4104	0.0394	221.9132	0.0045	8.7331	0.1145	7.8210	68.3016	31
32	28.2056	0.0355	247.3236	0.0040	8.7686	0.1140	7.9147	69.4007	32
33	31.3082	0.0319	275.5292	0.0036	8.8005	0.1136	8.0021	70.4228	33
34	34.7521	0.0288	306.8374	0.0033	8.8293	0.1133	8.0836	71.3724	34
35	38.5749	0.0259	341.5896	0.0029	8.8552	0.1129	8.1594	72.2538	35
40	65.0009	0.0154	581.8261	0.0017	8.9511	0.1117	8.4659	75.7789	40
45	109.5302	0.0091	986.6386	0.0010	9.0079	0.1110	8.6763	78.1551	45
50	184.5648	0.0054	1668.7712	0.0006	9.0417	0.1106	8.8185	79.7341	50
55	311.0025	0.0032	2818.2042	0.0004	9.0617	0.1104	8.9135	80.7712	55
60	524.0572	0.0019	4755.0658	0.0002	9.0736	0.1102	8.9762	81.4461	60

12.0%

	Single Payment		Equal Payment Series				Gradient Series		
	Compound Amount Factor	Present Worth Factor	Compound Amount Factor	Sinking Fund Factor	Present Worth Factor	Capital Recovery Factor	Gradient Uniform Series	Gradient Present Worth	
N	(F/P,i,N)	(P/F,i,N)	(F/A,i,N)	(A/F,i,N)	(P/A,i,N)	(A/P,i,N)	(A/G,i,N)	(P/G,i,N)	N
1	1.1200	0.8929	1.0000	1.0000	0.8929	1.1200	0.0000	0.0000	1
2	1.2544	0.7972	2.1200	0.4717	1.6901	0.5917	0.4717	0.7972	2
3	1.4049	0.7118	3.3744	0.2963	2.4018	0.4163	0.9246	2.2208	3
4	1.5735	0.6355	4.7793	0.2092	3.0373	0.3292	1.3589	4.1273	4
5	1.7623	0.5674	6.3528	0.1574	3.6048	0.2774	1.7746	6.3970	5
6	1.9738	0.5066	8.1152	0.1232	4.1114	0.2432	2.1720	8.9302	6
7	2.2107	0.4523	10.0890	0.0991	4.5638	0.2191	2.5515	11.6443	7
8	2.4760	0.4039	12.2997	0.0813	4.9676	0.2013	2.9131	14.4714	8
9	2.7731	0.3606	14.7757	0.0677	5.3282	0.1877	3.2574	17.3563	9
10	3.1058	0.3220	17.5487	0.0570	5.6502	0.1770	3.5847	20.2541	10
11	3.4785	0.2875	20.6546	0.0484	5.9377	0.1684	3.8953	23.1288	11
12	3.8960	0.2567	24.1331	0.0414	6.1944	0.1614	4.1897	25.9523	12
13	4.3635	0.2292	28.0291	0.0357	6.4235	0.1557	4.4683	28.7024	13
14	4.8871	0.2046	32.3926	0.0309	6.6282	0.1509	4.7317	31.3624	14
15	5.4736	0.1827	37.2797	0.0268	6.8109	0.1468	4.9803	33.9202	15
16	6.1304	0.1631	42.7533	0.0234	6.9740	0.1434	5.2147	36.3670	16
17	6.8660	0.1456	48.8837	0.0205	7.1196	0.1405	5.4353	38.6973	17
18	7.6900	0.1300	55.7497	0.0179	7.2497	0.1379	5.6427	40.9080	18
19	8.6128	0.1161	63.4397	0.0158	7.3658	0.1358	5.8375	42.9979	19
20	9.6463	0.1037	72.0524	0.0139	7.4694	0.1339	6.0202	44.9676	20
21	10.8038	0.0926	81.6987	0.0122	7.5620	0.1322	6.1913	46.8188	21
22	12.1003	0.0826	92.5026	0.0108	7.6446	0.1308	6.3514	48.5543	22
23	13.5523	0.0738	104.6029	0.0096	7.7184	0.1296	6.5010	50.1776	23
24	15.1786	0.0659	118.1552	0.0085	7.7843	0.1285	6.6406	51.6929	24
25	17.0001	0.0588	133.3339	0.0075	7.8431	0.1275	6.7708	53.1046	25
26	19.0401	0.0525	150.3339	0.0067	7.8957	0.1267	6.8921	54.4177	26
27	21.3249	0.0469	169.3740	0.0059	7.9426	0.1259	7.0049	55.6369	27
28	23.8839	0.0419	190.6989	0.0052	7.9844	0.1252	7.1098	56.7674	28
29	26.7499	0.0374	214.5828	0.0047	8.0218	0.1247	7.2071	57.8141	29
30	29.9599	0.0334	241.3327	0.0041	8.0552	0.1241	7.2974	58.7821	30
31	33.5551	0.0298	271.2926	0.0037	8.0850	0.1237	7.3811	59.6761	31
32	37.5817	0.0266	304.8477	0.0033	8.1116	0.1233	7.4586	60.5010	32
33	42.0915	0.0238	342.4294	0.0029	8.1354	0.1229	7.5302	61.2612	33
34	47.1425	0.0212	384.5210	0.0026	8.1566	0.1226	7.5965	61.9612	34
35	52.7996	0.0189	431.6635	0.0023	8.1755	0.1223	7.6577	62.6052	35
40	93.0510	0.0107	767.0914	0.0013	8.2438	0.1213	7.8988	65.1159	40
45	163.9876	0.0061	1358.2300	0.0007	8.2825	0.1207	8.0572	66.7342	45
50	289.0022	0.0035	2400.0182	0.0004	8.3045	0.1204	8.1597	67.7624	50
55	509.3206	0.0020	4236.0050	0.0002	8.3170	0.1202	8.2251	68.4082	55
60	897.5969	0.0011	7471.6411	0.0001	8.3240	0.1201	8.2664	68.8100	60

13.0%

	Single Payment		Equal Payment Series				Gradient Series		
	Compound Amount Factor	Present Worth Factor	Compound Amount Factor	Sinking Fund Factor	Present Worth Factor	Capital Recovery Factor	Gradient Uniform Series	Gradient Present Worth	
N	(F/P,i,N)	(P/F,i,N)	(F/A,i,N)	(A/F,i,N)	(P/A,i,N)	(A/P,i,N)	(A/G,i,N)	(P/G,i,N)	N
1	1.1300	0.8850	1.0000	1.0000	0.8850	1.1300	0.0000	0.0000	1
2	1.2769	0.7831	2.1300	0.4695	1.6681	0.5995	0.4695	0.7831	2
3	1.4429	0.6931	3.4069	0.2935	2.3612	0.4235	0.9187	2.1692	3
4	1.6305	0.6133	4.8498	0.2062	2.9745	0.3362	1.3479	4.0092	4
5	1.8424	0.5428	6.4803	0.1543	3.5172	0.2843	1.7571	6.1802	5
6	2.0820	0.4803	8.3227	0.1202	3.9975	0.2502	2.1468	8.5818	6
7	2.3526	0.4251	10.4047	0.0961	4.4226	0.2261	2.5171	11.1322	7
8	2.6584	0.3762	12.7573	0.0784	4.7988	0.2084	2.8685	13.7653	8
9	3.0040	0.3329	15.4157	0.0649	5.1317	0.1949	3.2014	16.4284	9
10	3.3946	0.2946	18.4197	0.0543	5.4262	0.1843	3.5162	19.0797	10
11	3.8359	0.2607	21.8143	0.0458	5.6869	0.1758	3.8134	21.6867	11
12	4.3345	0.2307	25.6502	0.0390	5.9176	0.1690	4.0936	24.2244	12
13	4.8980	0.2042	29.9847	0.0334	6.1218	0.1634	4.3573	26.6744	13
14	5.5348	0.1807	34.8827	0.0287	6.3025	0.1587	4.6050	29.0232	14
15	6.2543	0.1599	40.4175	0.0247	6.4624	0.1547	4.8375	31.2617	15
16	7.0673	0.1415	46.6717	0.0214	6.6039	0.1514	5.0552	33.3841	16
17	7.9861	0.1252	53.7391	0.0186	6.7291	0.1486	5.2589	35.3876	17
18	9.0243	0.1108	61.7251	0.0162	6.8399	0.1462	5.4491	37.2714	18
19	10.1974	0.0981	70.7494	0.0141	6.9380	0.1441	5.6265	39.0366	19
20	11.5231	0.0868	80.9468	0.0124	7.0248	0.1424	5.7917	40.6854	20
21	13.0211	0.0768	92.4699	0.0108	7.1016	0.1408	5.9454	42.2214	21
22	14.7138	0.0680	105.4910	0.0095	7.1695	0.1395	6.0881	43.6486	22
23	16.6266	0.0601	120.2048	0.0083	7.2297	0.1383	6.2205	44.9718	23
24	18.7881	0.0532	136.8315	0.0073	7.2829	0.1373	6.3431	46.1960	24
25	21.2305	0.0471	155.6196	0.0064	7.3300	0.1364	6.4566	47.3264	25
26	23.9905	0.0417	176.8501	0.0057	7.3717	0.1357	6.5614	48.3685	26
27	27.1093	0.0369	200.8406	0.0050	7.4086	0.1350	6.6582	49.3276	27
28	30.6335	0.0326	227.9499	0.0044	7.4412	0.1344	6.7474	50.2090	28
29	34.6158	0.0289	258.5834	0.0039	7.4701	0.1339	6.8296	51.0179	29
30	39.1159	0.0256	293.1992	0.0034	7.4957	0.1334	6.9052	51.7592	30
31	44.2010	0.0226	332.3151	0.0030	7.5183	0.1330	6.9747	52.4380	31
32	49.9471	0.0200	376.5161	0.0027	7.5383	0.1327	7.0385	53.0586	32
33	56.4402	0.0177	426.4632	0.0023	7.5560	0.1323	7.0971	53.6256	33
34	63.7774	0.0157	482.9034	0.0021	7.5717	0.1321	7.1507	54.1430	34
35	72.0685	0.0139	546.6808	0.0018	7.5856	0.1318	7.1998	54.6148	35
40	132.7816	0.0075	1013.7042	0.0010	7.6344	0.1310	7.3888	56.4087	40
45	244.6414	0.0041	1874.1646	0.0005	7.6609	0.1305	7.5076	57.5148	45
50	450.7359	0.0022	3459.5071	0.0003	7.6752	0.1303	7.5811	58.1870	50
55	830.4517	0.0012	6380.3979	0.0002	7.6830	0.1302	7.6260	58.5909	55
60	1530.0535	0.0007	11761.9498	0.0001	7.6873	0.1301	7.6531	58.8313	60

14.0%

	Single Payment		Equal Payment Series				Gradient Series		
	Compound Amount Factor	Present Worth Factor	Compound Amount Factor	Sinking Fund Factor	Present Worth Factor	Capital Recovery Factor	Gradient Uniform Series	Gradient Present Worth	
N	(F/P,i,N)	(P/F,i,N)	(F/A,i,N)	(A/F,i,N)	(P/A,i,N)	(A/P,i,N)	(A/G,i,N)	(P/G,i,N)	N
1	1.1400	0.8772	1.0000	1.0000	0.8772	1.1400	0.0000	0.0000	1
2	1.2996	0.7695	2.1400	0.4673	1.6467	0.6073	0.4673	0.7695	2
3	1.4815	0.6750	3.4396	0.2907	2.3216	0.4307	0.9129	2.1194	3
4	1.6890	0.5921	4.9211	0.2032	2.9137	0.3432	1.3370	3.8957	4
5	1.9254	0.5194	6.6101	0.1513	3.4331	0.2913	1.7399	5.9731	5
6	2.1950	0.4556	8.5355	0.1172	3.8887	0.2572	2.1218	8.2511	6
7	2.5023	0.3996	10.7305	0.0932	4.2883	0.2332	2.4832	10.6489	7
8	2.8526	0.3506	13.2328	0.0756	4.6389	0.2156	2.8246	13.1028	8
9	3.2519	0.3075	16.0853	0.0622	4.9464	0.2022	3.1463	15.5629	9
10	3.7072	0.2697	19.3373	0.0517	5.2161	0.1917	3.4490	17.9906	10
11	4.2262	0.2366	23.0445	0.0434	5.4527	0.1834	3.7333	20.3567	11
12	4.8179	0.2076	27.2707	0.0367	5.6603	0.1767	3.9998	22.6399	12
13	5.4924	0.1821	32.0887	0.0312	5.8424	0.1712	4.2491	24.8247	13
14	6.2613	0.1597	37.5811	0.0266	6.0021	0.1666	4.4819	26.9009	14
15	7.1379	0.1401	43.8424	0.0228	6.1422	0.1628	4.6990	28.8623	15
16	8.1372	0.1229	50.9804	0.0196	6.2651	0.1596	4.9011	30.7057	16
17	9.2765	0.1078	59.1176	0.0169	6.3729	0.1569	5.0888	32.4305	17
18	10.5752	0.0946	68.3941	0.0146	6.4674	0.1546	5.2630	34.0380	18
19	12.0557	0.0829	78.9692	0.0127	6.5504	0.1527	5.4243	35.5311	19
20	13.7435	0.0728	91.0249	0.0110	6.6231	0.1510	5.5734	36.9135	20
21	15.6676	0.0638	104.7684	0.0095	6.6870	0.1495	5.7111	38.1901	21
22	17.8610	0.0560	120.4360	0.0083	6.7429	0.1483	5.8381	39.3658	22
23	20.3616	0.0491	138.2970	0.0072	6.7921	0.1472	5.9549	40.4463	23
24	23.2122	0.0431	158.6586	0.0063	6.8351	0.1463	6.0624	41.4371	24
25	26.4619	0.0378	181.8708	0.0055	6.8729	0.1455	6.1610	42.3441	25
26	30.1666	0.0331	208.3327	0.0048	6.9061	0.1448	6.2514	43.1728	26
27	34.3899	0.0291	238.4993	0.0042	6.9352	0.1442	6.3342	43.9289	27
28	39.2045	0.0255	272.8892	0.0037	6.9607	0.1437	6.4100	44.6176	28
29	44.6931	0.0224	312.0937	0.0032	6.9830	0.1432	6.4791	45.2441	29
30	50.9502	0.0196	356.7868	0.0028	7.0027	0.1428	6.5423	45.8132	30
31	58.0832	0.0172	407.7370	0.0025	7.0199	0.1425	6.5998	46.3297	31
32	66.2148	0.0151	465.8202	0.0021	7.0350	0.1421	6.6522	46.7979	32
33	75.4849	0.0132	532.0350	0.0019	7.0482	0.1419	6.6998	47.2218	33
34	86.0528	0.0116	607.5199	0.0016	7.0599	0.1416	6.7431	47.6053	34
35	98.1002	0.0102	693.5727	0.0014	7.0700	0.1414	6.7824	47.9519	35
40	188.8835	0.0053	1342.0251	0.0007	7.1050	0.1407	6.9300	49.2376	40
45	363.6791	0.0027	2590.5648	0.0004	7.1232	0.1404	7.0188	49.9963	45
50	700.2330	0.0014	4994.5213	0.0002	7.1327	0.1402	7.0714	50.4375	50

15.0%

	Single Payment		Equal Payment Series				Gradient Series		
N	Compound Amount Factor (F/P,i,N)	Present Worth Factor (P/F,i,N)	Compound Amount Factor (F/A,i,N)	Sinking Fund Factor (A/F,i,N)	Present Worth Factor (P/A,i,N)	Capital Recovery Factor (A/P,i,N)	Gradient Uniform Series (A/G,i,N)	Gradient Present Worth (P/G,i,N)	N
1	1.1500	0.8696	1.0000	1.0000	0.8696	1.1500	0.0000	0.0000	1
2	1.3225	0.7561	2.1500	0.4651	1.6257	0.6151	0.4651	0.7561	2
3	1.5209	0.6575	3.4725	0.2880	2.2832	0.4380	0.9071	2.0712	3
4	1.7490	0.5718	4.9934	0.2003	2.8550	0.3503	1.3263	3.7864	4
5	2.0114	0.4972	6.7424	0.1483	3.3522	0.2983	1.7228	5.7751	5
6	2.3131	0.4323	8.7537	0.1142	3.7845	0.2642	2.0972	7.9368	6
7	2.6600	0.3759	11.0668	0.0904	4.1604	0.2404	2.4498	10.1924	7
8	3.0590	0.3269	13.7268	0.0729	4.4873	0.2229	2.7813	12.4807	8
9	3.5179	0.2843	16.7858	0.0596	4.7716	0.2096	3.0922	14.7548	9
10	4.0456	0.2472	20.3037	0.0493	5.0188	0.1993	3.3832	16.9795	10
11	4.6524	0.2149	24.3493	0.0411	5.2337	0.1911	3.6549	19.1289	11
12	5.3503	0.1869	29.0017	0.0345	5.4206	0.1845	3.9082	21.1849	12
13	6.1528	0.1625	34.3519	0.0291	5.5831	0.1791	4.1438	23.1352	13
14	7.0757	0.1413	40.5047	0.0247	5.7245	0.1747	4.3624	24.9725	14
15	8.1371	0.1229	47.5804	0.0210	5.8474	0.1710	4.5650	26.6930	15
16	9.3576	0.1069	55.7175	0.0179	5.9542	0.1679	4.7522	28.2960	16
17	10.7613	0.0929	65.0751	0.0154	6.0472	0.1654	4.9251	29.7828	17
18	12.3755	0.0808	75.8364	0.0132	6.1280	0.1632	5.0843	31.1565	18
19	14.2318	0.0703	88.2118	0.0113	6.1982	0.1613	5.2307	32.4213	19
20	16.3665	0.0611	102.4436	0.0098	6.2593	0.1598	5.3651	33.5822	20
21	18.8215	0.0531	118.8101	0.0084	6.3125	0.1584	5.4883	34.6448	21
22	21.6447	0.0462	137.6316	0.0073	6.3587	0.1573	5.6010	35.6150	22
23	24.8915	0.0402	159.2764	0.0063	6.3988	0.1563	5.7040	36.4988	23
24	28.6252	0.0349	184.1678	0.0054	6.4338	0.1554	5.7979	37.3023	24
25	32.9190	0.0304	212.7930	0.0047	6.4641	0.1547	5.8834	38.0314	25
26	37.8568	0.0264	245.7120	0.0041	6.4906	0.1541	5.9612	38.6918	26
27	43.5353	0.0230	283.5688	0.0035	6.5135	0.1535	6.0319	39.2890	27
28	50.0656	0.0200	327.1041	0.0031	6.5335	0.1531	6.0960	39.8283	28
29	57.5755	0.0174	377.1697	0.0027	6.5509	0.1527	6.1541	40.3146	29
30	66.2118	0.0151	434.7451	0.0023	6.5660	0.1523	6.2066	40.7526	30
31	76.1435	0.0131	500.9569	0.0020	6.5791	0.1520	6.2541	41.1466	31
32	87.5651	0.0114	577.1005	0.0017	6.5905	0.1517	6.2970	41.5006	32
33	100.6998	0.0099	664.6655	0.0015	6.6005	0.1515	6.3357	41.8184	33
34	115.8048	0.0086	765.3654	0.0013	6.6091	0.1513	6.3705	42.1033	34
35	133.1755	0.0075	881.1702	0.0011	6.6166	0.1511	6.4019	42.3586	35
40	267.8635	0.0037	1779.0903	0.0006	6.6418	0.1506	6.5168	43.2830	40
45	538.7693	0.0019	3585.1285	0.0003	6.6543	0.1503	6.5830	43.8051	45
50	1083.6574	0.0009	7217.7163	0.0001	6.6605	0.1501	6.6205	44.0958	50

16.0%

	Single Payment		Equal Payment Series				Gradient Series		
	Compound Amount Factor	Present Worth Factor	Compound Amount Factor	Sinking Fund Factor	Present Worth Factor	Capital Recovery Factor	Gradient Uniform Series	Gradient Present Worth	
N	(F/P,i,N)	(P/F,i,N)	(F/A,i,N)	(A/F,i,N)	(P/A,i,N)	(A/P,i,N)	(A/G,i,N)	(P/G,i,N)	N
1	1.1600	0.8621	1.0000	1.0000	0.8621	1.1600	0.0000	0.0000	1
2	1.3456	0.7432	2.1600	0.4630	1.6052	0.6230	0.4630	0.7432	2
3	1.5609	0.6407	3.5056	0.2853	2.2459	0.4453	0.9014	2.0245	3
4	1.8106	0.5523	5.0665	0.1974	2.7982	0.3574	1.3156	3.6814	4
5	2.1003	0.4761	6.8771	0.1454	3.2743	0.3054	1.7060	5.5858	5
6	2.4364	0.4104	8.9775	0.1114	3.6847	0.2714	2.0729	7.6380	6
7	2.8262	0.3538	11.4139	0.0876	4.0386	0.2476	2.4169	9.7610	7
8	3.2784	0.3050	14.2401	0.0702	4.3436	0.2302	2.7388	11.8962	8
9	3.8030	0.2630	17.5185	0.0571	4.6065	0.2171	3.0391	13.9998	9
10	4.4114	0.2267	21.3215	0.0469	4.8332	0.2069	3.3187	16.0399	10
11	5.1173	0.1954	25.7329	0.0389	5.0286	0.1989	3.5783	17.9941	11
12	5.9360	0.1685	30.8502	0.0324	5.1971	0.1924	3.8189	19.8472	12
13	6.8858	0.1452	36.7862	0.0272	5.3423	0.1872	4.0413	21.5899	13
14	7.9875	0.1252	43.6720	0.0229	5.4675	0.1829	4.2464	23.2175	14
15	9.2655	0.1079	51.6595	0.0194	5.5755	0.1794	4.4352	24.7284	15
16	10.7480	0.0930	60.9650	0.0164	5.6685	0.1764	4.6086	26.1241	16
17	12.4677	0.0802	71.6730	0.0140	5.7487	0.1740	4.7676	27.4074	17
18	14.4625	0.0691	84.1407	0.0119	5.8178	0.1719	4.9130	28.5828	18
19	16.7765	0.0596	98.6032	0.0101	5.8775	0.1701	5.0457	29.6557	19
20	19.4608	0.0514	115.3797	0.0087	5.9288	0.1687	5.1666	30.6321	20
21	22.5745	0.0443	134.8405	0.0074	5.9731	0.1674	5.2766	31.5180	21
22	26.1864	0.0382	157.4150	0.0064	6.0113	0.1664	5.3765	32.3200	22
23	30.3762	0.0329	183.6014	0.0054	6.0442	0.1654	5.4671	33.0442	23
24	35.2364	0.0284	213.9776	0.0047	6.0726	0.1647	5.5490	33.6970	24
25	40.8742	0.0245	249.2140	0.0040	6.0971	0.1640	5.6230	34.2841	25
26	47.4141	0.0211	290.0883	0.0034	6.1182	0.1634	5.6898	34.8114	26
27	55.0004	0.0182	337.5024	0.0030	6.1364	0.1630	5.7500	35.2841	27
28	63.8004	0.0157	392.5028	0.0025	6.1520	0.1625	5.8041	35.7073	28
29	74.0085	0.0135	456.3032	0.0022	6.1656	0.1622	5.8528	36.0856	29
30	85.8499	0.0116	530.3117	0.0019	6.1772	0.1619	5.8964	36.4234	30
31	99.5859	0.0100	616.1616	0.0016	6.1872	0.1616	5.9356	36.7247	31
32	115.5196	0.0087	715.7475	0.0014	6.1959	0.1614	5.9706	36.9930	32
33	134.0027	0.0075	831.2671	0.0012	6.2034	0.1612	6.0019	27.2318	33
34	155.4432	0.0064	965.2698	0.0010	6.2098	0.1610	6.0299	37.4441	34
35	180.3141	0.0055	1120.7130	0.0009	6.2153	0.1609	6.0548	37.6327	35
40	378.7212	0.0026	2360.7572	0.0004	6.2335	0.1604	6.1441	38.2992	40
45	795.4438	0.0013	4965.2739	0.0002	6.2421	0.1602	6.1934	38.6598	45
50	1670.7038	0.0006	10435.6488	0.0001	6.2463	0.1601	6.2201	38.8521	50

18.0%

	Single Payment		Equal Payment Series				Gradient Series		
	Compound Amount Factor	Present Worth Factor	Compound Amount Factor	Sinking Fund Factor	Present Worth Factor	Capital Recovery Factor	Gradient Uniform Series	Gradient Present Worth	
N	(F/P,i,N)	(P/F,i,N)	(F/A,i,N)	(A/F,i,N)	(P/A,i,N)	(A/P,i,N)	(A/G,i,N)	(P/G,i,N)	N
1	1.1800	0.8475	1.0000	1.0000	0.8475	1.1800	0.0000	0.0000	1
2	1.3924	0.7182	2.1800	0.4587	1.5656	0.6387	0.4587	0.7182	2
3	1.6430	0.6086	3.5724	0.2799	2.1743	0.4599	0.8902	1.9354	3
4	1.9388	0.5158	5.2154	0.1917	2.6901	0.3717	1.2947	3.4828	4
5	2.2878	0.4371	7.1542	0.1398	3.1272	0.3198	1.6728	5.2312	5
6	2.6996	0.3704	9.4420	0.1059	3.4976	0.2859	2.0252	7.0834	6
7	3.1855	0.3139	12.1415	0.0824	3.8115	0.2624	2.3526	8.9670	7
8	3.7589	0.2660	15.3270	0.0652	4.0776	0.2452	2.6558	10.8292	8
9	4.4355	0.2255	19.0859	0.0524	4.3030	0.2324	2.9358	12.6329	9
10	5.2338	0.1911	23.5213	0.0425	4.4941	0.2225	3.1936	14.3525	10
11	6.1759	0.1619	28.7551	0.0348	4.6560	0.2148	3.4303	15.9716	11
12	7.2876	0.1372	34.9311	0.0286	4.7932	0.2086	3.6470	17.4811	12
13	8.5994	0.1163	42.2187	0.0237	4.9095	0.2037	3.8449	18.8765	13
14	10.1472	0.0985	50.8180	0.0197	5.0081	0.1997	4.0250	20.1576	14
15	11.9737	0.0835	60.9653	0.0164	5.0916	0.1964	4.1887	21.3269	15
16	14.1290	0.0708	72.9390	0.0137	5.1624	0.1937	4.3369	22.3885	16
17	16.6722	0.0600	87.0680	0.0115	5.2223	0.1915	4.4708	23.3482	17
18	19.6733	0.0508	103.7403	0.0096	5.2732	0.1896	4.5916	24.2123	18
19	23.2144	0.0431	123.4135	0.0081	5.3162	0.1881	4.7003	24.9877	19
20	27.3930	0.0365	146.6280	0.0068	5.3527	0.1868	4.7978	25.6813	20
21	32.3238	0.0309	174.0210	0.0057	5.3837	0.1857	4.8851	26.3000	21
22	38.1421	0.0262	206.3448	0.0048	5.4099	0.1848	4.9632	26.8506	22
23	45.0076	0.0222	244.4868	0.0041	5.4321	0.1841	5.0329	27.3394	23
24	53.1090	0.0188	289.4945	0.0035	5.4509	0.1835	5.0950	27.7725	24
25	62.6686	0.0160	342.6035	0.0029	5.4669	0.1829	5.1502	28.1555	25
26	73.9490	0.0135	405.2721	0.0025	5.4804	0.1825	5.1991	28.4935	26
27	87.2598	0.0115	479.2211	0.0021	5.4919	0.1821	5.2425	28.7915	27
28	102.9666	0.0097	566.4809	0.0018	5.5016	0.1818	5.2810	29.0537	28
29	121.5005	0.0082	669.4475	0.0015	5.5098	0.1815	5.3149	29.2842	29
30	143.3706	0.0070	790.9480	0.0013	5.5168	0.1813	5.3448	29.4864	30
31	169.1774	0.0059	934.3186	0.0011	5.5227	0.1811	5.3712	29.6638	31
32	199.6293	0.0050	1103.4960	0.0009	5.5277	0.1809	5.3945	29.8191	32
33	235.5625	0.0042	1303.1253	0.0008	5.5320	0.1808	5.4149	29.9549	33
34	277.9638	0.0036	1538.6878	0.0006	5.5356	0.1806	5.4328	30.0736	34
35	327.9973	0.0030	1816.6516	0.0006	5.5386	0.1806	5.4485	30.1773	35
40	750.3783	0.0013	4163.2130	0.0002	5.5482	0.1802	5.5022	30.5269	40
45	1716.6839	0.0006	9531.5771	0.0001	5.5523	0.1801	5.5293	30.7006	45
50	3927.3569	0.0003	21813.0937	0.0000	5.5541	0.1800	5.5428	30.7856	50

20.0%

	Single Payment		Equal Payment Series				Gradient Series		
	Compound Amount Factor	Present Worth Factor	Compound Amount Factor	Sinking Fund Factor	Present Worth Factor	Capital Recovery Factor	Gradient Uniform Series	Gradient Present Worth	
N	(F/P,i,N)	(P/F,i,N)	(F/A,i,N)	(A/F,i,N)	(P/A,i,N)	(A/P,i,N)	(A/G,i,N)	(P/G,i,N)	N
1	1.2000	0.8333	1.0000	1.0000	0.8333	1.2000	0.0000	0.0000	1
2	1.4400	0.6944	2.2000	0.4545	1.5278	0.6545	0.4545	0.6944	2
3	1.7280	0.5787	3.6400	0.2747	2.1065	0.4747	0.8791	1.8519	3
4	2.0736	0.4823	5.3680	0.1863	2.5887	0.3863	1.2742	3.2986	4
5	2.4883	0.4019	7.4416	0.1344	2.9906	0.3344	1.6405	4.9061	5
6	2.9860	0.3349	9.9299	0.1007	3.3255	0.3007	1.9788	6.5806	6
7	3.5832	0.2791	12.9159	0.0774	3.6046	0.2774	2.2902	8.2551	7
8	4.2998	0.2326	16.4991	0.0606	3.8372	0.2606	2.5756	9.8831	8
9	5.1598	0.1938	20.7989	0.0481	4.0310	0.2481	2.8364	11.4335	9
10	6.1917	0.1615	25.9587	0.0385	4.1925	0.2385	3.0739	12.8871	10
11	7.4301	0.1346	32.1504	0.0311	4.3271	0.2311	3.2893	14.2330	11
12	8.9161	0.1122	39.5805	0.0253	4.4392	0.2253	3.4841	15.4667	12
13	10.6993	0.0935	48.4966	0.0206	4.5327	0.2206	3.6597	16.5883	13
14	12.8392	0.0779	59.1959	0.0169	4.6106	0.2169	3.8175	17.6008	14
15	15.4070	0.0649	72.0351	0.0139	4.6755	0.2139	3.9588	18.5095	15
16	18.4884	0.0541	87.4421	0.0114	4.7296	0.2114	4.0851	19.3208	16
17	22.1861	0.0451	105.9306	0.0094	4.7746	0.2094	4.1976	20.0419	17
18	26.6233	0.0376	128.1167	0.0078	4.8122	0.2078	4.2975	20.6805	18
19	31.9480	0.0313	154.7400	0.0065	4.8435	0.2065	4.3861	21.2439	19
20	38.3376	0.0261	186.6880	0.0054	4.8696	0.2054	4.4643	21.7395	20
21	46.0051	0.0217	225.0256	0.0044	4.8913	0.2044	4.5334	22.1742	21
22	55.2061	0.0181	271.0307	0.0037	4.9094	0.2037	4.5941	22.5546	22
23	66.2474	0.0151	326.2369	0.0031	4.9245	0.2031	4.6475	22.8867	23
24	79.4968	0.0126	392.4842	0.0025	4.9371	0.2025	4.6943	23.1760	24
25	95.3962	0.0105	471.9811	0.0021	4.9476	0.2021	4.7352	23.4276	25
26	114.4755	0.0087	567.3773	0.0018	4.9563	0.2018	4.7709	23.6460	26
27	137.3706	0.0073	681.8528	0.0015	4.9636	0.2015	4.8020	23.8353	27
28	164.8447	0.0061	819.2233	0.0012	4.9697	0.2012	4.8291	23.9991	28
29	197.8136	0.0051	984.0680	0.0010	4.9747	0.2010	4.8527	24.1406	29
30	237.3763	0.0042	1181.8816	0.0008	4.9789	0.2008	4.8731	24.2628	30
31	284.8516	0.0035	1419.2579	0.0007	4.9824	0.2007	4.8908	24.3681	31
32	341.8219	0.0029	1704.1095	0.0006	4.9854	0.2006	4.9061	24.4588	32
33	410.1863	0.0024	2045.9314	0.0005	4.9878	0.2005	4.9194	24.5368	33
34	492.2235	0.0020	2456.1176	0.0004	4.9898	0.2004	4.9308	24.6038	34
35	590.6682	0.0017	2948.3411	0.0003	4.9915	0.2003	4.9406	24.6614	35
40	1469.7716	0.0007	7343.8578	0.0001	4.9966	0.2001	4.9728	24.8469	40
45	3657.2620	0.0003	18281.3099	0.0001	4.9986	0.2001	4.9877	24.9316	45

25.0%

	Single Payment		Equal Payment Series				Gradient Series		
N	Compound Amount Factor (F/P,i,N)	Present Worth Factor (P/F,i,N)	Compound Amount Factor (F/A,i,N)	Sinking Fund Factor (A/F,i,N)	Present Worth Factor (P/A,i,N)	Capital Recovery Factor (A/P,i,N)	Gradient Uniform Series (A/G,i,N)	Gradient Present Worth (P/G,i,N)	N
1	1.2500	0.8000	1.0000	1.0000	0.8000	1.2500	0.0000	0.0000	1
2	1.5625	0.6400	2.2500	0.4444	1.4400	0.6944	0.4444	0.6400	2
3	1.9531	0.5120	3.8125	0.2623	1.9520	0.5123	0.8525	1.6640	3
4	2.4414	0.4096	5.7656	0.1734	2.3616	0.4234	1.2249	2.8928	4
5	3.0518	0.3277	8.2070	0.1218	2.6893	0.3718	1.5631	4.2035	5
6	3.8147	0.2621	11.2588	0.0888	2.9514	0.3388	1.8683	5.5142	6
7	4.7684	0.2097	15.0735	0.0663	3.1611	0.3163	2.1424	6.7725	7
8	5.9605	0.1678	19.8419	0.0504	3.3289	0.3004	2.3872	7.9469	8
9	7.4506	0.1342	25.8023	0.0388	3.4631	0.2888	2.6048	9.0207	9
10	9.3132	0.1074	33.2529	0.0301	3.5705	0.2801	2.7971	9.9870	10
11	11.6415	0.0859	42.5661	0.0235	3.6564	0.2735	2.9663	10.8460	11
12	14.5519	0.0687	54.2077	0.0184	3.7251	0.2684	3.1145	11.6020	12
13	18.1899	0.0550	68.7596	0.0145	3.7801	0.2645	3.2437	12.2617	13
14	22.7374	0.0440	86.9495	0.0115	3.8241	0.2615	3.3559	12.8334	14
15	28.4217	0.0352	109.6868	0.0091	3.8593	0.2591	3.4530	13.3260	15
16	35.5271	0.0281	138.1085	0.0072	3.8874	0.2572	3.5366	13.7482	16
17	44.4089	0.0225	173.6357	0.0058	3.9099	0.2558	3.6084	14.1085	17
18	55.5112	0.0180	218.0446	0.0046	3.9279	0.2546	3.6698	14.4147	18
19	69.3889	0.0144	273.5558	0.0037	3.9424	0.2537	3.7222	14.6741	19
20	86.7362	0.0115	342.9447	0.0029	3.9539	0.2529	3.7667	14.8932	20
21	108.4202	0.0092	429.6809	0.0023	3.9631	0.2523	3.8045	15.0777	21
22	135.5253	0.0074	538.1011	0.0019	3.9705	0.2519	3.8365	15.2326	22
23	169.4066	0.0059	673.6264	0.0015	3.9764	0.2515	3.8634	15.3625	23
24	211.7582	0.0047	843.0329	0.0012	3.9811	0.2512	3.8861	15.4711	24
25	264.6978	0.0038	1054.7912	0.0009	3.9849	0.2509	3.9052	15.5618	25
26	330.8722	0.0030	1319.4890	0.0008	3.9879	0.2508	3.9212	15.6373	26
27	413.5903	0.0024	1650.3612	0.0006	3.9903	0.2506	3.9346	15.7002	27
28	516.9879	0.0019	2063.9515	0.0005	3.9923	0.2505	3.9457	15.7524	28
29	646.2349	0.0015	2580.9394	0.0004	3.9938	0.2504	3.9551	15.7957	29
30	807.7936	0.0012	3227.1743	0.0003	3.9950	0.2503	3.9628	15.8316	30
31	1009.7420	0.0010	4034.9678	0.0002	3.9960	0.2502	3.9693	15.8614	31
32	1262.1774	0.0008	5044.7098	0.0002	3.9968	0.2502	3.9746	15.8859	32
33	1577.7218	0.0006	6306.8872	0.0002	3.9975	0.2502	3.9791	15.9062	33
34	1972.1523	0.0005	7884.6091	0.0001	3.9980	0.2501	3.9828	15.9229	34
35	2465.1903	0.0004	9856.7613	0.0001	3.9984	0.2501	3.9858	15.9367	35
40	7523.1638	0.0001	30088.6554	0.0000	3.9995	0.2500	3.9947	15.9766	40

30.0%

	Single Payment		Equal Payment Series				Gradient Series		
	Compound Amount Factor	Present Worth Factor	Compound Amount Factor	Sinking Fund Factor	Present Worth Factor	Capital Recovery Factor	Gradient Uniform Series	Gradient Present Worth	
N	(F/P,i,N)	(P/F,i,N)	(F/A,i,N)	(A/F,i,N)	(P/A,i,N)	(A/P,i,N)	(A/G,i,N)	(P/G,i,N)	N
1	1.3000	0.7692	1.0000	1.0000	0.7692	1.3000	0.0000	0.0000	1
2	1.6900	0.5917	2.3000	0.4348	1.3609	0.7348	0.4348	0.5917	2
3	2.1970	0.4552	3.9900	0.2506	1.8161	0.5506	0.8271	1.5020	3
4	2.8561	0.3501	6.1870	0.1616	2.1662	0.4616	1.1783	2.5524	4
5	3.7129	0.2693	9.0431	0.1106	2.4356	0.4106	1.4903	3.6297	5
6	4.8268	0.2072	12.7560	0.0784	2.6427	0.3784	1.7654	4.6656	6
7	6.2749	0.1594	17.5828	0.0569	2.8021	0.3569	2.0063	5.6218	7
8	8.1573	0.1226	23.8577	0.0419	2.9247	0.3419	2.2156	6.4800	8
9	10.6045	0.0943	32.0150	0.0312	3.0190	0.3312	2.3963	7.2343	9
10	13.7858	0.0725	42.6195	0.0235	3.0915	0.3235	2.5512	7.8872	10
11	17.9216	0.0558	56.4053	0.0177	3.1473	0.3177	2.6833	8.4452	11
12	23.2981	0.0429	74.3270	0.0135	3.1903	0.3135	2.7952	8.9173	12
13	30.2875	0.0330	97.6250	0.0102	3.2233	0.3102	2.8895	9.3135	13
14	39.3738	0.0254	127.9125	0.0078	3.2487	0.3078	2.9685	9.6437	14
15	51.1859	0.0195	167.2863	0.0060	3.2682	0.3060	3.0344	9.9172	15
16	66.5417	0.0150	218.4722	0.0046	3.2832	0.3046	3.0892	10.1426	16
17	86.5042	0.0116	285.0139	0.0035	3.2948	0.3035	3.1345	10.3276	17
18	112.4554	0.0089	371.5180	0.0027	3.3037	0.3027	3.1718	10.4788	18
19	146.1920	0.0068	483.9734	0.0021	3.3105	0.3021	3.2025	10.6019	19
20	190.0496	0.0053	630.1655	0.0016	3.3158	0.3016	3.2275	10.7019	20
21	247.0645	0.0040	820.2151	0.0012	3.3198	0.3012	3.2480	10.7828	21
22	321.1839	0.0031	1067.2796	0.0009	3.3230	0.3009	3.2646	10.8482	22
23	417.5391	0.0024	1388.4635	0.0007	3.3254	0.3007	3.2781	10.9009	23
24	542.8008	0.0018	1806.0026	0.0006	3.3272	0.3006	3.2890	10.9433	24
25	705.6410	0.0014	2348.8033	0.0004	3.3286	0.3004	3.2979	10.9773	25
26	917.3333	0.0011	3054.4443	0.0003	3.3297	0.3003	3.3050	11.0045	26
27	1192.5333	0.0008	3971.7776	0.0003	3.3305	0.3003	3.3107	11.0263	27
28	1550.2933	0.0006	5164.3109	0.0002	3.3312	0.3002	3.3153	11.0437	28
29	2015.3813	0.0005	6714.6042	0.0001	3.3317	0.3001	3.3189	11.0576	29
30	2619.9956	0.0004	8729.9855	0.0001	3.3321	0.3001	3.3219	11.0687	30
31	3405.9943	0.0003	11349.9811	0.0001	3.3324	0.3001	3.3242	11.0775	31
32	4427.7926	0.0002	14755.9755	0.0001	3.3326	0.3001	3.3261	11.0845	32
33	5756.1304	0.0002	19183.7681	0.0001	3.3328	0.3001	3.3276	11.0901	33
34	7482.9696	0.0001	24939.8985	0.0000	3.3329	0.3000	3.3288	11.0945	34
35	9727.8604	0.0001	32422.8681	0.0000	3.3330	0.3000	3.3297	11.0980	35

35.0%

	Single Payment		Equal Payment Series				Gradient Series		
N	Compound Amount Factor (F/P,i,N)	Present Worth Factor (P/F,i,N)	Compound Amount Factor (F/A,i,N)	Sinking Fund Factor (A/F,i,N)	Present Worth Factor (P/A,i,N)	Capital Recovery Factor (A/P,i,N)	Gradient Uniform Series (A/G,i,N)	Gradient Present Worth (P/G,i,N)	N
1	1.3500	0.7407	1.0000	1.0000	0.7407	1.3500	0.0000	0.0000	1
2	1.8225	0.5487	2.3500	0.4255	1.2894	0.7755	0.4255	0.5487	2
3	2.4604	0.4064	4.1725	0.2397	1.6959	0.5897	0.8029	1.3616	3
4	3.3215	0.3011	6.6329	0.1508	1.9969	0.5008	1.1341	2.2648	4
5	4.4840	0.2230	9.9544	0.1005	2.2200	0.4505	1.4220	3.1568	5
6	6.0534	0.1652	14.4384	0.0693	2.3852	0.4193	1.6698	3.9828	6
7	8.1722	0.1224	20.4919	0.0488	2.5075	0.3988	1.8811	4.7170	7
8	11.0324	0.0906	28.6640	0.0349	2.5982	0.3849	2.0597	5.3515	8
9	14.8937	0.0671	39.6964	0.0252	2.6653	0.3752	2.2094	5.8886	9
10	20.1066	0.0497	54.5902	0.0183	2.7150	0.3683	2.3338	6.3363	10
11	27.1439	0.0368	74.6967	0.0134	2.7519	0.3634	2.4364	6.7047	11
12	36.6442	0.0273	101.8406	0.0098	2.7792	0.3598	2.5205	7.0049	12
13	49.4697	0.0202	138.4848	0.0072	2.7994	0.3572	2.5889	7.2474	13
14	66.7841	0.0150	187.9544	0.0053	2.8144	0.3553	2.6443	7.4421	14
15	90.1585	0.0111	254.7385	0.0039	2.8255	0.3539	2.6889	7.5974	15
16	121.7139	0.0082	344.8970	0.0029	2.8337	0.3529	2.7246	7.7206	16
17	164.3138	0.0061	466.6109	0.0021	2.8398	0.3521	2.7530	7.8180	17
18	221.8236	0.0045	630.9247	0.0016	2.8443	0.3516	2.7756	7.8946	18
19	299.4619	0.0033	852.7483	0.0012	2.8476	0.3512	2.7935	2.9547	19
20	404.2736	0.0025	1152.2103	0.0009	2.8501	0.3509	2.8075	8.0017	20
21	545.7693	0.0018	1556.4838	0.0006	2.8519	0.3506	2.8186	8.0384	21
22	736.7886	0.0014	2102.2532	0.0005	2.8533	0.3505	2.8272	8.0669	22
23	994.6646	0.0010	2839.0418	0.0004	2.8543	0.3504	2.8340	8.0890	23
24	1342.7973	0.0007	3833.7064	0.0003	2.8550	0.3503	2.8393	8.1061	24
25	1812.7763	0.0006	5176.5037	0.0002	2.8556	0.3502	2.8433	8.1194	25
26	2447.2480	0.0004	6989.2800	0.0001	2.8560	0.3501	2.8465	8.1296	26
27	3303.7848	0.0003	9436.5280	0.0001	2.8563	0.3501	2.8490	8.1374	27
28	4460.1095	0.0002	12740.3128	0.0001	2.8565	0.3501	2.8509	8.1435	28
29	6021.1478	0.0002	17200.4222	0.0001	2.8567	0.3501	2.8523	8.1481	29
30	8128.5495	0.0001	23221.5700	0.0000	2.8568	0.3500	2.8535	8.1517	30

	Single Payment		Equal Payment Series				Gradient Series		
	Compound Amount Factor	Present Worth Factor	Compound Amount Factor	Sinking Fund Factor	Present Worth Factor	Capital Recovery Factor	Gradient Uniform Series	Gradient Present Worth	
N	(F/P,i,N)	(P/F,i,N)	(F/A,i,N)	(A/F,i,N)	(P/A,i,N)	(A/P,i,N)	(A/G,i,N)	(P/G,i,N)	N
1	1.4000	0.7143	1.0000	1.0000	0.7143	1.4000	0.0000	0.0000	1
2	1.9600	0.5102	2.4000	0.4167	1.2245	0.8167	0.4167	0.5102	2
3	2.7440	0.3644	4.3600	0.2294	1.5889	0.6294	0.7798	1.2391	3
4	3.8416	0.2603	7.1040	0.1408	1.8492	0.5408	1.0923	2.0200	4
5	5.3782	0.1859	10.9456	0.0914	2.0352	0.4914	1.3580	2.7637	5
6	7.5295	0.1328	16.3238	0.0613	2.1680	0.4613	1.5811	3.4278	6
7	10.5414	0.0949	23.8534	0.0419	2.2628	0.4419	1.7664	3.9970	7
8	14.7579	0.0678	34.3947	0.0291	2.3306	0.4291	1.9185	4.4713	8
9	20.6610	0.0484	49.1526	0.0203	2.3790	0.4203	2.0422	4.8585	9
10	28.9255	0.0346	69.8137	0.0143	2.4136	0.4143	2.1419	5.1696	10
11	40.4957	0.0247	98.7391	0.0101	2.4383	0.4101	2.2215	5.4166	11
12	56.6939	0.0176	139.2348	0.0072	2.4559	0.4072	2.2845	5.6106	12
13	79.3715	0.0126	195.9287	0.0051	2.4685	0.4051	2.3341	5.7618	13
14	111.1201	0.0090	275.3002	0.0036	2.4775	0.4036	2.3729	5.8788	14
15	155.5681	0.0064	386.4202	0.0026	2.4839	0.4026	2.4030	5.9688	15
16	217.7953	0.0046	541.9883	0.0018	2.4885	0.4018	2.4262	6.0376	16
17	304.9135	0.0033	759.7837	0.0013	2.4918	0.4013	2.4441	6.0901	17
18	426.8789	0.0023	1064.6971	0.0009	2.4941	0.4009	2.4577	6.1299	18
19	597.6304	0.0017	1491.5760	0.0007	2.4958	0.4007	2.4682	6.1601	19
20	836.6826	0.0012	2089.2064	0.0005	2.4970	0.4005	2.4761	6.1828	20
21	1171.3556	0.0009	2925.8889	0.0003	2.4979	0.4003	2.4821	6.1998	21
22	1639.8978	0.0006	4097.2445	0.0002	2.4985	0.4002	2.4866	6.2127	22
23	2295.8569	0.0004	5737.1423	0.0002	2.4989	0.4002	2.4900	6.2222	23
24	3214.1997	0.0003	8032.9993	0.0001	2.4992	0.4001	2.4925	6.2294	24
25	4499.8796	0.0002	11247.1990	0.0001	2.4994	0.4001	2.4944	6.2347	25
26	6299.8314	0.0002	15747.0785	0.0001	2.4996	0.4001	2.4959	6.2387	26
27	8819.7640	0.0001	22046.9099	0.0000	2.4997	0.4000	2.4969	6.2416	27
28	12347.6696	0.0001	30866.6739	0.0000	2.4998	0.4000	2.4977	6.2438	28
29	17286.7374	0.0001	43214.3435	0.0000	2.4999	0.4000	2.4983	6.2454	29
30	24201.4324	0.0000	60501.0809	0.0000	2.4999	0.4000	2.4988	6.2466	30

50.0%

	Single Payment		Equal Payment Series				Gradient Series		
	Compound Amount Factor	Present Worth Factor	Compound Amount Factor	Sinking Fund Factor	Present Worth Factor	Capital Recovery Factor	Gradient Uniform Series	Gradient Present Worth	
N	(F/P,i,N)	(P/F,i,N)	(F/A,i,N)	(A/F,i,N)	(P/A,i,N)	(A/P,i,N)	(A/G,i,N)	(P/G,i,N)	N
1	1.5000	0.6667	1.0000	1.0000	0.6667	1.5000	0.0000	0.0000	1
2	2.2500	0.4444	2.5000	0.4000	1.1111	0.9000	0.4000	0.4444	2
3	3.3750	0.2963	4.7500	0.2105	1.4074	0.7105	0.7368	1.0370	3
4	5.0625	0.1975	8.1250	0.1231	1.6049	0.6231	1.0154	1.6296	4
5	7.5938	0.1317	13.1875	0.0758	1.7366	0.5758	1.2417	2.1564	5
6	11.3906	0.0878	20.7813	0.0481	1.8244	0.5481	1.4226	2.5953	6
7	17.0859	0.0585	32.1719	0.0311	1.8829	0.5311	1.5648	2.9465	7
8	25.6289	0.0390	49.2578	0.0203	1.9220	0.5203	1.6752	3.2196	8
9	38.4434	0.0260	74.8867	0.0134	1.9480	0.5134	1.7596	3.4277	9
10	57.6650	0.0173	113.3301	0.0088	1.9653	0.5088	1.8235	3.5838	10
11	86.4976	0.0116	170.9951	0.0058	1.9769	0.5058	1.8713	3.6994	11
12	129.7463	0.0077	257.4927	0.0039	1.9846	0.5039	1.9068	3.7842	12
13	194.6195	0.0051	387.2390	0.0026	1.9897	0.5026	1.9329	3.8459	13
14	291.9293	0.0034	581.8585	0.0017	1.9931	0.5017	1.9519	3.8904	14
15	437.8939	0.0023	873.7878	0.0011	1.9954	0.5011	1.9657	3.9224	15
16	656.8408	0.0015	1311.6817	0.0008	1.9970	0.5008	1.9756	3.9452	16
17	985.2613	0.0010	1968.5225	0.0005	1.9980	0.5005	1.9827	3.9614	17
18	1477.8919	0.0007	2953.7838	0.0003	1.9986	0.5003	1.9878	3.9729	18
19	2216.8378	0.0005	4431.6756	0.0002	1.9991	0.5002	1.9914	3.9811	19
20	3325.2567	0.0003	6648.5135	0.0002	1.9994	0.5002	1.9940	3.9868	20

Answers to Selected Problems

Chapter 2

2.1	$1,000
2.2	Simple interest = $700, Compound interest = $967.15
2.6	$F = \$13,382.26$
2.9	(a) $F = \$13,948$
2.10	(a) $P = \$2,999$
2.11	(a) $P = \$4,083$
2.14	$P = \$23,055.58$
2.18	$20,734,618
2.21	$X = \$83,734$
2.25	(a) $F = \$28,613.16$
2.26	(a) $A = \$721.20$
2.30	(a) $A = \$6,262.50$
2.32	(a) $P = \$6,209.79$
2.37	$P = \$1,651.18$
2.40	$C = \$458.90$
2.43	(a) $A = \$923.88$, (b) $A = \$1,073.61$
2.47	$A = \$110.84$
2.50	$C = \$838.28$
2.53	$X = \$715.43$
2.55	(2), (4), and (5)
2.58	(b)

Chapter 3

3.1	(a) 18%, (b) 19.56%
3.5	(a) 12.5%, (b) 650%, (c) 45,601.60%
3.10	(a) 0.75%
3.13	(a) $F = \$8,376.74$
3.14	(a) $F = \$80,611$
3.15	(a) $A = \$446.59$
3.16	(a) $P = \$6,795$
3.18	(d)
3.21	$F = \$14,052.02$
3.25	$A = \$1,774.37$
3.28	(b)
3.30	(a) 13.87 years (b) 13.78 years, (c) 13.73 years

3.34	$A = \$730.26$
3.39	(c)
3.42	(a) $A = \$965.55$ (b) $A = \$1,025.31$
3.43	(a) $\$14,000(A/P, 0.75, 24)$ (b) $B_{12} = A(P/A, 0.75\%, 12)$
3.46	$A = \$1,477.11$
3.55	$A = \$453.43$

Chapter 4

4.1	165.45
4.4	$N = 10.24$ years, Rule of 72: 10.29 years
4.8	$658.35 in actual $, $595.85 in constant $
4.11	$45 in actual $, $28.23 in constant $
4.15	$A = \$2,174.52$ per month
4.18	(a) $f = 7.4312\%$, (c) $P = \$16,925$

Chapter 5

5.2	(a) 1.2 years, (b) 1.44 years
5.5	PW(9%) = $1,386.29
5.7	(d)
5.9	$P = \$7,913.16$
5.13	(a)
5.16	(c)
5.19	(a) $FW(15\%)_A = \$2,348.75$ (b) Accept all except Project E.
5.25	$3,034
5.29	$PW(12\%)_A = \$140.87$ $PW(12\%)_B = \$197.68$ Select B.
5.32	(a) Select A. (b) Select A at any interest rate.
5.36	Model A: $PW(12\%)_A = \$9,989$ Model B: $PW(12\%)_B = \$15,056$ Select Model B.
5.39.1	Method A: $CE(12\%)_A = \$43,117.50$; Model B: $CE(12\%)_B = \$75,312.50$
5.44	Bid A is better.
5.45	Option 1: PW(12%) = −$304,320; Option 2: PW(12%) = −$258,982

Chapter 6

6.2	(b)
6.4	(a)
6.6	$AE(12\%) = \$62.25$
6.8	$CR(14\%) = \$3,522.40$
6.11	$44,300
6.13	(a) $X = \$2,309.55$
	(b) $AE(13\%) = \$526.46 > 0$
6.15	$R = \$46,971$
6.17	$T = 1,018$ hours
6.19	$C = \$4.02$ per hour
6.22	$0.2401 per mile
6.25	667,140 rides
6.30	(a) Process A: $AE(15\%) = \$1,892.95$; Process B: $AE(15\%) = \$2,140.82$
	(b) Process A: $0.9465/hour; Process B: $1.0734/hour
6.34	Select Truck B with $AE(12\%)_B = \$6,910$.

Chapter 7

7.1	$i_a = 30.27\%$ per year
7.3	17.67%
7.5	Select (c).
7.8	(a) Simple investments: A, B, and D
	(b) Project A: 21.21%, Project B: 56.78%, Project C: 0.91% or 283.08%, and Project D: 154.76%
7.12	(a) $X = \$704$, (b) Accept the investment.
7.16	Select (c).
7.19	$i^*_{1-2} = 10\%$, select Option 1.
7.24	Select A2
7.27	Select A3.
7.30	Select Project B.

Chapter 8

8.2	$55,000
8.5	$31,000
8.7	$\alpha = 2\left(\dfrac{1}{8}\right) = 0.25$
	$D_1 = \$37,500$
	$D_2 = \$28,125$
	$D_3 = \$21,094$
8.11	(a) $\alpha = \left(\dfrac{1}{5}\right)1.5 = 0.3$
	(b) $D_1 = (0.3)(12,000) = \$3,600$
	(c) $B_4 = \$2,881.20$

8.14	$D_1 = \$6,713$
8.18	(a) Cost basis = $215,000
	(b) $D_1 = \$30,714$, $D_2 = 52,653$
8.25	(a) $D_1 = \$909$
	(b) $14,892
8.31	$7,893,500
8.35	(a) Income taxes = $195,500
	(b) Operating income = $389,400
	(c) Net cash flow = $444,500
8.37	(a) loss credit = $23,916
	(b) gain tax = $15,341
	(c) gain tax = $40,684.40

Chapter 9

9.4	(a) A = 400 units and B = 300 units
	(b) A = 371.43 units or 372 units
	(c) A = $5, B = $2
	(d) Product A
9.6	(a) tax rate w/o project = 39% tax rate with project = 34%
	(b) w/o project = 32.3%, with project = 34%
9.9	(a) $11,450
	(b) 38.17%
9.11	Net cash flow: −$235,000, $93,500, $104,216, $92,786, $85,927, $80,784, and $106,640.
9.14	Net cash flow: −$54,000, $27,830, $33,088, $30,929, $28,957, $23,627, and $22,539.
9.16	Net cash flow: −$50,000, $12,333, $14,279, $9092, $7,797, $6,500, and $6,500.
9.24	Net cash flow: −$100,000, $34,190, $47,090, $42,790, $90,490, $86,930, $82,460, $78,000, and $90,000
9.28	Net cash flow: $0, $15,170, $19,451, $15,050, $11,559, and $49,844
9.30	Option 1: −$695,968 Option 2: −$931,548 Option 3: −494,434
9.32	(a) −$21,670
	(b) −$21,071
	(c) Borrow-buy option
9.34	(a) $9,000
	(b) −$45,043
9.38	$PW(18\%) = \$30,885$ $IRR = 103.59\%$
9.44	$PW(20\%) = -\$345,989$ $AE(20\%) = -\$115,692$

9.48 (a) 9%
9.50 $k = 15.61\%$

Chapter 10

10.1 (a) PW(15%) = \$4,137
(b) $X = \$35,150$
10.3 5 floors at $i < 20\%$
2 floors at $20\% < i < 30\%$
10.6 (a) Model A: PW(10%) = $-\$7,152$;
Model B: PW(10%) = $-\$8,627$
(b) $X = \$793$
(c) Select Model A.
10.10 52.24%
10.15 24,867,011 copies
10.17 $E[\text{PW}] = \$694.21$
$V[\text{PW}] = 94,529$
10.23 (a) $E[\text{PW}]_1 = \$1,900$, $E[\text{PW}]_2 = \$1,850$,
and select 1.
(b) $V[\text{PW}]_1 = 1,240,000$, $V[\text{PW}]_2 = 2,492,500$, and select 2, as
$V[\text{PW}]_1 < V[\text{PW}]_2$, and $E[\text{PW}]_1 > E[\text{PW}]_2$.
10.27 Select (d).

Chapter 11

11.1 (a) \$9,000
(b) \$6,000
(c) $-\$6,450.58$
(d) AE(15%) = $-\$6,771.46$
11.3 (a) \$120,000
(b) Year 0: $-\$10,000$; Year 1–5: \$0
(c) \$23,868
11.7 (a) Economic service life = 4 years, with
AE(12%) = \$772
11.10 $\text{AEC}_{\text{challenger}} = \$4,959.04$
$\text{AEC}_{\text{defender}} = \$5,824.62$, replace the
defender now.

11.14 $\text{AEC}_{\text{defender}} = \$28,619$
$\text{AEC}_{\text{challenger}} = \$48,257$, do not replace the
defender for now.
11.18 $\text{AEC}_{\text{Option 1}} = \$35,210$
$\text{AEC}_{\text{Option 2}} = \$37,292$, select Option 1.
11.22 (a) Economic service life for defender = 2
years, with AE(12%) = \$4,748
(b) $\text{AE}(12\%)_{\text{Challenger}} = \$4,736$
(c) Keep for two more years.
11.24 (a) 8 years with AE(10%) = \$6,352
11.30 $X > \$91,215$

Chapter 12

12.3 (a) Design A: $I = \$400,000$,
$C' = \$427,974$, $B = \$85,000$
Design B: $I = \$300,000$,
$C' = \$684,758$, $B = \$85,000$
(b) Incremental analysis:
$\Delta \text{B}'\text{C}(8\%)_{\text{A-B}} = 2.56 > 1$, select Design A.
12.6 Design A: BC(10%) = 1.24
Design B: BC(10%) = 1.65
Design C: BC(10%) = 1.25
Incremental Analysis:
$\Delta \text{BC}(10\%)_{\text{C-B}} = -5.7$ and
$\Delta \text{BC}(10\%)_{\text{A-B}} = 0.37$, select Design B.
12.9 Select A4.
12.10 Select Site 1.

Chapter 13

13.2 (7), (8), (1), (11), (3), (9)
13.4 (a) working capital requirement = \$90,000
(b) \$680,000
(c) \$408,000
(d) \$318,000
13.7 Debt ratio = 0.33

Index

Categorization of Problems, Chapter Openers, and Examples by Engineering Disciplines

Civil & Environmental		Chemical & Petroleum		Electrical & Electronics		Industrial & Eng. Mngt.		Mechanical & Manufacturing			Others
p.5.27	p.12.4	p.5.47	p.11.35	CO.01	p.8.41	CO.05	p.9.29	EX.2.8	EX.8.11	p.9.52	EX.4.4 (Ag.)
p.5.28	p.12.6	p.5.48	p.12.1	EX.4.3	p.8.42	EX.5.6	p.9.30	EX.5.3	p.8.3	p.9.53	p.5.1. (Textile)
p.6.28	p.12.7	p.6.26		EX.4.6	p.9.5	p.5.45	p.9.32	EX.5.5	p.8.18	p.10.2	EX.7.3 (Ag.)
p.6.37	p.12.8	p.6.36		p.5.10	p.9.8	EX6.1	p.9.35	p.5.11	p.8.28	p.10.4	p.9.40 (Textile)
p.6.41	p.12.10	p.6.38		p.5.42	p.9.21	p.6.27	p.9.38	CO.06	p.8.30	p.10.11	p.10.7 (Textile)
p.8.2	p.12.11	p.7.14		p.5.44	p.9.33	p.6.33	p.9.42	EX.6.3	p.9.1	p.10.28	
p.8.16	p.12.12	p.7.33		p.5.46	p.10.5	p.6.34	p.9.44	EX.6.6	p.9.10	p.10.29	
p.8.31	p.12.13	p.7.34		p.5.49	p.10.9	p.6.40	EX.10.1	p.6.21	p.9.12	CO.11	
p.9.13		p.9.31		p.6.9	EX.11.5	EX.7.4	EX.10.2	p.6.22	p.9.18	EX.11.4	
p.9.14		p.9.39		p.6.17	p.11.3	EX.7.9	p.10.6	p.6.24	p.9.19	p.11.2	
p.9.28		p.9.45		p.6.23	p.11.26	p.7.13	p.10.10	p.6.35	p.9.20	p.11.4	
p.9.46		p.9.54		p.6.29		p.7.15	p.10.15	p.6.39	p.9.24	p.11.5	
p.10.1		p.9.55		EX.7.6		p.7.24	p.10.16	p.6.40	p.9.26	p.11.11	
p.10.3		CO.10		p.7.27		EX.8.10	p.10.22	EX.7.5	p.9.34	p.11.12	
p.10.13		p.11.14		p.7.28		p.8.32	p.11.1	EX.7.9	p.9.36	p.11.15	
p.12.2		p.11.17		p.7.28		CO.09	p.11.3	CO.08	p.9.41	p.11.36	
p.12.3		p.11.18		p.7.32		p.9.15	p.11.6	EX.8.8	p.9.47		

p. = end-of-chapter problem, CO = chapter opener, EX = example